MATHEMATICS OF MODELS: Continuous and Discrete Dynamical Systems

H. BRIAN GRIFFITHS
Emeritus Professor of Pure Mathematics,
University of Southampton

ADRIAN OLDKNOW
Head of Mathematics,
West Sussex Institute of Higher Education

ELLIS HORWOOD
NEW YORK LONDON TORONTO SYDNEY TOKYO SINGAPORE

First published in 1993 by
ELLIS HORWOOD LIMITED
Market Cross House, Cooper Street,
Chichester, West Sussex, PO19 1EB, England

A division of
Simon & Schuster International Group
A Paramount Communications Company

Printed and bound in Great Britain
by Redwood Books Limited, Trowbridge, Wiltshire

British Library Cataloguing in Publication Data

A catalogue record for this book is available from the British Library

ISBN 0–13–563800–3 cloth
0–13–563792–9 pbk

Library of Congress Cataloging-in-Publication Data

Available from the publisher

To Catherine and Jennie

Table of contents

Preface

Over the course of time, mathematics has become the principal tool for those seeking to make general statements about observed behaviour in the physical world. Such theoretical descriptions are nowadays called *models* of the behaviour, and of necessity, they do not tell us everything of interest about the behaviour. For example, Boyle's law $PV^\gamma = RT$ models the relationship between the pressure, volume, and temperature of a gas, but it doesn't tell you if the gas is poisonous or not. Concentrating on some interesting part of the behaviour, we may increase our physical knowledge of it and then try to incorporate this into a more accurate, but more complex, mathematical description; but this may then need new mathematics to handle it. In more recent years, and especially as a result of access to computing power, mathematical models have been produced in other disciplines such as economics and biology. Many interesting models now exist, and several have a similar *mathematical* structure: it's just that the terms mean different things in the different disciplines. Also, a model usually involves something corresponding to change in time (growth or decay)—either in discrete jumps or smoothly varying—and it is called *dynamical.* In this book, one of our aims is to describe several dynamical models for students, and to introduce ways in which a calculator or computer can be used for exploring both discrete dynamical models and discrete approximations to existing models. Another aim is to use these models to interest students in developing the relevant mathematics beyond the elementary (usually linear) stage.

We must make it quite clear that this book does not deal with testing the validity of such models, nor with the art of formulating new models. That is the business of *modelling*, on which there is an extensive literature, but which is essentially an activity for group discussion, and successive trials. The models that we discuss are the outcomes of past modelling activity; our hope is to give readers a good mathematical understanding of them, so that they may build on them later when they need to design powerful models.

Readers wishing to use this book will need first to understand its unorthodox arrangement. Contrary to a widespread view, mathematics is not a subject with a linear development: in practice we have to move sideways as well as back and forth, to use what we know, then fill in gaps in our theoretical understanding, then move on, and so on. That is why the book has two parts. The first part consists of Chapters 1–11 and is the main material for active reader participation; the second part, chapters A–F, is the theoretical support. Thus readers can immediately begin active mathematical work, and refer to Chapters A–F when they need to increase their knowledge of some theory or technique. (For location purposes, references such as theorem 7.2.1 or equation (B.1.2) will mean the theorem and equation, respectively, thus labelled in Chapter 7 and Chapter B.) However, there is much more to our arrangement than this, as we explain in Chapter 1.

The material of the book grows from three 'themes'. Theme 1 concerns some simple discrete and continuous mathematical models, which occur in the study of the growth of populations, and in mechanics and electrical circuits. These well-known classical models point to the need for extensions and generalizations, and we shall also have a non-traditional concern with them: to see the effect of changing certain 'control parameters' in a model (such as the constant R, or the exponent γ in Boyle's law mentioned in the first paragraph above). This is like moving from studying a particular curve in geometry, to examining a whole family of related curves at once, where we can see how one form evolves from another by passing through a 'singular' type. The singularities are traditionally set aside as unimportant, yet rare as they may be, they are necessarily present—like cell-division in biology—if radical change of form is to take place. Hence theme 2 develops further mathematical ideas and tools, via the geometry and tracing of plane curves; and there also we learn how to 'read' the resulting diagrams. We can then take up theme 3: models with greater complexity. These new models may consist of non-linear differential equations and their phase portraits as in Chapter 9, or the still controversial catastrophe models of Chapter 11; or the models from discrete non-linear dynamics in Chapter 10, which reveal the phenomenon of chaos and the related geometry of fractals, to which we give an introduction.

The two authors have worked within different teaching traditions, which we have here tried to combine. In the university tradition, mathematics tends to be explained to a passive audience as an organized body of knowledge, and students gain activity by working at old examination questions (and occasionally some course work). But there is a teacher–education tradition, in which formal theory takes second place to great emphasis on group work and active problem formulation and solving, using 'elementary' mathematics so that the students can develop the higher-level skills needed to allow them the confidence to become independently active. To reflect the virtues of each tradition, we have put the 'organized' theory in the later Chapters A–F, to support the 'activity' skills needed for Chapters 1–11. Some idea of their scope can be gained from the table of contents.

Of course, skill is needed to apprehend the theory, as well as the practice, so the reader is expected to work through the exercises, some based on the many short computer programs supplied. Others are adapted from university examinations and tutorial sheets. Thanks are due to the University of Southampton for permission to include questions taken from its past examination papers; these are indicated by the notation [SU]. Some of them ask for 'bookwork' that is already in the text, but we thought their inclusion reasonable, to allow the reader an element of self-testing. As to the tutorial questions, it is difficult to attribute all of them explicitly, because their creators are often anonymous, but we are especially indebted to colleagues in Southampton for permitting us to use their questions. Several sets of exercises are deliberately repetitive, because we think it is necessary for readers to *gain fluency by practising.*

We say more about our philosophy of mathematics education in the introductory Chapter 1. To conclude this Preface, we wish to thank Ellis Horwood and his staff, for encouraging us to write this book, and helping to bring it into production.

HBG, AJO
March 1993.

The programs listed in this book, together with a sampler version of TrueBASIC, are available in PC and Macintosh formats from:

UK:
Asher Research
16 Wellsworth Lane
Rowlands Castle
Hants PO9 6BY

Tel & Fax: 0705 412668

USA:
True Basic Inc.
12 Commerce Avenue
West Lebanon
New Hampshire, USA

Tel: 603 298 8517
Fax: 603 298 7015

1

Introduction: how to use this book

1.1 VIEWS OF MATHEMATICS

Although this first chapter is written in ordinary English prose, we hope that readers will not ignore it as 'waffle'; we think they will get more out of the book if they understand what we are about. The book began as a course of lectures in a university, but it has evolved into an unusual form. We therefore need to explain our intentions, which are moulded by our views on the practical teaching of mathematics.

In the early years of this century, a certain mathematics curriculum was developed with wide agreement (in Britain at least), as being suitable for a race of scientists and engineers. It led from elementary algebra and trigonometry to calculus, with much emphasis on interesting applications, especially in the physical sciences. It emphasized algorithms and was a good basis for the university mathematics courses of the time. (For details, see Griffiths and Howson (1974).) The approach changed very little and has become the British A-level tradition, but meanwhile university courses since that time have changed greatly, to allow for the rigour of analysis and the abstraction of algebra and topology: algorithms have often taken second place to conceptual mathematics. But the preliminary courses are still algorithmic, concentrating on being fluent in certain algorithms rather than on conceptual and critical matters. Students who have enjoyed the A-level course are frequently shocked into hostility when they come to mathematics in tertiary education: that is not their idea of 'mathematics'.

Indeed, mathematics has now become a rather strange study to embark upon. For many within traditional universities, it consists of a number of discrete courses with titles such as 'numerical analysis', 'linear algebra', 'probability theory', and so on, which are frequently studied with little interdependence. It is not unusual for a student to complete a course of three years' undergraduate study of the subject called 'mathematics' and yet to have little idea of what 'the subject' really is, except the name of a set of parts. This contrasts strongly with, say, the education of a doctor, who may well study discrete courses in anatomy, physiology, histology, and so on, but who already has a good idea of what a living human body is (being already in possession of one).

University mathematicians have tended to take the view that, at any given level, there is a set of basic skills and concepts which must be learned and practised *before* any engagement can be attempted with the actual practice of mathematics—which for them means the delight of discovery by research. To some extent this is not surprising as mathematics is one of the oldest of human intellectual activities, and it contains a vast body of knowledge and achievements. It then seems natural that a great amount has to be learnt to be able to come close to understanding some of the matters of current interest to research mathematicians. Perhaps the odd thing here is that so many students choose to study mathematics; though very few actually go on to practise what could be described as mathematics.

Within that university tradition, there is also a widespread single style of teaching and learning. Mathematics tends to be explained as an organized body of knowledge, in which students are largely passive, practising old, clearly formulated, and unambiguous questions for timed examinations. The large body of theory is found to be abstract and depends on unfamiliar language. These features are of course essential for the purposes of professional mathematicians, but they leave many students dispirited and bored, and their performance in more advanced courses is poor because the foundations are weak: the examiners are reduced to setting only bookwork or stereotyped questions, which can be remembered without becoming a vital part of the student.

For final year students who had been alienated by this less familiar mathematics, one of the authors developed an optional 'morale-building' course of advanced calculus. It was intended to be something like a course in an old-fashioned scholarship sixth form—using algorithms to do interesting things with relatively elementary mathematics. It was a modification of an earlier course on catastrophe theory, developed by colleagues (David Chillingworth and Leslie Lander); but their course was found difficult because the students lacked fluency and confidence in the deeper skills of 'elementary' calculus and both manipulative and linear algebra. The new version cried out for good graphics, which were supplied by the second author as an observer of the course. He normally teaches in a teacher–education tradition, in which formal theory takes second place to great emphasis on group work and active problem formulation and solving, using elementary mathematics so that the students can develop the higher-level skills that they need to become autonomously active. Their resulting competence could be said to be untidy, and not organized in the (idealized) university way. From that combined experience, this book evolved.

1.2 WHY MODELS?

For various reasons involving the system of timed examinations, the A-level tradition has in recent years tended to leave out many of the well-known applications, so it was natural for us to start with the classical models. It then seemed reasonable to draw on the current wide range of general linear and non-linear models, which are famous for their applications (with varying degrees of success) in fields such as physics, engineering, economics, biology, medicine, and elsewhere. Thinking of the future employment of the students, it is clear that mathematicians will be under growing pressure to use and develop these models further, as well as inventing new ones; so they must be acquainted with the standard models, their strengths and weaknesses. It also seems reasonable to let students gain experience with existing models before moving to the more difficult art of *modelling*. In any case, true modelling is not practical in those universities (gradually declining in number) that use only conventional timed examinations; although the latter can of course assess knowledge of models. The most interesting models usually allow for changes in time, and are nowadays called **dynamical systems**. When these do not involve probability, they are called **deterministic**; we have left aside the whole field of probabilistic models as beyond our scope (and competence).

This book, then, is about the mathematics of (deterministic) dynamical systems. It assumes that the reader wants to find out more about such systems and their properties, and hopes to achieve some understanding. We therefore introduce readers first of all to some classical models which usually display behaviour known as 'linear'. These can be analysed using relatively elementary mathematics, such as high-school algebra and calculus. After seeing some of their limitations, we develop further mathematics for handling more complicated non-linear models. These are capable of a range of complex behaviour, including limit cycles and chaos.

Besides introducing these new mathematical ideas, we have a deeper, educational, aim. Because students find mathematics difficult, some tertiary institutions seem to evade coming to grips with it, by substituting activity with computers and modelling instead. This can lead to a complacent hostility to mathematics, leaving students with an inadequate basis for the rest of their working lives. We are thinking of the student who has enjoyed the activity of 'doing' A-level mathematics, has then chosen to take a more advanced course in mathematics, but who will not take kindly to being told 'you must take a thorough study of X, Y, and Z before you can even start further activity'. This book, then, seeks to reverse that process. For, believing as we do that the first requisite for teaching mathematics is to arouse students' interest, we view many models as useful stepping stones, to let students see the point of learning new mathematics—either for understanding the model or to extend it further.

In order to handle these models, one needs to know a range of mathematical techniques, some classical and some modern. We also use geometric methods and imagery, and use computers to form such images, because the graphics are a strong stimulus for the imagination. So, analysis of the models requires us to bring together ideas from various parts of mathematics, and these ideas are laid out in a formal way in Chapters A–F, but readers are not expected to master those chapters first. Rather, we assume that readers have had a first course of calculus, with some algebra, trigonometry, and coordinate work (e.g., the British A-level course), and that they have access to computing machinery in the form of a personal computer and/or graphical calculator. Then it is best, we suggest, to read about the models and to take matters on trust, until one's existing mathematical equipment needs strengthening by a visit to the appropriate piece of theory. Practice in the use—the 'feel'—of a formal result is essential before one can understand its proof.

However, because the subject matter has such intimate connections with so many aspects of the mainstream of mathematical knowledge, this book provides (as far as possible) the necessary support for interested readers to extend or revise their knowledge of algebra (e.g., solution of differential equations), geometry (e.g., recognizing forms of ellipses, hyperbolae, etc.), numerical analysis (e.g., iterative methods for finding roots of equations), complex numbers, vectors, matrices, and so on. In order to understand the theory further, readers will need to develop more sophisticated tools, such as those associated with the mathematics of several variables, and those too are developed within the book.

Thus someone using the book to gain understanding of the mathematics of dynamical systems will, *en route*, be learning a large amount of mathematical theory which has been of much significance in mankind's attempt to explain the world in which it finds itself, and in attempting (for good or ill) to modify it. Within the world, subsystems change as certain controls are changed in value, and several of the figures in this book represent the evolution of such changes; skill is needed to 'read' them, to extract the information that they convey. To learn to create and use them is, we feel, an important part of education, because it can lead to the appreciation and understanding of change and form.

1.3 THREE THEMES

As explained in the Preface, the material of the book grows from three themes. Theme 1 concerns some simple discrete and continuous mathematical models, which occur in the study of the growth of populations, and in mechanics and electrical circuits. The models in mechanics go back to Newton's demonstration that the orbits of the planets are ellipses, and for which he and others had to develop the language of calculus. This should be within the range of readers equipped only with the 'first course', described above. But these models point to the need for extensions and generalizations, so theme 2 develops further mathematical ideas and tools, via the geometry and tracing of plane curves. These range from conics, cycloids, and so on, to the contours of two-variable functions. In the most interesting cases, they vary with a control parameter, as in the simple example of Fig. 1.3.1, which illustrates the changes that occur in the graph of the function $y = x^\alpha$ on the domain $x \geq 0$; here there is a graph for each value of the parameter α, as it ranges from $-\infty$ to $+\infty$, but there are only six significantly different types. Techniques and language are needed for describing these, for making graphical representations by computer, and for learning how to read the resulting diagrams. By assembling the diagrams appropriately, they can sweep out a surface in three-dimensional space, for which we need to develop our visual and spatial skills. Also, the need for control parameters means that we cannot confine ourselves to the often-heard demand 'please do a numerical example'; algebraic symbols may be 'theory' in one context, and completely concrete in another.

With this material, we can then take up theme 3: models with greater complexity. The new models we discuss are of three types: (i) non-linear differential equations and their phase-portraits; (ii) the still controversial catastrophe models; or (iii) models from discrete non-linear dynamics which were virtually inaccessible until the 1960s with the availability of fast computers. It was then that the phenomenon of chaos

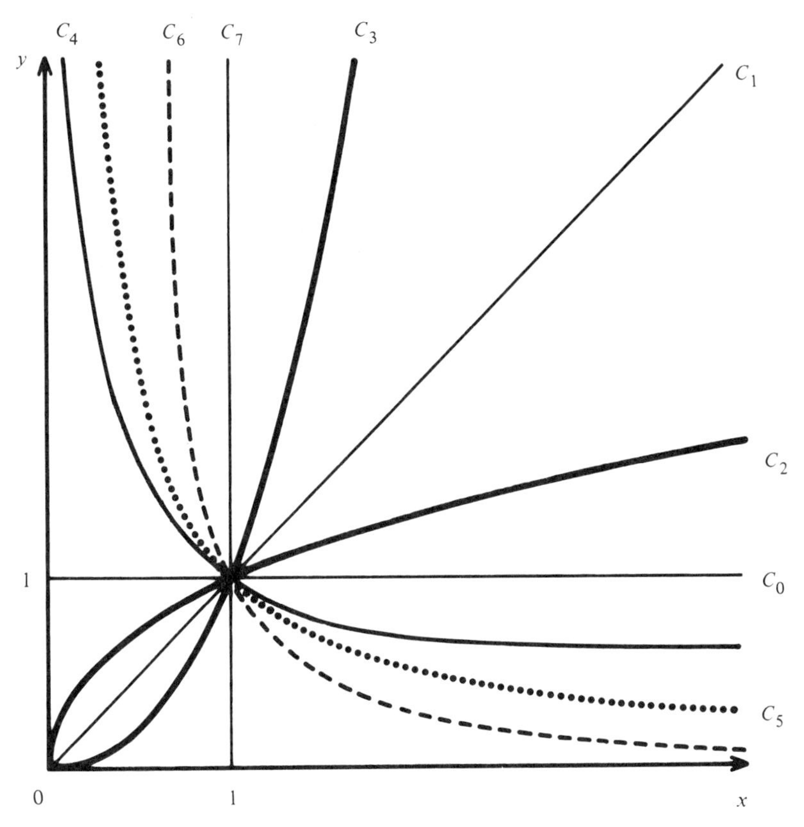

Fig. 1.3.1. All the possible shapes of the graph of the function $y = x^{\alpha}$, for x positive, are shown in one diagram, in which $C_0, C_1, \ldots, C_7$ correspond, respectively, to the cases $\alpha = 0$; $\alpha = 1$; $0 < \alpha < 1$; $\alpha > 1$; $0 < -\alpha < 1$; $\alpha = -1$; $-\alpha > 1$; ∞.

was discovered, with the related geometry of fractals, to which we give an introduction. Each of the three chapters within this theme concerns a field which is being rapidly developed by current research, and since complete books exist to cover each of them, we could not hope to be comprehensive. Nevertheless, we do not think that students will find the coverage superficial.

But that, too, is part of our own motivation for producing this book: we hope to bring readers to the stage at which they can read more specialized advanced texts.

1.4 ACTIVITY AND JOY

Our hope is that readers will acquire both approaches of the two traditions mentioned above, and even that the enjoyable activity will lead them to become active in producing new theory, either for its own sake, or because it is vital for new methods when old ones fail. We hope also that readers will not ignore the historical context: we often mention mathematicians who worked in the past, and it helps to know that they had to struggle to make progress, just like ordinary students, once they were interested in a problem. We usually give a mathematician's dates, but readers will get a better feeling for the mathematical timescale by asking who could have written letters to whom (for example, Kepler to Shakespeare, Euler to Mozart); mathematics is partly about people. For more detailed historical material throughout the book, we refer readers to the historical introduction of Boyer and Mersbach (1989). The material is interesting, and one's own mathematical understanding can be greatly helped by seeing how ideas gradually evolve from natural questions over the centuries, and building on the work of others, Knowing of this evolution, we can see the futility of quarrels about who discovered what; and as we struggle to learn mathematics we can be reassured that it is not just the product of a sequence of flashes of genius by whizz-kids.

We have also laid great stress on the need to practise the many exercises, to gain fluency—or in some cases to develop an idea further than in the text. Several exercises are structured so that some computer experience is gained before theoretical results are proved, but in others it seems best to do theory first. We sometimes use the computer for mathematical experimentation, in order to suggest what theory is needed, and Chapter A describes our programming conventions. We have not assumed that readers have access to facilities for doing symbolic manipulation by computer; even where those exist, we still think it is necessary to do some algebraic exercises by hand, in order to gain fluency in *reading* formulae, and to appreciate such matters as symmetry of form. Thus, an exercise like 'evaluate the determinant ...' is intended to be done first without the mechanical aid which might be good for checking the result. Also, it is still necessary (for avoiding tedium in later work) to remember some formulae, and writing these by hand will aid the memory; even then, the intelligent memorizer will look at the structure of a formula or definition, to see what makes it sensible and worthwhile.

Some readers may feel that it is dishonest to begin a problem before they understand the theory completely; but it is best not to be thinking of the theory at all when learning to carry out the moves of an algorithm: the thoughts interfere with the brain's handling of the instructions. Understanding often comes with the success of mastering the algorithm, and then the enjoyment is complete. The religious dispute between Abelard and Anselm in the twelfth-century University of Paris is still applicable within mathematics: 'I must understand in order to believe' versus 'I must believe in order to understand'; the same person may be a believer and understander at different periods of their mathematical growth.

Besides practising mathematics and reflecting about it, we all have to learn to fill in gaps in our knowledge by reading a range of books. This book, as any other, cannot cover everything, and we have frequently had to suggest books for further reading. With standard material like calculus (for example) there is no single 'best' book, although we may ourselves have a preference. Sometimes this will be for an older book that may only be obtainable from a large library, so we have also added a newer, more accessible, title in the list of references at the back. In mathematics, 'old' need not mean inferior; some 'old' books are very good indeed.

1.5 COMPUTING

We have already mentioned two particularly recent developments in mathematics, that we can characterize succinctly as chaos and computing. It is now time to look in more detail at their implications for the way in which mathematics is practised and learned. The first of these, chaos, arises in non-linear dynamical systems, both in the physical world and in the abstract deterministic models of mathematics: chaos is a seemingly random type of behaviour that can (paradoxically) be seen in deterministic systems. It has been rare indeed for mathematical developments to catch the public's eye and imagination, through the medium of mass communication; but chaos, and the related images of fractals, has certainly achieved 'star status'. This is not just because of its far-reaching significance but also because it can be related to our physical experience (such as a pendulum which apparently cannot make up its mind on which of a number of competing magnets to settle) and to quite elementary mathematics (like 'think of a number, square it, and subtract one and a half, and repeat the process on the result').

However, there would be something lacking in studying a subject concerned with movement and change, if we were restricted only to studying static symbols and illustrations on the printed page. Computing technology has developed so quickly in recent years that for less than the price of a pair of training shoes there is a range of pocket calculators, which are really computers, with graphic displays and built-in programming languages. Similarly, for less than the price of a bicycle there is a range of personal computers with power and speed in excess of those available only to researchers in the field of dynamical systems of twenty years ago (or less). Many college students would expect to have trainers and a bike, and we would contend that modern students of mathematics will be severely hampered if they do not also have ready access to powerful calculators and computing.

As with the range of readers' mathematical backgrounds, we have had to make assumptions about the capacity of their available computing machinery. We assume the minimum of a graphical display (for which colour is desirable but not essential) and an accessible programming language. The book contains a large number of computer programs, which are relatively short to type in at a keyboard, and which are intended to be self-explanatory. They have been written in a fairly simple convention and use the dialect of BASIC known as TrueBASIC, for which there are widely available interpreters for popular PCs such as IBM compatibles and Apple MACs. For convenience, all the programs are collected in Appendix B; and in Chapter A, we explain how they may be adapted for other versions of BASIC and for graphic calculators.

There exist of course many other items of computer software which can be profitably used as an adjunct to studying the material of this book (and mathematics in general), such as graph plotters, spreadsheets, symbolic manipulators, numerical toolkits, and so on. It is the very existence of such tools which makes it far more important for the modern mathematical practitioner to know what X (such as a matrix) is, how and why X is used, and when it is safe to do so, rather than be practised in carrying out manipulations on X by hand (such as finding a matrix inverse). This is not to gainsay our earlier remarks about doing exercises by hand to gain familiarity; but it forces a reappraisal of what types of techniques are important. For example, the known techniques for eliminating t from the parametric equations of an envelope are of only limited application, and now less important than the technique of writing down the necessary equations for a program to plot the curve. Also a learner of mathematics can have experimented with many instances of X (such as an envelope) and observed a great range of X's behaviour, in the time it would have taken to carry out a single lengthy manipulation of X.

To conclude this introductory chapter, then, we hope to have prepared readers to expect that a study of the mathematics of dynamical systems requires a synthesis of many parts of mathematics mentioned earlier—it is about applied mathematics since it uses models, taken from our environment (change in populations, temperature, radioactivity, etc.) and applies techniques from pure mathematics and computing in exploring and predicting (where possible) their range of behaviour. It uses geometry in displaying and analysing results and involves the techniques and concepts of calculus of several variables. It even involves probability in the generation of fractal images. So, a study of the mathematics of dynamical systems undertaken through reading this book, while not in itself a complete course of mathematics, will nevertheless have given readers an introduction to many of the main branches of mathematics which they may then choose to study further, with some idea of their content, approach, and relevance.

2

Discrete dynamics

This chapter starts by considering some simple but important models for the change over time (or **dynamics**) of a single variable such as the numbers in a human population or the money in a bank account. It goes on to extend the models to those in which there are two variables changing in a coupled way, such as the numbers of predator and prey in a natural system. Here the changes are considered to happen at regular distinct intervals of time, such as each year, month, or day. Such change is known as **discrete**, and is distinguished from the constant growth, such as that of the length of a plant, which is **continuous**. Discrete models can be easily simulated using a computer or programmable calculator. In some cases, especially the **linear** case, a general mathematical analysis is possible. For coupled systems the theory of matrices plays an important part.

2.1 THE MALTHUS MODEL

In 1798 the English clergyman T. J. Malthus published his *Essay concerning human population* in which he did something never previously attempted: he made a mathematical model of the human population in order to further rational discussion of basic political questions that were being widely discussed around that period. (Recall that the American Declaration of Independence was made in 1776 and was followed in 1779 by the upheaval of the French Revolution and the growing social stress of the Industrial Revolution; one outward sign of that distress was the rapid rise in the population of manufacturing towns, and the consequent overcrowding.)

Malthus's model was this: let $P(n)$ denote the number of people at the beginning of time period n, measured in years (say) from some base-year zero. Then the change from one year to the next is the 'increment' $P(n+1) - P(n)$, and is denoted by $\Delta P(n)$. Malthus suggested that $\Delta P(n)$ would be proportional to $P(n)$ itself, say

$$\Delta P(n) = b \cdot P(n) \tag{2.1.1}$$

where b is a constant that measures the proportionate excess of births over deaths. Hence

$$P(n+1) = P(n) + \Delta P(n) = P(n) + b \cdot P(n)$$

that is

$$P(n + 1) = k \cdot P(n), \qquad \text{where } k = 1 + b \tag{2.1.2}$$
$$= k \cdot (k \cdot P(n - 1)) = k^2 \cdot P(n - 1)$$
$$= (k^2) \cdot k \cdot P(n - 2) = k^3 \cdot P(n - 2),$$

and so on, until $P(n + 1) = k^{n+1} \cdot P(0)$.

For example, if $b = 0.1$ and $P(0) = 100$, the following are obtained:

n	0	1	2	3	4	5	6	7	8
$P(n)$	100	110	121	133.1	146.4	161.1	177.2	194.9	214.4
$\Delta P(n)$	10	11	12.1	13.3	14.7	16.1	17.7	19.5	

and the population has doubled in just 8 years.

Therefore, if $b > 0$ then $k > 1$ so k^n will increase to infinity with n; $P(n)$ increases 'geometrically', as Malthus put it. On the other hand, he argued that the quantity $F(n)$ of food available for year n could only be increased by extending cultivation to land of ever poorer quality; so at best he reckoned that

$$F(n + 1) = F(n) + c \tag{2.1.3}$$

where c is a constant. Thus $F(n + 1) = F(0) + (n + 1) \cdot c$, and $F(n)$ increases only 'arithmetically'.

Hence, although $F(n)$ increases to infinity with n, it does so far more slowly than $P(n)$ (see Fig. 2.1.1). From this, Malthus concluded that the human race would

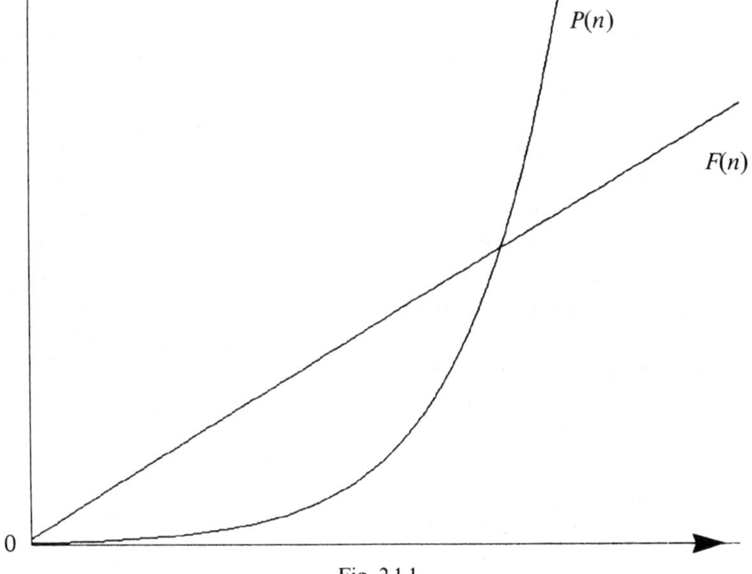

Fig. 2.1.1.

eventually starve to death, were it not for the sporadic reductions of population pressure caused by war, pestilence, and localized famine. Consequently these 'scourges' must have been provided by God to prevent total disaster, and therefore it would be wrong to alleviate these scourges unless people were to limit the size of their families.

Notice the different types of deduction: verbal arguments lead to the model equations (2.1.1), (2.1.3), from which mathematical arguments show that $P(n)$ must overtake $F(n)$; from this it is inferred that the 'scourges' act as a brake, and then a conclusion is drawn that had serious social effects, in that it was used by others as a serious argument to avoid mitigating the severe Poor Laws of the time. For many years, politicians acted on those arguments, and did avoid such mitigation, to the moral satisfaction of the proponents but not of the affected poor.

In formulating the mathematical model (2.1.1) we have chosen to use the notation $P(n)$ for the population at time step n, rather than using the more conventional subscript notion P_n, for two main reasons. First, our notation emphasizes that the population is a **function** of the number of time slots n, and secondly it allows easy conversion of the mathematical symbolism into a programming language, where $P(n)$ is the nth element of an **array** called P. Technically we need to specify the **domain** and **codomain** of the function P. Since n may only take the discrete values 0, 1, 2, ... the domain of P is the set of non-negative integers $\mathbb{Z}^+$. Because we are dealing with populations, whose size at any time n must also be a member of $\mathbb{Z}^+$, we would expect the codomain of P also to be $\mathbb{Z}^+$, and would write $P\colon \mathbb{Z}^+ \to \mathbb{Z}^+$. However, the parameter b in (2.1.1) is not necessarily an integer (a typical value would be between 0 and 1), and so the function P given by our mathematical model is $P\colon \mathbb{Z}^+ \to \mathbb{R}$, which again emphasizes the distinctions between reality and mathematical idealizations. We could adjust (2.1.1) to force $\Delta P(n)$ to be an integer by rounding or truncating, but then mathematical analysis becomes fearsome, if not impossible.

The Malthus model for population growth is found to be roughly correct for populations of such primitive organisms as yeast, when these are observed for limited periods in laboratories. His model (2.1.3) for increase of food production was based on prevailing knowledge of availability of land, and was temporarily invalidated by the opening of the American Middle West as the 'bread-basket' of the world. In our own time of course, the problem of how to increase food production to match the rapidly increasing world population is back with us again, and we know much more about world geography than did Malthus.

We know a type of population that really does obey the Malthus law, namely money in a bank account. If $P(n)$ denotes the sum in the account at the beginning of the year n, then $P(0)$ is called the **principal** and b the **interest rate**, usually expressed as a percentage $i\%$. Thus if the money is left in the account, then $P(n + 1)$ consists of $P(n)$ together with the accrued interest of $i \cdot P(n)/100$ (which we still write as $b \cdot P(n)$ for brevity), so

$$P(n + 1) = P(n) + b \cdot P(n) = k \cdot P(n)$$

as before.

This is an appropriate point to show how the situation can be modelled with a computer or calculator. The programs are given in TrueBASIC—see Chapter A for details of how to adapt them for other systems. If it is desired to keep a record of all the values $P(0), P(1), \ldots. P(n+1)$, then it is appropriate to use an **array**, otherwise it is wise to compute $\Delta P(n)$ directly and store it in a variable, `deltaP`, say:

```
REM Program 2.1.1a,Malthus
OPTION BASE 0
DIM P(50)
LET P(0) = 100
LET b = 0.1
LET k = 1 + b
FOR n = 0 TO 20
  PRINT n, P(n)
  LET P(n+1) = k*P(n)
NEXT n
END
```

```
REM Program 2.1.1b, Malthus
LET P = 100
LET b = 0.1
FOR n = 0 TO 20
  PRINT n, P
  LET deltaP = b*P
  LET P = P + deltaP
NEXT n
END
```

Such programs can be easily adapted to plot graphs.

```
REM Program 2.1.1c, Malthus graph
SET MODE "graphics"
SET WINDOW 0,20, 0,1000
LET P = 100
LET b = 0.1
FOR n = 0 TO 20
  PLOT n, P
  LET deltaP = b*P
  LET P = P + deltaP
NEXT n
END
```

(Note that the output of this program is a set of discrete points.)

Now suppose that, on the other hand, $P(0)$ is borrowed rather than lent, and a constant amount R is repaid at the end of each regular time period, such as each year or each month (as with a house mortgage). Then $P(n+1)$ is the amount owing at the start of time-period $n+1$, and consists of $P(n)$ plus the interest on $P(n)$, minus the repayment R; so

$$P(n+1) = k \cdot P(n) - R,$$

which reduces to (2.1.2) if $R = 0$ and to (2.1.3) if $k = 0$. In terms of the increment $\Delta P(n)$ this modifies (2.1.1) to

$$\Delta P(n) = b \cdot P(n) - R \tag{2.1.4}$$

which tells us that there is growth or decline according as $P(n)$ exceeds or is less than R/b, and $P(n)$ remains constantly equal to R/b if ever $P(m) = R/b$ for some m.

For example, with a loan $P(0) = 50\,000$ and a monthly interest rate of 1%, assuming (2.1.4) is applied monthly, then R must exceed 500 if the loan is to be paid off. Suppose $R = 750$; then the outstanding debt $P(n)$ at the end of each year, to the nearest integer, is given by:

n	12	24	36	48	60	72	84	96	108
$P(n)$	46829	43257	39230	34694	29583	23823	17332	10018	1777

and the loan is paid off during the ninth year.

We have a still more general form of equation if we allow the repayments to fluctuate (e.g., as the debtor's prosperity improves), so we consider the equation:

$$P(n+1) = k \cdot P(n) - R(n), \tag{2.1.5}$$

with $R(n)$ now a function of n. (In recent years, the interest rate has fluctuted as well—in contrast to nineteenth-century England, when it remained at 3% p.a.—but we shall keep k constant for simplicity.)

Usually, (2.1.5) is called a **first-order linear difference equation** and to solve it we **iterate** it to get

$$\begin{aligned} P(n+1) &= k \cdot (k \cdot P(n-1) - R(n-1)) - R(n) = k^2 . P(n-1) - k \cdot R(n-1) - R(n) \\ &= k^2 \cdot (k \cdot P(n-2) - R(n-2)) - k \cdot R(n-1) - R(n) \\ &= k^3 \cdot P(n-2) - [k^2 \cdot R(n-2) + k \cdot R(n-1) + R(n)], \end{aligned}$$

and so on, until

$$P(n+1) = k^{n+1} \cdot P(0) - \sum_{r=0}^{n} k^r \cdot R(n-r). \tag{2.1.6}$$

For example, if R is constant as in (2.1.4), then this becomes:

$$P(n+1) = k^{n+1} \cdot P(0) - R \cdot (1 + k + k^2 + \ldots + k^n)$$

and the sum of the geometric progression is $1 \cdot (k^{n+1} - 1)/(k-1) = (k^{n+1} - 1)/b$ so:

$$P(n+1) = k^{n+1} P(0) - \frac{R}{b} \cdot (k^{n+1} - 1). \tag{2.1.7}$$

Note that (2.1.7) can be rearranged to give

$$P(n+1) = \frac{R}{b} - k^{n+1} \cdot \left(\frac{R}{b} - P(0)\right) \tag{2.1.8}$$

which remains equal to R/b if $P(0) = R/b$, but otherwise steadily increases or decreases from $P(0)$ as described following (2.1.4).

Equation (2.1.8) can also be used to find the time taken to pay off a mortgage if $R > b \cdot P(0)$, by solving (2.1.8) for n with $P(n+1) = 0$. This can be done by taking logarithms, but the solution will usually give a non-integer value for n which must be interpreted in terms of the discrete model being used. For the numerical example above we need to solve

$$0 = \frac{750}{0.01} - 1.01^{n+1} \cdot \left(\frac{750}{0.01} - 50000\right)$$

which simplifies to $1.01^{n+1} = 3$ giving $n + 1 = 110.41$.

Of course this can also be arrived at numerically by adapting the programs 2.1.1a or 2.1.1b by replacing the counted `FOR..NEXT` loop with a tested `UNTIL` or `WHILE` loop. For example,

```
REM Program 2.1.2a, Mortgage
OPTION BASE 0
DIM P(100)
READ P(0), R , b
DATA 100 , 20,0.1
LET k = 1 + b
LET n = 0
DO
  PRINT n, P(n)
  LET P(n+1) = k*P(n) - R
  LET n = n + 1
LOOP UNTIL  P(n) < 0
END
```

```
REM Program 2.1.2b, Mortgage
READ P  , R , b
DATA 100, 20,0.1
LET n = 0
DO WHILE  P >= 0
  PRINT n, P
  LET deltaP = b*P - R
  LET P = P + deltaP
  LET n = n + 1
LOOP
END
```

Exercises 2.1.1

1. Calculate how long it takes for a sum of money to double in value if interest is compounded annually at the following rates: 1%, 2%, 5%, 7%, 9%, 13% (use 3 decimal places).
2. At a compound interest of r, $2r$, respectively, a sum of money doubles in n, m years. What is n/m? Is it less or more than 2? (Cambridge B.A., 1807)
3. Show that the time τ taken for $P(0)$ to double in the Malthus model is $\ln 2/\ln k$—without rounding. In 1990, the world population was at least 5 billion (1 billion is 1000 million). If the Malthus model were to apply, estimate the population for the year 2000 assuming values 1.05, 1.03, and 1.01 for k.
4. If the average age of death is d years then in a population of P people we can assume a death-rate of about P/d per year. In China, P and d are about 1 billion, and 65, respectively, while about 40 000 babies are born each day. Estimate k in this case. (Thus after a census lasting a few weeks, the population at the end is a good deal larger than at the beginning, but neither of them can be known accurately, with current technology. The World Health Organization calculated in 1992, that there are about 910 000 conceptions per day, throughout the world.)
5. Solve the equation $x(n+1) = k \cdot x(n) + n$. (It may be useful to look at $d/dk(1 + k + k^2 + \ldots + k^q)$.)
6. If, in the 'mortgage' equation, the rate of repayment is constant, find a formula for the number of years it will take to pay off a loan on a house costing £C, given a constant interest rate r. What happens if r changes to $r + \delta r$?
7. If my monthly salary is £s, which is better: to have a rise of r% in a year's time, or to have monthly rises of $(r/12)$%? What are the rises worth if the currency is devaluing at a rate c%?
8. Let T be the time taken to pay off a mortgage with an initial loan of P, an interest rate of b, and a fixed regular repayment R. Use a computer or calculator to investigate how:
   (a) T varies with R for fixed P and B;
   (b) R varies with b for fixed P and T;
   and any other similar relationships.
9. A loan of £L is to be repaid by annual instalments of £R per year. If the interest rate is r%, show that the sum owing after n years is £$L(n)$ where $L(n+1) = L(n) \cdot (1 + r/100) - R$,

and that the repayments decrease each $L(n)$ provided that $L < 100 \cdot R/r$. Show that the solution of the difference equation:

$$\lambda(n+1) = k\lambda(n) - c$$

is

$$\lambda(n) = k^n \cdot \left(\lambda(0) + \frac{c}{1-k}\right) - \frac{c}{1-k}$$

and use this to calculate the number of years N it will take to pay off the loan of £L. If r, L and N are fixed, calculate what R should be.

10. Suppose that $R(n) = A \cdot p^n$ in (2.1.5). Show that the solution (2.1.6) becomes in this case $P(n) = k^n . P(0) - A \cdot X(n)$, where $X(0) = 0$ and

$$X(n) = \frac{k^n - p^n}{k - p} \qquad \text{or} \qquad n \cdot k^{n+1}, \qquad n > 0,$$

according as $p \neq k$ or $p = k$. (Note: $k^r \cdot p^{n-r} = p^n \cdot (k/p)^r$.)

11. Writing now x_n for $x(n)$, solve the difference equations:

$$x_{n+1} = 3 \cdot x_n + 2 \qquad x(0) = 1$$

$$x_{n+1} = x_n/3 + 2 \qquad x(0) = 1$$

$$x_{n+1} = x_n/3 + n \qquad \text{(i) } x(0) = 0, \qquad \text{(ii) } x(0) = 1.$$

2.2 THE VERHULST MODEL

The Malthus model and its social consequences naturally generated much criticism, and the model was modified in 1840 by the Belgian (Flemish) demographer Jean François Verhulst. He felt (in the spirit of the time) that an individual's chance of survival would depend on his/her ability to cope with competitive clashes with other people. Now, if there are P people at a given time n, then the maximum number of pairings is $P(P-1)/2$, which is approximately $P^2/2$, so he suggested modifying (2.1.1) by using:

$$\Delta P(n) = b \cdot P(n) - c \cdot P(n)^2$$

to give:

$$P(n+1) = k \cdot P(n) - c \cdot P(n)^2 \tag{2.2.1}$$

where c is a constant that represents the competitive friction. He then approximated this model by one using calculus (see (3.1.3)), and it seemed more satisfactory than Malthus's model, especially since it predicted that the population would stabilize at a steady level. If that happened, much uncertainty about the future would go, and people could plan their lives more easily. That is no reason for accepting one model, rather than another, however! But there are other reasons why it is instructive to stick to (2.2.1) for a little while.

Program 2.1.1 can be easily adapted to get a feel for the numerical behaviour of this iteration. For example, start by setting $c = b$ with $P(0) = 0.2$ and see the effect of changing b, then repeat with other values for $P(0)$. If $c = b < 1$, say, with $P(0) < 1$ then the population $P(n)$ seems to grow initially in a Malthus fashion before levelling off to grow towards a steady value of 1. For example, if $b = c = 0.1$ and $P(0) = 0.2$ (where the units might now be tens of millions) then we can track a sample of values of $P(n)$:

n	1	2	10	28	36	59	81	103
$P(n)$	0.216	0.2329	0.3997	0.8044	0.9041	0.9906	0.9991	0.99991

starting with Malthus growth but then steadily levelling towards 1.

Similarly, if $P(0) > 1$ the population declines towards 1. This is numerical evidence for the stabilizing of the population. However, if quite large values of b and c are used (e.g., $b = c = 3$) then the population can fluctuate wildly with no stable state.

Now (2.2.1) is a **non-linear** equation, and there is no obvious formula that gives its solution in the manner of (2.1.6). With the advent of powerful computers in the 1970s, it became possible to investigate (2.2.1) numerically and this led to quite unexpected results and insights that led in turn to a new subject in mathematics—chaotic dynamics. We shall say more about this later (see Chapter 10).

One way to deal with the lack of a formula is to modify (2.2.1) into a more tractable form, and it was suggested that in the 'competition term', $P(n)^2$ could be replaced by $P(n) \cdot P(n+1)$ to give

$$P(n+1) = k \cdot P(n) - c \cdot P(n) \cdot P(n+1). \tag{2.2.2}$$

This can now be rearranged to give $P(n+1)$ in terms of $P(n)$:

$$P(n+1) = \frac{k \cdot P(n)}{1 + c \cdot P(n)} \tag{2.2.3}$$

which can be investigated numerically by adapting Program 2.1.1. The values for $b = c = 0.1$, $P(0) = 0.2$, closely match the example above.

Dividing (2.2.2) by $P(n+1) \cdot P(n)$ and rearranging gives:

$$\frac{1}{P(n+1)} = \frac{1}{k} \cdot \frac{1}{P(n)} + \frac{c}{k}$$

which can be converted into a linear recurrence equation, as in section 2.1, by setting $q(n) = 1/P(n)$ (assuming that the population $P(n)$ never becomes zero) to give:

$$q(n+1) = \frac{1}{k} \cdot q(n) + \frac{c}{k} \tag{2.2.4}$$

which is of the linear form (2.1.5) and so can be solved by the method of that section to give:

$$q(n+1) = \frac{c}{b} + \left(q(0) - \frac{c}{b}\right) \cdot k^{-(n+1)}. \tag{2.2.5}$$

So, if $b > 0$ then $k = 1 + b > 1$ and $k^{-(n+1)} \to 0$ as $n \to \infty$. Hence $q(n) \to c/b$, and thus $P(n)$ tends to a constant value $P_m = b/c$ which is the maximum sustainable population under this model, and which is independent of the initial population $P(0)$. If $P(0) = P_m$ then the population is constant; if $P(0) > P_m$ then it declines steadily towards the limiting value P_m; and similarly if $P(0) < P_m$ then it increases steadily towards P_m. Figure 2.2.1 shows these three types of behaviour.

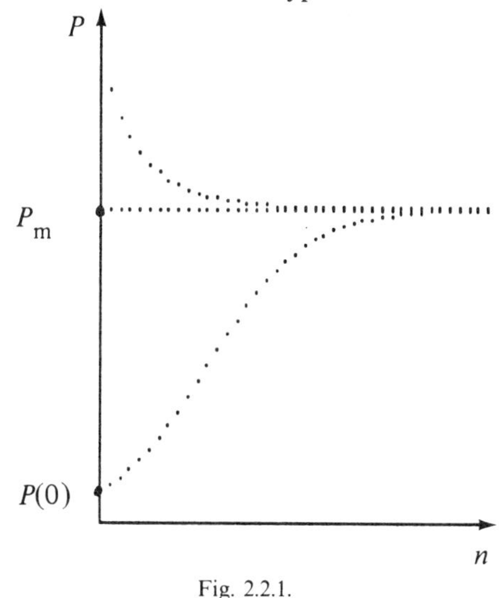

Fig. 2.2.1.

Substituting for p gives:

$$P(n+1) = \frac{P_m}{1 + (P_m/P(0) - 1)\cdot k^{-(n+1)}}. \tag{2.2.6}$$

Thus we can try to 'fit' this model to any point of the data generated from (2.2.1). For example, with $b = c = 0.1$ and $P(0) = 0.2$ we found that $P(1) = 0.216$. Here $P_m = b/c = 1$ and (2.2.6) can then be solved with $n = 0$ to give $k \cong 1.102$, modifying both k and c in (2.2.2).

The simple steady behaviour from this model for all values of the parameters b and c contrasts with the highly complex behaviour of equation (2.2.1) for certain parameter ranges (see the next exercise). Model (2.2.1) fluctuates wildly, in a **chaotic** state, whereas model (2.2.3) still shows a steady, albeit rapid, approach to the stable value P_m.

Thus the suggestion for changing the term involving $P(n)^2$, while apparently reasonable, is not justified mathematically.

Exercises 2.2.1

1. Try using a calculator or the programs of this section to compare the models (2.2.1) and (2.2.3) with $b = c = 3$ and $P(0) = 0.2$.
2. In many countries at present, over half the population is under 14 years of age. If the average age of death is 40 years, and you use the Verhulst model, what can you deduce about the excess of births over deaths?

2.3 SECOND-ORDER LINEAR DIFFERENCE EQUATIONS

The very first population problem to be considered was about rabbits, not humans, and was whimsical rather than realistic. It was proposed in 1202 by Leonardo of Pisa, known also as Fibonacci, and runs as follows. Each pair of rabbits begets a pair of young every month except the first, and each young pair takes a month to become adult. Starting with one newly born pair, how many pairs $p(n)$ will there be after n months if no rabbit dies?

To handle this problem, let $p(n)$ consist of $a(n)$ adult pairs and $b(n)$ baby pairs; thus $p(n) = a(n) + b(n)$. Next month, the former babies are now adults, and the $a(n)$ adults have produced $a(n)$ more babies. Hence

$$a(n+1) = a(n) + b(n), \qquad b(n+1) = a(n). \tag{2.3.1}$$

Our primary interest is in $p(n)$, so we get the famous Fibonacci equation

$$p(n+2) = p(n+1) + p(n) \tag{2.3.2}$$

since $p(n+2) = a(n+2) + b(n+2) = p(n+1) + a(n+1) = p(n+1) + p(n)$.

We are given that $p(0) = 1 = p(1)$, so (2.3.2) tells us that $p(2) = 2$. Feeding this back into (2.3.2) gives $p(3) = 3$, then $p(4) = 5$, $p(5) = 8$, and so on. The resulting sequence of positive integers is the 'Fibonacci sequence', and it has many fascinating properties that have given rise to an enormous literature.

As we have just seen, knowing the first two terms allows us eventually to calculate $p(n)$ for any n; and this can be automated in a program such as:

```
REM Program 2.3.1, Fibonacci
OPTION BASE 0
DIM p(30)
LET p(0) = 1
LET p(1) = 1
FOR n = 0 TO 20
  PRINT n, p(n), p(n+1)/p(n)
  LET p(n+2) = p(n+1) + p(n)
NEXT n
END
```

When the mathematics permits it is desirable to 'solve' such a system as (2.3.2) in the sense of finding a convenient formula for $p(n)$, and this we now consider.

Equation (2.3.2) is an example of a **second-order linear difference equation**. Many other problems lead to such equations, of which the general type is of the form

$$x(n+2) + a \cdot x(n+1) + b \cdot x(n) = f(n) \tag{2.3.3}$$

in which a, b are given constants and $f(n)$ is a given function of n. To 'solve' it is to find the sequence x of values $x(n)$ when n runs through the set $\mathbb{Z}^+$ of integers $0, 1, 2, \ldots$.

Note: it will avoid confusion if we carefully distinguish between 'the sequence x' and its set of values $x(n)$: each individual value is a number, not a sequence. Sometimes we emphasize the distinction by writing $x = \langle x(n) \rangle$; so, for example, if t is any number we can speak of 'the sequence $t \cdot x$' and mean the sequence which, at n, takes the value $t \cdot x(n)$ (in brief, $t \cdot x = \langle t \cdot x(n) \rangle$).

When f in (2.3.3) is the zero sequence (i.e., $f(n)$ is constantly zero, as in (2.3.2)) the equation is said to be **homogeneous**, because if a sequence x satisfies it, so does the sequence $t \cdot x$ for any number t; the factor t cancels through the equation.

We now look at a general method which will solve (2.3.3). First we introduce an operator E which changes each sequence y into 'y shifted one along', that is, Ey is the sequence which at n takes the value $y(n+1)$, so

$$Ey(n) = y(n+1). \tag{2.3.4}$$

For brevity, if t is any number we also write $E + t$ for the operator which converts y to the sequence $Ey + ty$; thus

$$(E + t)y = Ey + t \cdot y = \langle y(n+1) + t \cdot y(n) \rangle. \tag{2.3.5}$$

We can also operate with E on Ey to produce the sequence E^2y, that is, $E^2y(n) = y(n+2)$, so now we can write (2.3.3) as

$$(E^2 + a \cdot E + b)x = f \tag{2.3.6}$$

and think of the bracketed part as one operator. This suggests that we should extend (2.3.5) to get (for s a second number)

$$\begin{aligned}(E+s)(E+t)y &= E(E+t)y + s(E+t)y \\ &= E^2y + (t+s)Ey + st \cdot y.\end{aligned}$$

Hence, we have the same operator as in (2.3.6) if we choose s, t so that $a = s + t$ and $b = st$, which means that $-s, -t$ are the roots of what is known as the **auxiliary equation**

$$u^2 + au + b = 0. \tag{2.3.7}$$

Therefore, we change the signs of s, t and rewrite equation (2.3.6) as

$$(E - s)(E - t)x = f \tag{2.3.8}$$

where now s, t are the roots of (2.3.7). We then split (2.3.8) into two first-order equations that we have seen before:

$$\text{(i) } (E - s)y = f \qquad \text{where} \qquad \text{(ii) } (E - t)x = y.$$

Here (i) is just equation (2.1.5) with k and $R(n)$ there replaced by s and $-f$, so y is given by (2.1.7) with $P(0)$ replaced by $y(0)$. But we can use (ii) to see that $y(0) = x(1) - t \cdot x(0)$; and now (ii) is of the form (2.1.5) with y as the known function. We can therefore use (2.1.7) to solve (ii) for x and we are finished.

As a concrete example, let us solve the Fibonacci equation (2.3.2) by this method. We first solve the auxiliary equation (2.3.7) which here is

$$u^2 - u - 1 = 0, \qquad \text{with roots } u = \frac{1 \pm \sqrt{(1 - (-4))}}{2}$$

so

$$s = \frac{1 + \sqrt{5}}{2}, \qquad t = \frac{1 - \sqrt{5}}{2} \tag{2.3.9}$$

with approximate decimal expansions

$$s \cong 1.61803399, \qquad t \cong -0.618033989.$$

(The number s is often called the **golden section** for historical reasons: see exercise 2.3.1(6) below.)

The equations (i), (ii) become, respectively,

$$(E - s)y = 0, \qquad (E - t)x = y \tag{2.3.10}$$

and the first has solution $y(n) = y(0) \cdot s^n$ where, using the given initial conditions, we have $y(0) = x(1) - t \cdot x(0) = 1 - t = s$ by (2.3.9). Thus we are left with solving the second equation, which is of the form (2.1.5) with $k = t$ and $R(n) = -y(n) = -y(0) \cdot s^n$. Since $s \neq t$ then by exercise 2.1.1(10) we get

$$x(n) = t^n \cdot x(0) - y(0) \cdot \frac{t^n - s^n}{t - s}$$

which is of the form

$$x(n) = A \cdot t^n + B \cdot s^n. \tag{2.3.11}$$

With $A = 1 - B$ and $B = s/\sqrt{5}$. It is rather surprising that this formula should always give $x(n)$ as an integer, but the square roots cancel nicely. The approximate values given above for s, t show that the term $A \cdot t^n$ soon becomes negligible, so that $x(n) \cong B \cdot s^n$ and the ratio $x(n + 1)/x(n)$ of successive terms is approximately s. Thus the growth is approximately that of the Malthus model.

Except for substituting the particular values of s, t, and $x(0)$, the above working would have been valid with any homogeneous equation (i.e., (2.3.6) with $f = 0$), provided the roots s, t of the auxiliary equation (2.3.7) are unequal, to give a solution of the form (2.3.11). When $s = t$ we refer again to exercise 2.1.1(10) to obtain $x(n)$ in the form

$$x(n) = (nA + t \cdot x(0)) \cdot t^{n-1}, \qquad (n > 0). \tag{2.3.12}$$

Knowing these forms, it is often quicker to start with them and to find the two constants by using the initial conditions. For example, if $s \neq t$ we use (2.3.10) to get

$$x(0) = A + B, \qquad x(1) = A \cdot t + B \cdot s \tag{2.3.13}$$

and we can solve to find A, B. But for a non-homogeneous equation (2.3.6), there are not usually such short cuts.

Example 2.3.1

Consider the homogeneous equation: $x(n + 2) + 2 \cdot x(n + 1) - 3 \cdot x(n) = 0$. Comparing this with (2.3.3) the auxiliary equation (2.3.7) is: $u^2 + 2 \cdot u - 3 = 0$; so the given equation is: $(E + 3)(E - 1)x = 0$, with general solution: $x(n) = A \cdot (-3)^n + B \cdot (1)^n$ by

(2.3.11). If $x(0)$ and $x(1)$ are specified, say $x(0) = 2$, $x(1) = 7$, then by (2.3.13): $2 = A + B$, $7 = -3 \cdot A + B$; whence $A = -5/4$, $B = 13/4$. Hence the solution is $x(n) = [13 - 5 \cdot (-3)^n]/4$, which oscillates wildly.

Exercises 2.3.1

(Again x_n is used as an alternative for $x(n)$.)

1. Find the complete solution of each of the following equations:
   (a) $x_{n+2} + 7 \cdot x_{n+1} + 12 \cdot x_n = 0$
   (b) $x_{n+2} + 6 \cdot x_{n+1} + 9 \cdot x_n = 0$
   (c) $x_{n+2} + 2 \cdot x_{n+1} + 2 \cdot x_n = 0$
   (d) $x_{n+1} - 3 \cdot x_n = 4^n + n$
   (e) $x_{n+2} + 4 \cdot x_n = \cos n$
   (f) $x_{n+2} - 4 \cdot x_{n+1} + 4 \cdot x_n = 2^n$
2. Solve the difference equations:
   (a) $x_{n+2} - 5 \cdot x_{n+1} + 6 \cdot x_n = 0$ $\quad x_0 = 1, \quad x_1 = 2$
   (b) $5 \cdot x_{n+2} - 14 \cdot x_{n+1} - 3 \cdot x_n = 0$ $\quad x_0 = -1, \quad x_1 = 6$
   (c) $10 \cdot x_{n+2} + 3 \cdot x_{n+1} - x_n = 0$ $\quad x_0 = 1, \quad x_1 = -1$
   (d) $x_{n+2} + 3 \cdot x_{n+1} + 2 \cdot x_n = 5$ $\quad x_0 = 1, \quad x_1 = 2$
   (e) $x_{n+2} + 2 \cdot x_{n+1} + x_n = n$ $\quad x_0 = 2, \quad x_1 = 1$
3. By assuming that $(2x + 7)/(x^2 + 5x + 6)$ has an expansion in ascending powers of x in the form $a_0 + a_1 \cdot x + a_2 \cdot x^2 + \cdots$, prove that $6 \cdot a_{n+2} + 5 \cdot a_{n+1} + a_n = 0$ $(n \geq 0)$. Solve this difference equation to find the coefficient of x^n in the expansion.
4. A plant is such that each of its seeds when one year old produces 21-fold and produces 44-fold when two years old or more. A seed is planted and, as soon as a new seed is produced, it is planted. Taking $u(n)$ to be the number of seeds produced at the end of the nth year, show that:

   $$u(n + 1) = 21 \cdot u(n) + 44 \cdot (1 + u(1) + u(2) + \ldots + u(n - 1)).$$

   Hence show that $u(n + 2) - 22 \cdot u(n + 1) - 23 \cdot u(n) = 0$ and find $u(n)$.
5. If $u(n) = \int_0^\pi \cos nx/(5 - 3 \cos x)\, \mathrm{d}x$ (where $n \geq 0$ is an integer), prove that $3 \cdot u(n + 2) - 10 \cdot u(n + 1) + 3 \cdot u(n) = 0$ and deduce that $u(n) = (1/3)^n \cdot \pi/4$.
6. A rectangle has length l and breadth b. These dimensions are said to be in the 'divine proportion', or 'golden section' if, when a square is removed as shown on Fig. 2.3.1, a rectangle (shaded) remains whose proportions are again in the same ratio. By considering a sequence of rectangles with dimensions $l(n)$, $b(n)$ each obtained from the next by the above process of removal, find a second-order difference equation for $b(n)$ and hence find the value of the golden section.

2.4 PAIRS OF POPULATIONS

In the Fibonacci problem, it was natural to consider the population as split into two populations, one of adults and one of young. We then derived the Fibonacci recurrence equation (2.3.2) from the pair of simultaneous first-order equations (2.3.1), and in terms of the column vector $\begin{pmatrix} a(n) \\ b(n) \end{pmatrix}$ we can rewrite these in matrix form

$$\mathbf{c}(n + 1) = F \cdot \mathbf{c}(n), \qquad \mathbf{c}(n) = \begin{pmatrix} a(n) \\ b(n) \end{pmatrix}$$

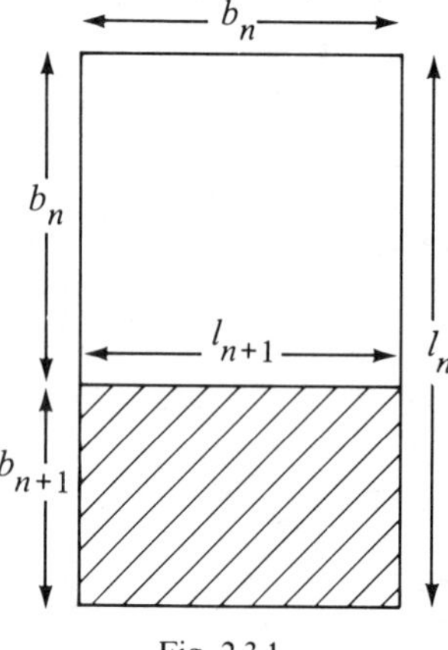

Fig. 2.3.1.

where

$$F = \begin{pmatrix} 1 & 1 \\ 1 & 0 \end{pmatrix}.$$

The matrix equation can be solved by iteration, as with the Malthus equation (2.1.1), and we have the solution analogous to (2.1.2):

$$\mathbf{c}(n) = F^n \cdot \mathbf{c}(0)$$

which is simple in form but leaves us with the problem of calculating the powers F^n of F.

Again it is easy to use a computer or calculator to study the behaviour of $\mathbf{c}(n)$ or that of F^n for a particular matrix F. We now introduce a matrix technique for the general mathematical analysis of such behaviour. This involves finding the **eigenvalues** of the matrix F. Readers unfamiliar with matrix theory may prefer to read Chapter 7 before following the details of this section, and of section 6. Following the technique of (7.4.4) we first calculate the eigenvalues of F; these are the roots of its **characteristic equation**

$$0 = u^2 - u - 1$$

since F has trace and determinant equal to 1 and -1, respectively. This quadratic equation is the same as the auxiliary equation with roots s, t in (2.3.9); and since s and t are real and unequal then by a theorem (proved as theorem 7.4.1), there is an invertible matrix R such that $F = R^{-1}DR$, where D is a diagonal matrix with diagonal entries s, t, that is $D = \begin{pmatrix} s & 0 \\ 0 & t \end{pmatrix}$. Hence

$$\mathbf{c}(n) = R^{-1}D^nR \cdot \mathbf{c}(0) = R^{-1}\begin{pmatrix} s^n & 0 \\ 0 & t^n \end{pmatrix} R \cdot \mathbf{c}(0) \tag{2.4.1}$$

from which we can obtain $a(n)$, $b(n)$ by multiplying out the matrix product. But we can understand this solution better if we write (2.4.1) in the form

$$\mathbf{d}(n) = D^n \cdot \mathbf{d}(0), \qquad \mathbf{d}(n) = R \cdot \mathbf{c}(n),$$

and study the track of the points $d(1), \mathbf{d}(2), \ldots, \mathbf{d}(n) \ldots$ as n increases. This track is called the **orbit** of $\mathbf{d}(0)$, and R maps each point $\mathbf{c}(n)$ of the orbit of $\mathbf{c}(0)$ onto the point $\mathbf{d}(n)$ of the orbit of $\mathbf{d}(0)$. For the Fibonacci matrix F the **eigenvectors**, given by:

$$\mathbf{v}_\lambda = \begin{pmatrix} s \\ 1 \end{pmatrix} \quad \text{and} \quad \mathbf{v}_\mu = \begin{pmatrix} t \\ 1 \end{pmatrix}$$

are **orthogonal** since $\mathbf{v}_\lambda \cdot \mathbf{v}_\mu = st + 1 = 0$ and in this case the matrix R is that of a rotation. Hence the orbits of **c** and **d** are identical in shape because one can be rotated through about 32° ($= \tan^{-1}(1/s)$) to coincide with the other.

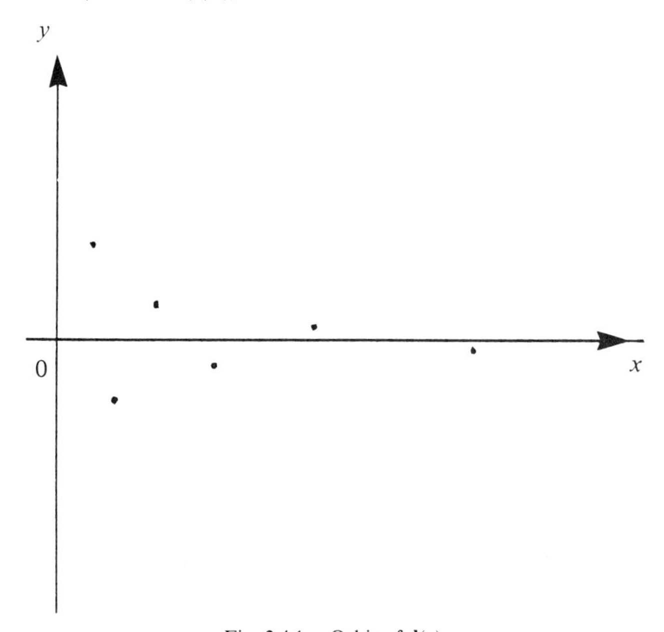

Fig. 2.4.1. Orbit of $\mathbf{d}(n)$.

As observed following (2.3.10), the powers of t soon become negligible. Hence the orbit of $\mathbf{d}(0)$ gets closer and closer to the x-axis as $\mathbf{d}(n)$ is driven further and further away from **0** owing to the rapid increase in the magnitude of its x-coordinate. In this example we have used a two-dimensional model of the Fibonacci problem, and by observing orbits we can get more insight into the actual growth pattern. More complicated behaviour can be observed in other problems, as in the next section. Meanwhile, we note here that the second-order equation (2.3.2) gave rise to a two-population problem (2.3.10); conversely a homogeneous problem

$$\mathbf{z}(n+1) = A \cdot \mathbf{z}(n), \qquad A = \begin{pmatrix} a & b \\ c & d \end{pmatrix}, \qquad \mathbf{z}(n) = \begin{pmatrix} x(n) \\ y(n) \end{pmatrix},$$

leads to a second-order homogeneous equation; for we are given

$$(E - a)x(n) = b \cdot y(n), \qquad (E - d)y(n) = c \cdot x(n),$$

whence, eliminating as in ordinary algebra gives,

$$(E - d)(E - a)x(n) = b(E - d)y(n) = bc \cdot x(n)$$

that is,

$$(E^2 - T \cdot E + \Delta)x(n) = 0,$$

where T and Δ are the trace $a + d$ and determinant $ad - bc$ of A, respectively. We obtain a similar equation for $y(n)$.

The following program should help in investigating orbits in this and the next two sections.

```
REM Program 2.4.1, Linear orbits
OPTION BASE 0
DIM x(21), y(21)
READ a  , b, c, d  , xmax, ymax
DATA 1.2, 0, 0, 0.8, 15  , 10
SET MODE "graphics"
SET WINDOW -xmax,xmax, -ymax,ymax
DO
     INPUT PROMPT "x0 , y0 :  " : x(0),y(0)
     IF x(0) = 0 AND y(0) = 0 THEN EXIT DO
     FOR n = 0 TO 20
         PLOT x(n),y(n)
         LET x(n+1) = a*x(n) + b*y(n)
         LET y(n+1) = c*x(n) + d*y(n)
     NEXT n
LOOP
END
```

Exercises 2.4.1

1. Show that the homogeneous equation $(E - a)(E - b)x = 0$ is equivalent to the equation $\mathbf{z}(n + 1) = A \cdot \mathbf{z}(n)$, where $A = \begin{pmatrix} b & 1 \\ 0 & a \end{pmatrix}$.
2. Find the corresponding matrix B for each homogeneous equation in exercise 2.3.1. Show that the equation $x_{n+2} + px_{n+1} + qx_n = 0$ is also equivalent to the equation $\mathbf{z}_{n+1} = B \cdot \mathbf{z}_n$, where $B = \begin{pmatrix} 0 & 1 \\ -q & -p \end{pmatrix}$. (Put $y_n = x_{n+1}$.)

2.5 THE 'BATTLE' PROBLEM

Next we consider a very simplified problem concerning two warring armies. It is due to the Quaker pacifist L. F. Richardson. For further details see (Newman 1956).

We suppose that the numbers in the armies after n units of time are $x(n)$ and $y(n)$, and each soldier in the x-army can inflict casualties on the y-army at the rate of a per time interval. Thus, for the increment $\Delta y(n)$, we have

$$\Delta y(n) = y(n+1) - y(n) = -a \cdot x(n), \qquad (a > 0).$$

Similarly, we have a positive rate b for the soldiers of the y-army, and $\Delta x(n) = -b \cdot y(n)$. Thus we have a **coupled pair** of recurrence equations:

$$x(n+1) = x(n) - b \cdot y(n)$$

$$y(n+1) = y(n) - a \cdot x(n)$$

which can be used directly in a program.

```
REM Program 2.5.1, Battle
OPTION BASE 0
DIM x(100), y(100)
READ a    , b    , x(0), y(0)
DATA 0.02, 0.01, 1000, 2000
LET n = 0
DO
  PRINT n, x(n), y(n)
  LET x(n+1) = x(n) - b*y(n)
  LET y(n+1) = y(n) - a*x(n)
  LET n = n + 1
LOOP UNTIL  x(n) <= 0  OR  y(n) <= 0
END
```

The last value of n printed gives the duration of the battle. Note that in general the values of $x(n)$ and $y(n)$ are not integers—and so the model is clearly an idealization. The program can easily be adapted to plot graphs of any or all of: (i) $x(n)$ against n; (ii) $y(n)$ against n; (iii) $y(n)$ against $x(n)$. In the last case the time, represented by n, is a parameter, and the resulting curve is the orbit of the population which looks like (but is not quite) an arc of an hyperbola.

To analyse the situation mathematically we combine the two linked recurrence equations into a single form using the notation and techniques of matrices and vectors. Thus set $\mathbf{p}(n)$ as the column vector $\begin{pmatrix} x(n) \\ y(n) \end{pmatrix}$ representing the population at time n to obtain a difference equation

$$\mathbf{p}(n+1) = A \cdot \mathbf{p}(n), \qquad \text{with } A = \begin{pmatrix} 1 & -b \\ -a & 1 \end{pmatrix}, \tag{2.5.1}$$

and so

$$\mathbf{p}(n) = A^n \cdot \mathbf{p}(0). \tag{2.5.2}$$

Here, A has trace 2 and determinant $(1 - ab)$ and, as with the Fibonacci problem, we examine the eigenvalues of A. These are the roots λ, μ of the characteristic equation of A,

$$u^2 - 2u + (1 - ab) = 0$$

so, by the quadratic formula (6.1.1), $u = (2 \pm \sqrt{(4 - 4(1 - ab))})/2$ giving $\lambda = 1 + r$, $\mu = 1 - r$, ($r = +\sqrt{(ab)}$), and these are real, positive, and unequal if (as is reasonable to assume) a and b are small but non-zero. Therefore, by theorem 7.5.1 there is an invertible matrix R such that $A = R^{-1}DR$, where D is a diagonal matrix for which we use the self-explanatory notation $D = \mathrm{diag}(\lambda, \mu)$. In this case the eigenvectors of A are $\mathbf{v}_\lambda = \begin{pmatrix} r \\ -a \end{pmatrix}$ and $\mathbf{v}_\mu = \begin{pmatrix} r \\ a \end{pmatrix}$ giving $R^{-1} = \begin{pmatrix} r & r \\ -a & a \end{pmatrix}$ and $R = \dfrac{1}{2ar}\begin{pmatrix} a & -r \\ a & r \end{pmatrix}$. In this case the eigenvectors will only be orthogonal if $a = 0$ or $a = b$, so in general R is not a rotation.

By (2.4.1), we can now find $\mathbf{p}(n)$ as:

$$\mathbf{p}(n) = R^{-1} \cdot \mathrm{diag}(\lambda^n, \mu^n) \cdot R \cdot \mathbf{p}(0),$$

which is of the form: (matrix) $\cdot\, \mathrm{diag}(\lambda^n, \mu^n) \cdot$ (vector). Hence we see that the coordinates $x(n)$, $y(n)$ of $\mathbf{p}(n)$ are of the form

$$x(n) = B \cdot \lambda^n + C \cdot \mu^n, \qquad y(n) = E \cdot \lambda^n + F \cdot \mu^n$$

for some constants B, C, E, F. To evaluate these, we have

$$x(0) = B + C, \qquad x(1) = B \cdot \lambda + C \cdot \mu = x(0) - b \cdot y(0)$$

by (2.5.1), and on solving these equations we find

$$B = \frac{x(0) \cdot \alpha - y(0) \cdot \beta}{2\alpha}, \qquad C = \frac{x(0) \cdot \alpha + y(0) \cdot \beta}{2\alpha}$$

where $\alpha = \sqrt{a}$, $\beta = \sqrt{b}$. To find E, F we interchange x and y, and a and b, so

$$E = -B \cdot s, \quad F = C \cdot s \qquad \text{where } s = \sqrt{\frac{a}{b}}.$$

Thus C and F are positive. Hence if $B > 0$ then always $x(n) > 0$ but $y(n)$ becomes negative when $-Bs\lambda^n + Cs\mu^n < 0$, which is to say: as soon as n exceeds $\ln(C/B)/\ln(\lambda/\mu)$ the y-army will have been annihilated while the x-army still has some soldiers left. Similarly, if $B < 0$ then E and F are positive, so $y(n)$ then stays positive and the x-army is annihilated.

The condition that $B > 0$ says that

$$a \cdot x(0)^2 - b \cdot y(0)^2 > 0 \tag{2.5.3}$$

so by computing B we can decide the outcome of the battle without a shot being fired.

As a pacifist, Richardson felt that the opposing commanders needed only to know the numbers a, b, $x(0)$, $y(0)$ to calculate B, and no blood need be shed: discussion was the only sensible way out. However, one commander might instead use the model to decide that the value of B was in his favour, and choose that opportunity to wipe out the enemy. Indeed, it has been argued that Nelson might have known this 'square law' instinctively (rather than as a piece of mathematics) because at Trafalgar he split the French fleet into two parts, beating the first according to this law, and using his remnants to beat the second. But then, Nelson was a thoughtful commander, and it is possible that the above mathematics could give pause to any later commander who normally thought only with verbal models about complex situations.

We now consider the shape of the orbits of $\mathbf{p}(n)$ in (2.5.2). In the next chapter we shall see that the analysis of a similar model using continuous change leads to orbits which are arcs of hyperbolae. We shall now show that for the discrete case the orbit 'jumps' between neighbouring points on a family of hyperbolae with common asymptotes. This is akin to the way in which some numerical techniques for solving differential equations (such as the simple Euler method) 'slip' at each step onto a neighbouring solution curve.

To study the orbit of $\mathbf{p}(n)$ in (2.5.1) we look at the expression analogous to (2.5.3) and use (2.5.1) to calculate:

$$\begin{aligned} a\cdot x(n+1)^2 - b\cdot y(n+1)^2 &= a\cdot(x(n) - b\cdot y(n))^2 - b\cdot(-a\cdot x(n) + y(n))^2 \\ &= (1-ab)(a\cdot x(n)^2 - b\cdot y(n)^2). \end{aligned}$$

So, if we write $Z(n) = a\cdot x(n)^2 - b\cdot y(n)^2$ we have $Z(n+1) = (1-ab)\cdot Z(n)$ and so $Z(n) = (1-ab)^n\cdot Z(0)$, that is

$$a\cdot x(n)^2 - b\cdot y(n)^2 = (1-ab)^n\cdot Z(0). \qquad (2.5.4)$$

Now

$$\begin{aligned} Z(0) &= a\cdot x(0)^2 - b\cdot y(0)^2 \\ &= (\sqrt{a}\cdot x(0) - \sqrt{b}\cdot y(0))\cdot(\sqrt{a}\cdot x(0) + \sqrt{b}\cdot y(0)) \\ &= 2\cdot\sqrt{a}\cdot(\sqrt{a}\cdot x(0) + \sqrt{b}\cdot y(0))\cdot B = k\cdot B, \end{aligned}$$

say, and thus $\mathbf{p}(n)$ lies on the **hyperbola**:

$$H(n): \qquad a\cdot x^2 - b\cdot y^2 = k\cdot B\cdot(1-ab)^n.$$

These hyperbolae all share the same asymptotes ($a\cdot x^2 - b\cdot y^2 = 0$) and lie in the same V-shaped region between them. Also as n increases, $\mathbf{p}(n)$ jumps to $\mathbf{p}(n+1)$ from $H(n)$ to $H(n+1)$. The orbit is best seen by means of computer graphics (Fig. 2.5.1), for which we refer the reader back to Program 2.5.1.

Returning to the theory, suppose next that our two armies replace casualties by recruiting new soldiers at rates proportional to their numbers. Then the increments leading to (2.5.1) are changed to:

$$\Delta x(n) = p\cdot x(n) - a\cdot y(n), \qquad \Delta y(n) = -b\cdot x(n) + q\cdot y(n) \qquad (2.5.5)$$

where p, q are positive rates of recruitment. As we shall see after (2.6.3) below, it turns out that the military outcome is essentially the same as before. Without further analysis, one might expect that the fortunes of the battle might oscillate, with first one army in the ascendant and then the other. Mathematically, this could happen; but first we remark that such behaviour can occur—with rather more general mathematics—in other types of population, as we now discuss.

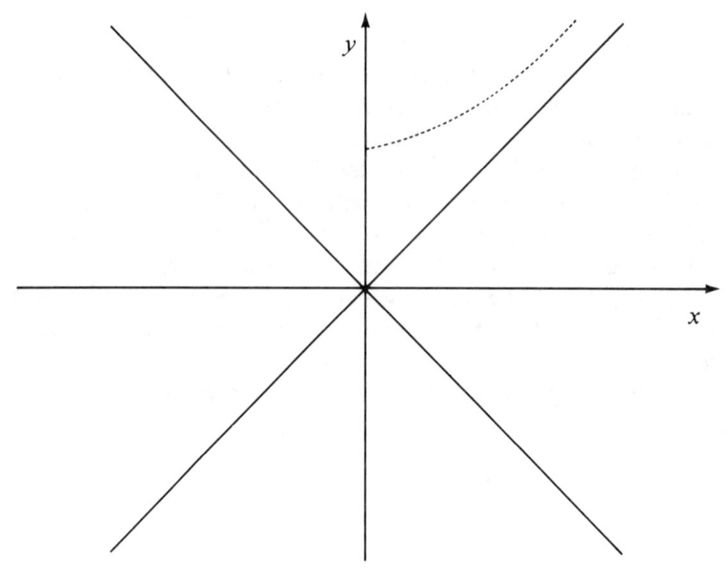

Fig. 2.5.1.

Often in nature, one observes pairs of populations interacting, but not like armies. For example, we observe predator and prey, or host and parasite, in each of which it is in the interest of at least one population to keep the other thriving. A standard example is that of sharks and minnows in an ocean: if there are many minnows the sharks will thrive, but they will then kill off minnows at a great rate, so that eventually the sharks begin to die off as food becomes scarce, and the minnows have a safer environment in which to rear their young. As they flourish once more, they provide easier food supplies for the few remaining sharks and the previous cycle begins again. This description is itself a model (albeit crude) which uses literary, rather than mathematical, language; it is nevertheless strong enough to sound reasonable and to suggest what type of mathematical solution we should expect.

A simple mathematical model is given by a pair of equations like (2.5.5), and when we carry $x(n)$ and $y(n)$ across from the increments we can write the result in a more standard form:

$$x(n) = a \cdot x(n) + b \cdot y(n), \qquad y(n) = c \cdot x(n) + d \cdot y(n) \tag{2.5.6}$$

wherein the constants a, b, c, d have meanings appropriate to the particular problem that we happen to be modelling. If $a = 1 = d$, then we have the original 'battle' problem when the other constants are negative. We have a richer situation if, for example, x refers to the predator population and y to the prey; in that case we would take a and b to be positive, and assume that x will increase in the 'Malthus' way due to its own birth-rate, and that it also increases with its food supply. But we expect c to be negative, because increase in prey leads to greater danger for y; while d is taken positive for the same reason as with x.

If, however, x is a parasite, then its presence may be good for the host y, and c would be positive. The reader is invited to consider other pairings of populations in the real world, such as cars and petrol stations, managers and workers, and so on, and to decide the signs of the constraints. It may be objected that the Malthus term, for example, should not be linear, but as in the Verhulst model. To that we reply that we are keeping the mathematics as simple as possible at first, and we shall complicate it later. There are also biological objections; for example, when a predator is short of prey it may change to another type of food. Another type of objection is that we are in any case using deterministic mathematics to model a situation where the data can only be obtained subject to certain probabilities, which is to say that we are using the wrong sort of mathematics to make our models. But different types of model can show hitherto unsuspected relationships, and at this stage, we are keeping things as simple as possible just to open up the possibilities. Mathematicians should also be aware that what they may consider to be a good explanation of a phenomenon (in the sense of helping them to understand it) can leave a biologist cold: the biologist may 'understand' better with a computer simulation, for example, without any mathematical theory.

As mathematicians, we are interested here in analysing the general model given by (2.5.6), which brings us to the next section.

2.6 LINEAR DYNAMICAL SYSTEMS ON THE PLANE

The general model (2.5.6) can be written in the standard matrix form

$$\mathbf{z}(n+1) = A \cdot \mathbf{z}(n), \qquad \mathbf{z}(n) = \begin{pmatrix} x(n) \\ y(n) \end{pmatrix}, \qquad A = \begin{pmatrix} a & b \\ c & d \end{pmatrix}. \tag{2.6.1}$$

Our approach is geometrical rather than algebraic, in as much as we study the orbit of the initial point $P = \mathbf{z}(0)$ under iteration by the matrix A. That is to say, we want to work out the general shape of the set of points $\mathbf{z}(0), \mathbf{z}(1), \mathbf{z}(2), \ldots. \mathbf{z}(n), \ldots,$ these being given by the rule (2.6.1). 'Shape' will involve such matters as whether the orbit spirals out to infinity, or converges inwards toward the origin $\mathbf{0}$.

The model (2.6.1) is an example of what is called a **dynamical system on the plane** (since $\mathbf{z}(n)$ stays in $\mathbb{R}^2$), and the system is **linear** because $\mathbf{z}(n+1)$ in equation (2.6.1) is a linear function of $\mathbf{z}(n)$. The 'dynamics' refers to the motion of $\mathbf{z}(n)$ along its orbit. Of course, if P starts at the origin $\mathbf{0}$, it will stay there because $A \cdot \mathbf{0}$ is always zero. Therefore we assume that $\mathbf{z}(0)$ is not zero in what follows.

In the simplest case suppose A is a **similitude** of the form $A = \begin{pmatrix} a & 0 \\ 0 & a \end{pmatrix}$. If $a = 0$, then A is the zero matrix, and $\mathbf{z}(0)$ is moved by the dynamics to $\mathbf{0}$, there to remain forever. If $a \neq 0$, then $\mathbf{z}(n+1) = a \cdot \mathbf{z}(n) = \ldots = a^n \cdot \mathbf{z}(0)$, so each $\mathbf{z}(n)$ lies on the line L from $\mathbf{0}$ through $\mathbf{z}(0)$, and there are several possibilities. For example:

(i) $a = 1$; (ii) $0 < a < 1$; (iii) $a > 1$.

If (i) then $\mathbf{z}(n) = \mathbf{z}(0)$ and A is the identity matrix, and $\mathbf{z}(0)$ remains stuck. If (ii) then the powers a^n converge to zero so $\mathbf{z}(n)$ moves along L from $\mathbf{z}(0)$ and converges to $\mathbf{0}$. If (iii) and $a > 1$ then a^n increases without limit so $\mathbf{z}(n)$ moves along L away from $\mathbf{0}$

and off to infinity. But if (iv) $a < -1$ then a^n oscillates with ever larger swings from positive to negative, so $\mathbf{z}(n)$ oscillates on L across $\mathbf{0}$, and its distance away becomes infinite. (Readers who are unsure about convergence and limits may find it helpful to do some computer experiments with program 2.4.1. They should also investigate the cases $a = -1$ and $-1 < a < 0$ to see how the orbit oscillates.)

Next, we suppose that A is not a similitude. Then the description of the orbit is going to depend on the eigenvalues of A. At this point the reader will need to know the statement of a theorem, which we later prove as theorem 7.5.1. It tells us that there is an invertible matrix V such that $V^{-1}AV = S$, where S is a matrix of one of the three standard forms:

$$\text{(i)} \begin{pmatrix} \lambda & 0 \\ 0 & \mu \end{pmatrix}; \quad \text{(ii)} \begin{pmatrix} \lambda & 1 \\ 0 & \lambda \end{pmatrix}; \quad \text{(iii)} \begin{pmatrix} p & -q \\ q & p \end{pmatrix} \tag{2.6.2}$$

depending on whether the eigenvalues are (i) real and distinct, (ii) real and equal, or (iii) complex conjugates $p \pm jq$. We therefore make the transformation $\mathbf{z} = V\mathbf{w}$, so that (2.6.1) becomes $V \cdot \mathbf{w}(n+1) = AV \cdot \mathbf{w}(n)$; then since V is invertible, this gives

$$\mathbf{w}(n+1) = V^{-1}AV \cdot \mathbf{w}(n) = S \cdot \mathbf{w}(n), \qquad \mathbf{z}(n) = V \cdot \mathbf{w}(n).$$

Now V maps the orbit of $\mathbf{w}(0)$ to that of $\mathbf{z}(0)$, so we shall know the general features of the one if we know those of the other.

The orbit of $\mathbf{w}(0)$ consists of the set of points $\mathbf{w}(0)$, $\mathbf{w}(1) = S \cdot \mathbf{w}(0)$, $\mathbf{w}(2) = S^2 \cdot \mathbf{w}(0), \ldots,$ $\mathbf{w}(n) = S^n \cdot \mathbf{w}(0)$. To get more precise information, we now look at the possible forms that S can take, and these are expressed in terms of the eigenvalues λ, μ of A. Using the three forms (i)–(iii) above we must examine the corresponding orbits as follows.

Perhaps the most interesting occurs when S is of type (iii), when we can then write $S = r \cdot R_\theta$, for some rotation matrix R_θ of angle θ, where $r = |\lambda|$ (λ is complex so this is its **modulus**); that is,

$$S = \begin{pmatrix} r\cos\theta & -r\sin\theta \\ r\sin\theta & r\cos\theta \end{pmatrix} = r \cdot \begin{pmatrix} \cos\theta & -\sin\theta \\ \sin\theta & \cos\theta \end{pmatrix}.$$

Hence $\mathbf{w}(n) = r^n \cdot R_{n\theta}$, and if $r = 1$ then $\mathbf{w}(n)$ moves on the unit circle with centre $\mathbf{0}$ and moves through an anticlockwise angle θ to $\mathbf{w}(n+1)$. Now $\mathbf{z}(n) = V \cdot \mathbf{w}(n)$, and the effect of an invertible matrix V on points of a circle centre $\mathbf{0}$ is to map them to points on an ellipse with centre $\mathbf{0}$. Thus $\mathbf{z}(n)$ moves steadily round that ellipse, always in the same direction. In this case the point $\mathbf{0}$ is known as a **centre**.

If $r \neq 1$ then $\mathbf{w}(n)$ winds round $\mathbf{0}$ anticlockwise on a spiral which converges towards $\mathbf{0}$ if $r < 1$, and which expands outwards to infinity if $r > 1$. (The reader might like to experiment with this on computer graphics by taking various values for r.) Thus the orbits either all converge to $\mathbf{0}$ (and we then call $\mathbf{0}$ a **spiral sink**) or they all diverge to infinity (and we then call $\mathbf{0}$ a **spiral source**). This is illustrated in Fig. 2.6.1.

If S is of type (i) above, then $S = \text{diag}(\lambda, \mu) = \begin{pmatrix} \lambda & 0 \\ 0 & \mu \end{pmatrix}$ with λ, μ real and distinct.

Hence $\mathbf{w}(n) = S^n \cdot \mathbf{w}(0)$ and $S^n = \text{diag}(\lambda^n, \mu^n)$; so, for example, $\mathbf{w}(n)$ converges to $\mathbf{0}$ if λ, μ are each of modulus < 1. In this case, all orbits converge to $\mathbf{0}$, and $\mathbf{0}$ is a **sink**;

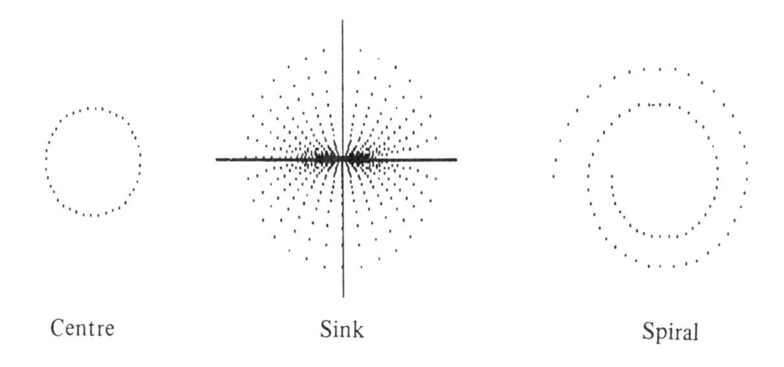

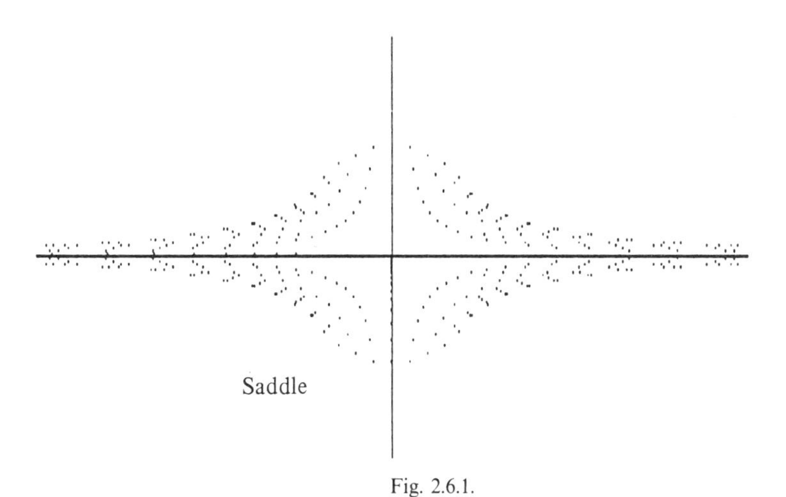

Fig. 2.6.1.

if λ, μ each exceed 1 then all orbits diverge to infinity, and **0** is a **source**. There are several other cases to consider: $\lambda = 1 < \mu$, $\lambda < -1 < \mu < 1$, and so on, and readers will find it more enlightening (and requiring less stamina) to investigate these cases personally, than to read the details at length. The case when $(\lambda - 1)(\mu - 1) < 0$ occurs rather often, and **0** is then called a **saddle**, because of the shape of the orbits in, for example, the battle problem (2.5.1).

Finally, if S is of type (ii) then it can be shown that (e.g., in exercise 7.5.1(2)),

$$S^n = \begin{pmatrix} \lambda^n & (n-1)\lambda^{n-1} \\ 0 & \lambda^n \end{pmatrix} \tag{2.6.3}$$

so that, for example, $a(n)$ converges to **0** if $\lambda < 1$ (in which case **0** is called a **degenerate sink**). Again we have various other cases depending on the size and sign of λ, and it is best for readers to try these for themselves, perhaps with the aid of program 2.4.1.

The case (i), with real distinct eigenvalues, is that of the battle problem (2.5.1) with recruitment; for in that notation, the discriminant D of A is $(p - q)^2 + 4ab$ which is positive; and since $0 < \mu < 1 < \lambda$ we have in fact a saddle. The oscillations we mentioned can only occur by allowing just one of a, b to be negative—and in 'army' terms this would correspond to one army having growth proportional to the size of the other. Such a situation is unlikely to occur in an actual battle without constant desertion, to replenish the losers; but it does not happen in an arms race, when the size of the assumed 'threat' is used by the military lobby as an argument to get more funds.

We end this section with a warning: this analysis of these models shows that they are inadequate for most real-life situations, for the following reason. If $\mathbf{z}(n)$ represents a pair of populations then, except for case (iii) above with $r = 1$, the orbit of $\mathbf{z}(n)$ always either decreases to zero or grows infinite so the individual populations can only die out or explode—both populations die out if $\mathbf{0}$ is a source, and one of them dies out if $\mathbf{z}(n)$ becomes negative or $\mathbf{0}$ is a saddle. The exceptional case is unstable, as we shall discuss later. This paucity of types of behaviour arises from the linear nature of the dynamics, and if we wish to capture more realistic behaviour, then we are forced to look at some non-linear models (in which this linear mathematics nevertheless plays a vital part). The lack of understanding of this point has caused some users of mathematics to make dogmatic assertions about the future behaviour of systems, in economics especially, thinking that only 'linear' behaviour is possible. An example of the possible complexity of behaviour was seen in exercise 2.2.1(1) with the Verhulst model, therefore we shall need to develop more mathematics before we can return to look at non-linear discrete models in Chapter 10. Meanwhile we now proceed to look at a calculus approach.

Exercises 2.6.1

1. Continue exercise 2.4.1(2) by finding the standard form of each matrix and describing the dynamics of the planar system. (Does it display a saddle, sink, source, centre, etc?)
2. Let P denote a real invertible 2×2 matrix. Show that if $\mathbf{z}$ lies on the circle $x^2 + y^2 = r^2$ then $P\mathbf{z}$ lies on an ellipse. (Hint: the matrix equation $\mathbf{z}^T Q \mathbf{z} = c^2$ represents an ellipse if, and only if, $\det Q > 0$.)

   Show that if A has complex eigenvalues, then orbits of the difference equation $\mathbf{w}(n+1) = A \cdot \mathbf{w}(n)$ lie on circles or spirals according as $\det(A) = 1$ or not. Given the difference equation $\Delta\mathbf{z}(n) = B \cdot \mathbf{z}(n)$ on $\mathbb{R}^2$, where $\Delta\mathbf{z}(n) = \mathbf{z}(n+1) - \mathbf{z}(n)$ and B is a real 2×2 matrix, show that the points $\mathbf{z}(0), \mathbf{z}(1), \ldots, \mathbf{z}(n)$ all lie on an ellipse if:

   $$\mathrm{T} + \Delta = 0 \qquad \text{and} \qquad -4 < T < 0$$

   where T, Δ denote the trace and determinant of B, respectively. [SU]
3. In exercise 2.4.1, (1) and (2) give us two different ways of associating a matrix C with the equation $x_{n+2} + px_{n+1} + qx_n = 0$. Show that the two matrices, thus obtained, each have the same eigenvalues, and hence if one displays a sink, saddle, ..., etc, so does the other.
4. Describe the dynamics of the system (2.6.1) when A is one of the matrices

   $$\begin{pmatrix} 1 & 2 \\ 1 & 1 \end{pmatrix}, \quad \begin{pmatrix} 2 & 1 \\ -2 & 4 \end{pmatrix}, \quad \begin{pmatrix} 2 & 1 \\ 1 & 2 \end{pmatrix}, \quad \begin{pmatrix} 0.4, & 0.2 \\ 0.1, & 0.3 \end{pmatrix}$$

3

Growth and decay; some calculus models

In this chapter we consider how many of the discrete models from the last chapter can be tackled by the methods of the calculus when they are reformulated as models of continuous change.

3.1 THE MALTHUS MODEL AGAIN

Suppose we have a population which at time t has x members. In Chapter 2 we considered this problem when t was observed at discrete moments of time, and the law of growth was expressed in terms of the increment Δ in x in one unit of time. For many forms of change this increment is a crude measure of the average change in x per unit time. If x were to vary continuously, we could use smaller and smaller time intervals, and the resulting time averages could be expected to converge to the rate of change in the sense of calculus. The problem of predicting future values of x can then be handled by the powerful techniques of calculus.

For mathematical simplicity therefore, we now make an idealization, and assume that x varies continuously with t. This is not unreasonable when x is very large, as with populations of atoms, bacteria, memory cells, or even (in some economies) money. We are therefore working with a 'continuous' model, and apply to it the methods of calculus; naturally we assume that x is a differentiable function $x(t)$ of t. Then the Malthus model (2.1.2) becomes here a differential equation:

$$\frac{\mathrm{d}x}{\mathrm{d}t} = kx \tag{3.1.1}$$

where k is a constant which depends on the nature of the population. For example, a population of radium atoms will decay, and then $k < 0$; but a population of bacteria will grow, and then $k > 0$. The population may be totally static, in which case $k = 0$. Our aim is to predict what x will be at any time t after the start, given the initial value x_0 of x at time $t = 0$. Thus we must solve equation (3.1.1) to find what function x of t satisfies it.

Knowledge of the exponential function (for details see Chapter C) allows us to recognize at once that the solution is (for all times t):

$$x = x_0 \cdot e^{kt} \tag{3.1.2}$$

and this covers the case $k = 0$ since $e^0 = 1$. There are then three types of graph for x as a function of t: a horizontal line when $k = 0$, a steeply rising curve when $k > 0$, and a falling one when $k < 0$. These are shown in Fig. 3.1.1.

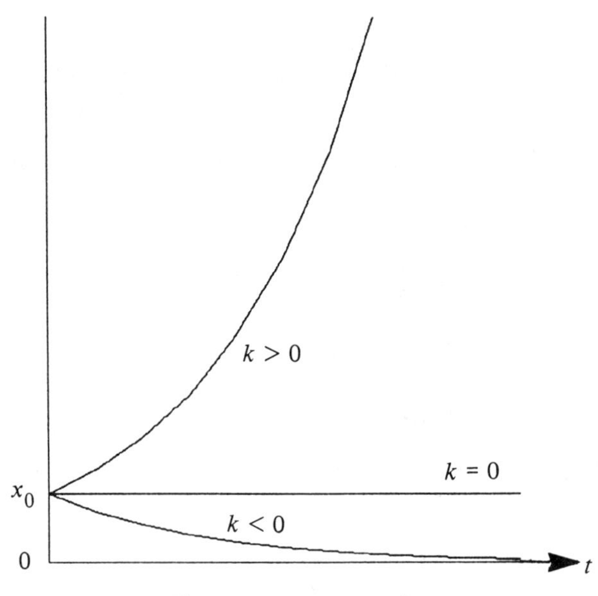

Fig. 3.1.1. Graphs of $x_0 e^{kt}$

With radium, it is important to know the half-life of x, which is the time τ by which $x(\tau) = \frac{1}{2}x_0$; thus by (3.1.2) we must have

$$\frac{1}{2} = e^{k\tau} \qquad \text{whence } \tau = -\frac{1}{k} \cdot \ln 2.$$

Since $k < 0$ and $\ln 2 > 0$ then τ is positive. Similarly, when $k > 0$ we can ask when x will double in size, and obtain a positive value:

$$\tau = \frac{1}{k} \ln 2$$

These values should be compared with those given by the discrete model in Chapter 2, the ratio of the calculus to the discrete values being

$$\frac{\ln(1 + k)}{k} = 1 + \text{error}$$

where (since $0 < k < 1$) the error is of magnitude $< \frac{1}{2}k$. Thus any dangers of handling such radioactive material, predicted from either model, will be about the same. The difference might be significant, though, to an archaeologist using methods of carbon dating.

As in (2.2.1) we next consider the Verhulst model that embodies a term representing competition for resources, and translate it into the language of calculus. Writing $\dot{x}$ for $\mathrm{d}x/\mathrm{d}t$, we obtain the differential equation $\dot{x} = b \cdot x - c \cdot x^2$, and this can be written as:

$$\dot{x} = bx(1 - ax) \tag{3.1.3}$$

where b and a are positive constants, and, for algebraic convenience, we have written the competition coefficient as ba rather than c.

From the equation itself, we see that if $x_0 < 1/a$ then $\dot{x}$ is positive at $t = 0$, so x increases towards a maximum population of size $1/a$; similarly if $x_0 > 1/a$ then x decreases towards $1/a$. These observations are confirmed when we integrate the equation. To simplify the calculus techniques we use a trick similar to that used in the discrete case after (2.2.3) and we write $x = 1/y$ so that

$$\dot{x} = -\frac{\dot{y}}{y^2}$$

which we substitute in equation (3.1.3) to get

$$-\frac{\dot{y}}{y^2} = \frac{b}{y} - \frac{ba}{y^2}$$

whence $\dot{y} + by = ba$, which is now a linear equation with solution:

$$y = a + (y_0 - a) \cdot \mathrm{e}^{-bt} \qquad \text{where } y_0 = \frac{1}{x_0}. \tag{3.1.4}$$

(Readers unfamiliar with techniques for solving differential equations should satisfy themselves, for now, that y is indeed a solution by differentiating it and substituting in the equation. The next chapter includes techniques for solving differential equations.)

Since $b > 0$, then $y \to a$ as $t \to \infty$, so $x \to 1/a$ as $t \to \infty$. Note that if $x_0 = 1/a$, then $y_0 = a$ and (by (3.1.4)) y remains at a, so x remains at $1/a$. Writing $x_{\mathrm{m}} = 1/a$, as in (2.2.6) we arrive at the following expression for x:

$$x = \frac{x_{\mathrm{m}}}{1 + (x_{\mathrm{m}}/x_0 - 1) \cdot \mathrm{e}^{-bt}}. \tag{3.1.5}$$

There are therefore three types of solution, as shown in Fig. 3.1.2, and these correspond to the three cases $x_0 = x_{\mathrm{m}}$, $x_0 < x_{\mathrm{m}}$, $x_0 > x_{\mathrm{m}}$. When $x_0 < x_{\mathrm{m}}$, the graph is called the **logistic curve** and the differential equation is called the **logistic equation**. Here, we have nothing corresponding to the chaos we saw with the discrete model (2.2.1), but we do have very similar behaviour to that of the modified model (2.2.3).

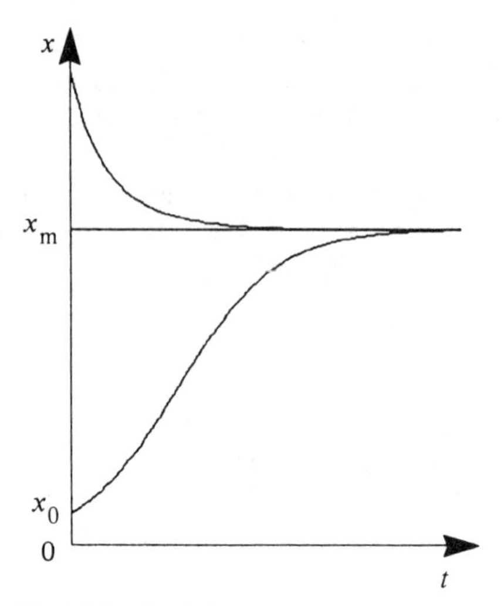

Fig. 3.1.2. Logistic curve and its relatives.

```
REM Program 3.1.1, Logistic curve
READ a, b  ,  x0, tmax
DATA 1, 0.1, 0.2, 100
LET xm = 1/a
SET MODE "graphics"
SET WINDOW 0,100 , 0,2*xm
FOR t = 0 TO tmax
   PLOT t , xm/(1 + (xm/x0 - 1)*EXP(-b*t)) ;
NEXT t
PLOT
END
```

Using this result, Verhulst predicted (in 1840) the populations of various countries, one hundred years ahead. He was remarkably accurate with the population of the United States, but highly inaccurate with that of his own country, Belgium; but in that span of time, Belgium changed from a rural country to a highly industrialized one, while the United States suffered a civil war, as well as accepting vast numbers of immigrants from Europe. Thus the constants b and c were not modelling the same things (such as birth-rates) either for each country or for the whole of the century.

Exercises 3.1.1

1. The death-rate of an ant colony is proportional to the number present. If no births were to take place, the population at the end of one week would be reduced by one-half. However, because of births, the rate at which it changes is also proportional to the population present, and the ant population doubles in two weeks. Determine the birth-rate of the colony per week. Compare the calculus solution with a model using difference equations and a suitable unit of time.
2. A tank contains 50 gallons of brine in which 25 pounds of salt are dissolved. Water begins to run into this tank at a constant rate of 2 gallons per minute, and the well-stirred mixture

flows out at the same rate (and starting at the same time) through a second tank initially containing 50 gallons of pure water. Show that the second tank contains the maximum concentration of salt after 25 minutes.

3. A nuclear scientist isolated one gramme of a radioactive element which she called Laurelium. It decayed at a rate proportional to the square of the amount present. After one year, half a gramme remained. Determine the mass of Laurelium present after t years. A rival scientist found a substance, Hardium, which decays at a rate proportional to the square root of the amount present. Half the original mass was present after one year. Show that Hardium decays entirely within a finite time T, and calculate T. Equal masses of Laurelium and Hardium are mixed together. Each element continues to obey its own law of decay, irrespective of the presence of the other element. Show that after 4 years, precisely 10% of the original mass of the mixture remains. Give a rough sketch to show the amounts of the two elements present during the decay process. [SU]
4. Suppose that the population of Erewhon is x at time t, and that the rate of change of the population satisfies the differential equation: $\dot{x} = Ax - Bt$, for $x > 0$, $t > 0$, where A and B are positive constants. Find x as a function of t, given that $x = x_0$ when $t = 0$ $(x_0 > 0)$. Deduce that either:
   (a) the population of Erewhon increases without limit; or
   (b) the population increases up to a certain maximum, then decreases until it becomes extinct.

   State conditions in terms x_0, A, and B for each of (a) and (b) to occur. It is found that $A = 0.025$, $B = 100$, and that the population has doubled after 50 years. Determine the initial population of Erewhon to 4 significant figures and show that the Erewhonians will become extinct after 130 years. [SU]
5. Using program 2.1.1c, plot the discrete output from the Malthus model for a fixed value of b (and hence k). Superimpose the graphs of curves of the form $x = x_0 \cdot e^{kt}$ and hence find an appropriate exponent k to fit the data.
6. Suppose Verhulst's equation (3.1.3) is changed to: $\dot{x} = bx(1 - ax^2)$. Use the same method as for (3.1.4) above to find the solution, and compare it with (3.1.5). (Hint: $2y\dot{y} = \mathrm{d}/\mathrm{d}t\,(y^2)$.) What if x^2 is changed to x^n? The integration becomes too difficult, but see program 4.6.1 below.

3.2 THE BATTLE PROBLEM

Next consider a problem with two populations, that of the two warring armies, of which we gave a discrete version in Chapter 2. The calculus model consists of a pair of simultaneous differential equations:

$$\dot{x} = -by, \qquad \dot{y} = -ax, \qquad (a, b > 0) \tag{3.2.1}$$

in which x, y are the numbers in each army at time t, and a, b are measures of wastage rates in the sense that each soldier in the x-army is able to destroy soldiers in the other army at an average rate of a per second, with a similar interpretation for b. Thus a, b are each positive. The assumption that x, y are continuous functions of t is of course unrealistic, but we pursue it to show how the mathematics simplifies.

One view of 'solving' these equations is to find functions x, y of t which satisfy the equations. Thus, if we differentiate the first equation in (3.2.1), we then obtain a single differential equation, but of the second order, which can be solved by the algebraic algorithm of the next chapter; but this gives little insight into the dynamical process that is going on.

Instead we use a small trick again, and multiply the first equation by ax and the second by by. Then we subtract and obtain

$$ax\dot{x} - by\dot{y} = 0$$

that is,

$$\frac{\mathrm{d}}{\mathrm{d}t}(ax^2 - by^2) = 0$$

whence

$$\begin{aligned} ax^2 - by^2 &= \{\text{value at } t = 0\} \\ &= ax_0^2 - by_0^2. \end{aligned} \tag{3.2.2}$$

Compare this with the solution of the discrete case in (2.5.4). Now, the state of the battle at time t is given by the pair of numbers $x(t)$, $y(t)$, which together form the coordinates of the point $P(t)$ in the coordinate plane $\mathbb{R}^2$; so as time goes on, $P(t)$ traces out a curve called its **trajectory**.

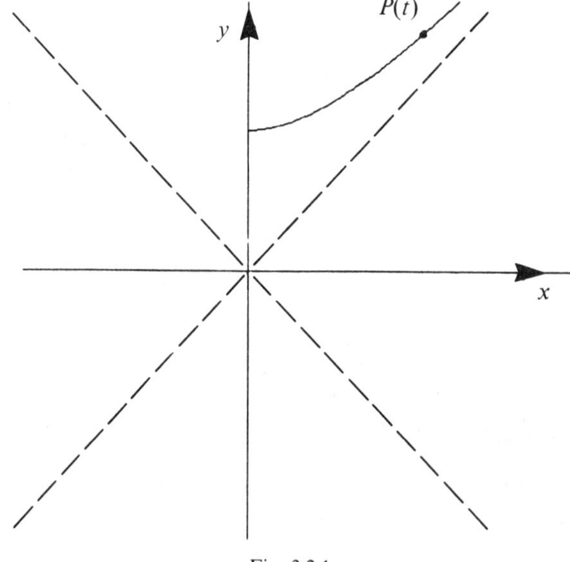

Fig. 3.2.1.

As we have seen, x and y have to satisfy (3.2.2), so the trajectory is a section of the hyperbola H, of which the asymptotes have equation

$$ax^2 - by^2 = 0$$

and the two lines have equations: $x\sqrt{a} \pm y\sqrt{b} = 0$. Also H lies above or below these. To decide which of these alternatives holds, let (in (3.2.2))

$$c = ax_0^2 - by_0^2. \tag{3.2.3}$$

Then H lies above the asymptotes if $c < 0$ and below if $c > 0$.

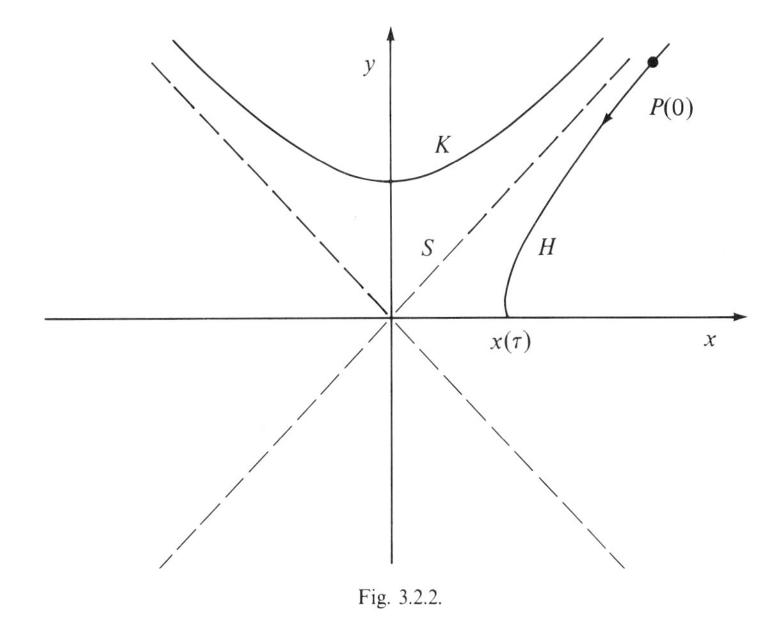

Fig. 3.2.2.

Therefore, we can decide the outcome of the battle, without a shot being fired, by reading Fig. 3.2.1 as follows. Suppose $c > 0$; then $P(0)$ lies on the branch of H shown separately in Fig. 3.2.2. Since a, b are each positive, then by (3.2.1) both x and y are decreasing. Hence $P(t)$ moves down the branch H until it hits the x-axis, which it does at a point with coordinates of the form $x = x(\tau)$, $y = 0$, and $x(\tau)$ is still positive. Therefore the x-army still has some soldiers left by time τ, whereas the y-army has been annihilated. If, instead, $c < 0$ then P would start on the branch K in Fig. 3.2.2, and would then run downwards until it hits the y-axis at a point of the form $x = 0$, $y = y(\tau)$. In this case $P(\tau)$ represents the end of a battle in which the y-army wins. With this model, as with the discrete one, the opposing generals only need to know the numbers a, b, x_0, y_0 and to calculate c, and then no bloodshed need follow: everything could be settled purely by discussion. Of course, the same practical considerations apply with this model as well.

Some points to notice about the above analysis are these. First, we have not needed to solve the equations (3.2.1) to find $x(t)$, $y(t)$ as explicit functions of t. If we do solve them, then we can find how long the battle will last. In fact, if $c > 0$ then

$$\tau = -\frac{1}{\sqrt{ab}}\sinh^{-1}\left(-\frac{\sqrt{b}}{\sqrt{c}}\,y_0\right)$$

and then:

$$x = \frac{\sqrt{c}}{\sqrt{a}}\cosh\left((t-\tau)\sqrt{ab}\right) \qquad y = -\frac{\sqrt{c}}{\sqrt{b}}\sinh\left((t-\tau)\sqrt{ab}\right)$$

with similar results if $c < 0$ with sinh and cosh interchanged.

Second, Fig. 3.2.2 shows that the positions of P are 'stable' in the sense that if small errors were made in calculating c in (3.2.3) when actual numbers are substituted, then we will remain on the same branch of the curve, and the outcome of the battle is the same. However, if $c = 0$, then $P(0)$ lies on the asymptote S between the two branches, and $P(t)$ simply moves along the line down to $\mathbf{0}$; the armies are annihilated simultaneously (though not in a finite time!). But if $P(t)$ were to wobble slightly onto K, for example, it would stay on K and the y-army would win. Such a wobble can easily be produced by rounding errors in computer arithmetic, so computer output in such unstable cases must be viewed with suspicion (or appropriate formulae must be build in, as in later exercises).

Exercises 3.2.1

1. If in (3.2.1), $a = -\frac{1}{4}$, $b = -\frac{1}{9}$, sketch the corresponding hyperbola. Decide who wins when (x_0, y_0) is (in thousands) (10, 7), (7, 10). In each case, calculate by how much the losing army would have needed to increase its fire-power in order to have won. Next suppose $a = -\frac{1}{8}$. With $x_0 = 10$ calculate y_0 so that (x_0, y_0) lines on the asymptote S. Substitute in the formula (3.2.3) for c to see whether your calculator gives $c = 0$. Investigate the internal arithmetic of your calculator, by deciding the sign of c for different values of x_0. (Because of this unstable situation, your calculator might mislead a commander as to the likely outcome of the battle.)
2. If the x-army is outnumbered 2:1 by the y-army, that is, $y_0 = 2 \cdot x_0$, show that the x-army can only win the battle if $a > 4 \cdot b$. Prove Lanchester's square law (1914) that if $y_0 = m \cdot x_0$ then the x-army can only win if $a > m^2 b$.

Returning to the theory, suppose next that our two armies replace casualties by recruiting new soldiers at rates proportional to their numbers. Then the differential equations (3.2.2) are changed to:

$$\dot{x} = px - by, \qquad \dot{y} = -ax + qy \tag{3.2.4}$$

and formulae for their solution are derived in the next chapter. Such formulae are difficult to penetrate, and more insight can be gained from a geometrical approach like the preceding one. We shall carry it out later on; suffice it to say for now that just as with the discrete model the military outcome is essentially the same as before. One might expect that the fortunes of the battle might oscillate, with first one army in the ascendant and then the other. Mathematically, this could happen if a, b in (3.2.1) are of opposite sign, for then instead of hyperbolae we obtain ellipses as trajectories. On these, $P(t)$ goes round and round, and the situation repeats endlessly with a fixed period. We shall not give the details, because we might as well look at what can occur with the other types of population that we discussed in the last chapter.

The calculus analogue of the model in (2.5.5) is given by a pair of equations like (3.2.4), which we now write in a more standard form:

$$\dot{x} = ax + by, \qquad \dot{y} = cx + dy \tag{3.2.5}$$

wherein the constants a, b, c, d have meanings appropriate to the particular problem that we happen to be modelling. If $a = 0 = d$, then we have the battle problem (3.2.1) already considered, where the other constants are negative. We have a richer situation if, for example, x refers to a predator population and y to its prey, in which case we would take a, b, and d positive, and c negative for the same reasons we used in the discrete model.

Exercises 3.2.2

1. Consider the equations (3.2.5) when $a = d$, and $c = -b$. (This seemingly special case is actually very important, as we shall see in the next section.) It is helpful to convert to polar coordinates by writing

$$r^2 = x^2 + y^2, \qquad \tan\theta = \frac{y}{x}.$$

Now differentiate each side to obtain $r\dot{r} = x\dot{x} + y\dot{y}$; substitute for $\dot{x}$, $\dot{y}$ from (3.2.5) to show that $\dot{r} = ar$. Also show that

$$\sec^2\theta \cdot \dot{\theta} = \frac{x\dot{y} - y\dot{x}}{x^2} = -b\frac{r^2}{x^2},$$

whence $\dot{\theta} = -b$. Thus θ increases clockwise or anticlockwise, according to $b > 0$ or $b < 0$, suppose $b < 0$. Since $\dot{r} = ar$, then r increases or decreases according as $a > 0$, or $a < 0$. Armed with this information draw rough sketches for different values of the pair (a, b), to see that the trajectories are spirals as in Fig. 3.2.3. There the arrow shows the spiral winding inwards, and we call **0** a **spiral sink**. When the spiral moves outwards as r decreases (but time increases), we call **0** a **spiral source**.

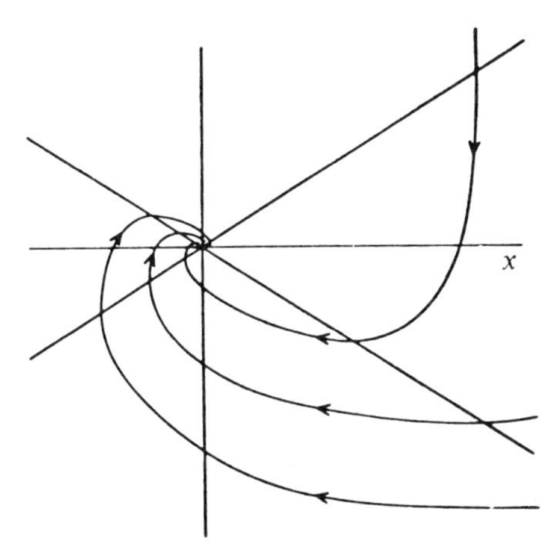

Fig. 3.2.3.

The case when $a = 0$ is very special, for then $\dot{r} = 0$. Show that $P(t)$ always moves on a circle of radius $r_0 = \sqrt{(x_0^2 + y_0^2)}$ so all trajectories are circles (that of **0** having zero radius). We then say that **0** is a **centre**.

If $b = 0$, show that the trajectories consist of all straight lines through **0** (which is then called a **degenerate focus**).

2. Now consider the equations (3.2.5) when $c = 0$ and $a = d$. Find y and insert this in the $\dot{x}$-equation to integrate and get

$$x(t) = (x_0 + tby_0)\cdot e^{at}.$$

Show that the x-axis consists of two trajectories, with both either converging to, or diverging from, **0**. At first sight it is not easy to work out the shape of the other trajectories, so it will be best for the reader to use a graphical calculator or graph plotting software to display some trajectories from their parametric equations, graphically. These should look like those in Fig. 3.2.4, and we call **0** a **degenerate node**. (The arrows point towards **0** if $a < 0$ and we have a **sink**; if $a > 0$ they reverse, and we have a **source**.)

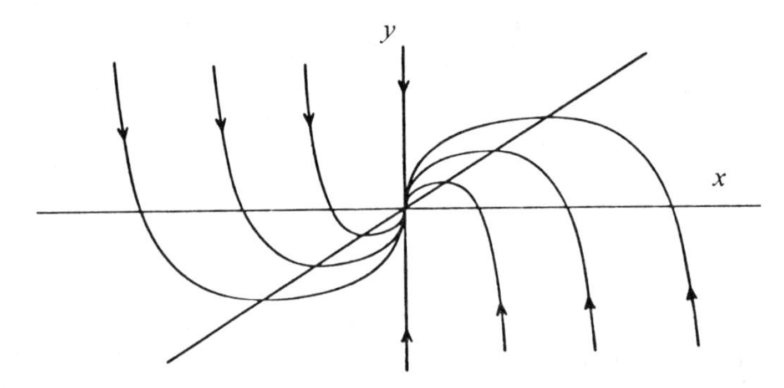

Fig. 3.2.4. Degenerate nodal sink.

3.3 SOME PORTRAITS OF SYSTEMS

Using matrix notation, we can write (3.2.5) in matrix form as:

$$\dot{\mathbf{z}} = A\mathbf{z} \tag{3.3.1}$$

where $\mathbf{z}$ is the column vector $\mathbf{z} = \begin{pmatrix} x \\ y \end{pmatrix}$, and $\dot{\mathbf{z}} = \begin{pmatrix} \dot{x} \\ \dot{y} \end{pmatrix}$, while A is the real 2×2 matrix $A = \begin{pmatrix} a & b \\ c & d \end{pmatrix}$.

We call (3.3.1) a **planar system**, more precisely a system of linear differential equations on the plane $\mathbb{R}^2$. To analyse the system, we again use the three standard types of matrix of (2.6.2). Already the systems analysed in exercises 3.2.2(1),(2) have A of types (iii), (ii), respectively, and the names 'spiral', etc., attached to them conform with those attached to the discrete case of section 2.6. When A is of type (i), however, there is more to analyse as we shall soon see.

At the outset, we must state what our general plan is. Given a general system of the form (3.3.1), we wish to describe the trajectory of a point when it starts at an initial position $P(0)$ and moves subject to the equation (3.3.1). Also, we want to have a sketch of all the various types of trajectory; this sketch is called the **portrait** of the equation (3.3.1). For example, if the portrait looks like that for (3.2.1) then we have a saddle, and the above exercises have displayed types of spiral sources and sinks, as well as of centres and a degenerate node. We shall work out mathematically all other possible types of portrait, and it turns out that there is only a small number of them.

First, we use a general simplifying principle: by (2.6.2), we know that given A in (3.3.1), there is an invertible matrix R such that $A = R^{-1}BR$, where B is a matrix of type (i), (ii), or (iii). Then we make the transformation $\mathbf{z} = R\mathbf{w}$ so that $\dot{\mathbf{z}} = R\dot{\mathbf{w}}$ (since R is constant), whence (3.3.1) becomes

$$\dot{\mathbf{w}} = B\mathbf{w}. \tag{3.3.2}$$

The trajectory of any point $\mathbf{w}_0$ in this new system (3.3.2) is mapped by R onto the trajectory of $\mathbf{z}_0 = R\mathbf{w}_0$, and if (for example) one is a spiral, so is the other. Hence the general features of the portrait of (3.3.2) are all transferred by the linear isomorphism R to become those of the original system (3.3.1): if one displays a source, etc., so does the other, and we regard the two portraits as being 'the same'—in a qualitative sense, because R will distort minor features such as angles and curvature.

Therefore, we need only investigate the possible portraits of (3.3.2) when B is of type (i)–(iii), and we have already dealt with (ii) and (iii) in the exercises above. Hence our next task is to give a systematic analysis of the case when B is diagonal (i.e., $b = 0 = c$), so B is either a similitude $\begin{pmatrix} a & 0 \\ 0 & a \end{pmatrix}$ or of type (i) $\begin{pmatrix} \lambda & 0 \\ 0 & \mu \end{pmatrix}$. It will be helpful first to investigate them by microcomputer. The computer pictures we generate will suggest some possibilities, but we need to know whether some may have been missed. Computer investigation is an experimental science, and cannot guarantee the completion of results as mathematics can. Of course there are situations when mathematical analysis is not feasible and computer simulation is the only available tool.

We must first think about what the equation (3.3.1) is telling us. At each point (u, v) of the plane, it provides us with a direction, namely that of the vector $\begin{pmatrix} ax + by \\ cx + dy \end{pmatrix}$. We can calculate this whenever we know the matrix A and the point (u, v). If we perform this calculation at each point X of a regular grid, we can then draw a small arrow at X in the direction of the calculated vector. There results a picture that resembles the distribution of iron filings when they are scattered on a sheet of paper that lies in a magnetic field. Our eye then tends to make us see the arrows (or filings) lined up along certain quite pronounced curves, to which they are tangent. These curves are trajectories, threading their way through the **vector field** created. The following program can be used for such investigations.

```
REM Program 3.3.1, Compass needle
READ a, b , c, d, w
DATA 2, -1, 1, 3, 0.05
READ xmin, xmax, ymin, ymax, xstep, ystep
DATA -3  , 3   , -2  , 2   , 0.25 , 0.25
SET MODE "graphics"
SET WINDOW xmin,xmax, ymin,ymax
PLOT xmin,0; xmax,0
```

Program 3.3.1 continued

```
PLOT 0,ymin; 0,ymax
FOR x = xmin TO xmax STEP xstep
  FOR y = ymin TO ymax STEP ystep
    LET dx = a*x + b*y
    LET dy = c*x + d*y
    LET s = SQR(dx*dx + dy*dy)
    IF s>0 THEN PLOT x-w*dx/s,y-w*dy/s; x+w*dx/s,y+w*dy/s
  NEXT y
NEXT x
END
```

and a sample output is shown in Fig. 3.3.1.

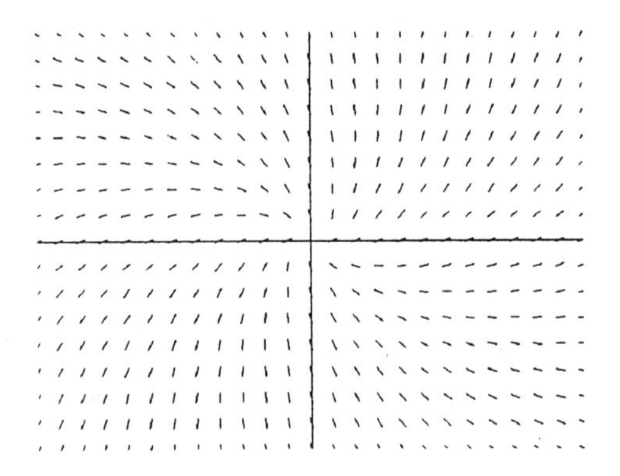

Fig. 3.3.1.

3.4 UNCOUPLED SYSTEMS

If B in (3.3.2) is a diagonal matrix, then the separate equations are of the form

$$\dot{x} = ax, \qquad \dot{y} = dy, \qquad (a^2 + d^2 \neq 0) \tag{3.4.1}$$

the condition in brackets saying that B is not the zero matrix. Since each equation can be integrated as in (3.1.2), the system is said to be **uncoupled**. Here, we call a, d the **control parameters** because as we change them the portrait of the system (3.4.1) will also change. Indeed our object is to obtain a full sample of all possible portraits as in Fig. 3.4.1; but it contains so much information that a good deal of explanation will be necessary to make the diagram intelligible to the reader. The pair (a, d) of parameters can run through the entire **control space** $\mathbb{R}^2$ except **0**, and it is best, first, to 'normalize' the system so as to have a smaller control space. In fact, we can restrict $\begin{pmatrix} a \\ d \end{pmatrix}$ to lie on the unit circle S, as we now show. This allows the circular arrangement of Fig. 3.4.1, and we can imagine the portraits changing as we turn a circular knob through an angle corresponding to the parameter value. The figure is complicated and we explain it as follows.

Fig. 3.4.1.

Clearly, it makes no difference to the trajectory of a point P whether or not P traverses it quickly, so we can rescale our unit of time by writing $u = kt$ with k constant. By the chain rule

$$\frac{\mathrm{d}x}{\mathrm{d}u} = \frac{\mathrm{d}x}{\mathrm{d}t} \cdot \frac{\mathrm{d}t}{\mathrm{d}u} = \frac{\dot{x}}{k}$$

and similarly for y; therefore if we choose $k = \sqrt{(a^2 + d^2)}$, then (3.4.1) becomes

$$\frac{\mathrm{d}x}{\mathrm{d}u} = sx, \qquad \frac{\mathrm{d}y}{\mathrm{d}u} = cy, \qquad \left(s = \frac{a}{k}, \quad c = \frac{d}{k}\right) \tag{3.4.2}$$

and now the control parameters s, c are the coordinates of the point $(\sin\theta, \cos\theta)$ with $\tan\theta = a/d$, lying on the circle S as required. As in (3.1.2) the solutions are

$$x = x_0 \cdot \mathrm{e}^{su}, \qquad y = y_0 \cdot \mathrm{e}^{cu}, \tag{3.4.3}$$

from which we see that if x_0, y_0 are both positive (and so lie in the positive quadrant of $\mathbb{R}^2$) then x, y remain in that quadrant. Further, if we change the sign of x_0, then we change the sign of x, so the trajectory is the reflection in the y-axis of the previous trajectory. Similarly, when (x_0, y_0) lies in the other quadrants, the resulting trajectory is a reflection in some coordinate axis of a trajectory in the first quadrant. We may suppose, therefore, that (x_0, y_0) lies in the first quadrant.

Consider first the special case when $s = 0$, so $c = \pm 1$; say $c = 1$. Then in (3.4.3), x remains constant at x_0, while y increases from y_0 to infinity. Thus the trajectory is a portion of the vertical line in $\mathbb{R}^2$ with equation $x = x_0$, and we can think of it as flowing away from the x-axis, each point of which is a source. The resulting portrait looks like Γ_4 in Fig. 3.4.1, and we call it a **gate**. If, instead, $c = -1$, we have the same portrait except that the flow is toward the x-axis, each point of which is now a sink. Similarly, at the other two points of the compass ($s = \pm 1, c = 0$), we obtain gates with horizontal trajectories and a vertical line of sources or sinks.

The same argument applies for any s, c if x_0 or y_0 is zero, so *the semi-axes are trajectories in every portrait*. At this point we interpose a type of self-testing exercise.

Exercises 3.4.1

1. If A is the zero matrix, show that the trajectory of each point consists only of that point; so the portrait consists of all the separate points of $\mathbb{R}^2$.
2. Sketch the portrait for the equation (3.4.1) when $a = 3$, $b = -2$, $c = 6$, $d = -4$. Does your solution agree with the theory of the next exercise (3)?
3. In (3.2.5) suppose the first equation is a constant multiple k of the second. Show that the portrait of (3.2.5) consists of the set of all parallel straight lines of gradient $1/k$ (if $k = 0$ this means the lines are vertical). If L is one of these lines, it is cut at Q, say, by the line $ax + by = 0$. Show that the flow is always towards Q or always away from Q analogously to the cases considered above. Hence the portraits are also called gates. (See Fig. 3.4.2.)

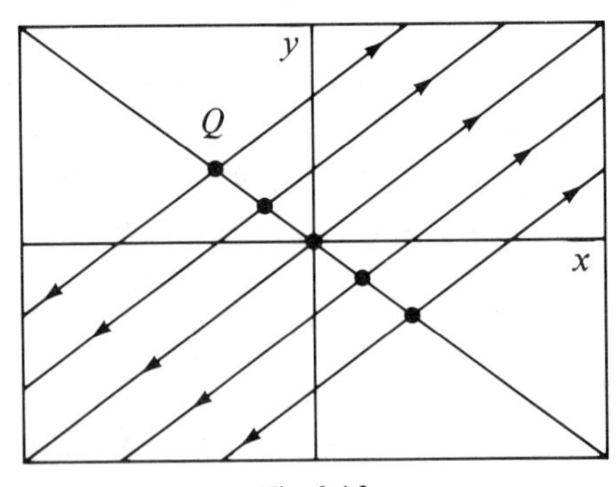

Fig. 3.4.2.

Continuing the deductions begun before the last exercises, we next suppose that s, c in (3.4.2) are both non-zero. Since x_0, y_0 are positive then we can raise the x-equation to the cth power, and the y-equation to the sth power, and obtain

$$\left(\frac{x}{x_0}\right)^c = (e^{su})^c = e^{scu} = \left(\frac{y}{y_0}\right)^s,$$

whence

$$y = k_0 \cdot x^p, \tag{3.4.4}$$

where

$$p = \frac{c}{s} = \cot\theta, \qquad k_0 = \frac{y_0}{x_0^p}.$$

The trajectory through (x_0, y_0) is therefore the graph of the function (3.4.4). Recalling Fig. 3.3.1, we can then assemble the portrait for the angle θ of the control variable (s, c); the flow at (x_0, y_0) on any trajectory is determined by the vector $\begin{pmatrix}\dot{x}\\ \dot{y}\end{pmatrix}$ at that point, and is for example north-easterly or south-westerly if s, c are both positive or both negative. Except for the gates, we say that the portrait displays a **focus**. Note that the arrows always flow either towards $\mathbf{0}$ or away from it, and accordingly we call $\mathbf{0}$ a **sink** or a **source**.

An 'adequate' sample of all the possible portraits is sketched in Fig. 3.4.1. Here, the portraits alternate between the stable and the unstable, in the sense that if we move slightly away from a stable portrait, we still have one that resembles it, whereas immediately on either side of an unstable portrait we have different types of stable portrait. The stable portraits depend continuously on the control parameter, whereas the unstable ones do not. (Always, however, the trajectory near any point $P \neq \mathbf{0}$ changes continuously.)

Readers should observe closely how the portrait Γ_i changes as θ varies steadily around the circle S. For example, with Γ_{14}, the value of p in (3.4.4) is -1 so the trajectories are rectangular hyperbolae, while on either side the trajectories of Γ_{13} and Γ_{15} resemble (but are not) hyperbolae; because of the resemblance to the portrait for the 'battle' problem, we say that each of these portraits displays a **saddle**. The same holds for portraits Γ_5, Γ_6, and Γ_7. As θ increases, Γ_7 changes into the unstable gate Γ_8, which then allows it to change to the stable sink Γ_9, with 'horizontal' parabolae which became 'vertical' in Γ_{11}, by first changing into the straight rays of the unstable Γ_{10}. These three portraits display $\mathbf{0}$ as a focus, as do $\Gamma_1, \Gamma_2, \Gamma_3$, but, in the latter, $\mathbf{0}$ is a source. Observe, too, how, by these changes, the sink Γ_9 changes eventually into the source Γ_1.

Exercise 3.4.2

Anticipating results about the solutions of linear differential equations in the next chapter we exhibit a computer program to plot the portraits shown in Fig. 3.4.1. For fixed values of a, b, c, d a set of solution curves can be plotted for a selection of initial values `(x0, y0)` to give a portrait. Try this with various a, b, c, d.

```
REM Program 3.4.1, Linear portraits
READ a, b , c, d
DATA 0, -1, 1, 0
READ xmin, xmax, ymin, ymax, tmax, tstep
DATA -6  , 6   , -4  , 4   , 10,   0.1
SET MODE "graphics"
SET WINDOW xmin,xmax, ymin, ymax
PLOT xmin,0 ; xmax,0
PLOT 0,ymin ; 0,ymax
LET trace = a + d
LET del = a*d - b*c
LET m2 = trace^2 - 4*del
DO
   INPUT PROMPT " x0, y0 : " : x0,y0
   IF x0 = 0 AND y0 = 0 THEN EXIT DO
   IF m2=0 THEN CALL equal
   IF m2>0 THEN CALL real
   IF m2<0 THEN CALL complex
LOOP
STOP

SUB equal
   LET u = trace/2
   FOR t = 0 TO tmax STEP tstep
     LET x = (x0 + ((a-u)*x0 + b*y0)*t)*EXP(u*t)
     LET y = (y0 + (c*x0 + (d-u)*y0)*t)*EXP(u*t)
     PLOT x,y;
   NEXT t
   PLOT
END SUB

SUB real
   LET m = SQR(m2)
   LET u = (trace + m)/2
   LET v = (trace - m)/2
   LET P = ((a-v)*x0 + b*y0)/m
   LET Q = ((u-a)*x0 - b*y0)/m
   LET R = ( c*x0 + (d-v)*y0)/m
   LET S = (-c*x0 + (u-d)*y0)/m
   FOR t = 0 TO tmax STEP tstep
     LET x = P*EXP(u*t) + Q*EXP(v*t)
     LET y = R*EXP(u*t) + S*EXP(v*t)
     PLOT x,y;
   NEXT t
   PLOT
END SUB


SUB complex
   LET g = trace/2
   LET h = SQR(-m2)/2
   LET P = x0
   LET Q = ((a-g)*x0 + b*y0)/h
```

Program 3.4.1 continued

```
        LET R = y0
        LET S = (c*x0 + (d-g)*y0)/h
        FOR t = 0 TO tmax STEP tstep
          LET x = (P*COS(h*t) + Q*SIN(h*t))*EXP(g*t)
          LET y = (R*COS(h*t) + S*SIN(h*t))*EXP(g*t)
          PLOT x,y;
        NEXT t
        PLOT
     END SUB
     END
```

3.5 SUMMARY: THE ALGORITHM

As explained in section 3.3, by having completed our analysis of the possible portraits of system (3.3.2) we have also completed the investigation of system (3.3.1). But readers may have been bemused by the considerable amount of detail we have met, and it remains to organize our conclusions into an intelligible whole that can be memorized.

Since the matrix A in (3.3.1) contains the four numbers a, b, c, d then the control space is the four-dimensional Euclidean space $\mathbb{R}^4$. Nevertheless, instead of a potential infinity of portraits, we have seen that there is only a finite, and small, number of 'qualitatively different' portraits. The previous theory can all be summarized in a simple algorithm that tells us, just by testing A, which portrait applies in any given case. The quantitative detail of the portrait can be supplied by a computer, if we should need it. We emphasize that we count two portraits as being of the same type if one can be mapped onto the other by a linear transformation. The algorithm uses the trace T, determinant Δ, and discriminant D of the matrix A in (3.3.1), which are given by:

$$T = a + d, \qquad \Delta = ad - bc, \qquad D = T^2 - 4\Delta.$$

Algorithm 3.5.1
The steps of the algorithm are then as follows.

(1) If $\Delta = 0$ then either
  (i) $A = 0$ (the zero matrix), or
  (ii) $A \neq 0$, in which case one row of A is a multiple of the other.
  If (i) then the trajectory of each point consists only of that point; so the portrait consists of all the separate points of $\mathbb{R}^2$.
  If (ii) suppose that the first row is a constant multiple k of the second row. Then the portrait of equation (3.3.1) (in this case) consists of the set of all straight lines in the plane, which have gradient $1/k$ (if $k = 0$ this means that the lines are vertical); the lines are therefore all parallel and the portrait is a **gate**.

(2) If $\Delta \neq 0$ then either (i) $\Delta < 0$ or (ii) $\Delta > 0$.
  If (i) then the portrait is a **saddle**.
  If (ii) then work out the discriminant D.
  If $T = 0$ then $D < 0$ and the portrait is a **centre** (and all trajectories are ellipses).

If $T \neq 0$ then the portrait is linearly isomorphic to a **spiral node**, a **focus**, or a **degenerate node** according as $D > 0$, $D < 0$, or $D = 0$. Each of these is a **source** or a **sink** according as $T > 0$, or $T < 0$.

This is the end of the algorithm

The reader may feel that there is an enormous amount to remember in this algorithm, but intelligent use of memory drives us through in a rather natural way, once we see that everything follows from deciding the signs of the three numbers T, Δ, and D, and asking such routine questions as: 'is $\mathbf{0}$ a source or sink?', and so on.

The validity of the algorithm has been established in the various arguments given earlier, and readers should cast the details from their mind while gaining familiarity with the algorithm by practising the following exercises.

Exercises 3.5.1

1. Draw phase portraits for the system $\dot{\mathbf{z}} = A\mathbf{z}$ when A is:

$$\begin{pmatrix} 1 & 0 \\ 0 & 2 \end{pmatrix}, \quad \begin{pmatrix} -1 & 0 \\ 0 & 2 \end{pmatrix}, \quad \begin{pmatrix} 3 & 1 \\ -1 & 3 \end{pmatrix}, \quad \begin{pmatrix} 2 & 1 \\ 0 & 2 \end{pmatrix}, \quad \begin{pmatrix} 1 & 0 \\ 0 & 0.5 \end{pmatrix}$$

$$\begin{pmatrix} -3 & 0 \\ 0 & -3 \end{pmatrix}, \quad \begin{pmatrix} 0 & 2 \\ -2 & 0 \end{pmatrix}, \quad \begin{pmatrix} 3 & 0 \\ 0 & -1 \end{pmatrix}, \quad \begin{pmatrix} 3 & 0 \\ 0 & 3 \end{pmatrix}, \quad \begin{pmatrix} 0 & -2 \\ 2 & 0 \end{pmatrix}.$$

2. For each of the matrices A in (i)–(iv) below, find a real matrix P for which $P^{-1}AP$ is in standard form:

$$\text{(i)} \begin{pmatrix} 1 & 2 \\ 1 & 1 \end{pmatrix}, \quad \text{(ii)} \begin{pmatrix} 1 & 2 \\ 2 & 4 \end{pmatrix}, \quad \text{(iii)} \begin{pmatrix} 3 & -1 \\ 1 & 1 \end{pmatrix}, \quad \text{(iv)} \begin{pmatrix} 2 & 1 \\ -2 & 4 \end{pmatrix}.$$

   In each case, state the form of the equilibrium exhibited by the origin in the planar system $\dot{\mathbf{z}} = A\mathbf{z}$.

   In cases (i) and (ii) sketch the corresponding phase diagram, and solve the differential equation to find the trajectory through $\mathbf{z}_0 = \begin{pmatrix} 5 \\ 7 \end{pmatrix}$. [SU]

3. Let S_c, $c \in \mathbb{R}$, denote the dynamical system

$$\dot{x} = cx + 2y, \qquad \dot{y} = -3x + y$$

   with control parameter c. Find the set of c for which S_c does not have a stable portrait. Using this information, sketch portraits of S_c when $c = -2, -\frac{3}{2}, -1, -\frac{1}{2}$. [SU]

4. For each of the matrices A in exercise 2 describe the dynamics of the system $z(n+1) = Az(n)$ in (2.6.1), to compare it with the system $\dot{\mathbf{z}} = A\mathbf{z}$. Do a similar comparison for the matrices in exercise 1.

So far, we have not actually 'solved' the equations (3.2.5) in the sense of expressing x and y as functions of t. We do that in section 8 of the next chapter.

4

Calculus and some classical models

4.1 INTRODUCTION: THE DERIVATIVE

In the last chapter we assumed that readers would have some knowledge of calculus to investigate the models suggested by the nineteenth-century questions about populations. But calculus itself arose from attempts in the seventeenth century to model the hitherto vague notions of the area of a curved region, and the tangent to a curve. Once some precision was attained, mathematicians could go on to model such processes as the motion of planets, oscillations of electrical and mechanical systems, and the cooling of hot bodies. Later work required the use of multi-variable calculus, but in this chapter we restrict ourselves to one variable. Further theory will be found in Chapters C and D.

We assume that readers have at least some acquaintance with calculus, in particular with the standard rules for calculating derivatives of products of functions, and of finding integrals as antiderivatives. In this form, calculus appears as a merely algorithmic subject, whereas it was largely intended by its seventeenth-century developers (such as Newton) as a tool for penetrating the secrets of nature. Hard thought was necessary (they would aver) for making intelligent guesses about these secrets, which could then be translated into the language of calculus, to be analysed by application of its routines. (A 'calculus' is a set of routines to be carried out automatically, which frees the mind from the labour of calculating and allows it to dwell instead on really serious work. 'The' calculus has come to mean the particular calculi involving differentials and integrals.)

First we need to gather together various bits of calculus that we shall need later. This chapter will therefore be revision for many readers, but perhaps our point of view may be new to them (although its spirit is that of the 'Perry reforms' of the early 1900s, and the classical English textbooks that followed (see, e.g., Griffiths and Howson, 1974).

4.2 WHAT IS A TANGENT TO A GRAPH?

When asked, students nowadays remember being taught calculus as a set of arbitrary rules such as

$$\text{'If } y = x^n\text{, then } \frac{\mathrm{d}y}{\mathrm{d}x} = nx^{n-1},\ n = 1, 2, \ldots\text{'} \tag{4.2.1}$$

and then told to start differentiating polynomials and so forth. (To be fair to their teachers, it has to be said that students often remember badly, and dismiss as waffle anything that will not appear in their examinations.) We now remind such students why any intelligent person should concoct such a rule in the first place. It arose during attempts, some 300 years ago, to give good working definitions for the then vague notions of the tangent to a curve, and of velocity and acceleration. Such definitions would be mathematical *models* of the vague notions, and would allow precise deductions to be made.

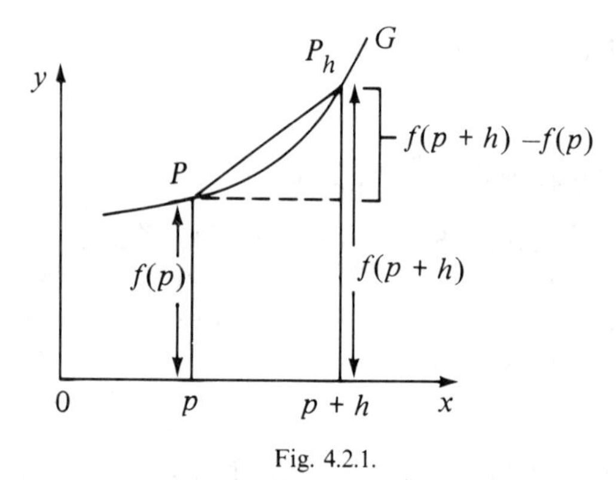

Fig. 4.2.1.

Success came from contemplating the basic Fig. 4.2.1, wherein G is the graph of a function $y = f(x)$, and P is a fixed point upon it with coordinates $(p, f(p))$. We then consider the chord through P and a nearby point $P_h = (p + h, f(p + h))$; its gradient is

$$\frac{f(p + h) - f(p)}{h} \tag{4.2.2}$$

which we think of as a function of h, say $u(h)$, because we are holding p fixed. The diagram suggests, and calculations with simple functions confirm, that as h gets smaller and smaller, $u(h)$ gets closer and closer to a fixed value v. (Briefly we write: $\lim u(h) = v$ as h tends to 0, or $\lim_{h \to 0} u(h) = v$.) Whenever v exists, we say that the function f is **differentiable** at p with **differential coefficient** $f'(p) = v$. If f is differentiable at each point p of some 'open' interval $\{a < p < b\}$, then the **derivative** of f is the function f' whose value at p is $f'(p)$. Of course, $f'(p)$ is often written in the familiar, but curious, form $\mathrm{d}f/\mathrm{d}x$ as a reminder that $u(h)$ in (4.2.2) is a quotient that was written $\delta f/\delta x$ when people thought in terms of 'infinitesimals'.

It was then agreed by the mathematical community that the tangent to G at P should be deemed to be the line through P with gradient $f'(p)$: there is no proof here—we are simply modelling the 'intuitive' notion of tangent by something that has shown itself to work satisfactorily over the course of time. The equation of the tangent is then

$$y = f(p) = f'(p)\cdot(x - p). \tag{4.2.3}$$

It would be very laborious if, every time we wished to differentiate a function, we had to calculate a limit. Fortunately, it is possible to derive, from (4.2.2), the usual rules of differentiation which give a purely algebraic algorithm for handling the process. Thus it can be shown that for any differentiable functions f, g their sum, product, and quotient are differentiable with derivatives

$$(f+g)' = f' + g'; \qquad (fg)' = f'g + fg';$$
$$\left(\frac{f}{g}\right)' = \frac{f'g - fg'}{g^2}, \qquad \text{if } g \neq 0. \tag{4.2.4}$$

If $g(x)$ is in the domain of f, then we can form $f(g(x))$, a function denoted by $f\circ g$, and called the **composition** of f with g. Usually $f\circ g \neq g\circ f$ (e.g., try $f(x) = x^2$ and $g(x) = x + 1$). We then have the **chain rule**:

$$(f\circ g)' = (f'\circ g)\cdot g' \tag{4.2.5}$$

which might be more familiar to readers in the form

$$\frac{\mathrm{d}}{\mathrm{d}x}(f(g(x)) = \frac{df}{\mathrm{d}g}\cdot\frac{\mathrm{d}g}{\mathrm{d}x},$$

which is surely not so intelligible, or suitable for computation, as

$$(f\circ g)'(x) = (f'(g(x))\cdot g'(x)$$

obtained from (4.2.5) by evaluating at x.

If $f =$ constant in (4.2.2), then $u(h) = 0$, so $f' = 0$ in this case. If f is the **identity function** (i.e., for all $x, f(x) = x$), then $u(h) = 1$, so f' is the constant function 1.

Now let us show that the 'weird' equation (4.2.1) follows quite naturally from the second rule in (4.2.4): we use the method of mathematical induction and assume that (4.2.1) holds for powers up to n (since it holds when $n = 1$). Then by definition of taking powers, x^{n+1} is the product $x^n\cdot x$, so by the product rule in (4.2.4), and writing $f(x) = x^n$ and $g(x) = x$, we have

$$(fg)' = f'g + fg' = nx^{n-1}\cdot x + n^n\cdot 1 = nx^n + x^n$$

by our inductive hypothesis. The last term is $(n + 1)\cdot x^n$ as required.

By using the third rule in (4.2.4), it is easily shown that (4.2.1) holds for all integers n, positive or negative. For a general differentiable function f it may well happen that f' is also differentiable; so we can form $(f')'$, written f'' or $\mathrm{d}^2f/\mathrm{d}x^2$. Similarly for the higher derivatives, denoted variously in books as $f^{(n)}$, $\mathrm{d}^nf/\mathrm{d}x^n$, or D^nf.

Exercises 4.2.1

1. Use a graphical calculator or graph-plotting software to study the behaviour of the function $u(x)$ in (4.2.2) for a variety of functions $f(x)$ and holding h fixed as a small real number. For example, consider the function $f(x) = \tan x$ (with x measured in radians). The graph is a set of disconnected arcs each of which moves steadily upwards through all possible real numbers. Thus its gradient is always positive (in going from left to right on an arc you are always climbing) and so its derivative will always take a positive value, where it is defined. The gentlest climb is at the mid-point of an arc, for example, $x = 0$, and this corresponds to the minimum value of the derivative. The climbs either side of the mid-point are similar and so the derivative should show symmetry. This qualitative approach should enable you to sketch the shape of the derivative $f'(x)$. Now you can use your computing kit to draw the graph of the function: $u(x) = (\tan(x + 0.1) - \tan(x))/0.1$ say, in which h has been taken as 0.1, and check whether it resembles your sketch. Now, knowing that $\tan x = \sin x/\cos x$ and that $(\sin x)' = \cos x$ and $(\cos x)' = -\sin x$ you can use the quotient rule of (4.2.4) to derive the analytic form $(\tan x)' = 1 + \tan^2 x$ and check that its graph closely aligns with that of $u(x)$. Try these techniques with functions such as: $1/x$, $\sqrt{x}$, $\ln x$, e^x, $\sin^2 x$, abs(x), int(x), sec x, etc.

   Use the techniques to test results for products, quotients, and functions of functions. The following program may also help in visualizing derivatives.

```
REM Program 4.2.1, Chords
FUNCTION f(x) = TAN(x)
FUNCTION u(x,h) = (f(x+h)-f(x))/h
READ xmin, xmax, xstep, ymax
DATA -3  , 3   , 0.05 , 10
SET MODE "graphics"
SET WINDOW xmin,xmax, -ymax,ymax
FOR x = xmin TO xmax STEP xstep
     LET y = f(x)
     PLOT x, y;
     IF ABS(y)>ymax THEN PLOT
NEXT x
PLOT
DO
       INPUT PROMPT "value of h :  " : h
       IF h = 0 THEN EXIT DO
       FOR x = xmin TO xmax STEP xstep
           LET y = u(x,h)
           PLOT x,y;
           IF ABS(y)>ymax THEN PLOT
       NEXT x
       PLOT
LOOP
END
```

2. Prove the rules (4.2.4), (4.2.5). (Hint: For $(fg)'$, test (4.2.2) by showing that

$$(f(p+h)\cdot g(p+h) - f(p)\cdot g(p))/h = f(p+h)\cdot v(h) + g(p)\cdot u(h),$$

where

$$v(h) = \frac{g(p+h) - g(p)}{h} \qquad \text{and} \qquad u(h) = \frac{f(p+h) - f(p)}{h}$$

and now use $f(p+h) = f(p) + h \cdot u(p)$ by (4.2.2), and take limits as h tends to zero.) The others are done similarly, and the proofs are 'rigorous' (in the sense of the subject called mathematical analysis, provided we have a model of limits in which the natural rules are valid; such a model is described in many analysis texts).

3. Carry out the proof, mentioned above, that (4.2.1) holds for negative integers.
4. Assuming (as can be shown) that the derivative of the sine function is the cosine, that is, $(\sin)' = \cos$, prove that $(\cos)' = -\sin$. (Use the chain rule with $\cos(x) = \sin(\pi/2 - x)$.)

   Now use the rules (4.2.4) to work out the derivatives of the other trigonometrical functions: tan, sec, and so on.
5. Show that the higher derivatives of sine are given by:

$$\sin^{(2n)} = (-1)^n \cdot \sin \qquad \sin^{(2n+1)} = (-1)^n \cdot \cos \qquad \text{when } n = 0, 1, \ldots,$$

   and find similar formulae for the higher derivatives of cos. Show that $\tan' = 1 + \tan^2$ and find a similar formula for $\tan''$ in terms of tan and $\tan^3$. Derive an iterative scheme for finding the higher derivatives of tan and show that the derivatives of even order only involve odd powers of tan, and that a similar result holds for derivatives of odd order.
6. Prove Leibniz's theorem: the nth derivative $(D^n)fg$ of the product of f and g is given by the sum $\sum_{r=0}^{n} {}^nC_r \cdot D^r f \cdot D^{n-r} g$ where ${}^nC_r$ is the binomial coefficient $n!/(n-r)!r!$. (Use induction on n, but first prove the identity: ${}^nC_r = {}^{n-1}C_{r-1} + {}^{n-1}C_r$.)

4.3 LOCAL MAXIMA AND MINIMA

If $f'(p)$ is neither strictly positive nor strictly negative, then $f'(p)$ must be zero. We then call p a **critical point** of f; and it is among the critical points that we look for maxima and minima. Older books call p a *stationary* point, and $f(p)$ a stationary *value*.

We say that f has a **local maximum** at p if $f(p+h) < f(p)$ for all sufficiently small $h \neq 0$. At such a point p, $f'(p) = 0$, and the tangent is horizontal. Similarly for a local minimum. This explains why we seek maxima and minima among the zeros of f'; but if, for example, $f(x) = x^3$, then $f' = 0$ at $x = 0$, yet 0 is neither a local maximum nor a local minimum of f (Fig. 4.3.1).

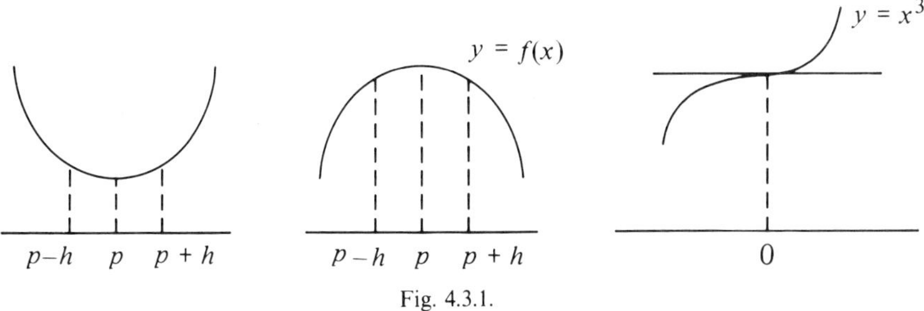

Fig. 4.3.1.

Note: if $f(x) = x^2(x-1)(x+1)$ then $f' = 0$ when $x = 0$ or $\pm 1/\sqrt{2}$, and 0 is a local maximum but not *the* maximum value of f. What about $x = 1/\sqrt{2}$? (See Fig. 4.3.2.)

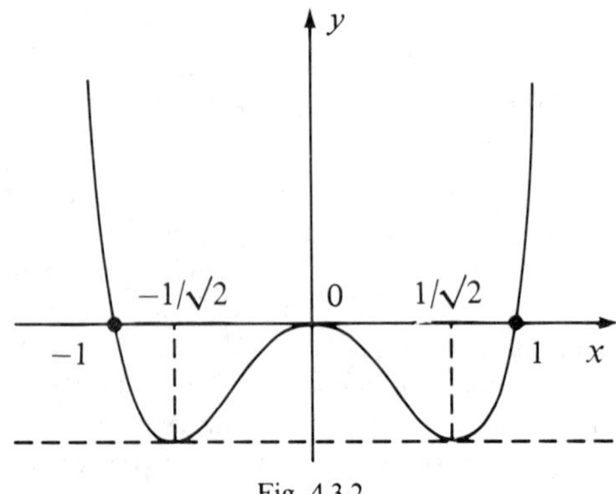

Fig. 4.3.2.

For brevity, we shall write 'max' for 'local maximum' and similarly for 'min'. The sketches in Fig. 4.3.1 indicate that, at a max, we have

$$f'(p-h) > 0 > f'(p+h), \tag{4.3.1}$$

with the inequalities reversed for a min. But then at a max f' is decreasing, so $f''(p) < 0$. Therefore if $f''(p) < 0$ then (4.3.1) holds; and similarly for a min when $f''(p) > 0$. In practice it is often easiest to use (4.3.1), unless $f''(p)$ can be evaluated easily. Note that it is possible to have $f''(p) = 0$ at a max or min: consider $f(x) = +x$. A point q at which $f'' = 0$ is called a **point of inflexion**, provided q is not a max or min of f.

Exercises 4.3.1

1. Classify the critical points (if any) of the functions:

   $$\sin x, \qquad x \cdot \sin x, \qquad x^{-1} \cdot \sin x, \qquad x^3 + 3x - 1, \qquad x^3 - 3x - 1, \qquad x^3.$$

   Let $f(x) = x^3 + ax^2 + bx + c$ with a, b, c real constants. Show that f has two critical points or none, according as $a > 3b$ or $a < 3b$; and in the former case there is one max and one min.
2. On the domain $-\pi/2 < x < \pi/2$, find the critical points (if any) of the functions:

   $$\sin x + \tan x, \qquad 2\sin x + 3\tan x, \qquad 3\sin x + 2\tan x,$$

   and use the information to sketch the graphs. Now show how the function $f(x) = \sin x + k \cdot \tan x$ changes as k changes. (First show that f has two critical points or none, according as $|k| < 1$ or $|k| \geq 1$.)
3. Show that the maximum and minimum values of $r^2 = x^2 + y^2$, as (x, y) moves on the curve $ax^2 + 2hxy + by^2 = 1$, are given by $r^{-2} = (T \pm \sqrt{D})/2$, where $T = a + b$ and $D = (a - b)^2 + 4h^2$.

4.3.1 Light

In many problems, minimum solutions are of great significance. As an example, we consider models of the theory of light. The first comes from the period of classical Greek mathematics which lasted from 700 BC to AD 500, and was at its greatest around 400 BC (with Euclid and the greater geometers of whom he wrote) and again around 250 BC (with Archimedes). (The woman mathematician, Hypatia, was Librarian at Alexandria and was murdered by Christians in AD 415; not long after, the school there fell into decay.)

To the classical Greeks, rays of light were straight lines to which they could apply the geometrical theory of Euclid. Around AD 100, Heron used this to explain the observation that light is reflected from a flat mirror according to the law: *the angle of incidence equals the angle of reflection*, where these angles are made with the normal to the surface (see Fig. 4.3.3(a)). Heron suggested that there was a more basic law, that the length of the path taken by a ray of light always was to be a minimum; and then he deduced the law of reflection by a geometrical argument that we sketch in the exercises below. But this theory did not account for the refraction that is observed when (for example) a rod is inserted into water. Therefore in the seventeenth century, Fermat and others changed the model: instead of saying that rays of light *were* straight lines, they said that they *travelled* in straight lines. This apparently trivial change of words suggests that light has a velocity, which will depend on the medium (air, water, glass) through which the light is passing. Eventually this idea gave rise to Fermat's **principle of least time**: that the path of a ray of light between two points is the one that takes the least time to traverse, at the velocities permitted by the various media that have to be crossed. Clearly (although it is not simple to prove mathematically) if the points are in the same medium, then the time is least if the path is straight. With refraction at a surface, we can use calculus to find the minimum time, and easily deduce Snell's law of refraction as well as the law of reflection; the details are left to the reader in the exercises.

This improved model would still not explain other phenomena such as the splitting of light into colours, or interference fringes, and for this, Huyghens and others developed the wave model for light. Here, Fermat's principle still featured, as a very general minimum principle of 'least action' that played a great role in the eventual enunciation of such all-pervading principles of physics as the rule that a physical system tends to that state in which its energy is least. Strategies for mathematical theorems, too, are often based on the search for a point at which some function takes a minimum value, and we shall see examples below. Not all such problems can be solved by calculus; for a very interesting survey, see Courant and Robbins (1984), Chapter 7.

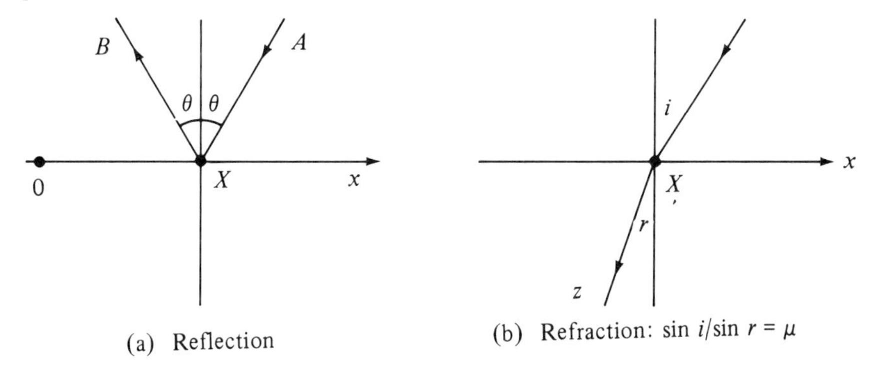

(a) Reflection

(b) Refraction: $\sin i/\sin r = \mu$

Fig. 4.3.3.

Exercises 4.3.2

1. Give Heron's proof of the reflection principle as follows. Take the x-axis as the mirror, and suppose a ray of light passes through $A = (a, c)$ with $c > 0$, to be reflected at $X = (x, 0)$ and then to pass through $B = (b, c)$; suppose $b < x < a$. Let $C = (b, -c)$ and let the line AC cut $\mathbf{0}x$ in Y; show that, since AC is shorter than $AX + XC$, then $AY + YB$ is shorter than $AX + XB$ unless $X = Y$. Hence by symmetry, AX and BX make equal angles with $\mathbf{0}y$, as required.

   Next give a calculus proof, by finding x such that $f(x)$ is a minimum, where $f(x) = AX^2 + XB^2$. (Why is it enough to use f, rather than $g(x) = AX + XB$?)
2. In the previous question, suppose that $\mathbf{0}$x is not a mirror but the boundary between two transparent media, in which light has velocities u (above $\mathbf{0}x$) and v (below $\mathbf{0}x$). Suppose the ray from A penetrates $\mathbf{0}x$ at X, and passes through $Z = (z, -c)$, and that it makes angles i, r of incidence and refraction as in Fig. 4.3.3(b). Then the time taken is

$$T(x) = \frac{1}{u} \cdot AX + \frac{1}{v} \cdot XB.$$

   Show that $T'(x) = 0$ when (and only when)

$$\frac{a-x}{u} \cdot AX = \frac{x-b}{v} \cdot XB,$$

   that is, when $\sin(i)/\sin(r) = u/v$; and that x is then a local minimum of T. (Snell found that $\sin(i)/\sin(r)$was constant, apparently by patiently scrutinizing the experimental data! He did not give this mathematical proof.)
3. It is required to build a road from A to B on opposite sides of a straight motorway, and the cost of the bridge is independent of its position. If the cost per kilometre of road on one side of the motorway is £x, and £y on the other, find where the bridge should be for minimum total cost.
4. When a billiard ball is projected against the cushion of a billiard table, it rebounds off the cushion according to the law of reflection (if we neglect frictional effects). If it keeps on rebounding from one cushion to another, when will it return to the start? Investigate this problem by means of a computer program. What happens with tables of different shapes?
5. A projectile is fired from a point $\mathbf{0}$ on the ground.
   (a) If the ground is level, when and where will the projectile land? It is usual to write $u = U \cos \alpha$, $v = U \sin \alpha$, (so $\tan \alpha = \dot{y}/\dot{x}$ when $t = 0$). Show that if U is fixed and α varies, then the maximum range is $L = U^2/g$, and the time taken is $U^2/g\sqrt{2}$.
   (b) If $\mathbf{0}$ is on top of a cliff of height d above sea-level, how far off shore will the projectile hit the water?
   (c) Find an equation in x and y (but *not* t) which describes the path of the projectile.
   (d) Show that the maximum range R in (b) is $(L^2/g)\sqrt{(1 + 2d/L)}$ and then $\tan \alpha = R/L$.
6. The length plus girth of a cylindrical parcel must not exceed c units of length. Show that the volume V has one max, and it is the largest possible value, $2c^3/27\pi$. If, instead, the cross-section is square, show that the maximum volume is $c^3/108$, smaller than before.
7. A cylindrical vessel with a flat lid has volume V, and is made from sheet metal. Show that the area of metal is least when the radius r is $(V/2\pi)^{1/3}$ and the height is $2r$. If, instead, the lid is omitted, show that the area is least when the radius equals the height. What happens to these optimal dimensions if we cost both the quantity of metal, and the soldering along a vertical line to form the cylinder and along its rim?

8. A triangle T has one side of length 1, and perimeter $1 + p$. Show that the area is greatest when T is isosceles.
9. Let P be a convex polygon in $\mathbb{R}^2$, and for each $X \in \mathbb{R}^2$ let $f_p(X)$ denote the sum of the squares of the distance of X from the sides of P.
   (a) Take P to be an isosceles triangle T of height h and base a, symmetrical about $\mathbf{0}y$. If X runs on $\mathbf{0}y$, show that the least value of f is $2a^2h/(h^2 + 3a^2)$.
   (b) If, instead, P is the trapezium formed by removing from T a similar triangle of height $k < h$, and again X runs on $\mathbf{0}y$, find the least value of f, and show that (depending on the size of k/h) it can occur when X lies outside T.
10. A rectangle R has its vertices on the ellipse $x^2/a^2 + y^2/b^2 = 1$; show that its sides are parallel to the axes. If one vertex is at $(a\cos\theta, b\sin\theta)$ find θ for which R has (a) greatest area, (b) greatest perimeter. What if we change 'greatest' to 'least'?
11. Taking $\mathbb{R}^2$ as a model of the sea, suppose that a ship sails with constant speed u on the line $y = k$ from (a, k) to (b, k) with $a < b$. As it starts, a boat sails from $\mathbf{0}$ with speed v, along the line $y = mx$ so as to meet the ship. Show that the boat cannot meet the ship at all unless $a/k \leq m \leq b/k$; and find the least possible time of meeting.
12. Find the straight line of quickest descent from a given point to a given vertical straight line, assuming that the time of sliding a distance s from rest at an inclination θ to the horizontal is

$$\left(\frac{2s}{g\sin\theta}\right)^{\frac{1}{2}}.$$

(The minimum time is $2(a/g)^{\frac{1}{2}}$, corresponding to $\theta = \frac{1}{4}\pi$, where a is the horizontal distance of the point from the given straight line.)

13. Find at what height on the wall of a room a source of light must be placed in order to produce the greatest brightness at a point on the floor at a given distance a from the wall. (Assume that the brightness of a surface varies inversely as the square of the distance from the source, and directly as the cosine of the angle which the rays make with the normal to the surface.)
14. If the power required to propel a steamer through the water varies as the cube of the speed, show that the most economical rate of steaming against a current will be at a speed equal to $1\frac{1}{2}$ times that of the current.
15. The velocity of waves of length λ on deep water is proportional to

$$\sqrt{\left(\frac{\lambda}{a} + \frac{a}{\lambda}\right)},$$

where a is a constant. Prove that the velocity is a minimum when $\lambda = a$.

16. The inclination of a pendulum to the vertical, when the resistance of the air is taken into account, is given by the formula

$$\theta = ae^{-kt}\cos(nt + \varepsilon).$$

Prove that the greatest elongations occur at equal intervals π/n of time, and that they form a series diminishing in geometrical progression.

(The last five problems are taken from the large collection in Lamb (1907).)

4.4 THE MEAN VALUE THEOREM

It is proved in analysis texts that if f is a differentiable function on an interval I, and $a < b$ in I, then there is some point c between a and b such that:

$$f(b) - f(a) = (b - a)\cdot f'(c). \tag{4.4.1}$$

This result is called the **mean value theorem**, and is illustrated in Fig. 4.4.1(a). Because of it, we have an important result:

$$\textit{If } f' = 0 \textit{ on } I \textit{ then } f = \textit{constant}. \tag{4.4.2}$$

For then $f'(c) = 0$, so $f(a) = f(b)$; thus f takes equal values at any two points of I, and so f is constant, as required. Note that we could not use this argument unless the domain I is an interval, as the following argument shows.

Suppose that we remove the middle third from I to leave a domain J consisting of two separate intervals. Then we could take f to be 1 on one of these, and to be -1 on the other—still we have $f' = 0$ on J but f is not constant.

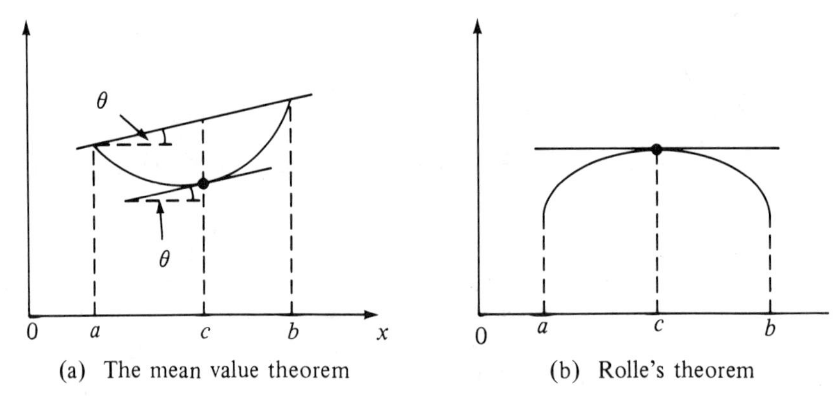

(a) The mean value theorem (b) Rolle's theorem

Fig. 4.4.1.

Exercises 4.4.1

1. With f as above, use the mean value theorem (abbreviated as MVT) to derive **Rolle's theorem**: if $f(a) = f(b)$ then f' must be zero somewhere between a and b; see Fig. 4.4.1(b). Conversely, show that if we assume Rolle's theorem then the MVT follows. (Rotate Fig. 4.4.1 until the chord is horizontal.)

   These two theorems can be used to locate roots of equations, if they are combined with the **intermediate value theorem**, one form of which is this: if in (4.4.1), $f(a)$ and $f(b)$ are of opposite signs, then f is zero somewhere between a and b. For example, if $f(x) = x^3 + 2x + 1$ then f has at least one real zero; for if $|x|$ is very large, $f(x)$ is positive if $x > 0$, and negative if $x < 0$. But if f had two zeros, say at a, b, then by Rolle's theorem, f' has a zero between them. This is impossible since $f'(x) = 3x^2 + 2 > 0$, so f has only one zero.

   (a) By writing $f(x) = x^3(1 + 2/x + 1/x^2)$, estimate how large x must be to ensure that $f(x) > 0$. Similarly for $f(x) < 0$.

   (b) Show that the equation $f(x) \equiv x^3 + 3x^2 + 5x + 1 = 0$ has only one real root.

2. Show that, for all values of k, the equation $x^3 - 3x + k = 0$ has just one root in the interval $0 \leq x \leq 1$.
3. If $f(x) = x^3 + ax + 7$ has at least two real zeros, show that $a < 0$, in which case f has no more than three real zeros.
4. Show that the equation $x = \cos x$ has just one root in the set $X = \{0 < x < \pi/2\}$. (Use the MVT and then Rolle's theorem.) For what values of k has the equation $kx = \sin x$ a root in X? (Draw a sketch to show where the line $y = kx$ meets the curve $y = \sin x$.) Deal similarly with the equation $kx = \tan x$. To find these roots, one can use Newton's method: see exercise 10.4.2(1).
5. The function $f(x)$ with domain $-\pi/2 < x < 3\pi/2$, $x \neq \pi/2$, is given by

$$f(x) = 8 \sin x - \tan x.$$

   Show that $f(x)$ has two stationary values (i.e. with $f'(x) = 0$). Find the corresponding critical points of the graph, $y = f(x)$, of the function and determine their nature. Sketch the graph.
   From your graph, determine the number of solutions for the equation $8 \sin x - \tan x = a$ with $-\pi/2 < x < 3\pi/2$, when (i) $0 < a < 3\sqrt{3}$, (ii) $a = 3\sqrt{3}$, and (iii) $a > 3\sqrt{3}$. [SU]
6. Apply the mean value theorem to $\ln(1 + x)$ to show that $x/(1 + x) < \ln(1 + x) < x$ when $-1 < x < 0$ and when $x > 0$. [SU]

4.5 INTEGRATION

For thousands of years, mathematicians had been fascinated by the problem (which they found difficult) of calculating the area enclosed by such plane curves as the circle. (One approximate technique involved cutting the shape from a piece of wood and weighing it.) Once the notion of a graph had emerged, it was seen that it would suffice to be able to calculate the area under the graph of a function. Although particular areas had been calculated before him (see Boyer and Merzbach (1989)), Isaac Newton (1642–1727) discovered a systematic method for finding the area under the graph of a function f on a 'closed' interval $I = \{a \leq x \leq b\}$. His argument was roughly this: let $A(p)$ denote the area from a to p, as in Fig. 4.5.1. Then the shaded area in the figure is $A(p + h) - A(p)$, and if h is small this area is approximately the rectangle R of height $f(p)$ and base h. Thus

$$\frac{A(p + h) - A(p)}{h} \cong f(p)$$

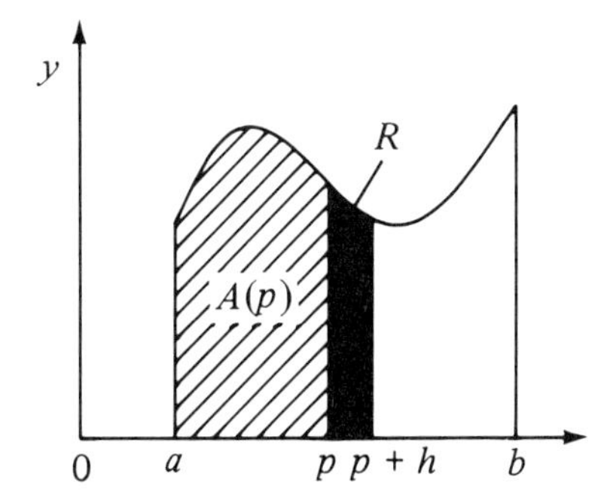

Fig. 4.5.1.

so letting h tend to zero we have, as in section 4.2,

$$A'(p) = f(p). \tag{4.5.1}$$

Hence f is the derivative of A (so A is called an **antiderivative** of f), and A is that function such that $A(a) = 0$. This is why we find areas by looking for antiderivatives; the area under the graph is then $A(b) - A(a)$. Since the area can also be thought of as approximately a sum of thin rectangles like R above, the integral sign $\int$ was introduced as an elongated 'S' (for 'sum') to write

$$A(p) = \int_a^p f(x)\,\mathrm{d}x. \tag{4.5.2}$$

Here, the x is redundant in that we can change it to any other symbol provided we change it at both appearances:

$$\int_a^p f(x)\,\mathrm{d}x = \int_a^p f(t)\,\mathrm{d}t. \tag{4.5.3}$$

The fact that A is an antiderivative of f can now be written as the 'fundamental theorem of calculus':

$$\frac{\mathrm{d}A}{\mathrm{d}p} = \frac{\mathrm{d}}{\mathrm{d}p}\int_a^p f(x)\,\mathrm{d}x = f(p). \tag{4.5.4}$$

This argument of Newton's assumes that the area exists and is a differentiable function. Later work showed that this will still apply if f is a 'continuous' function (which need not necessarily be differentiable; for example, $f(x) = \mathrm{abs}(x)$).

We assume that readers have had some practice in evaluating specific integrals, so we now give some theoretical properties that will codify what the reader already knows but perhaps has not actually been 'said' in a formal way.

First the integral *respects inequalities* in the sense that if g is a second function then

$$\textit{If } g \leq f, \textit{ then } \int_a^p g(x)\,\mathrm{d}x \leq \int_a^p f(x)\,\mathrm{d}x. \tag{4.5.5}$$

Here, $g \leq f$ means that for each x, $g(x) \leq f(x)$; the graph of g lies below that of f, so the area is less. This is useful when we wish to estimate the size of an integral we cannot evaluate exactly. For example, we may know that $|f|$ never exceeds some constant M, whence, for each x, $-M \leq f(x) \leq M$; if now we integrate and use (4.5.5) we obtain the estimate (when $a > 0$):

$$-M \cdot a \leq \int_0^a f(x)\,\mathrm{d}x \leq M \cdot a. \tag{4.5.6}$$

Similarly, if $f(p), f(q)$ are the least and greatest values of f on I, then from (4.5.5) we obtain

$$f(p)\cdot(b-a) \le \int_a^b f(x)\,\mathrm{d}x \le f(q)\cdot(b-a),$$

so the integral (J, say) lies between two values of the continuous function $g(x) = f(x)\cdot(b-a)$; hence by the intermediate value theorem (see section C.4), J must itself be a value of g, so:

There is some point c in I such that $f(c)\cdot(b-a) = \int_a^b f(x)\,\mathrm{d}x.$ (4.5.7)

This result is called the (first) **mean value theorem** for integrals, and says that the area under the graph of f is equal to that of a rectangle of height $f(c)$ for some c. This also is useful for getting rough estimates of the size of an integral.

As a young man in the 1660s, Isaac Newton saw that a vital notion for his project of explaining the natural world was that of the 'rate of change' of a process, with respect to time. Thus he was able to translate such concepts as velocity (v) and acceleration (a) into calculus language as

$$v = \frac{\mathrm{d}s}{\mathrm{d}t}, \qquad a = \frac{\mathrm{d}v}{\mathrm{d}t}, \tag{4.5.8}$$

where s is the distance covered in time t. (Instead of Newton's own 'fluxion' notation $\dot{s}$, we are using Leibniz's $\mathrm{d}s/\mathrm{d}t$.) Therefore, by the above theory, the distance s is given by the area under the velocity–time graph, so 'distance is the integral of velocity'. Also, Newton (following Galileo) formulated one of his laws of motion as 'the rate of change of momentum of a body equals the force acting on the body'. Here, momentum is defined to be mv, where m is the mass of the body and v is its velocity; so the law translates into calculus as

$$\frac{\mathrm{d}}{\mathrm{d}t}(mv) = \text{Force} \tag{4.5.9}$$

from which many consequences can be deduced, as we shall see below. Once such laws are formulated, various mathematical questions arise. For example, if the rate of change of a function f is f itself, can we find f? We deal with some of these below.

Exercises 4.5.1

1. Show that A in (4.5.2) is unique. (Use the mean value theorem to show that any two antiderivatives of f on the same interval can differ only by a constant.)
2. A particle covers 1 m in 1 sec (along a straight line). Prove that somewhere its acceleration or deceleration exceeds $4\ \mathrm{ms}^{-2}$. (Assume the contrary; then show that the velocity–time graph would be contained in an isosceles triangle with too small an area.)
3. Show by integration that a circular disc of radius a has area πa^2; and if two radii OA, OB make an angle 2θ, the area of the sector between them is θa^2, while the area of the 'cap' between AB and the circular arc is $a^2(\theta - \sin\theta)$. If now the circle is replaced by the ellipse $E: x^2/a^2 + y^2/b^2 = 1$, and AB cuts E at $(a\cos\theta, b\sin\theta)$ show that the cap has area $ab(\theta - \sin\theta)$. (Compare the area of a vertical strip of E at $x = a\cos\phi$ with that of the strip of the circle $x^2 + y^2 = a^2$.)

4. *The Gentleman's Mathematical Companion* (or as it was originally called *A Companion to the Gentleman's Diary or a Preparative to that useful Work*) was published annually from about 1798 to 1826. Issues contained a Questions section, with Answers published the next year. In Issue No. 19 (dated 1816) the following question was proposed by Mr. John Baines, Jnr.

   'Admitting I have an elliptical enclosure of an acre of ground, whose transverse and conjugate diameters are to each other as five to four; how long must a chord be, which fastened to the fence at the extremity of the transverse diameter, so as to allow a horse tethered to the same to eat half an acre?'
5. Use integration to establish the following results. In each case you should obtain first an approximation of the form

   $$F(p+h) - F(p) \simeq h \cdot f(p)$$

   and then pass to an equation like (4.5.1).
   (a) The curved surface of a right circular cylinder of height h and radius a is $2\pi ah$;
   (b) The volume of the same cylinder is $\pi a^2 h$;
   (c) The curved surface of a right circular cone of height h, base-radius a, and slant side l is πal;
   (d) The volume of the same cone is $\frac{1}{3}\pi a^2 h$;
   (e) The surface of a sphere of radius a is $4\pi a^2$;
   (f) The volume of the same sphere is $\frac{4}{3}\pi a^3$.
6. Evaluate $\int_0^2 x^p/(1 + x^2)^q \, dx$ in each of the following cases:
   (a) $p = 1, q = 1$; (b) $p = 1, q = 2$;
   (c) $p = 3, q = 1$; (d) $p = 3, q = 2$.
7. Given $I_n = \int_0^{\pi/2} \sin^n\theta \cos^2\theta \, d\theta$ show that $I_1 = \frac{1}{3}$. Establish the recurrence relation

   $$I_n = ((n-1)/(n+2)) \cdot I_{n-2}.$$ [SU]

   Hence evaluate I_3 and I_5.
   (Try writing $\sin^n\theta \cdot \cos^2\theta = \sin^{n-1}\theta \cdot (\sin\theta \cdot \cos\theta)$ and integrate by parts.)
8. Show that

   $$\int_0^{\pi} \sin^5\theta \cos^2\theta \, d\theta = \frac{16}{105}$$

   and that

   $$\int_0^{2\pi} \sin^2\phi \cos^2\phi \, d\phi = \frac{\pi}{4}.$$

9. Use program 4.5.1 to investigate finite approximations to the area function using trapezoidal slices.

```
REM Program 4.5.1, Area by slices
FUNCTION f(x) = SIN(x)
FUNCTION slice(x,h) = h*(f(x) + f(x+h))/2
READ xmin  , xmax , xstep, ymax
DATA -6.283, 6.283, 0.05 , 2
SET MODE "graphics"
SET WINDOW xmin, xmax, -ymax, ymax
FOR x = xmin TO xmax STEP xstep
    LET y = f(x)
    PLOT x, y;
```

Program 4.5.1 continued

```
        IF ABS(y)>ymax THEN PLOT
    NEXT x
    PLOT
    DO
        INPUT PROMPT "value of h : " : h
        IF h = 0 THEN EXIT DO
        LET a = 0
        LET x = xmin
        DO
            PLOT x,a;
            IF ABS(a)>ymax THEN PLOT
            LET a = a + slice(x,h)
            LET x = x + h
        LOOP UNTIL x > xmax
        PLOT
    LOOP
    END
```

4.6 SOME FIRST-ORDER DIFFERENTIAL EQUATIONS

The notion of rate of change, mentioned with (4.5.8), allowed Newton to formulate his 'law of cooling' into calculus as:

$$\frac{\mathrm{d}\theta}{\mathrm{d}t} = -k\cdot(\theta - S)$$

where t denotes time, θ the temperature of a cooling body, S that of its suroundings, and k is a constant. This says: the body loses heat at a rate proportional to the difference between its temperature and that of the surroundings. Since the body is losing heat, we take k to be positive. It is convenient to rearrange the equation in the form:

$$\frac{\mathrm{d}\theta}{\mathrm{d}t} + k\theta = kS. \tag{4.6.1}$$

This is an example of a **first-order differential equation**, and the problem is to 'solve' it, that is, to find the unknown function $\theta(t)$. The trick is to multiply by e^{kt}, to get (by the rules of differentiation)

$$\frac{\mathrm{d}}{\mathrm{d}t}(\mathrm{e}^{kt}\cdot\theta) = \mathrm{e}^{kt}\cdot(kS). \tag{4.6.2}$$

Now integrate between 0 and T to get

$$[\mathrm{e}^{kt}\cdot\theta]_0^T = k\cdot\int_0^T \mathrm{e}^{kt}\cdot S\,\mathrm{d}t,$$

so

$$\theta(T) = \mathrm{e}^{-kT}\cdot(\theta_0 + k\cdot\int_0^T \mathrm{e}^{kt}\cdot S\,\mathrm{d}t). \tag{4.6.3}$$

This tells us that if there is a solution of (4.6.1) then it has to be (4.6.3); but if we differentiate (4.6.3) and insert into equation (4.6.1), then it is a solution; so (4.6.3) is *the* solution. Therefore we have proved an 'existence and uniqueness' theorem—a solution exists, and it is the only solution if we specify the 'initial condition' that θ at time zero must have the value θ_0.

The solution (4.6.3) allows for S to be fluctuating, perhaps depending on the central heating in the laboratory. Usually, however, S is constant, and then (4.5.3) simplifies to

$$\theta(T) = S + \mathrm{e}^{-kT} \cdot (\theta_0 - S) \tag{4.6.4}$$

from which we note that as time goes on, $\theta(T)$ decreases towards S, as we would expect from experience, because $\theta_0 \geq S$. This is a check on the algebra, if we believe that the model of the physical situation is a good one. Mathematically, of course, e^{-kT} is never zero, but it becomes immeasurably small very rapidly, and we cannot tell with instruments whether the body is really at room temperature or always slightly above it.

Equation (4.6.1) appears in many models of quite different physical processes, sometimes in disguised form. For example, consider the motion of a body falling through the air, subject to air resistance. Suppose the body has constant mass m. Then by the law of motion (4.5.9), we have

$$m \cdot \frac{\mathrm{d}v}{\mathrm{d}t} = mg - R, \tag{4.6.5}$$

where g is the acceleration due to gravity, and R is the upward force of the air resistance. If R is proportional to the velocity, say $R = mkv$, then the last equation becomes

$$\frac{\mathrm{d}v}{\mathrm{d}t} = -k(v - V), \qquad V = \frac{g}{k}, \qquad k > 0. \tag{4.6.6}$$

This is equation (4.6.1) with solution given by (4.6.4):

$$v = V + \mathrm{e}^{-kt} \cdot (v_0 - V) = V \cdot (1 - \mathrm{e}^{-kt}),$$

if we assume the initial velocity v_0 is zero. Thus, v increases from 0 toward 'terminal' velocity V, which it never attains in finite time.

Another deduction can be made from (4.6.6) if we use the chain rule to write

$$\frac{\mathrm{d}v}{\mathrm{d}t} = \frac{\mathrm{d}v}{\mathrm{d}s} \cdot \frac{\mathrm{d}s}{\mathrm{d}t} = v \cdot \frac{\mathrm{d}v}{\mathrm{d}s}, \tag{4.6.7}$$

for then, by (4.6.6), we have

$$\frac{1}{v - V} \cdot v \cdot \frac{\mathrm{d}v}{\mathrm{d}s} = -k, \tag{4.6.8}$$

from which we can find to what height h the body has fallen from its initial height H, by integrating with respect to s. The result gives h in terms of v, which is not a very interesting relationship. We leave the calculation as practice for the reader.

Next, suppose the air resistance is proportional to v^2. Then using (4.6.7) in (4.6.5) we have

$$\frac{\mathrm{d}}{\mathrm{d}s}\left(\frac{v^2}{2}\right) = v \cdot \frac{\mathrm{d}v}{\mathrm{d}s} = g - c \cdot v^2$$

for a suitable positive constant c. Writing $w = v^2$, we get

$$\frac{\mathrm{d}w}{\mathrm{d}s} = -2c \cdot (w - W), \qquad W = \frac{g}{c}$$

and again have an equation of the form (4.6.1) but with s playing the role of the time. From (4.6.4) we can express $v = \sqrt{w}$ in terms of s:

$$v = \sqrt{(W(1 - \mathrm{e}^{-2cs}))} < \sqrt{W}.$$

By (4.5.8), $v = \mathrm{d}s/\mathrm{d}t$, but if we now try to do the analogue of (4.6.8) we cannot perform the integration to find the relation between s and t. These different cases can be combined by assuming R to be a quadratic function of v, so that (4.6.5) takes the form

$$\frac{\mathrm{d}v}{\mathrm{d}t} = g - kv - c \cdot v^2, \qquad (g, k, c \geq 0). \tag{4.6.9}$$

If $c = 0 \neq k$ this is just $\dot{w} = -kw$ with $w = v - g/k$, so we assume $c \neq 0$. In that case, the equation resembles Verhulst's equation (3.1.3) for population growth; and it becomes Verhulst's itself if we write $v = w + h$, where h is a suitably chosen constant. For then

$$\frac{\mathrm{d}v}{\mathrm{d}t} = \frac{\mathrm{d}w}{\mathrm{d}t} = g - k(w + h) - c(w + h)^2 = (g - kh - ch^2) + w(-k - 2ch) - cw^2.$$

Now, the roots of the equation $g - kh - ch^2 = 0$ are p, q, where

$$p = -\frac{(D + k)}{2c} < 0 < q = \frac{D - k}{2c}, \qquad D = \sqrt{(k^2 + 4cg)}$$

and D is real since $k^2 + 4cg > 0$. We take h to be p, so that

$$\frac{\mathrm{d}w}{\mathrm{d}t} = \beta w(1 - \alpha w), \qquad \beta = D > 0, \qquad \alpha = c/\beta > 0.$$

This is Verhulst's equation (3.1.3) with a, b taken to be α, β. The solution is given by (3.1.5).

The discussion following (3.1.4) shows that as $t \to \infty$, then $w \to 1/\alpha = D/c$; so v tends to the limiting velocity $1/\alpha + p = q$. As a function of t, $v(t)$ can be obtained as $v(t) = w(t) + p$, with $w(t)$ given by (3.1.5).

If we wish to find v as a function of s, then we use (4.6.7) to get

$$v\frac{\mathrm{d}v}{\mathrm{d}s} = (w + p)\frac{\mathrm{d}w}{\mathrm{d}s} = \beta w(1 - \alpha w)$$

so

$$\int \frac{(w + p)\,\mathrm{d}w}{w(1 - \alpha w)} = \beta t + \text{constant}$$

when we integrate from 0 to t (and from w_0 to $w(t)$). The division by $1 - \alpha w$ is legitimate since w never attains the limiting value $1/\alpha$.

Exercises 4.6.1

1. Complete the above integration, with $w < 1/\alpha$. What happens if $w_0 > 1/\alpha$?
2. Solve the equation (4.6.9) when k, c may be negative and $k^2 + 4cg > 0$. Can you think of a physical process that the equation then models?
3. A particle moves in a vertical plane, having been projected from **0** with speed u at an angle θ with **0**x. It s subject to an air resistance, so that its x- and y-coordinates are given by the differential equations.

$$\ddot{x} = -cu\cos\theta, \qquad \ddot{y} = g - cu\sin\theta$$

where c is a non-zero constant. Show that, if $p(t) = 1 - \mathrm{e}^{-ct}$, then

$$x = \frac{u}{c}\cos\theta\cdot p(t), \qquad y = \frac{u}{c}\sin\theta\cdot p(t) + \frac{g}{c^2}(p(t) - ct).$$

Show that, if $c = 0$ in the differential equations, then the path of the particle is a parabola. (When $c \neq 0$, it can be investigated by the computer program 5.1.1 in the next chapter.)

After (4.6.9), take $h = -k/2c$, so that

$$\frac{\mathrm{d}w}{\mathrm{d}t} = c\cdot(u^2 - w^2), \qquad u = D/\sqrt{c},$$

from which we can see that, if the body is to increase its velocity from zero then, as above, $u \geq v$. Rearranging the equation, and integrating with respect to time we get

$$\int \frac{\mathrm{d}w/\mathrm{d}t}{u^2 - w^2}\,\mathrm{d}t = ct + a,$$

where a is a constant; the integral is

$$\frac{1}{2u}\int\left(\frac{1}{u - w} + \frac{1}{u + w}\right)\mathrm{d}v = \frac{1}{2u}\cdot\ln\left(\frac{u + w}{u - w}\right)$$

so a can be evaluated since $v = 0$ at $t = 0$. If we now take exponentials, we can solve algebraically to find v as a function of t, and again we leave the details to the reader.

Clearly, these tricks of integration are not going to be very successful with most other possible forms of resistance, such as when R is proportional to v^p with p a fractional power. But with a computer, we can investigate the general picture by taking suitable samples.

```
REM Program 4.6.1, Resisted fall
READ c  , p
DATA 0.1, 1.5
READ t, tmax, dt , v, vmax, g
DATA 0, 10  , 0.1, 0, 40  , 9.81
SET MODE "graphics"
SET WINDOW t,tmax, v,vmax
DO
  PLOT t, v
  LET dv = (g - c*v^p)*dt
  LET v = v + dv
  LET t = t + dt
LOOP UNTIL t >= tmax
END
```

(Adapt the program to confirm the solutions of (4.6.9).)

This program uses a very simple step-by-step approximation for solving the differential equation, known as the **Euler method**; in fact it replaces the differential equation by a corresponding difference equation. For small time steps, dt, this technique is simple, fast, and gives a good idea of the nature of a solution. However, it is continually drifting away from the correct solution curve onto neighbouring curves. When it is important to minimize the effects of any such algorithmic errors, a more sophisticated numerical technique is required. The next exercise shows how one such technique, the **fourth-order Runge–Kutta method**, provides such an improvement.

Exercises 4.6.2

1. Adapt program 4.6.2 to plot an approximate solution to Newton's model of cooling (4.6.1) when $k = 1$, $S = 1$, and $\theta_0 = 10$. See the effect of altering the value of the time increment dt on the solution curve. Adapt the program to superimpose the solution given by (4.6.4) to see how the approximate solution tends to the exact solution as dt gets smaller.
2. The following program uses the fourth-order Runge–Kutta technique instead of the simpler Euler method.

```
REM Program 4.6.2, Runge-Kutta
FUNCTION  f(t,x) = -k*(x-S)
READ k, S, x0, t, tmax, dt
DATA 1, 1, 10, 0, 10  , 0.25
DECLARE FUNCTION RK4
SET MODE "graphics"
SET WINDOW 0,tmax, 0,x0
LET x = x0
DO
  PLOT t,x
  LET x = RK4(x,t,dt)
  LET t = t + dt
LOOP UNTIL t>tmax
FOR t = 0 TO tmax STEP dt
  LET x = S + EXP(-t)*(x0 - S)
  PLOT t,x;
NEXT t
STOP
```

Program 4.6.2 continued

```
FUNCTION  RK4(x,t,dt)
  LET h = dt/2
  LET d1x = f(t,x)*dt
  LET d2x = f(t + h ,  x + d1x/2)*dt
  LET d3x = f(t + h ,  x + d2x/2)*dt
  LET d4x = f(t + dt,  x + d3x)*dt
  LET RK4 = x + (d1x + 2*d2x + 2*d3x + d4x)/6
END FUNCTION
END
```

Try adjusting the value of the time increment dt. Adapt it to the air resistance models above.

3. A pilot always keeps the nose of the plane pointed towards a city T due west of the starting point. When there is no wind blowing, the speed of the plane is v miles per hour. Show that the differential equation of the plane's path when a wind is blowing from the South at the rate of w miles per hour is:

$$\frac{dy}{dx} = \frac{y - k\sqrt{(x^2 + y^2)}}{x}$$

where $k = w/v$ and (x, y) is the position of the plane at time t. Find the equation of the plane's path when it starts from a position a miles from T (0, 0) at time $t = 0$. Comment on your result.

4. The Ricatti equation is given by $y' = P(x)y^2 + Q(x)y + R(x)$. Prove that if one solution of this equation, say $y_1(x)$, is known, the general solution can be found by using the transformation $y = y_1 + 1/u$, when u is a new independent variable. Show that a solution to $y' = x^2 - 2 + 2xy + y^2$ is $y = 1 - x$ and find a general solution. ($y' = dy/dx$.)

5. If a first-order differential equation can be easily solved for y in terms of x and $v = y'$, that is, $y = F(x, v)$, then by differentiating with respect to x we obtain $v = \phi(x, v, v')$. If this last equation can be solved for v in terms of x, say $v = \phi(x, c)$, the solution to the original equation can be obtained as $y = F(x, \phi)$. Illustrate this method by obtaining solutions to:
(a) $y = xy' + (y')^2$, (b) $y = 2xy' + \ln(yy')$.

6. Suppose that in (4.6.5) the mass is a function $m(t)$. Use Newton's law (4.5.9) to change (4.6.5) to $\dot{m}v + m\dot{v} = mg - R$.

   Now suppose the falling body is a spherical raindrop, of radius $r(t)$ which increases its mass by condensation, at a rate proportional to its surface area. Show that $r = r_0 + at$, for some constant a; and then $\dot{v} + r(t)(v - g)/3a = -R/br^2$, where b is another constant.

   Investigate the motion when R is (a) constant, (b) proportional to v. What are some limitations on this model? (Note: $\dot{v} + f(t)v = e^{-g}\, d/dt\,(e^g v)$, where $g = \int f\, dt$. Other problems of variable mass, involving rockets burning fuel, etc., can be found in Mechanics books.)

7. A particle moves in one dimension under the action of a force which gives it an acceleration:

$$\frac{d^2x}{dt^2} = -k\left(\frac{dx}{dt}\right)^3,$$

where k is a positive constant. At time $t = 0$ the particle has a speed v_0.
(a) Find the distance travelled by the particle when its speed has fallen to a value v.

(b) Show that, at any time t, the speed of the particle $v(t) \equiv \mathrm{d}x/\mathrm{d}t$ is given by

$$\frac{1}{v^2} \equiv \frac{1}{v_0^2} + 2kt.$$

(c) Hence show that the distance travelled from its starting position in time t is given by

$$x = \frac{(2ktv_0^2 + 1)^{\frac{1}{2}} - 1}{kv_0}.$$

(d) Expand $x(t)$ in powers of t to obtain the first two non-vanishing terms. (See section C.2) [SU]

8. A particle moves with coordinate y and velocity $v = \mathrm{d}y/\mathrm{d}t$ under the action of a force that gives it an acceleration

$$\frac{\mathrm{d}^2y}{\mathrm{d}t^2} = -\frac{gc^2}{(c + y)^2},$$

where g and c are positive constants. If the initial conditions are

$$y = 0 \qquad \text{and} \qquad v = \sqrt{2gc} \qquad \text{at } t = 0$$

show that

$$v^2 = \frac{2gc^2}{c + y}.$$

Use this to write down an expression for $\mathrm{d}y/\mathrm{d}t$ and integrate to show that the value of y at time t is

$$y = \left[c^{3/2} + \frac{3}{2}c\sqrt{2g}\,t\right]^{2/3} - c.$$

[SU]

9. Suppose that $\mathrm{d}x/\mathrm{d}t = f(x)$, where $f(x)$ is never zero on the interval $I = [a, b]$. Then

$$\int_a^x \frac{\mathrm{d}u}{f(u)} = \int_a^x 1\,\mathrm{d}u = x - a.$$

Use this method to solve the following problems.

(a) $\dfrac{\mathrm{d}y}{\mathrm{d}x} = y(y - 1)(y - 2)$ with $y(0) = \frac{1}{2}$.

(b) A chemical reaction is modelled by the equation

$$\frac{\mathrm{d}x}{\mathrm{d}t} = k(a - x)(b - x),$$

where a, b, k are positive constants, which satisfies the initial condition $x = 0$ when $t = 0$. Show that

$$\frac{b - x}{a - x} = \frac{b}{a}\mathrm{e}^{(b-a)kt}.$$

10. Show that if $y' + y \cdot p(x) = q(x)$ then

$$\frac{\mathrm{d}}{\mathrm{d}x}(\mathrm{e}^{P(x)} \cdot y) = \mathrm{e}^{P(x)}q(x)$$

where $P(x) = \int_a^x p(t)\,\mathrm{d}t$. Use this method to solve the equations:

(a) $y' + 2(\tan x)\cdot y = \sin x$, $\quad y(\pi/3) = 0$;

(b) $\cos^2 x \cdot y' - \frac{1}{2} y \sin 2x = \sec x$, $\quad y(0) = 1$;

(c) $xy' - y = xe^{y/x}$, $\quad y(e) = 1$;

(d) $x\dfrac{\mathrm{d}y}{\mathrm{d}x} - 5y = x^4$, $\quad y(1) = 1$.

4.7 SIMPLE HARMONIC MOTION: PLANETARY ORBITS

Heat does not oscillate: it oozes away. As we have seen, the first-order differential equations of section 4.6 describe processes that are governed by a real exponential term, which either dies out, or grows infinite. But many physical processes oscillate, and the simplest of these is **simple harmonic motion** (briefly: SHM) described as follows.

Suppose a particle P, of mass m, moves on the x-axis and is subject to a force that is proportional to the distance x of P from $\mathbf{0}$, and directed towards $\mathbf{0}$. (Such motion could be approximated physically if P moved in a smooth horizontal groove, and was attached to an elastic spring.) Then the acceleration a of P is given by $a = \mathrm{d}v/\mathrm{d}t = \mathrm{d}/\mathrm{d}t\,(\mathrm{d}x/\mathrm{d}t)$, denoted for brevity by $\ddot{x}$; and the description of the force allows us to translate Newton's law to:

$$m\ddot{x} = -kx, \qquad \text{or} \qquad \ddot{x} = -\omega^2\cdot x, \tag{4.7.1}$$

where k is the (positive) constant of proportionality, and $\omega^2 = k/m$. To solve equation (4.7.1), we use the trick of (4.6.7) to write the equation as:

$$v\frac{\mathrm{d}v}{\mathrm{d}x} = -\omega^2 x, \qquad (v = \dot{x})$$

and integrate with respect to x to get

$$\omega^2 x^2 = v_0^2 - v^2,$$

where v_0 is the velocity at $x = 0$; the last equation shows that $v^2 \leq v_0^2$ since the left-hand side is positive. Hence

$$v = \frac{\mathrm{d}x}{\mathrm{d}t} = \sqrt{(v_0^2 - \omega^2 x^2)},$$

and we can integrate with respect to t to obtain

$$\sin^{-1}\left(\frac{\omega x}{v_0}\right) = \omega\cdot(t + c), \qquad c = \text{constant} \tag{4.7.2}$$

whence $x(t) = A\cdot\sin(\omega t)$, $A = v_0/\omega > 0$, assuming $x = 0$ when $t = 0$. Thus $x(t)$ oscillates like the sine function, repeating each value every $2\pi/\omega$ seconds. It therefore has **period** $2\pi/\omega$. In particular, $x(t)$ varies between A and $-A$ inclusive, and reaches these values at times $(2n + \frac{1}{2})\,\pi/\omega$ and $(2n - \frac{1}{2})\,\pi/\omega$, respectively, for each integer n. We call A the **amplitude** of the motion. It is not necessary that $x = 0$ at time $t = 0$, in which case we can write (4.7.2) as $x(t) = A\sin(\omega t + \varepsilon)$, where $\varepsilon = \omega c$. Using the addition formula to expand $\sin(\omega t + \varepsilon)$, we obtain the general solution

$$x(t) = A\sin(\omega t + \varepsilon) = P\sin(\omega t) + Q\cos(\omega t), \tag{4.7.3}$$

where P, Q are constants. If we know both x and $\dot{x}$ at $t = 0$, then we have two equations to find P and Q.

Exercises 4.7.1

1. Find the general solution of the equation $\ddot{x} = -\omega^2 x + c$, where c is a constant. (Consider $y = x + c/\omega^2$.)
2. The simple numerical technique (Euler's method) for solving first-order differential equations, used in program 4.6.1, can be easily extended for second- and higher-order equations. Set $v = \dot{x}$ in (4.7.1) to give $\dot{v}$ as a function of x (in more general examples $\dot{v}$ may be a function of any or all of t, x, v). Now we know that $\dot{x} = v$ and so we have a pair of coupled equations:

   $$\dot{x} = f_1(t, x, v) = v \quad \text{(in this case)}$$
   $$\dot{v} = f_2(t, x, v) = -\omega^2 \cdot x \quad \text{(in this case)}$$

   and so the approximate increases dx, dv in x and v over a small finite time step dt are: $dx = v \cdot dt$, $dv = -w_2 \cdot x \cdot dt$ where $w_2 = \omega \cdot \omega$. Adapt program 4.6.1 to produce points (t, x) on an approximate solution curve for the SHM equation (4.7.1) and compare the result with the known solution given by (4.7.3).
3. As with program 4.6.2 it is also possible, and, in this case, highly desirable, to have a more accurate solution, and the fourth-order Runge–Kutta process provides one such means. The following program shows the technique.

```
REM Program 4.7.1, S.H.M.
READ x0, v0, w, t, tmax, dt , xmax
DATA 1 , 0 , 1, 0, 10  , 0.1, 1.2
FUNCTION  f1(t,x,v) = v
FUNCTION  f2(t,x,v) = -w^2*x
SET MODE "graphics"
SET WINDOW t,tmax, -xmax,xmax
LET x = x0
LET v = v0
LET vv = 0
LET xx = 0
DO
  PLOT t,x
  CALL RK4(v,x,t,dt,vv,xx)
  LET v = vv
  LET x = xx
  LET t = t + dt
LOOP UNTIL t>tmax
FOR t = 0 TO tmax STEP dt
  LET x = COS(t)
  PLOT t,x;
NEXT t
STOP
```

Program 4.7.1 continued

```
SUB RK4(v,x,t,dt,vv,xx)
  LET h = dt/2
  LET d1x = f1(t,x,v)*dt
  LET d1v = f2(t,x,v)*dt
  LET d2x = f1(t + h, x + d1x/2, v + d1v/2)*dt
  LET d2v = f2(t + h, x + d1x/2, v + d1v/2)*dt
  LET d3x = f1(t + h, x + d2x/2, v + d2v/2)*dt
  LET d3v = f2(t + h, x + d2x/2, v + d2v/2)*dt
  LET d4x = f1(t + dt, x + d3x, v + d3v)*dt
  LET d4v = f2(t + dt, x + d3x, v + d3v)*dt
  LET xx = x + (d1x + 2*d2x + 2*d3x + d4x)/6
  LET vv = v + (d1v + 2*d2v + 2*d3v + d4v)/6
END SUB
END
```

(Note that the variables xx, vv are used to return values from the procedure RK4.) The approximate solution should agree well with the analytic one. See how this varies with dt.

4.7.1 Planetary orbits

Although simple harmonic motion was defined in terms of motion in a line, we meet it (somewhat surprisingly) when we study the orbits of planets, as we now show. For this, it is necessary for readers (and indeed all educated persons, arts graduates or not) to know a little of the history of this problem, which has had an immense effect on our modern lives and on world culture.

Gradually, people came to believe that the planets of the solar system rotated about the Sun; but it was assumed that their orbits were circular, since that was the most perfect shape. Nowadays, we do not regard that as a good reason, but it seemed totally natural in the intellectual climate of the sixteenth century, when Kepler was forced to challenge it because of the accurate observations made by his teacher Tycho Brahe. An immense intellectual wrench was needed for Kepler to change his manner of thinking from that of an astrologer (by which he made his living) to that of a modern astronomer: things move as they do, not as some traditional belief says they 'ought'. He derived some support from his correspondence with Galileo, who was also having new thoughts on mechanics.

Eventually, Kepler announced his three laws, of which the most shocking to the orthodox was the one saying that *each planet moves in an elliptical orbit with the Sun at one focus.* But with Galileo's work, this raised the new question: what force acted on the planet's to make them move in this way? No longer was it satisfactory just to answer 'The hand of God'—the whole nature of what is meant by an explanation was now in question, and this could lead to subversive thoughts. (Even to burning at the stake, as happened to the natural philosopher Giordano Bruno in 1600.)

More than half a century later, Newton provided an answer, with all its later effects on the practice of science and much social thinking. Following the guesses of other contemporaries, he postulated that the Sun exerts a force of attraction on each planet, acting along the radius between them, and of magnitude

$$F = \frac{PMG}{r^2} \qquad \text{(the \textbf{inverse square law})}.$$

Here P and M are the masses of the planet and Sun, and G is the 'universal constant of gravitation', $\cong 6.672 \times 10^{-11}\ \mathrm{m}^3\,\mathrm{s}^{-2}\,\mathrm{kg}^{-1}$. Newton then used his law to obtain a differential equation, from which he deduced the three laws that Kepler had found by observation (and hard toil). Moreover, he showed that the inverse square law is the only one that would lead to these laws (within his general model). We shall content ourselves here with showing that the orbit is an ellipse, as a comprehensive account is very long.

Thus, suppose that the motion takes place in the coordinate plane with the Sun fixed at $\mathbf{0}$, and suppose that at time t the planet is at the position which we can represent by the complex number $z(t) = r\mathrm{e}^{j\theta}$, so r and θ are functions of t. Then the planet has velocity $\dot{z}(t)$ and acceleration $\ddot{z}(t)$ given by:

$$\dot{z}(t) = (\dot{r} + jr\dot{\theta})\mathrm{e}^{j\theta}, \qquad \ddot{z}(t) = ((\ddot{r} - r\dot{\theta}^2) + j(r\ddot{\theta} + 2\dot{r}\dot{\theta}))\mathrm{e}^{j\theta}.$$

Hence, the inverse square law says that the force is the vector $-(PMG/r^2)\cdot \mathrm{e}^{j\theta}$, and by Newton's law this is $P\ddot{z}$, so we may cancel $\mathrm{e}^{j\theta}$ and equate real and imaginary parts to get

$$-\frac{MG}{r^2} = \ddot{r} - r\dot{\theta}^2, \qquad r\ddot{\theta} + 2\dot{r}\dot{\theta} = 0. \tag{4.7.4}$$

The second equation says: $1/r \cdot \mathrm{d}/\mathrm{d}t\,(r^2\dot{\theta}) = 0$, so we have $r^2\dot{\theta} = h$, where h is a constant, and this is equivalent to Kepler's second law that *planets sweep out equal areas in equal times.* We can substitute for $\dot{\theta}$ in the first equation, but we will get a very hard equation to solve. In many cases like this, it is helpful to try putting $r = 1/u$, and also to regard r as a function of θ; to do all this in the seventeenth century was very clever! We then have

$$\dot{\theta} = hu^2 \qquad \text{and} \qquad \dot{r} = -\frac{\dot{u}}{u^2} = -h\frac{\mathrm{d}u}{\mathrm{d}\theta} \tag{4.7.5}$$

since

$$\frac{\mathrm{d}u}{\mathrm{d}\theta} = \frac{\mathrm{d}u}{\mathrm{d}t}\Big/\frac{\mathrm{d}\theta}{\mathrm{d}t} = \frac{\dot{u}}{hu^2}.$$

Hence

$$\ddot{r} = \frac{\mathrm{d}\dot{r}}{\mathrm{d}t} = \frac{\mathrm{d}\dot{r}}{\mathrm{d}\theta}\frac{\mathrm{d}\theta}{\mathrm{d}t} = -h\frac{\mathrm{d}^2u}{\mathrm{d}\theta^2}\cdot(hu^2) = -h^2u^2\frac{\mathrm{d}^2u}{\mathrm{d}\theta^2}.$$

Inserting this in the first equation of (4.7.4), we get

$$\frac{MG}{h^2} = \frac{\mathrm{d}^2u}{\mathrm{d}\theta^2} + u, \qquad \text{or} \qquad \frac{\mathrm{d}^2v}{\mathrm{d}\theta^2} = -v,$$

where $v = u - MG/h^2$. Thus we have a simple harmonic equation, with $\omega = 1$ and with θ playing the role of time; and by (4.7.3) its solution is of the form $v = A\sin(\theta + \varepsilon)$. Taking $\varepsilon = -\pi/2$ gives $v = -A\cos\theta$, so finally r is given by

$$r = \frac{1}{u} = p/(1 - e\cdot\cos\theta), \qquad p = \frac{h^2}{MG}, \qquad e = Ap. \tag{4.7.6}$$

Equation (4.7.6), giving the distance r as a function of the angle θ is the *polar equation* of a curve. As θ (in radians) goes through the angles 0, $\pi/2$, π, $3\pi/2$, 2π, so r takes the values $p/(1-e)$, p, $p/(1+e)$, p, $p/(1-e)$, and the path of the point (r, θ) in polar coordinates is an ellipse of eccentricity e $(0 < e < 1)$ with one of its foci as centre **0**. (This is proved in (7.2.5).) The extreme values of r: $p/(1+e)$ and $p/(1-e)$, are known as the **perihelion** (= nearest the Sun) distance and the **aphelion** (= furthest from the Sun) distance. The value p is known as the **mean distance** from the planet to the Sun. A sketch of the graph of r against θ shows that $(0, p/(1+e))$ and $(0, p/(1-e))$ are local minima and maxima, and hence that $dr/d\theta = 0$ at $\theta = n\pi$ (where n is an integer).

Now, for the planet Earth, at $\theta = 0$, the aphelion distance r_0 is 1.521×10^{11} m, and this corresponds to the Earth's minimum linear velocity of $V_0 = 2.929 \times 10^4\ \mathrm{ms}^{-1} = r_0\dot{\theta}_0$ (since here $dr/d\theta = 0$, and, since $\dot{\theta} \neq 0$, $\dot{r}_0 = 0$) which gives the initial angular velocity $\dot{\theta}_0 = 1.925 \times 10^{-7}\ \mathrm{rads\ s}^{-1}$. We can check this for size with the known period of the Earth's rotation about the Sun of 365.3 days, and so the mean angular velocity $\dot{\theta} = 1.991 \times 10^{-7}\ \mathrm{rads\ s}^{-1}$. Now h is given by $h = r_0^2\dot{\theta}_0 = r_0 V_0 = 4.455 \times 10^{15}\ \mathrm{m}^2\,\mathrm{s}^{-1}$. The values of M, G are: $M = 1.989 \times 10^{30}$ kg, $G = 6.672 \times 10^{-11}\ \mathrm{m}^3\,\mathrm{s}^{-2}\,\mathrm{kg}^{-1}$ and so $p = 1.496 \times 10^{11}$ m. Now $v = u - MG/h^2$ which becomes: $-A\cos\theta = 1/r - 1/p$, so when $\theta = 0$ this gives $A = 1/p - 1/r_0 = 1.118 \times 10^{-13}\ \mathrm{m}^{-1}$. Similarly $e = Ap = 1 - p/r_0 = 0.01672$, which means the orbit is very nearly circular, and we calculate the perihelion distance as $p/(1+e)$ to get 1.471×10^{11} m, and this, and the value of p, agree with known results. The next exercises include programs which simulate orbits under the inverse square law of attraction, and which plot curves from their polar equations.

Exercises 4.7.2

1. The following program plots polar curves of the form $r = f(\theta)$.

```
REM Program 4.7.2, Polar plot
FUNCTION fr(theta) = p/(1 - e*COS(theta))
READ p          , e         , xmax, ymax, tmax, tstep
DATA 1.496E11, 0.01672, 3E11, 2E11, 6.3 , 0.05
SET MODE "graphics"
SET WINDOW -xmax,xmax, -ymax,ymax
FOR theta = 0 TO tmax STEP tstep
    LET r = fr(theta)
    PLOT r*COS(theta),r*SIN(theta);
NEXT theta
END
```

Check that it gives the expected orbit with data as above. The following table gives the aphelion distance r_0 (in m), and the minimum linear velocity V_0 (in ms^{-1}) for the other planets.

Calculate the values of p and e in each case, and plot their orbits. Calculate $\dot{\theta}_0$ in each case, and so estimate the period of rotation about the Sun for each planet as a multiple of an Earth year.

Planet	r_0	V_0
Mercury	6.982×10^{10}	3.886×10^4
Venus	1.089×10^{11}	3.478×10^4
Mars	2.492×10^{11}	2.197×10^4
Jupiter	8.159×10^{11}	1.245×10^4
Saturn	1.507×10^{12}	9.119×10^3
Uranus	3.003×10^{12}	6.491×10^3
Neptune	4.542×10^{12}	5.378×10^3
Pluto	7.383×10^{12}	3.675×10^3

2. Deduce the other Kepler laws as follows. First, the radius vector sweeps out equal areas in equal times. For, if the radius, of length r, moves through a small angle, $\delta\theta$ then the area δA of the thin triangle swept out is approximately $\frac{1}{2}r^2\delta\theta$, whence the area swept out between angles θ and ϕ is $\frac{1}{2}\int_0^\phi r^2\,d\theta = \frac{1}{2}\int_0^\phi r^2\dot{\theta}\,dt$. Now use equation (4.7.5). Next, note that the length of a year on a planet is its **period** T, that is, the time of one revolution about the Sun. Kepler's third law states that *if a planet moves with period T, on an ellipse of major axis a, then $T^2/a^3 = 4\pi^2/MG$.* But the work for the first law shows that the area E of the ellipse is $hT/2$, and $E = \pi ab$ if b is the minor axis. The length p in (4.7.6) is b^2/a; hence $T^2/a^3 = 4\pi^2 \cdot p/h^2$, and the law follows.
3. The orbit of a particle under an inverse square law of attraction can be simply, albeit crudely, simulated by a difference equation technique like that in Chapter 2. The following program assumes that a particle is at position (x, y) at time t with velocity vector $(vx \quad vy)$ and that the Sun is at the origin **0** (0, 0).

```
REM Program 4.7.3, Central orbits
READ x        , y, vx, vy     , dt     , mu        , xmax, ymax, tmax
DATA 1.521E11, 0, 0 , 2.929E4, 8.64E4, 1.327E20, 3E11, 2E11, 32E6
SET MODE "graphics"
SET WINDOW -xmax,xmax, -ymax,ymax
LET t = 0
DO
  PLOT x,y
  LET r = SQR(x*x + y*y)
  LET c = x/r
  LET s = y/r
  LET f = -mu/(r*r)
  LET vx = vx + f*c*dt
  LET vy = vy + f*s*dt
  LET x = x + vx*dt
  LET y = y + vy*dt
  LET t = t + dt
LOOP UNTIL  t > tmax
END
```

Investigate the behaviour of the orbit for different initial conditions. How can you estimate the period of the motion? Use this with the data for the other planets in exercise 1. For an improved method see if you can model the solution of the differential equations using the Runge–Kutta technique.

The polar equation in (4.7.6) gives the form of the orbit as an ellipse and the polar plot program 4.7.2 plots points at equally spaced values of the polar angle θ. But these do not correspond to positions taken by the planet at *equal time steps.* So, an important physical calculation is, given the position P_1 of a planet at time t_1, to predict its position P_2 at time t_2. If we denote the time step by Δt then $\Delta t = t_2 - t_1$. Then we know the area ΔA swept out in this time is given by

$$\Delta A = \frac{1}{2} h \Delta t = \frac{1}{2} \int_{\theta_1}^{\theta_2} r^2 \, d\theta \tag{4.7.7}$$

where θ_1 and θ_2 are the polar angles of the points P_1 and P_2.

Unfortunately the corresponding integral $\int d\theta/(1 - e\cos\theta)^2$ is not simple to evaluate, and so a change of parameter is worth trying. In the following chapter we show in example 5.1.1(b) that an ellipse with centre $\mathbf{0}$ and eccentricity e can be parametrized in the form: $x(\phi) = a\cos\phi$, $y(\phi) = b\sin\phi$, where $2a$ is the longest diameter, the distance between the centre and a focus is ae, and b may be found from $b^2 = a^2(1 - e^2)$. So, for a planet, $2a$ is the sum of the perihelion and aphelion distances: $2a = p/(1+e) + p/(1-e)$ which gives $a = p/(1-e^2)$ and so $b = p/\sqrt{(1-e^2)}$. Since the centre of the elliptical orbit is the Sun, which is at one focus, then the centre of the ellipse is at $(ae, 0)$ and the corresponding parametric forms are:

$$x(\phi) = a(\cos\phi + e) \qquad y(\phi) = b\sin\phi \tag{4.7.8}$$

and these are connected to r and θ by:

$$x^2(\phi) + y^2(\phi) = r^2, \qquad \frac{y(\phi)}{x(\phi)} = \tan\theta. \tag{4.7.9}$$

Differentiating the second of these gives:

$$\frac{xy' - x'y}{x^2} = (1 + \tan^2\theta)\frac{d\theta}{d\phi} = \frac{x^2 + y^2}{x^2}\frac{d\theta}{d\phi}$$

where $x' = dx/d\phi$ and $y' = dy/d\phi$, and so (4.7.7) becomes:

$$\Delta A = \frac{1}{2}\int_{\theta_1}^{\theta_2} r^2 \, d\theta = \frac{1}{2}\int_{\phi_1}^{\phi_2} (x(\phi)y'(\phi) - x'(\phi)y(\phi)) \, d\phi. \tag{4.7.10}$$

The last integral is known as 'Leibniz's sector formula'.

Substituting from (4.7.8) gives the integrand as $ab(1 + e\cos\phi)$ and hence the equation corresponding to (4.7.7) becomes:

$$\mu = [\phi + e\sin\phi]_{\phi_1}^{\phi_2}, \qquad \mu = \frac{2}{ab}\Delta A. \tag{4.7.11}$$

If the planet starts at P_0 with $\phi_0 = 0$ at time $t = 0$, then to find the parameter ϕ_1 corresponding to time Δt means solving the equation $\mu = \phi_1 + e\sin\phi_1$, which is not separable. However the rearrangement

$$\phi = \mu - e\sin\phi \tag{4.7.12}$$

gives an iterative scheme which quickly converges.

For the Earth, suppose we take $\Delta t = 1/12$ year (i.e., roughly a calendar month). Then $\mu \cong 0.5235$ and we know $e \cong 0.01672$, and for any starting value of ϕ (e.g., $\phi = 0$) the iteration $\mu - e \sin \phi$ quickly converges to $\phi_1 \cong 0.5153$. Using this in (4.7.8) gives the next position P_1 of the planet. Similarly, from (4.7.11) the next equation to be solved is $2\mu = \phi_2 + e \sin \phi_2$, and so on. The reader can now attempt the following exercise.

Exercises 4.7.3

1. Check that the solution to $n\mu = \phi_n + e \sin \phi_n$ for $n = 12$ is approximately 2π as expected. (The behaviour of the related iteration, which has interesting dynamics for a range of values of e, will be considered in Chapter 10.) Here, the plot of ϕ_n versus n is nearly linear, but we get different behaviour for Halley's comet. Thus, repeat the above with $\mu = 2\pi/12$ as before, but with $e \cong 0.967$, and observe large changes in ϕ_n around $n = 6$. Repeat for smaller divisions of the comet's year.
2. (T. Crilly) Many particles move with acceleration given by $a = f(x)$ for some function f. To integrate this equation we write it as

$$v\frac{dv}{dx} = f(x) \text{ so } v^2 - v_0^2 = \int_0^a f(x)\,dx. \text{ With SHM, } f(x) = -\omega^2 x.$$

Prove that the time taken to reach 0 from rest at $x(0) = a > 0$ is $\pi/2\omega$, independent of a.

But if $f(x) = -\omega^2/x$, the origin acts as a 'black hole'. With the same initial conditions, prove that when $x > 0$, $v = -\omega\sqrt{2 \ln(a/x)}$.

Now show that the time T taken to reach $x = s < a$ is given by

$$\omega\sqrt{2}\cdot T = \int_s^a \frac{dx}{\sqrt{\ln(a/x)}} = \int_0^Y y^{1/2}\,e^{-y}\,dy$$

where $y = \ln(a/x)$ and $Y = y(a)$. When s tends to zero show that Y tends to infinity; and T tends to $a\Gamma(\frac{1}{2})/\omega\sqrt{2}$, where

$$\Gamma(z) = \int_0^\infty y^{z-1}\,e^{-y}\,dy$$

is the gamma function. It is known that $\Gamma(\frac{1}{2}) = \sqrt{\pi}$ so the time taken for our particle to reach 0 is $a/\omega\sqrt{(\pi/2)}$, and this depends on a. Next try the cases $f(x) = -\omega^2 x^p$ when $p > -1$ and $p < -1$. The results involve the Beta function.

4.8 LINEAR DIFFERENTIAL EQUATIONS OF THE SECOND ORDER

In a geniune physical situation like that of a particle in a groove, we would need to incorporate a friction term in (4.7.1), and it is traditional to suppose that the frictional force is proportional to velocity $\dot{x}$. There may also be a driving force that fluctuates with time, but is independent of x, so (4.7.1) is modified to the form

$$\ddot{x} + r\cdot\dot{x} + \omega^2\cdot x = f(t), \tag{4.8.1}$$

and we give various physical examples below. These include the basic electrical circuit shown in Fig. 4.8.1, in which a source of voltage E supplies energy through a resistor of value R ohms, a capacitor of value C farads, and an inductor of value L henries.

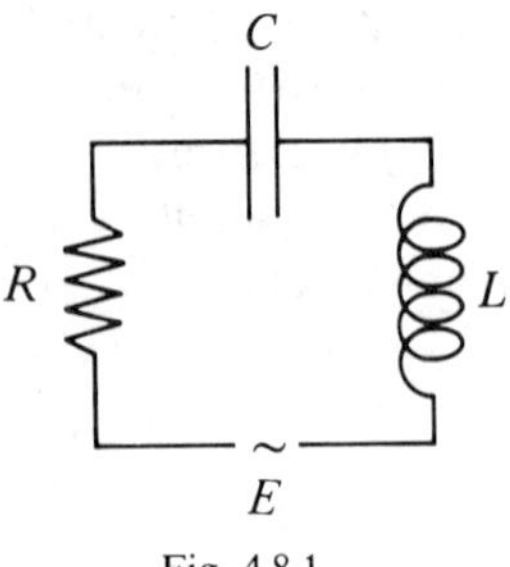

Fig. 4.8.1.

The value of E at time t is $f(t)$ volts, say, and electrons flow through the circuit to leave a charge of $q(t)$ coulombs on C; the rate of flow, $\mathrm{d}q/\mathrm{d}t$ is the **current** $i(t)$. The voltage drops across the components are V_R, V_C, V_L and physical laws state that:

(i) $E = V_R + V_C + V_L$; (ii) $V_C = qC$;

(iii) $V_R = iR$ (Ohm's Law); (iv) $V_L = L\dfrac{\mathrm{d}i}{\mathrm{d}t}$ (Lenz's law).

Hence, from (i) we obtain the equation $L\ddot{q} + R\dot{q} + Cq = f(t)$ which is essentially (4.8.1). Indeed, this is found to describe electrical systems more accurately than (4.8.1) describes mechanical ones, perhaps because the friction term $k\dot{x}$ is often only a crude approximation to reality.

The equation (4.8.1) is said to be **linear**, because x and its derivatives do not appear as powers other than the first; and it is of **second order**, because the highest derivative that occurs is the second. Once a system has been modelled by an equation like (4.8.1), with r, ω, and f assumed known, we have to 'solve' the equation by finding a function $x(t)$ that satisfies both it and the initial conditions. The latter conditions are assumed given, and in practice usually tell us the values of x and $\dot{x}$ at $t = 0$.

There exist several trial-and-error methods of solution, some of which can work speedily in special situations, such as those illustrated in the programs of this section; but now we prefer to explain a systematic method, which is more general. Readers will find it easiest if, before starting the systematic exposition, we summarize the method in the following algorithm, which may be partly familiar.

Algorithm 4.8.1

Given the initial conditions $x_0, \dot{x}_0$, the solution $x(t)$ of (4.8.1) is the sum of two functions:

$$x(t) = C(t) + P(t), \tag{4.8.2}$$

where $C(t)$ and $P(t)$ are called respectively the **complementary function**, and the **particular integral**, and are computed as follows.

(1) (For $C(t)$). Find the roots u, v of the **auxiliary equation**

$$k^2 + r\cdot k + \omega^2 = 0 \tag{4.8.3}$$

(these need not be real numbers). Let $m = u - v$.

(i) if $m = 0$ then $C(t) = e^{ut} \cdot (x_0 + (\dot{x}_0 - u \cdot x_0) \cdot t)$,
(ii) if $m \neq 0$ then $C(t) = A \cdot e^{ut} + B \cdot e^{vt}$, where

$$A = \frac{\dot{x}_0 - vx_0}{m}, \qquad B = \frac{ux_0 - \dot{x}_0}{m}. \tag{4.8.4}$$

If u, v are not real then they are complex conjugates

$$u = a + jb, \qquad v = a - jb, \qquad b \neq 0$$

and we express $C(t)$ in 'real form' as
(iii) $C(t) = e^{at} \cdot (A \cdot \cos(bt) + B \cdot \sin(bt))$, where

$$A = x_0, \qquad B = \frac{\dot{x}_0 - ax_0}{b}.$$

(2) (For $P(t)$). Form the integral (which it may or may not be possible to simplify further):

$$F(t) = e^{ut} \cdot \int_0^t e^{-us} \cdot f(s)\, ds.$$

Then

$$P(t) = e^{vt} \cdot \int_0^t e^{-vs} \cdot F(s)\, ds.$$

Note that if u, v are real and negative, then $C(t)$ dies away as t increases; similarly when they are complex conjugates, $C(t)$ oscillates and dies away if $a < 0$. In these cases, $C(t)$ is said to be a 'transient', because its practical effect lasts for only a short time: the long-term motion is due to $P(t)$.

Example 4.8.1

Suppose $\ddot{x} + 5\dot{x} + 6 = f(t)$, $x_0 = 1$, $\dot{x}_0 = -4$, where $f(t)$ is a 'jump' function which is zero when $t < 0$ and 1 otherwise. The auxiliary equation (4.8.3) is $k^2 + 5k + 6 = 0$ with unequal real roots $u = -3$, $v = -2$. Then $m = -1$ so by (ii) of the algorithm, $C(t) = Ae^{-3t} + Be^{-2t}$, where $A = (-4 - (-2))/(-1) = 2$, and $B = (-3 - (-4))/(-1) = -1$. To find $P(t)$ we have (4.8.4) to form

$$F(t) = e^{-3t} \int_0^t e^{+3s} f(s)\, ds$$

which is zero if $t < 0$; while if $t \geq 0$ then

$$F(t) = e^{-3t} \int_0^t e^{3s} \cdot 1\, ds = (1 - e^{-3t})/3.$$

Hence, if $t > 0$, then

$$P(t) = e^{-2t} \int_0^t e^{2s}(1 - e^{-3s})\, ds/3$$

$$= (1 + e^{-3t} - 2e^{-t})/6$$

and $P(t) = 0$ otherwise. Thus all terms in the solution $x(t) = C(t) + P(t)$ are transient except the 1/6, and $x(t)$ approaches this steady value as t increases.

Exercises 4.8.1

1. Use the algorithm to find the general solutions to the following differential equations:
   (a) $y'' + 4y' + 4y = x^2$ (Here, as often, y' means dy/dx)
   (b) $y'' - 5y' + 6y = e^{3x}$
   (c) $y'' - 2y' - 3y = \sin 2x + \cos 2x$
   (d) $\dfrac{d^2y}{dx^2} - 4\dfrac{dy}{dx} + 3y = 0$
   (e) $y'' + 4y' + 3y = 15e^{-2x}$
   (f) $y'' + 4y' + 4y = x$
   (g) $y'' + 4y = \sin x$.
   In each case write down the appropriate form of A and B as given by algorithm 4.8.1 in terms of y_0, y_0'.
2. Find the solution of the differential equation

$$\frac{d^2y}{dx^2} + 4\frac{dy}{dx} + 4y = x + e^{-2x}$$

   which satisfies the conditions $y = 2$ and $dy/dx = -2$ when $x = 0$. [SU]
3. Solve the equations

   (a) $\dfrac{d^2y}{dx^2} + 2\dfrac{dy}{dx} + 5y = 0$ with $y(0) = 1$ and $y'(0) = 0$

   (b) $y'' - y' - 2y = 60\, e^t \sin 2t$, with $y(0) = y'(0) = 0$. [SU]
4. Find the solution to

$$x^2\frac{d^2y}{dx^2} - 4x\frac{dy}{dx} + 6y = 2x$$

   satisfying $y(1) = y(-1) = 0$. Put $x = e^\theta$, and show that $xy' = dy/d\theta$, $x^2y'' = d^2y/d\theta^2$; what if $x < 0$?)
5. Find the solution to the differential equation

$$\frac{d^2y}{dx^2} + 6\frac{dy}{dx} + 10y = 0$$

   which satisfies the conditions $y = 0$ and $dy/dx = -1$ when $x = \pi/2$.
6. The motion of a damped oscillatory system with a periodic disturbing force is described by the differential equation

$$\frac{d^2x}{dt^2} + 2\frac{dx}{dt} + 3x = 4 \sin \omega t.$$

Find the general solution of this equation and state which term in the solution is the transient term and which the steady-state term.

Find also the maximum value of the amplitude of the steady-state solution that may be obtained by varying ω, and show that the corresponding value of the phase shift of the steady-state solution relative to the forcing term is $\pi/4$. [SU]

7. Find the equation satisfied by $u(x)$ if $e^{-3x}u(x)$ is a solution of

$$y'' + 6y' + 9y = e^{-3x} f(x).$$

Hence find the solution of

$$y'' + 6y' + 9y = xe^{-3x}$$

for which $y(0) = 1$, $y'(0) = 1$.

8. Solve the equation

$$\frac{d^2y}{dx^2} + \frac{2dy}{dx} + y = \begin{matrix} 0 & x < 0 \\ -x & 0 \le x \le 1 \\ -1 & x > 1 \end{matrix} .$$

$y(0) = y'(0) = 0$.

9. Solve the equation

$$(D - 2)(D^2 + 1)x = e^{3t},$$

with $y(0) = 1$, $\dot{y}(0) = 2$, $\ddot{y}(0) = -1$.

10. If, in (4.8.4), $u = a + jb$ with $a < 0$, then e^{ut} is bounded for all time t. Show that this will not be true of $F(t)$ there, if $f(s)$ is either $e^{at} \sin bt$ or $e^{at} \cos bt$; and hence show that $P(t)$ will be unbounded. (This is the phenomenon of **resonance**, when the forcing term happens to have the same amplitude and frequency as the system it is forcing; and why mechanical systems can shake themselves to destruction.)

11. Leakages in a transmission line can be investigated if it is modelled as a sequence of n simple circuits. Fig. 4.8.2 shows the case $n = 2$, and all resistances are neglected. Derive for this case the equations:

$$L\dot{I}_j = V_{j-1} - V_j \; (j = 1, 2); \qquad C\dot{V}_j = I_j - I_{j+1} \; (j = 0, 1, 2, \quad I_3 = 0).$$

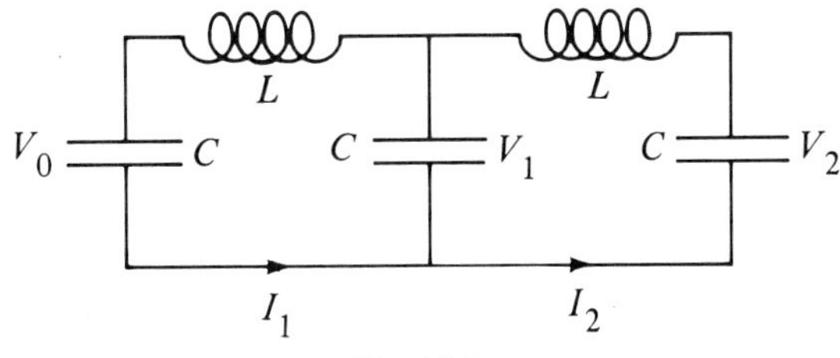

Fig. 4.8.2.

Here the capacitor at V_2 is such that $I_2 = f(V_1)$. Rescale by writing $I_j = \lambda u_j$, $V_j = \mu v_j$, $t = \alpha\tau$, to show that when $\alpha = \mu/\lambda = \sqrt{(L/C)}$, the equations become:

$$\dot{u}_j = v_{j-1} - v_j \; (j = 1, 2); \qquad \dot{v}_j = u_j - u_{j+1} \; (j = 0, 1); \qquad v_2 = \int u^2 \, dt$$

with $u_2 = I_2/\lambda = f(\mu v_1/\lambda) = g(v_1)$, say. If g is assumed linear, say $g(v_1) = kv_1$ with k constant, the equations form a linear system $\dot{\mathbf{z}} = A\mathbf{z}$, with $\mathbf{z} = (u_1, v_1, u_2, v_2)$ in $\mathbb{R}^4$. Write down the 4×4 matrix A. How would you analyse the geometry of the equilibrium point at $\mathbf{0}$? Working algebraically instead, obtain a fourth-order differential equation for one of the coordinates, and solve it. How does your solution depend on k? (If g is non-linear, the hardware for the circuit can demonstrate 'chaotic' behaviour in the sense of Chapter 10; this was studied in the undergraduate project of Ross (1990).)

4.9 PROOF OF THE ALGORITHM

In this section, we prove that algorithm 4.8.1 in section 4.8 actually delivers what it claims to do. We first explain some algebra of the D-operator.

The process of obtaining the derivative $\dot{x}$ of the function $x(t)$ can be thought of as an operator (denoted by D) which converts x into $\dot{x}$. Similarly, if u is any number the operator $\mathrm{D} + u$ converts x into the function $\dot{x} + ux$; thus we write

$$(\mathrm{D} + u)x = \dot{x} + ux, \tag{4.9.1}$$

and for example, equation (4.6.1) can be written $(\mathrm{D} + k)\theta = kS$. If now v is a second number, we can operate with $(\mathrm{D} + v)$ on $(\mathrm{D} + u)x$ to get

$$\begin{aligned}(\mathrm{D} + v)((\mathrm{D} + u)x) &= (\mathrm{D} + v)(\dot{x} + ux) = (\mathrm{D} + v)\dot{x} + u(\mathrm{D} + v)x \\ &= \ddot{x} + (u + v)\dot{x} + uvx.\end{aligned}$$

This is the same as the left-hand side of (4.8.1) if $u + v = r$, and $uv = \omega^2$, that is, if $-u$, $-v$ are the roots of the auxiliary equation (4.8.3). Hence we can write (4.8.1) as an equation about functions of t: $(\mathrm{D} - u)(\mathrm{D} - v)x = f$. We can now write this as a pair of simultaneous first-order equations:

$$(\mathrm{D} - u)y = f, \qquad (\mathrm{D} - v)x = y, \tag{4.9.2}$$

of which the initial conditions are that at $t = 0$, $x(0) = x_0$ as before, and (by the second equation in (4.9.2)),

$$y_0 = y(0) = \dot{x}(0) - vx(0) = \dot{x}_0 - vx_0. \tag{4.9.3}$$

By (4.6.3), the first equation in (4.9.2) has the unique solution:

$$y(t) = \mathrm{e}^{ut} \cdot (y_0 + \int_0^t \mathrm{e}^{-us} \cdot f(s)\,\mathrm{d}s) = \mathrm{e}^{ut} \cdot y_0 + \mathrm{F(t)},$$

say, where we can say nothing more in general about F, without knowing more about f. We can then substitute for y in the second equation of (4.9.2), and apply (4.6.3) again, to get the unique solution

$$\begin{aligned}x(t) &= \mathrm{e}^{vt} \cdot (x_0 + \int_0^t \mathrm{e}^{-vs} \cdot y(s)\,\mathrm{d}s) \\ &= \mathrm{e}^{vt} \cdot x_0 + \mathrm{e}^{vt} \cdot \int_0^t \mathrm{e}^{-vs} \cdot y(s)\,\mathrm{d}s\end{aligned}$$

$$= e^{vt} \cdot x_0 + e^{vt} \cdot \int_0^t e^{-vs} \cdot (e^{us} \cdot y_0 + F(s))\, ds$$

$$= e^{vt} \cdot x_0 + e^{vt} \cdot h(t) + P(t) \tag{4.9.4}$$

where $h(t) = \int_0^t e^{-vs} \cdot e^{us} \cdot y_0\, ds$ and $F(t)$, $P(t)$ are as in (4.8.4). Thus (4.9.4) accounts for (4.8.2) if

$$C(t) = e^{vt} \cdot (x_0 + h(t)).$$

Now $h(t)$ has two different forms, according as $m = u - v$ is zero or not; and these are given by (i) and (ii) of algorithm 4.8.1 above. Finally, if u, v are complex conjugates, then the form (iii) in algorithm 4.8.1 is obtained from (ii) by using the fact that, for example,

$$e^{ut} = e^{at} \cdot (\cos(bt) + j \sin(bt)) = e^{at} \operatorname{cis}(bt),$$

and then collecting up terms in sin and cos, which turn out to have coefficients $j(A - B)$, $(A + B)$, respectively (where $j = \sqrt{-1}$). These simplify to the real forms given in the algorithm.

This establishes the algorithm, and also justifies the claim that we now have **the** solution, because we have constructed it in terms of the unique solutions to the first-order equations (4.9.2). Again, then, we have proved an existence and uniqueness theorem; it is this which justifies the various trial-and-error methods that are often taught. Those usually add together various functions to produce a solution that fits the conditions, but they do not explain why there is nothing more to be added in, apart from an appeal to the (not clearly relevant) existence of two 'arbitrary constants'. A brief explanation follows.

When f is the zero function in (4.8.1) the equation is called **homogeneous**, because then, if x is a solution, so is any constant multiple of x. Also, if y is another solution (ignoring initial conditions) then so is $x + y$, as the reader can easily check. Readers who know about vector spaces will therefore recognize that the set of all possible solutions of the homogeneous equation forms a real vector space V. We have shown above that $C(t)$ can always be expressed in terms of two basis elements of V, namely (if u, v are real and distinct) the functions e^{ut}, e^{vt}, with analogous remarks in the other two cases. Thus V has dimension 2, so every solution is a **unique** linear sum of the two functions in any basis. Also, if we have any two solutions of (4.8.1) that satisfy the same initial conditions, it is easy to check that their difference z satisfies the homogeneous equation, but now with $z_0 = \dot{z}_0 = 0$. Since the zero function also has these properties, then by uniqueness, $z = 0$. Hence the two solutions of (4.8.1) must really have been the same. This rather sophisticated explanation therefore justifies the usual trial-and-error methods of solving (4.8.1).

Exercises 4.9.1

1. Show that the homogeneous differential equation $\ddot{x} + k\dot{x} + \omega^2 x = 0$ gives the dynamical system $S_{k,\omega}$: $\dot{\mathbf{z}} = B\mathbf{z}$, where $B = \begin{pmatrix} 0 & 1 \\ -\omega^2 & -k \end{pmatrix}$. Hence, use your knowledge of linear systems

to describe the behaviour of the differential equation for different values of k and ω by labelling the regions X of $\mathbb{R}^2$ corresponding to which, if $(k, \omega) \in X$, then $S_{k,\omega}$ exhibits a nodal source, spiral sink, etc.

2. Show how to write the differential equation $\ddot{x} + b\dot{x} + 25x = 0$ as a first-order linear system S_b on $\mathbb{R}^2$. Sketch phase portraits of S_b for the cases
(i) $b = -8$, (ii) $b = 0$, (iii) $b = 8$, (iv) $b = 26$.
Calculate the values of b for which the system S_b differs qualitatively from the system
T_b: $\dot{x} = 10^{-6}x + y$, $\dot{y} = -25x - by$. [SU]

3. Express the circuit equation $L\ddot{q} + R\dot{q} + Cq = 0$ in the form $\dot{\mathbf{z}} = A\mathbf{z}$ where L, R, C are real constants and A is a 2×2 matrix. Show that, if L, R, C are all positive, then the resulting planar system S has a phase portrait displaying a sink at $\mathbf{0}$, which is nodal or spiral according as $R^2 - 4LC$ is positive or negative. Let S_h denote the planar system $\dot{\mathbf{z}} = B\mathbf{z}$, where $B = A + hI$, and I is the 2×2 identity matrix, with $h > 0$. Show that A and B have the same discriminant. Now fix h and C, to show that R and L can be tuned to display a variety of behaviour, as follows. In the (R, L) plane, sketch the parabola Π: $L = R^2/4C$, and lines l: $L = R/2h$, m: $L = R/h - C/h^2$. These curves divide the positive quadrant into various regions. Find the region, say X_1, for which if $(R, L) \in X_1$ then S_n has a saddle at $\mathbf{0}$. Label X_1 'saddles', and similarly label each of the other regions (e.g., 'spiral sinks'), giving reasons in each case. [SU]

4. Given the plane dynamical system $\dot{\mathbf{z}} = A\mathbf{z}$ with $A = \begin{pmatrix} a & b \\ c & d \end{pmatrix}$, show that x and y satisfy the equation $\ddot{q} - (a + d)\dot{q} + (ad - bc)q = 0$. (Note that $(\mathrm{D} - a)x = by$, $(\mathrm{D} - d)y = cx$ and operate on the first equation by $\mathrm{D} - d$.)

5. (a) Solve the simultaneous differential equations:
$$\dot{x} + 5x - 2y = t \qquad \dot{y} + 2x + y = 0.$$
for $x(t)$ and $y(t)$, given that $x(0) = y(0) = 0$. (Obtain a second-order equation in x.)
(b) Similarly solve:
$$\ddot{x} = 2\dot{y} \qquad \ddot{y} = 8 - 2\dot{x}$$
where $x(0) = y(0) = \dot{x}(0) = \dot{y}(0) = 0$.

6. Solve the simultaneous differential equations
$$\dot{x} + x + \dot{y} = \mathrm{e}^t$$
$$3\dot{x} - \dot{y} + y = \mathrm{e}^{-t},$$
where $x(0) = y(0) = 0$.

7. Let $E_{p,q}$ denote the differential equation
$$\ddot{x} + p\dot{x} + qx = 0, \qquad (p, q) \in \mathbb{R}^2,$$
which describes the behaviour of an oscillator with resistance, if p, q are positive. Express this as a two-dimensional system $S_{p,q}$: $\dot{\mathbf{z}} = A\mathbf{z}$ where A is a 2×2 matrix. Label the regions X of $\mathbb{R}^2$ corresponding to which, if $(p, q) \in X$, then $S_{p,q}$ exhibits a nodal source, spiral sink, etc.

Let $T_{k,\varepsilon}$: $\dot{\mathbf{z}} = B\mathbf{z}$ denote a two-dimensional system with matrix $B = \begin{pmatrix} \varepsilon, & 1 \\ -1, & -k \end{pmatrix}$.

Show that if $\mathbf{z} = \begin{pmatrix} x \\ y \end{pmatrix}$ then x satisfies the equation $E_{k-\varepsilon, 1-k\varepsilon}$, and indicate a type of behaviour that the latter can exhibit, which the equation $E_{p,1}$ cannot, however small $\varepsilon > 0$ may be. [SU]

8. Obtain the solution to the differential equation

$$\frac{\mathrm{d}y}{\mathrm{d}x} = \frac{3y + x}{x - y}$$

for which $y = 0$ when $x = 1$. (Hint: $\mathrm{d}y/\mathrm{d}x = \dot{y}/\dot{x}$.)

The material of the next two sections could hardly be called revision, but it is a rather natural extension of our previous work. It may be omitted at a first reading.

4.10 MODELLING MATHEMATICS WITHIN MATHEMATICS

The use of the D operator in section 4.9 is very reminiscent of the E operator in section 2.3, and readers will doubtless be curious as to why eigenvalues appear in the two cases in the same way. When, in mathematics, we see a vague analogy between two pieces of work, it is often fruitful to burrow further. (Indeed, the Chicago mathematician E. H. Moore in 1910 said it is our *duty* to do so.) We cannot go fully into the matter here, but the following sketch should lay the secret sufficiently bare.

When solving linear differential equations, we are most usually interested in functions $f\colon \mathbb{R} \to \mathbb{R}$ which have derivatives of all orders. It is usual to denote the set of all such functions by C^∞, because they are 'infinitely differentiable'. Thus, if $f \in C^\infty$, then also $\mathrm{D}f \in C^\infty$. Hence we can associate with f an infinite sequence, denoted by $\mathscr{S}_f$, where

$$\mathscr{S}_f = (f(0), f'(0), f''(0), \ldots, f^{(n)}(0), \ldots).$$

This sequence lies in the set S of all sequences of real numbers, and S is the setting for difference equations. As examples:

(a) $f(t) = \mathrm{e}^{ct}$; $\quad \mathscr{S}_f = (1, c, c^2, \ldots, c^n, \ldots)$;
(b) $f(t) = \sin(ct)$; $\quad \mathscr{S}_f = (0, c, 0, -c^3, 0, c^5, 0, \ldots)$;
(c) $f(t) = \cos(ct)$; $\quad \mathscr{S}_f = (1, 0, -c^2, 0, c^4, 0, \ldots)$.

Since D is linear, the same is true of $\mathscr{S}$, which means that for any numbers a, b and functions f, g in C^∞, we have

$$\mathrm{D}(af + bg) = a\mathrm{D}f + b\mathrm{D}g; \qquad \mathscr{S}_{(af+bg)} = a\mathscr{S}_f + b\mathscr{S}_g,$$

because the $(n + 1)$th term of the sequence $\mathscr{S}_{(af+bg)}$ is $af^{(n)}(0) + bg^{(n)}(0)$, which agrees with that of the sum of the two sequences $a\mathscr{S}_f$ and $b\mathscr{S}_g$. Thus if we add and then map, the result is the same as mapping and then adding—and we leave the reader to verify that this is usually not true when f and g are multiplied, at least if sequences are multiplied term by term.

Just as D operates linearly on the set C^∞, we have the linear operator E acting on S by the 'move on one' rule (2.3.4), so

$$E\mathscr{S}_f = (f'(0), f''(0), \ldots, f^{(n+1)}(0), \ldots).$$

But this is the sequence we get from $\mathrm{D}f$, whence

$$E\mathscr{S}_f = \mathscr{S}_{\mathrm{D}(f)},$$

which means that if we use $\mathscr{S}$ to map f into S and then operate with E, we get the same result as operating with D and then mapping by $\mathscr{S}$.

Thus the process of obtaining a derivative at 0 is modelled by the use of E in S. Also if in C^{∞} we have an equation such as $Df = cf$ (with c a number) then it is modelled by S in using the mapping $\mathscr{S}$ to give $c\mathscr{S}_f = \mathscr{S}_{Df} = E\mathscr{S}_f$; so the model of $Df = cf$ is the difference equation $Ex = cx$, where $x = \mathscr{S}_f$. Its solution is the sequence $\langle c_n \rangle$, the model of e^{cx} in (a) above. It is now easy to verify that the model of the differential equation $(D - a)f = g$ is the difference equation $(E - a)x = y$ where $x = \mathscr{S}_f$ and $y = \mathscr{S}_g$, since

$$y = \mathscr{S}_g = \mathscr{S}_{(D-a)f} = \mathscr{S}_{Df} - \mathscr{S}_{af} = Ex - ax = (E - a)x.$$

Similarly, the model of the second-order equation $(D - a)(D - b)f = g$ is the difference equation $(E - a)(E - b)x = y$, since

$$\mathscr{S}_{(E-a)(E-b)f} = (E - a)\mathscr{S}_{(E-b)f} = (E - a)(E - b)\mathscr{S}_f.$$

When $g = 0$, the general solution of the differential equation is

$$f(t) = Ae^{at} + Be^{bt},$$

which is modelled by the sequence $A\langle x^n \rangle + B\langle y^n \rangle$. Thus *the model of the solution is the solution of the model.* Similarly for equations with complex roots, when we express the solutions in terms of sin and cos and use (b) and (c) above. We hope that with this explanation, and the following exercises, readers will have gained enough insight to assuage their curiosity about this particular modelling process.

Exercises 4.10.1

1. If f, g are the functions in (b), (c) above, calculate $\mathscr{S}_h$, when h is the product $f \cdot g$ (so $h(t) = f(t)g(t)$). Show that this is not the term-by-term product of the sequences $\mathscr{S}_f$ and $\mathscr{S}_g$. How would we have to define the product of two sequences in order to force $\mathscr{S}$ to be multiplicative? (Consider Leibniz's formula in exercise 4.2.1(6).)
2. Solve the equation $(D - c)f = g$ and compare $\mathscr{S}_f$ with the solution of (2.1.7). You will need to calculate $\mathscr{S}_P$, where $P(t) = e^{ct} \cdot G(t)$ and $G(t) = \int_0^t e^{-cs} \cdot g(s)\,ds$; first show that $\dot{P}(t) = g(t) + c \cdot P(t)$, and then iterate to find $P^{(n)}(0)$. This shows how certain integrals are modelled by sums.
3. Suppose the sequence $\langle a_n \rangle$ is such that $\sum a_n \cdot x^n/n!$ is convergent when $|x| < 1$. Show that $\mathscr{S}_f = \langle a_n \rangle$.
4. For each difference equation of the form $au_{n+2} + bu_{n+1} + cu_n = f(n)$ in exercises 2.3.1, compare its solution with that of the differential equation $a\ddot{x} + b\dot{x} + cx = f(t)$.

We conclude this section by remarking that with most models, some information may be lost when passing from the original to the model. This happens when we use $\mathscr{S}$ to model the function

(d) $f(t) = e^{-1/t^2}$ if $t \neq 0$; $f(0) = 0$,

which can be shown to lie in C^{∞}, yet $\mathscr{S}_f$ is the zero sequence. The trouble here is that f cannot be approximated by its Taylor expansion at $t = 0$. Such functions can be cast from C^∞ if we insist on using only 'analytic' functions (which are those with a power series expansion at $t = 0$). Such functions include most of the familiar ones, like those in the examples (a)–(c) above.

4.11 THE EXPONENTIAL OF A MATRIX

The differential equations

$$\dot{x} = kx, \qquad x(0) = x_0 \qquad \text{on } \mathbb{R}, \tag{4.11.1}$$

and

$$\dot{\mathbf{z}} = A\mathbf{z}, \qquad \mathbf{z}(0) = \mathbf{z}_0 \qquad \text{on } \mathbb{R}^2 \tag{4.11.2}$$

have the same algebraic notation, which suggests that there ought to be a way of solving (4.11.2) which is as neat as the solution $x = x_0 e^{kt}$ of (4.11.1). We shall show that there is indeed a matrix, e^{At}, such that the solution of $\dot{\mathbf{z}} = A\mathbf{z}$ in (4.11.2) is the vector $e^{At}\mathbf{z}_0$; and the matrix satisfies the 'law of indices' but with a restriction. This result holds for an $m \times m$ matrix A, and readers may assume in the following exposition that $m = 2$ if that makes the material seem less forbidding.

Our first task is then to show that, given the $m \times m$ matrix A, there is an $m \times m$ matrix E which is a function of t such that, for all $t \in \mathbb{R}$,

$$\frac{dE}{dt} = AE, \tag{4.11.3}$$

where we obtain the matrix dE/dt by differentiating each entry of E. For example d/dx of the matrix $\begin{pmatrix} x & x^2 \\ \sin x & e^{2x} \end{pmatrix}$ is $\begin{pmatrix} 1 & 2x \\ \cos x & 2e^{2x} \end{pmatrix}$. It turns out that one solution of equation (4.11.3) is a matrix constructed by analogy with the ordinary exponential function; thus we define e^{At} using the familiar exponential series and write

$$e^{At} = \sum_{n=0}^{\infty} \frac{(At)^n}{n!}, \tag{4.11.4}$$

which is a matrix of the same size as A, and such that its (i, j)th entry $p_{ij}(e^{At})$ is the limit, as N tends to infinity

$$\lim p_{ij}\left(\sum_{n=0}^{N} \frac{(At)^n}{n!}\right) = \lim \left(\sum_{n=0}^{N} p_{ij} \frac{(At)^n}{n!}\right).$$

By the methods of Mathematical Analysis, it can be proved that the last limit exists, so e^{At} is well defined; and that for any $m \times m$ matrix B,

$$B \cdot e^{At} = \sum_{n=0}^{\infty} \frac{B(At)^n}{n!}, \tag{4.11.5}$$

with a similar result for multiplication on the right. Thus, for example, if A has all its eigenvalues λ_i distinct (but possibly complex) then $A = PLP^{-1}$ with L the matrix $\text{diag}(\lambda_1, \ldots, \lambda_m)$ and we can calculate e^{At} easily; for

$$\begin{aligned} Pe^{At}P^{-1} &= \sum_{n=0}^{\infty} \frac{P(At)^n P^{-1}}{n!} = \sum_{n=0}^{\infty} \frac{PA^nP^{-1}t^n}{n!} \\ &= \sum_{n=0}^{\infty} \frac{L^n t^n}{n!} = \sum_{n=0}^{\infty} \frac{(Lt)^n}{n!}, \\ &= \sum_{n=0}^{\infty} \frac{\text{diag}((\lambda_1 t)^n, \ldots, (\lambda_m t)^n)}{n!}, \\ &= \text{diag}(e^{\lambda_1 t}, \ldots, e^{\lambda_m t}) = \text{D (say)} \end{aligned} \tag{4.11.6}$$

so $e^{At} = P^{-1}DP$.

As an example consider the matrix $A = \begin{pmatrix} \lambda & 1 \\ 0 & \lambda \end{pmatrix}$ which is of type (ii) in (2.6.2). Then $A^2 = \begin{pmatrix} \lambda^2 & 2\lambda \\ 0 & \lambda^2 \end{pmatrix}$, $A^3 = \begin{pmatrix} \lambda^3 & 3\lambda^2 \\ 0 & \lambda^3 \end{pmatrix}, \ldots$ and thus

$$\begin{aligned} e^{At} &= \begin{pmatrix} 1 + \lambda t + \lambda^2 t^2/2 + \lambda^3 t^3/6 + \ldots & 0 + t + 2\lambda t^2/2 + 3\lambda^2 t^3/6 + \ldots \\ 0 + 0 + 0 + 0 + \ldots & 1 + \lambda t + \lambda^2 t^2/2 + \lambda^3 t^3/6 + \ldots \end{pmatrix} \\ &= \begin{pmatrix} e^{\lambda t} & te^{\lambda t} \\ 0 & e^{\lambda t} \end{pmatrix} = E, \end{aligned}$$

so

$$\frac{dE}{dt} = \begin{pmatrix} \lambda e^{\lambda t} & e^{\lambda t} + t\lambda e^{\lambda t} \\ 0 & \lambda e^{\lambda t} \end{pmatrix} = AE.$$

Exercises 4.11.1

1. Calculate e^{At} when A is:
   (a) either of the other standard forms in (2.6.2);
   (b) each of the matrices in exercise 7.5.1(3).
2. If each number λ_s in (4.11.6) is written in the form $re^{j\theta}$, show that L itself is e^{Bt}, where $B = \text{diag}(x_1, \ldots, x_m)$ and

   $$x_s = \ln(r) + j\theta.$$

   Hence show that A has a 'logarithm', viz. PBP^{-1}.
3. When $m = 2$, we can use the matrices Q_1, Q_2 of exercise 7.4.1(9) to simplify e^{At}. Thus show that $e^{At} = e^{\lambda t} \cdot Q_1 + e^{\mu t} \cdot Q_2$ when λ, μ are the distinct eigenvalues of A. What happens if $\lambda = \mu$?
4. Use (4.11.6) to show that $\det e^{Atp} = e^{\Delta t}$, where $\Delta = \det A$. Do you think the equation holds even if the eigenvalues are not all distinct?
5. Calculate e^{At} when $A = \begin{pmatrix} -2 & 2 \\ -4 & -2 \end{pmatrix}$.

Returning to (4.11.4), it can be shown further that the derivative of e^{At} is given (as we would expect) by differentiating the series term by term:

$$\frac{d}{dt}(e^{At}) = \sum_{n=0}^{\infty} \frac{d}{dt}\frac{(At)^n}{n!}. \tag{4.11.7}$$

To simplify this, we need two remarks on the differentiation of matrices. First, if C is a constant matrix, then Ct^n is formed by multiplying each entry of C by t^n, so $d/dt\,(Ct^n) = nCt^{n-1}$. Second, if Z is a product of XY of matrices, then by the rule for multiplying matrices we obtain, just as for scalars,

$$\frac{dZ}{dt} = \frac{dX}{dt}\cdot Y + X\cdot\frac{dY}{dt}.$$

The first of these remarks shows that in (4.11.5), $d/dt\,(e^{At})$ is given by

$$\frac{d}{dt}(e^{At}) = \sum_{n=0}^{\infty} \frac{nA^n t^{n-1}}{n!} = A\sum_{n=1}^{\infty} \frac{A^{n-1}t^{n-1}}{(n-1)!} = A\sum_{k=0}^{\infty} \frac{(At)^k}{k!};$$

where $k = n - 1$, and then the last expression is $A\cdot e^{At}$. This shows that e^{At} satisfies (4.11.3). Hence, if $\mathbf{z}_0$ is a fixed vector and $\mathbf{z} = e^{At}\cdot\mathbf{z}_0$, then, by the above rule for differentiating a product, we have $d\mathbf{z}/dt = A\mathbf{z}$, so $\mathbf{z}$ is certainly one solution of this differential equation which satisfies the 'initial condition' that $\mathbf{z}$ shall be $\mathbf{z}_0$ at time $t = 0$.

We shall now prove that this solution is unique. The proof will also have, as a by-product, the following proposition.

Proposition 4.11.1

e^{At} *is always an invertible matrix, with inverse* e^{-At}.

Proof
Suppose that $\mathbf{w}(t)$ is a second vector satisfying the initial condition $\mathbf{w}(0) = \mathbf{z}_0$, and

$$\frac{d\mathbf{w}}{dt} = A\mathbf{w}. \tag{4.11.8}$$

Multiplying by e^{-At}, we get

$$0 = e^{-At}\dot{\mathbf{w}} - e^{-At}A\mathbf{w} = \frac{d}{dt}(e^{-At}\mathbf{w})$$

so $e^{-At}\mathbf{w}$ is constantly equal to its initial value. Now by (4.11.3), e^{-At} at $t = 0$ is the identity matrix I, whence for all t and all solutions of (4.11.6) we have

$$e^{-At}\mathbf{w}(t) = \mathbf{z}_0. \tag{4.11.9}$$

In particular, this is true when $\mathbf{w}$ is the solution $\mathbf{z} = e^{At}\mathbf{z}_0$, so we have $e^{-At}e^{At}\mathbf{z}_0 = \mathbf{z}_0$. Therefore, the matrix $X = e^{-At}e^{At}$ is such that, for all $\mathbf{z}_0 \in \mathbb{R}^m$, $X\mathbf{z}_0 = \mathbf{z}_0$. This can happen only if $X = I$. Now change A to $-A$, and we have

$$e^{-At}e^{At}\mathbf{z}_0 = I = e^{At}e^{-At}\mathbf{z}_0,$$

which tells us that e^{At} is an invertible matrix and its inverse is e^{-At}. This proves proposition 4.11.1.

Finally, to establish the uniqueness of solutions to (4.11.8), we multiply (4.11.9) by e^{At} to get

$$\mathbf{w} = I\mathbf{w} = (e^{At}e^{-At})\mathbf{w} = e^{At}(e^{-At}\mathbf{w}) = e^{At}\mathbf{w}_0,$$

so $\mathbf{w}$ is the solution we found at first. This proves uniqueness.

Next, let us justify the exponential notation by proving that an index rule holds. For this, we remind readers that the matrices A, B are said to **commute** if $AB = BA$. In particular, B commutes with any power A^n, and hence (by 4.11.5), B commutes with e^{At}.

Proposition 4.11.2

If the square matrices A, B commute, then $e^{At}e^{Bt} = e^{(A+B)t}$.

Proof

Let $x = e^{At}e^{Bt}$. Then $dX/dt\ \mathbf{z}_0 = (Ae^{At})e^{Bt}\mathbf{z}_0 + e^{At}(Be^{Bt})\mathbf{z}_0 = (A + B)X\mathbf{z}_0$ since B commutes with A and hence with e^{At}. Therefore $X\mathbf{z}_0$ and $e^{(A+B)t}\mathbf{z}_0$ are two solutions of the equation $\dot{\mathbf{z}} = (A + B)\mathbf{z}$ such that each is $\mathbf{z}_0$ at time $t = 0$. Hence by the uniqueness proved above, these two solutions are equal for every $\mathbf{z}_0 \in \mathbb{R}^m$, so $X = e^{(A+B)t}$ and the proof is complete.

It can be checked that when $k = 2$, the solution $e^{At}\mathbf{z}_0$ gives us precisely the same ones as in algorithm 4.8.1 of section 4.8. But it allows a similar algorithm to be worked out for higher dimensions, and it is also very useful in theoretical discussions. In particular, it relates also to the discrete dynamics of Chapter 2, for it gives us a special mapping $T: \mathbb{R}^m \to \mathbb{R}^m$ defined by letting $T(\mathbf{x})$ be the 'time 1' solution $e^A\mathbf{x}$, for each $\mathbf{x} \in \mathbb{R}^m$. Hence the iterates of $\mathbf{x}$ under T are given by $T^n(\mathbf{x}) = (e^{An})\mathbf{x}$ (i.e., the positions of $\mathbf{x}$ at time $1, 2, 3, \ldots$); and the orbit $(\mathbf{x}, T\mathbf{x}, T^2\mathbf{x}, \ldots, T^n\mathbf{x}, \ldots)$ lies in the 'A-trajectory' of $\mathbf{x}$, which consists of all points $e^{At}\mathbf{x}$ as t runs from $-\infty$ to ∞. Consider the special case when the eigenvalues of A are distinct and positive. Then by exercise 4.11.1, $A = e^{Bt}$ for some real matrix B, so we see that the orbit of $\mathbf{x}$ under A is embedded in the B-trajectory through the time 1 map. If one converges to ∞, so does the other; if one forms a spiral, so does the other. This will suggest why certain results in Chapter 3 are analogous to corresponding results in Chapter 2.

Exercises 4.11.2

1. Use matrix methods to find the general solution of the simultaneous first-order differential equations:

$$\frac{dx_1}{dt} = -x_1 + x_2, \qquad \frac{dx_2}{dt} = -x_1 + x_2 - 2x_3, \qquad \frac{dx_3}{dt} = x_1 - x_2 + x_3.$$

2. In a chemical reaction, a chemical A forms a chemical B which in turn forms a chemical C. If x, y, z are the concentrations of A, B, C, respectively, at time t, then

$$\frac{\mathrm{d}x}{\mathrm{d}t} = -\lambda_1 x, \qquad \frac{\mathrm{d}y}{\mathrm{d}t} = \lambda_1 x - \lambda_2 y, \qquad \frac{\mathrm{d}z}{\mathrm{d}t} = \lambda_2 y,$$

where λ_1, λ_2 are equal constants. Given that $x = a$, $y = z = 0$, when $t = 0$, obtain expressions for x, y, z in terms of t.

3. If $B = \begin{pmatrix} 0 & 1 & 0 \\ 1 & 0 & 1 \\ 0 & -1 & 0 \end{pmatrix}$, calculate B^2, B^3, e^B, e^{-B}. Verify that $\mathrm{e}^B \cdot \mathrm{e}^{-B} = I$. Define $\cos B$ and $\sin B$ by

$$\cos B = \sum_{n=0}^{\infty} (-1)^n \frac{B^{2n}}{(2n)!}, \qquad \sin B = \sum_{n=0}^{\infty} (-1)^n \frac{B^{2n+1}}{(2n+1)!}.$$

Evaluate the series to obtain $\cos B$, $\sin B$ as 3×3 matrices, and verify directly that $\cos^2 B + \sin^2 B = I$. [SU]

5

Plane curves

So far, the models we have investigated have needed just a few plane curves for their description. It is now time for the repertoire to be widened, and in this chapter we use an approach that will give readers practice in handling derivatives. After all, calculus was developed partly for the express purpose of analysing plane curves. Of course, the graphical output of computer or calculator can also help a lot in forming images.

5.1 WHAT IS A PLANE CURVE?

In common parlance, when we say that a point P 'moves along a curve in the coordinate plane', we mean that at each instant t of time, the coordinates of P are functions $x(t)$, $y(t)$ of t; moreover, as we observe the motion of P during a time interval I, a curve C is traced out by P. Now, there is a type of motion, called 'Brownian', in which a microscopic particle moves in a liquid and is buffeted by collisions with individual molecules; its path is then extremely jagged. To exclude such random motion, we shall assume that the functions $x(t)$, $y(t)$ are differentiable, which means that P has a **velocity vector**

$$\mathbf{vel}(P) = (\dot{x}(t) \quad \dot{y}(t)), \tag{5.1.1}$$

with direction along the tangent to its path. (For convenience velocity vectors are written as row vectors in this chapter.) Thus if $\mathbf{vel}(p) = \mathbf{0}$ there need not be a tangent to the curve at P.

Because higher derivatives are sometimes needed, it is briefer to call $x(t)$, $y(t)$ **smooth** functions if they have as many derivatives as are necessary (e.g., if they lie in the class C^{∞}), without spelling out the number required.

Rather more formally we shall *define* a 'plane curve' C to be the set of all points $(x(t), y(t))$, where t runs through an interval I, and the functions x, y are smooth functions of t. This gives what is known as the **parametric form** of the curve C, which is particularly suitable for plotting with a graphical calculator or computer. The definition therefore prevents the track of a Brownian motion from being called a plane curve. It is convenient to use the 'arrow' notation and write $x: I \to \mathbb{R}$ to indicate that the function x is defined on the domain I, and values $x(t)$ lie in $\mathbb{R}$ for each t in I.

Examples 5.1.1

We shall now give some simple examples, and the reader should sketch the velocity vector at typical points (in different quadrants). If you have access to a graphical calculator or graphing software, then find out how to use it to plot curves described in parametric form. The following program should also prove useful in exploring curves in parametric form.

```
REM Program 5.1.1, Parametric plot
FUNCTION x(t) = a*t^2
FUNCTION y(t) = 2*a*t
READ a, tmin, tmax, tstep, xmin, xmax, ymin, ymax
DATA 1, -3   , 3    , 0.1  , -1   , 3    , -3   , 3
SET MODE "graphics"
SET WINDOW xmin,xmax, ymin,ymax
FOR t = tmin TO tmax STEP tstep
    PLOT x(t),y(t);
NEXT t
END
```

(If the semicolon is omitted from the PLOT statement then a sample of discrete points on the curve is displayed. Successive points are generated at equal steps of the parameter t. The distance between successive points is a measure of the magnitude of the average velocity vector between them; thus a 'good' parametrization for curve plotting 'slows down' at points near which the curve is quite 'bendy', where detail is important, and can travel quite fast on the 'straighter bits'.)

(a) The standard **parabola** with equation $y^2 = 4ax$ is a plane curve according to the definition, because it is the set of all points of the form $(at^2, 2at)$. Here the required functions $x(t)$, $y(t)$ are given by $x(t) = at^2$, $y(t) = 2at$, and the interval I is the whole number line $\mathbb{R}$. Note that $y(t)$ can take all real values, but that $x(t)$ cannot be negative. The velocity vector is $(2at \quad 2a)$. Its velocity has a minimum magnitude of $|2a|$ at the apex $t = 0$.

(b) The standard **ellipse** with equation $x^2/a^2 + y^2/b^2 = 1$ is a plane curve: take $x(t) = a \cdot \cos(t)$, and $y(t) = b \cdot \sin(t)$, with I the interval $\{0 \le t \le 2\pi\}$. If $a > b > 0$ then the magnitude of the velocity vector $(-a \sin(t) \quad b \cos(t))$ is greatest $(=a)$ when $\cos(t) = 0$, and least $(=b)$ when $\sin(t) = 0$. (Thus this is a 'good' parametrization in the above sense.)

(c) Each of the two branches of the standard **hyperbola** with equation $x^2/a^2 - y^2/b^2 = 1$ is a plane curve: take $x(t) = a \cdot \sec(t)$, $y(t) = b \cdot \tan(t)$ with I the interval $\{-\pi/2 < t < \pi/2\}$ for one branch, and $\{\pi/2 < t < 3\pi/2\}$ for the other. An alternative is to parametrize with the hyperbolic functions and, for one branch, to take $x(t) = a \cdot \cosh(t)$, $y(t) = b \cdot \sinh(t)$, with $I = \mathbb{R}$, and to change, say, to $x(t) = -a \cdot \cosh(t)$ for the other.

(d) The **semi-cubical parabola** has equation $4x^3 + y^2 = 0$ with the parametrization $x(t) = -t^2$, $y(t) = 2t^3$, and the interval I is the number line $\mathbb{R}$. Here, the velocity vector is $(-2t \quad 6t^2)$. Note that, although the functions $x(t)$, $y(t)$ are smooth, the *curve* is 'rough' at $t = 0$ (i.e., at the origin $\mathbf{0}$ in $\mathbb{R}^2$) because it has the sharp point of the cusp at $t = 0$. See Fig. 5.1.1.

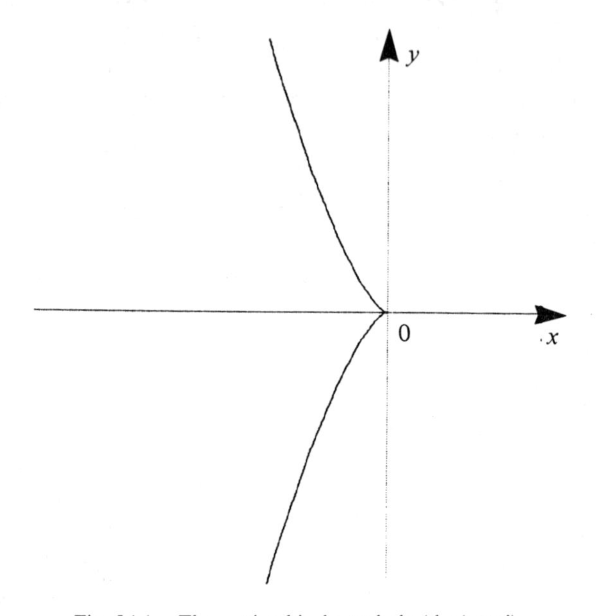

Fig. 5.1.1. The semi-cubical parabola (the 'cusp').

(e) The graph of a smooth function $f: I \to \mathbb{R}$ is a plane curve, because it is the set of all points $(t, f(t))$ as t runs through I; so here $x(t)$ is t, and $y(t)$ is $f(t)$. The velocity vector is $(1 \quad f'(t))$. Indeed, a plane curve can be built from pieces of graphs, but it is useful to have formulae which are symmetrical in the functions x and y. (Later, we shall have examples of plane curves that may have 'cross-over' points, as on a figure eight.)

For brevity we say that the plane curve C is **parametrized** by $\gamma: I \to \mathbb{R}^2$ when, for each $t \in I$, $\gamma(t)$ is the point $(x(t), y(t))$.

Exercise 5.1.1

Sketch the plane curve C parametrized by $\gamma: \mathbb{R} \to \mathbb{R}^2$ when

1. $\gamma(t) = (t, e^t)$
2. $\gamma(t) = (t^3, t^4)$
3. $\gamma(t) = (t - t^2, t + t^2)$
4. $\gamma(t) = (t - \sin t, 1 - \cos t)$
5. $\gamma(t) = (e^{kt} \cos t, e^{kt} \sin t) \qquad (k \neq 0)$
6. $\gamma(t) = (t^2 - 1, t^3 - t)$

In each case calculate the velocity vector. (If you have difficulties, read the next section.)

5.2 REGULARITY

The curve C is said to be **regular** if the velocity vector is nowhere zero. Thus, regularity is a property of the parametrization; for example, if L is a straight line, it could have parametrization $x(t) = at$, $y(t) = bt$, with velocity vector $(a \quad b)$; so we have regularity if either a or b is non-zero. In this form, L has equation $bx - ay = 0$, and so we could also parametrize L by the functions $x(t) = at^3$, $y(t) = bt^3$, and here the velocity vector is $3(at^2 \quad bt^2)$ which is zero when $t = 0$. With this parametrization, then, the curve is not regular, even though it has a tangent at each point. In example 5.1.1(d) above, we saw that lack of regularity could suggest the presence of a 'bad' point like the cusp.

As we have seen it is easy to see the shape of a plane curve by plotting on a calculator or computer screen. However, it is useful to have ways of verifying the output to check for errors. To make a rough sketch, by hand, we can start at a convenient point P on the curve, say with parameter $t = 0$, and then imagine that we are steering a car from P with the velocity vector $\mathbf{vel}(P)$ in (5.1.1), to trace out the curve in the following way.

For this purpose, it is not so much the magnitude as the *direction* of $\mathbf{vel}(P)$ that matters; for if it points north-west (NW), then we steer NW until either $\dot{x}$ or $\dot{y}$ is zero. If $\dot{y}(t) = 0$ at some first time t, then the velocity must have veered gradually until $\mathbf{vel}(P(t))$ points due west; at that moment, $\dot{y}$ either changes sign (so we steer SW) or becomes positive again (so we steer NW, after an inflection in our track), or—rarely—$\dot{y}$ remains zero for a while (so we steer due west). If, instead, it was $\dot{x}$ that was zero, then the track would have veered gradually to the north, probably ready to turn NE. Similarly with other points of the compass. Having got the rough shape for positive time, we can go back to base at P, and then steer backwards along the velocity vector.

It might happen that our steering causes us to cross a piece of track that we had already traced. This can occur only if there are two different times, t_1, and t_2, such that

$$x(t_1) = x(t_2), \qquad y(t_1) = y(t_2).$$

This will have to be verified analytically, possibly with help from the computer to finds approximate values of t_1 and t_2.

Exercises 5.2.1

1. Decide which of the curves in example 5.1.1 are regular.
2. If $\mathbf{0} \neq P = (x(p), y(p))$ lies on C, the unit vector at P is given by $\mathbf{u}(p) = P/c$ where $c = \sqrt{(x^2 + y^2)}$. Show that $\dot{\mathbf{u}}(p)$ is the vector $(-y(p) \quad x(p))$ multiplied by the scalar $(x\dot{y} - y\dot{x})/c^3$ evaluated at p. (Hence $\dot{\mathbf{u}}(p)$ is orthogonal to $\mathbf{u}(p)$.)
3. The **cycloid**, illustrated in Fig. 5.2.1, is the locus of a point on a circle that rolls without slipping on a line. If the line is taken as the x-axis $\mathbf{0}x$, the point starts at the origin $\mathbf{0}$, and, after the circle has turned through an angle t about its centre C, it is at the point P with coordinates $(x(t), y(t))$, prove that: $x(t) = r(t - \sin t)$, $y(t) = r(1 - \cos t)$. Find the velocity vector and the values of t for which it is zero.

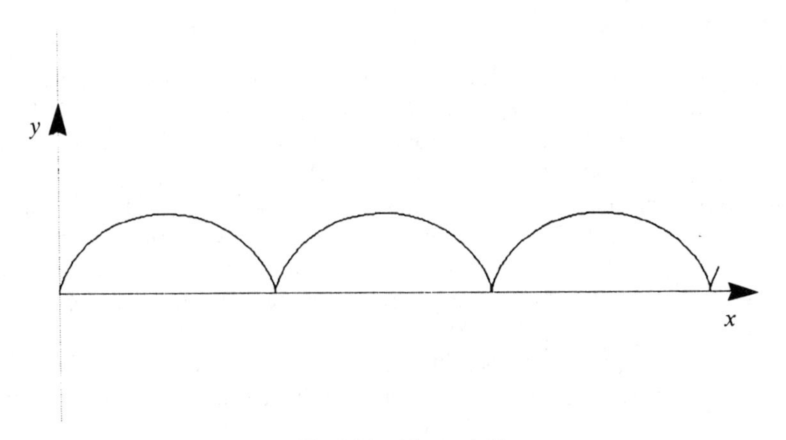

Fig. 5.2.1. The cycloid.

4. Investigate the **three-cusped hypocycloid**:
   $x = -c(2\cos t + \cos 2t) \qquad y = c(2\sin t - \sin 2t) \qquad c > 0$
5. If $x(t) = (1 - t^2)/(1 + t^2)$, $y(t) = 2t/(1 + t^2)$ then find $x^2(t) + y^2(t)$ and identify the plane curve which has the parametrization. What is the appropriate interval I? Hence suggest alternative parametrizations for the ellipse and hyperbola in example 5.1.1(b) and (c) above.
6. Find out how to plot curves in parametric form on your graphics calculator, or graph-plotting software. How does the machine know what to take as I? Explore the parametric forms of the curves introduced above. What is the snag in using the forms of exercise 5 above?

5.3 PARALLEL CURVES

Consider a curve C as before, but now suppose that C is regular, in the sense of section 5.2. For each real number r, we shall construct a curve C_r, which is 'parallel' to C in an obvious sense. Such curves are needed in engineering, when it is required to use a circular cutting tool to cut a curve like C in metal; the centre of the tool must be steered along a curve parallel to C.

For the construction, we note that if P is on C, again with parameter p, then by regularity the velocity vector $\mathbf{vel}(P)$ at P is always non-zero; hence it has a direction, and this lies along the tangent at P. Its magnitude (or 'norm') is called the **speed** $\sigma(P)$, and is given by

$$\sigma(P) = \sqrt{(\dot{x}^2 + \dot{y}^2)}, \tag{5.3.1}$$

evaluated at $t = p$. Besides the tangent, we also have the normal to C at P, which is orthogonal to $\mathbf{vel}(P)$ and hence parallel to any vector of the form $c\mathbf{w}$, where c is any scalar, and $\mathbf{w} = (-\dot{y} \quad \dot{x})$. Here, the length of $\mathbf{w}$, denoted by $\|\mathbf{w}\|$, is also $\sigma(P)$. Therefore the normal is parallel to the **unit normal vector** $\mathbf{n}(P)$ given by

$$\mathbf{n}(P) = \frac{\mathbf{w}}{\|\mathbf{w}\|} = \frac{1}{\sigma(P)}(-\dot{y} \quad \dot{x}). \tag{5.3.2}$$

To construct the new curve C_r, we define it parametrically by the vector equation

$$(X(p) \quad Y(p)) = \mathbf{P} + r \cdot \mathbf{n}(P), \tag{5.3.3}$$

which says: to go to the new point (X, Y), start at P and move a distance r along the normal (see Fig. 5.3.1). The sign of r determines on which side of C we move along the normal. Clearly, if $r = 0$, then C_0 is C itself.

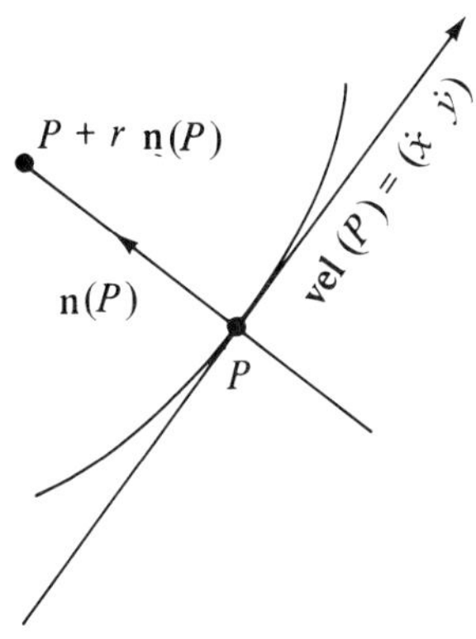

Fig. 5.3.1.

Exercises 5.3.1

1. Sketch families of curves parallel to the curves in examples 5.1.1 and exercise 5.2.1.
2. The following computer program draws a set of curves parallel to an ellipse for a variety of input values of r. Explore the shapes of such curves. Adapt it to sketch typical parallel curves C_r when C is one of the curves in examples 5.1.1 and exercise 5.2.1.

```
REM Program 5.3.1, Parallel curves
FUNCTION     x(t) =  a*COS(t)
FUNCTION     y(t) =  b*SIN(t)
FUNCTION xdot(t) = -a*SIN(t)
FUNCTION ydot(t) =  b*COS(t)
READ a, b, tmin, tmax, tstep
DATA 5, 3, 0    , 6.3 , 0.1
FUNCTION speed(x,y) = SQR(x*x + y*y)
SET MODE "graphics"
SET WINDOW -2*a,2*a, -2*b,2*b
FOR t = tmin TO tmax STEP tstep
  PLOT x(t),y(t);
NEXT t
PLOT
DO
  INPUT PROMPT " r : " : r
  IF r = 0 THEN EXIT DO
  FOR t = tmin TO tmax STEP tstep
    LET dx = xdot(t)
    LET dy = ydot(t)
    LET s = speed(dx,dy)
    PLOT x(t) - r*dy/s , y(t) + r*dx/s ;
  NEXT t
  PLOT
LOOP
END
```

```
r:-2.5
r:2.5
r:
```

Sample output

3. What do the curves parallel to a cycloid look like? For other examples, see Giblin (1982).

Even though C is regular, it may happen that C_r is not regular for certain values of r. To find such values we differentiate equation (5.3.2); here $X = x - r\dot{y}/\sigma$, and because of the square root we have

$$\dot{\sigma} = \frac{\dot{x}\ddot{x} + \dot{y}\ddot{y}}{\sigma}, \qquad \dot{X} = \dot{x} - r\cdot\left[\frac{\sigma\ddot{y} - \dot{y}\dot{\sigma}}{\sigma^2}\right]$$

and the square bracket simplifies to $\dot{x}\cdot\kappa(t)$, where

$$\kappa(t) = \frac{\dot{x}\ddot{y} - \dot{y}\ddot{x}}{\sigma^3}. \tag{5.3.4}$$

Similarly for the second coordinates, so

$$(\dot{X} \quad \dot{Y}) = (\dot{x} \quad \dot{y})\cdot(1 - r\cdot\kappa(t)).$$

Hence, C_r fails to be regular at any t for which

$$r = 1/\kappa(t). \tag{5.3.5}$$

Exercises 5.3.2

1. Calculate C_r for each of the curves in examples 5.1.1 and exercise 5.2.1. Find for which values of r irregularity occurs, and check your findings by modifying program 5.3.1.
2. For each real r, explain what is meant by the curve Γ_r parallel to the curve Γ with parametrization $(p(t), q(t))$. If $r > 0$, show that the two parallel curves Γ_r, Γ_{-r} constitute the envelope of the family of circles

$$(x - p(t))^2 + (y - q(t))^2 - r^2 = 0.$$

Let Γ be the hyperbola with $p(t) = t$, $q(t) = 1/t$. Indicate on a diagram the sense in which $P = (t, 1/t)$ moves on Γ as t increases, and hence show that if $r > 0$ then Γ_r has two branches of which one lies wholly in the positive quadrant. Show that Γ_r has parametric equations of the form $x = f(t)$, $y = f(1/t)$ and hence show that Γ_r has just two irregular points, both on the 'positive' branch of Γ_r. [SU]

5.4 ARC-LENGTH

Let C be a plane curve, parametrized by smooth functions $x, y\colon I \to \mathbb{R}$, and not necessarily regular. Consider a typical point P on C with parameter p, that is, $P = (x(p), y(p))$; and let us choose some point Q on C as a 'reference' point, with parameter q. What is the length of the arc of C from Q to P?

Our answer to this question will involve the quantity $\sigma(t)$ that was used in (5.3.1). Firstly the required length is a number that depends on p; it is usually denoted by $s(p)$, and we then have a function $s\colon I \to \mathbb{R}$. We shall prove the important formula

$$s(p) = \int_q^p \sigma(t)\,\mathrm{d}t = \int_q^p \surd[\dot{x}^2 + \dot{y}^2]\,\mathrm{d}t. \tag{5.4.1}$$

If this be granted for the moment, then since the derivative of an integral is the integrand (by the fundamental theorem of calculus) we find that s has derivative

$$\dot{s}(p) = \sigma(p) = \sqrt{[\dot{x}^2 + \dot{y}^2]}, \tag{5.4.2}$$

and (5.4.1) says that the arc-length is the integral of the speed (as we would expect, having seen (4.5.8)).

Before the proof, it is probably best for readers to learn how to handle the formula (5.4.1); so let us apply it to example 5.1.1(b), in the special case that $a = b$ (so we have a circle). Then $\dot{x}(t) = -a \cdot \sin(t)$, $\dot{y}(t) = a \cdot \cos(t)$, and substituting in (5.4.1) we get (since $\sin^2(t) + \cos^2(t) = 1$):

$$s(p) = \int_q^p a \, \mathrm{d}t = a \cdot (p - q),$$

and if we take the reference point to be at $q = 0$, then $s(p) = ap$, and the length of the entire circumference is $2\pi a$.

Exercises 5.4.1

1. Calculate the arc-length for each of the curves in examples 5.1.1 above. (In the case of the ellipse with $a \neq b$, the integral cannot be evaluated in terms of elementary functions, and gives rise to a new function, called an 'elliptic integral'. There is a vast theory of these, to be found in books on elliptic functions.)
2. The **catenary** is the curve formed by a heavy rope that hangs under gravity when its ends are fixed at the same height. Books on statics show that the equation of the curve is given by $y = a \cosh x/a$ where a is constant. Sketch the curve and show that the arc-length, measured from the lowest point, is $a \sinh x/a$.
3. In 4, 5 of exercise 5.1.1, calculate the arc-length from $(0, 0)$ to $(2\pi, 0)$ and $(1, 0)$ to $(e^{2\pi k}, 0)$, respectively.
4. In each case sketch the curve γ and find its total path length when its parametrization is given by

   (a) $\gamma(t) = (2 \cos t + \cos 2t, 2 \sin t - \sin 2t)$, $\quad 0 \leq t \leq 2\pi$;
   (b) $\gamma(t) = (\cos^3 t, \sin^3 t)$, $\quad 0 \leq t \leq 2\pi$.

Proof of (5.4.1)
Let Q be the point on C with parameter $p + h$; then, by Pythagoras' theorem, $PQ^2 = [x(p + h) - x(p)]^2 + [y(p + h) - y(p)]^2$. As t changes from p to $p + h$, the arc-length $s(t)$ changes from $s(p)$, and this is approximately the length of the straight segment PQ. By the above equation, therefore, we have

$$\frac{s(p + h) - s(p)}{h} \cong \sqrt{\left(\left[\frac{x(p + h) - x(p)}{h}\right]^2 + \left[\frac{y(p + h) - y(p)}{h}\right]^2\right)}$$

and, when h tends to zero, we obtain, as in (5.4.2),

$$\dot{s} = \sqrt{[\dot{x}^2 + \dot{y}^2]},$$

everything being evaluated at p. Now integrate $\dot{s}(t)$ with respect to t and obtain (5.4.1) as required.

A pedantic point

In the above theory, we assume that we knew what it mean to speak of the 'length' of a curve, as if it was as natural as the length of a straight segment. One can imagine approximating a curve by means of a polygon, and taking the sum of the lengths of its edges to give a rough measure of what the length of the curve ought to be.

This approach can be carried through, but not without several technical difficulties. Instead, one can simply define 'length' to mean the number given by the integral (5.4.1). This is not the place to labour this point, but simply to draw the reader's attention to something that the developers of calculus treated as obvious (because for them the interesting question was how to calculate the length that 'obviously' existed). In the nineteenth century, further progress would not have been possible without seriously investigating the nature of several 'obvious' things like length, area, and volume.

5.5 CURVATURE

With our curve C as before, we now ask: 'what is the curvature at P?' Roughly speaking, this requires us to find that circle which most closely approximates C at P, and its radius ρ is called the **radius of curvature** at P, while the number $1/\rho$ is traditionally called the **curvature** at P. We shall prove that the curvature is the function $\kappa(t)$ that we met in (5.3.4). Granted this for the moment, we can give a formula for ρ. For, we saw in (5.4.2) that in the denominator of (5.3.4), the function σ is $\dot{s}$, so at P with parameter p, the radius of curvature is given by the formula

$$\rho = \frac{\dot{s}^3}{(\dot{x}\ddot{y} - \dot{y}\ddot{x})}, \tag{5.5.1}$$

when all the terms are evaluated at $t = p$. If the denominator is zero, we say that ρ is infinite.

The formula (5.5.1) gives the right result for the circle with $x(t) = a \cdot \cos(t)$, $y(t) = a \cdot \sin(t)$; for then $\dot{s} = a$, and the denominator in (5.5.1) is

$$(-a)^2 \cdot [\sin(t) \cdot \sin(t) + \cos(t) \cdot \cos(t)] = a^2,$$

whence $\rho = a^3/a^2 = a$, the radius of the circle. The right-hand side of (5.5.1) has already been seen in (5.3.5), from which we shall deduce the following about parallel curves C_r.

Proposition 5.5.1

C_r is not regular at exactly those points which are centres of curvature for C itself.

Proof

By (5.3.5), we see that on the parallel curve C_r, we have $(\dot{X}, \dot{Y}) = \mathbf{0}$ for those parameter values t for which ρ(t) equals r. Since the corresponding point $(X(t), Y(t))$ on C_r is on the normal to C at P, it must be the centre of the circle of curvature at P. This completes the proof.

Before we prove the formula (5.5.1), we supply the following exercises, to enable readers to gain familiarity with the formula.

Exercises 5.5.1

1. Let C be the graph $y = f(x)$ of a function. Show that

$$\rho(x) = (1 + f'(x)^2)^{3/2}/f''(x).$$

2. In the above exercise, suppose that $\mathbf{0}$ lies on the graph, which touches the x-axis at $\mathbf{0}$. (Thus $f(0) = 0 = f'(0)$.) Show that $\rho(0) = 1/f''(0)$. Now let Z be the circle with centre at $(0, 1/f''(0))$, and radius $1/f''(0)$. Express it in the form $y = g(x)$ near $\mathbf{0}$, and show that f and g agree 'to third order' at $\mathbf{0}$, in that the Taylor expansion of $f - g$ at $\mathbf{0}$ is of the form $ax^3 + bx^4 + \cdots$.
3. Calculate ρ for each of the curves in examples 5.1.1. Show that for the catenary $y = a \cosh x/a$, $\rho = y^2/a$.
4. In (5.3.4) suppose that r is not constant, but a smooth function of t. Show that the resulting curve will be regular provided $(1 - r/\rho)^2 + (\dot{r}/\dot{s})^2$ is non-zero.
5. Which curves have constant curvature? (If you get stuck, see the next exercises.)
6. Assuming the usual formula for the radius ρ of curvature, show that the centre of curvature, corresponding to $P = (p(t), q(t))$ on Γ, has coordinates

$$x = p - \frac{\sigma^2 \dot{q}}{Q}, \qquad y = q + \frac{\sigma^2 \dot{p}}{Q}$$

where $Q = \dot{p}\ddot{q} - \dot{q}\ddot{p}$ and is assumed to be non-zero. Suppose Γ is the cycloid (with r a positive constant)

$$p(t) = r(t - \sin t) \qquad q(t) = r(1 - \cos t).$$

Show that the radius of curvature $\rho = -4 \sin^2 t/2$. [SU]

Proof of (5.5.1)

When the tangent at P is produced in length, it cuts the x-axis in a point M, and at an angle θ (see Fig. 5.5.1). It is traditional then to define the curvature (rather mysteriously) to be $d\theta/ds$ so

$$\rho = \frac{1}{\text{curvature}} = \frac{\dot{s}}{\dot{\theta}}, \tag{5.5.2}$$

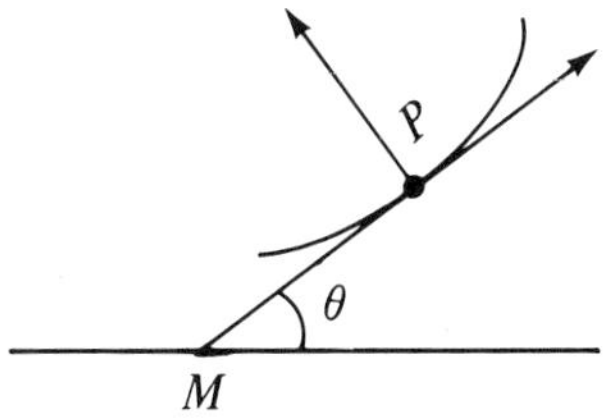

Fig. 5.5.1.

and we shall follow this line for a moment, before pointing out a snag. Since $\tan \theta = \dot{y}/\dot{x}$, then by differentiating we find

$$(1 + \tan^2\theta)\cdot\dot{\theta} = \frac{E}{\dot{x}^2}, \qquad E = \dot{x}\ddot{y} - \ddot{x}\dot{y},$$

so $\dot{\theta} = E/[\dot{x}^2(1 + (\dot{y}/\dot{x})^2)]$ whence $\rho = \dot{s}/\dot{\theta} = \dot{s}^3/E$ as required.

Now for the snag. Apart from the mystery, there is a problem about saying what the angle θ (rather than $\tan\theta$) is: in Fig. 5.5.1 it is $\tan^{-1}(\dot{y}/\dot{x})$, which lies in the interval $J = \{-\pi/2 < \theta \le \pi/2\}$.

Thus θ cannot be the obtuse angle marked in Fig. 5.5.2, and in any case what happens about differentiability of θ when $\theta = \pi/2$? These difficulties can be overcome with some trouble but there is an easier way.

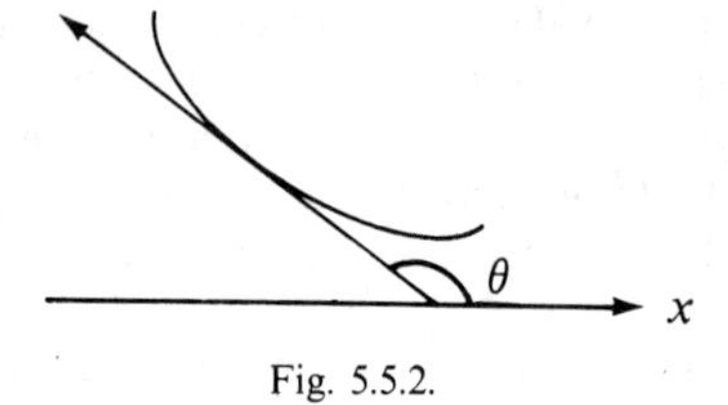

Fig. 5.5.2.

In the simple case of Fig. 5.5.1, we see that $\dot{\theta}$ exists as the rate of change of the direction of the tangent at P. This direction is $\mathbf{w}(p) = \mathbf{vel}(p)/\dot{s}(p)$, since $\mathbf{vel}(p)$ itself has magnitude $\dot{s}(p)$. We must assume $\dot{s}(p) \neq 0$ for this to make sense.) By differentiating each coordinate of $\mathbf{w}$, we find (with E as above):

$$\dot{\mathbf{w}} = c\cdot(-\dot{y} \quad \dot{x}), \qquad c = E/\dot{s}^3$$

evaluated at $t = p$. Now recall the unit normal vector $\mathbf{n}(P)$ in (5.3.2); we have just shown that $\dot{\mathbf{w}} = (c\dot{s})\cdot\mathbf{n}(P)$, which means that $\dot{\mathbf{w}}$ is a vector with direction $\mathbf{n}(P)$ and magnitude $c\dot{s}$. We therefore take this scalar as a substitute for the scalar $\dot{\theta}$ in the 'traditional' argument above. Hence, corresponding to the traditional definition (5.5.2), we now have

$$\rho = \frac{\dot{s}}{\dot{\theta}} = \frac{1}{c} = \frac{\dot{s}^3}{E}$$

as required by (5.5.1). This completes our proof.

To summarize: we have shown that for each point P on C with parameter p, there is a circle of curvature $1/f$ which touches C at P and has radius $\rho(p)$. Its centre lies on the normal at P, and by (5.3.3) has coordinates

$$(X(p), Y(p)) = P + \rho(p)\cdot\mathbf{n}(P). \tag{5.5.3}$$

Of course, as P runs along C, then $(X(p), Y(p))$ will sweep out a plane curve; this curve is called the **evolute** of C, and is traditionally defined as the 'locus of centres of curvature' of C.

Exercises 5.5.2

1. Use the calculations of ρ in exercise 5.5.1(3) to find the corresponding centres of curvature, as given by (5.5.4).
2. In exercise 5.5.1(1), plot a graph of ρ against x when f is (i) $\sin x$, (ii) $\cos x$. Also find the coordinates of the centre of curvature, in each case.

3. In questions 1 and 2, locate the centre of curvature at sample points, to check whether the formula (5.5.3) gives answers that lie on the side of the curve that you would expect.
4. Show that the radius of curvature of the curve

$$x^4 + y^2 = 2a(x + y),$$

at the origin **0** is $2a\sqrt{2}$. Calculate the coordinates of the centre of curvature at **0**.
5. For each of the curves in exercise 5.1.1(1–6), find an expression for the curvature $\kappa(t)$ of the point $\gamma(t)$; indicate any points where the curvature is not defined. Find the points where the curvature is a (local) maximum or minimum (so-called 'vertices'), and indicate them also on the curve.
6. If the curvature ρ is assumed constant over a small arc PQ of a curve C then the normals at P and Q will be radii of the corresponding circle, and their intersection R will be the centre of the circle, and hence a point of the evolute of C. The computer program 5.3.1 was used to plot points on the parallel curve to C by marking off points S at a fixed distance r along the normal to C at points P. Adapt this program to draw a set of normals PS of length r to the curve. By making r sufficiently large you should see another curve being enveloped by this set of lines—can you explain why this curve is the evolute of C?
7. The computer program 5.3.1 can also be easily adapted to plot the evolute directly from the definition of the radius of curvature ρ. Instead of using a constant value of r we substitute a value `rho` which is a function of the parameter t. Use the following program 5.5.1 to explore the forms of the evolutes of the curves met so far in this chapter.

```
REM Program 5.5.1, Evolute
READ a, b, tmin, tmax, tstep
DATA 5, 3, 0   , 6.3 , 0.1
FUNCTION   x(t) =  a*COS(t)
FUNCTION   y(t) =  b*SIN(t)
FUNCTION  x1(t) = -a*SIN(t)
FUNCTION  y1(t) =  b*COS(t)
FUNCTION  x2(t) = -a*COS(t)
FUNCTION  y2(t) = -b*SIN(t)
FUNCTION   s(t) =  SQR(x1(t)^2 + y1(t)^2)
FUNCTION rho(t) =  s(t)^2/(x1(t)*y2(t) - y1(t)*x2(t))
SET MODE "graphics"
SET WINDOW -2*a,2*a, -2*b,2*b
FOR t = tmin TO tmax STEP tstep
  PLOT x(t),y(t);
NEXT t
PLOT
  FOR t = tmin TO tmax STEP tstep
    LET r = rho(t)
    PLOT x(t) - r*y1(t) , y(t) + r*x1(t) ;
  NEXT t
  PLOT
END
```

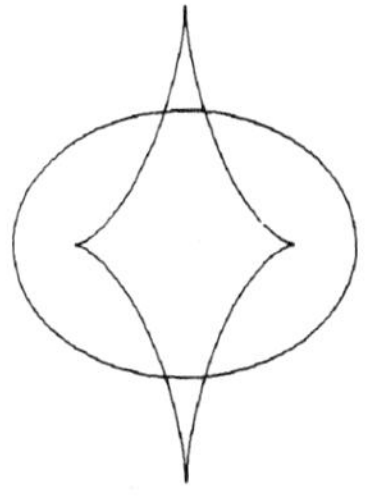

Sample output

8. Find the evolute of the catenary in exercise 5.4.1(2).
9. If you could not do exercise 5.5.1(5), try using Proposition 5.5.1 to show that if C has constant (finite) curvature, then C is a circle.
10. Use the implicit function theorem (8.1.1) to show that if p lies on the smooth curve $f(x, y) = 0$, then

$$\frac{1}{\rho} = \frac{f_{11}f_2^2 + f_{22}f_1^2 - 2f_1f_2f_{12}}{|\text{grad}\, f|^{3/2}}$$

where f_1, f_{12}, etc., are each evaluated at p, and $\text{grad}\, f|_p \neq 0$.
11. Fig. 5.5.3 shows how a draughtsman's approximation to an ellipse is constructed from a rhombus $ABCD$. The sides are bisected to obtain P, Q, R, S. PD intersects BS at L, and QD intersects RB at M. Circular arcs are drawn as follows: centre L from S to P, centre D from P to Q, centre M from Q to R, and centre B from R to S, to obtain a continuous curve. If the angle $\angle BAD = 60^\circ$ show that the curve has a tangent everywhere but the equations of the tangent are not always differentiable. Find the form of an ellipse which passes through $PQRS$ with common tangents to the draughtsman's 'ellipse'. Find its radius of curvature at P and show this is equal to $(LP + DP)/2$. Can you adapt the technique for more general cases?

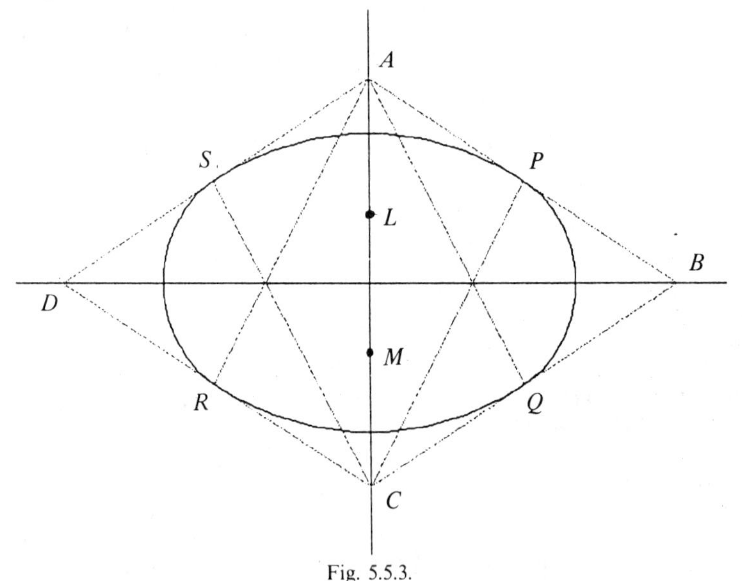

Fig. 5.5.3.

5.6 ENVELOPES

If we see a family of curves drawn on a sheet of paper, our eyes often detect another curve that touches each curve of the family. This 'ghost' curve is called the **envelope**. Many decorative embroidery patterns use this principle, and the family of curves is then usually a family of straight lines; the computer programs in the exercises below will give many other striking examples. In the present section, we aim to give the theory of these envelopes, which will introduce us to more plane curves.

In practice, the curves that occur in families do not usually present themselves as plane curves (i.e., complete with parametrizing functions $x(t)$, $y(t)$). Rather, each curve appears as a 'contour' of some function $f(x, y)$, just as (for example) the standard ellipse often occurs most naturally in the form $f(x, y) = 1$ where f is the function given by $f(x, y) = x^2/a^2 + y^2/b^2$. The ellipse is then the **contour** at level 1, and the name is suggested by geographical maps wherein the contour of level 100 (say) shows all points P at which the height h satisfies the equation $h(P) = 100$.

Now suppose we have a family of functions $f_t(x, y)$, one for each number t that varies in some interval I. From these, we then obtain a family of contours C_t, where for each t in I,

$$C_t = \{P \in \mathbb{R}^2 : f_t(P) = 0\}. \tag{5.6.1}$$

For example, let f_t be the function such that

$$f_t(x, y) = (x - t)^2 + y^2 - 1, \tag{5.6.2}$$

so that C_t is a circle of radius 1 and centre at $(t, 0)$ on the x-axis. If we sketch this family, our eye picks out the two parallel lines $y = \pm 1$, as being touched by each circle of the family (see Fig. 5.6.1). These two lines form the envelope in this case, as we shall justify shortly.

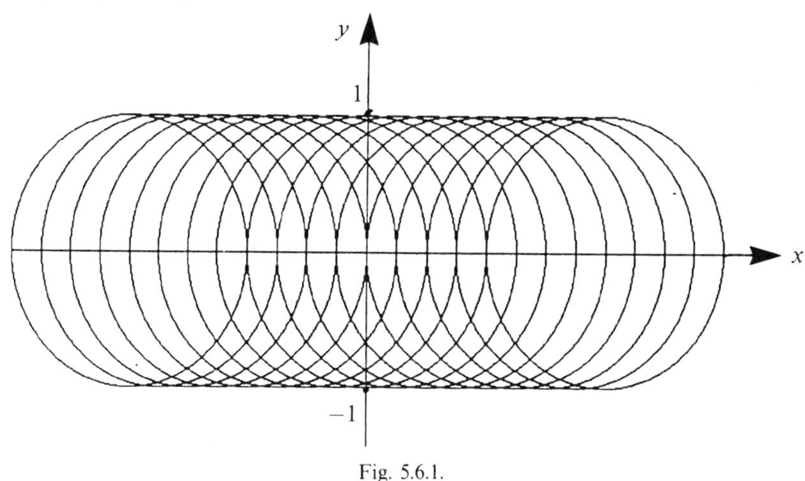

Fig. 5.6.1.

Firstly, however, we need to agree on a workable mathematical statement of what we mean by an 'envelope'. With such a problem, it is always best to look at what our historical predecessors did. Until the nineteenth century, mathematicians agreed that the envelope is the 'locus of points of intersection of consecutive curves' of the family. Here, the Latin word 'locus' simply means 'set', but the trouble lies with the word 'consecutive' which implies that the curves come one after another—but the numbers t do not. The difficulty arises from the fact that most people (including the greatest mathematicians) carry in their psychology two models of the number line: one in which numbers are strung about like separate dots, and another (less natural?) wherein numbers are thought of as flowing like ink. This second model is more like the modern 'Weierstrass' model $\mathbb{R}$, in which there is no 'next' number after (say) $t = \frac{1}{2}$.

However, as we shall see in the exercises, because a modern computer program uses a 'strung out' number system as a model of $\mathbb{R}$, the nineteenth century definition of envelope is ideally suited to a computer! (Our computers were not invented until the 1940s.)

Secondly, a notational point: in (5.6.1) we are using t as an index or label to distinguish between the various functions, and it is more convenient to regard t as a variable. Thus, we write $f(x, y, t)$ for $f_t(x, y)$, and the contour C_t is the set of points $(x, y) \in \mathbb{R}^2$ for which $f(x, y, t) = 0$.

In spite of the difficulty about 'consecutive', the traditional approach led to sensible results, and was as follows. Consider a particular t in I, and a nearby value $t + h$. Then, if (a, b) is a point of intersection of the curves C_t and C_{t+h}, we have

$$f(a, b, t) = 0 = f(a, b, t + h)$$

so

$$\frac{f(a, b, t + h) - f(a, b, t)}{h} = 0.$$

Now let h tend to zero (here our mathematical predecessors abandoned the 'consecutive' idea!). Then (a, b) will (we hope) tend to a limit (a^*, b^*) and must satisfy the simultaneous equations

$$f(a^*, b^*, t) = 0 = \frac{\partial f(a^*, b^*, t)}{\partial t}. \tag{5.6.3}$$

where we use the operator $\partial/\partial t$ to mean that we differentiate f as if only t varied. (We do not need the theory of partial differentiation here.)

We therefore define the **envelope** of the family $\{C_t\}$ to be the set of all points (a^*, b^*) that satisfy (5.6.3). This is a perfectly good definition, even if we were guided to it by a shaky model of $\mathbb{R}$. As we shall see, it gives satisfying results, although in some cases there may be points on the envelope which do not lie on any curve C_h.

Other definitions of envelope exist in the literature (see Bruce and Giblin (1981), and they all agree in the simpler cases, but they give different results with certain families. For our purposes here, the most suitable definition is that using (5.6.3).

Let us try out the definition on the family (5.6.2). We have to solve the equations

$$(x - t)^2 + y^2 - 1 = 0 = \frac{\partial f}{\partial t} = -2\cdot(x - t).$$

The last equation gives $x = t$, and when this is substituted in the first, we get $y^2 = 1$; thus the envelope consists of the set of all points (x, y) that satisfy $y^2 = 1$, and this set is the pair of parallel lines $y = \pm 1$ that we earlier 'saw' to be right.

A more interesting example concerns the family of normals to the parabola $P\colon y^2 = 4ax$ (see Fig. 5.6.2). The typical point A on the curve is $(at^2, 2at)$, at which the gradient is $\dot{y}/\dot{x} = 1/t$. Hence the normal to P at A has gradient $-t$, and equation

$$y - 2at = (-t)\cdot(x - at^2); \tag{5.6.4}$$

this is the contour at level 0 of f, where

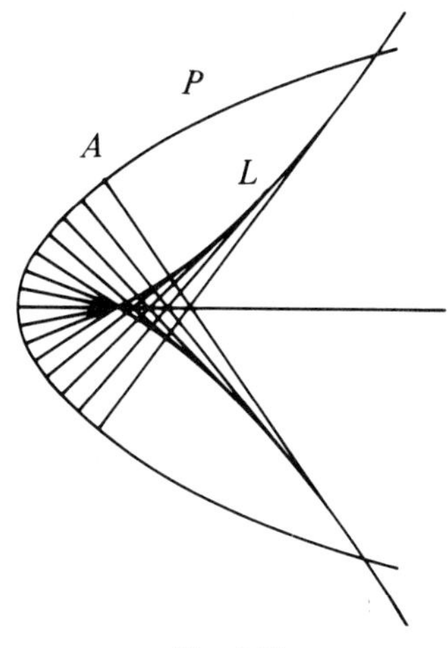

Fig. 5.6.2.

$$f(x, y, t) = y + xt - 2at - at^3$$

Hence, to find the envelope of the family $\{C_t\}$, we must solve the equations corresponding to (5.6.3), viz:

$$y + xt - 2at - at^3 = 0 = x - 2a - 3at^2.$$

The last equation gives $X = x - 2a = 3at^2$, and substitution in the first gives $y = -2at^3$. Therefore with a shift of origin to $x = 2a$, $y = 0$, the locus we require is a plane curve L, given in (X, y) coordinates) by

$$X = 3at^2, \qquad y = -2at^3, \qquad 27ay^2 = 4X^3,$$

which is a semi-cubical parabola. Further examples are given in the following exercises, which readers should try before continuing with the theory.

Exercises 5.6.1

1. Prove that the envelope of the family of lines: $x/\alpha + y/\beta = 1$, when

   (a) $\alpha + \beta = a$ (a constant), is the parabola $\sqrt{x} + \sqrt{y} = \sqrt{a}$;
   (b) $\alpha\beta = a^2$, is the astroid $x^{2/3} + y^{2/3} = a^{2/3}$.

   State the geometrical meaning of the conditions to which the parameters α, β are subject.
2. Prove that the envelope of the family of ellipses: $x^2/\alpha^2 + y^2/\beta^2 = 1$, when

   (a) $\alpha\beta = a^2$, is the two hyperbolas $2xy = \pm a^2$;
   (b) $\alpha + \beta = a$, is the astroid $x^{2/3} + y^{2/3} = a^{2/3}$.

   State the geometrical meaning of the conditions to which the parameters α, β are subject.
3. A straight line moves so that (i) the product, (ii) the sum of the squares of the perpendiculars drawn to it from two fixed points $(c, 0)$, $(-c, 0)$ is constant. Show that in each case the envelope is a central conic.

4. Let ABC be an equilateral triangle of side 2 and centroid G, with A at the origin of coordinates and the base BC below A and parallel to $\mathbf{0}x$. Let S, T be points on AB, AC such that

$$AS/AB = s, \qquad AT/AC = t, \qquad 0 \le s, t \le 1.$$

By using the formula $\Delta = \frac{1}{2}\, bc \sin A$, or otherwise, show that ST bisects the area of the triangle ABC if, and only if, $st = \frac{1}{2}$; and that ST then has the equation:

$$y(1 + 2s^2) + \sqrt{3}x(1 - 2s^2) + 2\sqrt{3}s = 0.$$

Show that the envelope of ST, as s varies, is the portion of the hyperbola H: $y^2 - 3x^2 = 3/2$ that is bounded by the lines GB, GC, each of which is tangent to H.

Hence show that the envelope of the family of all lines that bisect the area of ABC is a closed curve in the form of a triangle with curved sides. Indicate why a similar result must hold for *any* triangle. [SU]

5. Two particles P, Q move in a plane on the lines $x = 0$, $x = h$ (with $h > 0$), respectively. Particle P has constant velocity v, while Q has initial velocity u and constant acceleration a; and they start to move upwards simultaneously when each is on the x-axis. Show that the envelope of the line PQ is a branch of the hyperbola $\frac{1}{2}(wx - vh)^2 + ahxy = 0$ where $w = v - u$. Sketch this branch when $u = v$. [SU]

6. Show that the envelope of the family of lines:

$$L_t\colon At^2 + 2Bt + C = 0 \qquad (A, B, C, t \in \mathbb{R})$$

is the surface $B^2 = AC$ in (A, B, C)-space. Explain why this surface also relates to the condition for a quadratic equation to have equal roots. [SU]

7. A particle is projected from $\mathbf{0}$ at an elevation θ so that its coordinates are related by: $y = x \tan\theta - \frac{1}{4}x^2 \sec^2\theta$. Show that its envelope, as θ varies, is a parabola P with axis vertical and focus at 0. (It is called the **enveloping** parabola. Draw a sketch to show that the particle can never hit anything outside P.)

8. Find the envelope E of the family of lines $y + xt - f(t) = 0$ where $f\colon \mathbb{R} \to \mathbb{R}$ is a smooth function. Show that E is regular everywhere, except at the set Z of those values of t for which $f'' = 0$. If u is fixed, show that the envelope (in the (u, v)-plane of the lines:

$$t^4 + ut^2 + vt + w = 0 \tag{5.6.5}$$

is always regular if $u > 0$ and has just two irregular points if $u < 0$. Sketch the envelope when $u = 1$, and when $u = -1$.

Suppose t_0 is a simple root of the equation (5.6.5). Use the implicit function theorem 8.1.1 to explain how t_0 varies with u, v, w; and calculate its derivatives $\partial t_0/\partial u$, $\partial t_0/\partial v$.

9. Verify that the differential equation $y = xy' - (y')^3$ has the family of straight lines $y = cx - c^3$ as solution curves. Show that the envelope of the family $y = cx - c^3$ is $27y^2 = 4x^3$ and that this is also a solution of the differential equation.

10. In example 5.1.1(e), let G be the graph of f. Show that the envelope of the tangents to G is G itself. (The case when G is a straight line will need to be considered separately.) [Try the problem first when $f(x) = x^3$, then when G is a portion of a parabola, then the general case.] Let L_t denote the line joining $(t, f(t))$ to $(f(t), 0)$. Investigate the envelope of the lines L_t as t varies.

11. The ends X, Y of a ladder L of length a have coordinates $(a\cos\theta, 0)$, $(0, a\sin\theta)$ respectively, and $Z = (p, q)$ lies on L. Let $P = (p, 0)$, $Q = (0, q)$. As θ varies, find the envelope of XY, the locus of Z, and the envelope of PQ when (i) $p = ua\cos\theta$, $q = (1 - u)a\sin\theta$, for a fixed u with $0 < u < 1$; (ii) $\mathbf{0}Z$ is perpendicular to XY. In case (ii) display the envelope as a

'Maltese Cross' with eight cusps and six 'butterflies'. (The envelope of XY is the astroid of exercise 2.)

We shall now prove an important result.

Theorem 5.6.1
The envelope of a family touches every curve of the family.

(At the beginning of this section, we remarked that our eyes see the envelope as a curve which appears to touch each curve of the family. The theorem adds further confidence in this theoretical model of our perceptions. Our proof requires a little of the material of Chapter D, which we shall call on as needed. Readers may prefer to leave the proof until they have read that chapter.)

Proof of theorem 5.6.1
Recall that the envelope is constructed as follows. To find a point (a, b) on the envelope, we solve the equations (5.6.3) and find an interval I, a point s in I, and smooth functions $u, v: I \to \mathbb{R}$, such that $u(s) = a$, $v(s) = b$, and for all t in I,

$$f(u(t), v(t), t) = 0 = \frac{\partial f(u(t), v(t), t)}{\partial t}. \qquad (5.6.6)$$

Precise conditions for such solutions to exist are given in Chapter D, but we have seen this situation in all the examples considered so far.

We can therefore differentiate the first equation, with respect to t, and use the chain rule to get

$$\frac{\partial f}{\partial x} \cdot \dot{u}(t) + \frac{\partial f}{\partial y} \cdot \dot{v}(t) + \frac{\partial f}{\partial t} = 0$$

the partials being evaluated at $(u(t), v(t), t)$. The last term is zero, by the second equation in (5.6.6), so we have a scalar product

$$\left(\frac{\partial f}{\partial x} \quad \frac{\partial f}{\partial y}\right) \cdot (\dot{u}(t) \quad \dot{v}(t)) = 0.$$

To interpret this equation, let $\mathbf{T}$ be the tangent vector to the envelope E at $P = (a, b)$, and let $\mathbf{N}$ be the normal at P to the contour C_s. Then the last equation says that when $t = s$, $\mathbf{T}$ and $\mathbf{N}$ are orthogonal to each other. (This uses the fact that the gradient vector $(\partial f/\partial x \quad \partial f/\partial y)$ always points along the normal to a contour.) But $\mathbf{N}$ is also orthogonal to the tangent to C_s at P, so E and C_s have the common tangent $\mathbf{T}$. Thus, C_s touches E, and the proof is complete.

Next, we shall copy the example of the parabola in (5.6.4), but for a general curve, in order to display the structure. This will allow us to prove a result about the evolute in (5.5.3):

Theorem 5.6.2
The envelope of normals to a plane curve is the evolute; and each normal is a tangent to the evolute.

Proof

Suppose C is a plane curve, parametrized by smooth functions $u, v: I \to \mathbb{R}$. At $P = (u(t), v(t))$ on C, the normal has gradient $m = -\dot{u}/\dot{v}|_t$ and equation $f(x, y, t) = 0$, where

$$f(x, y, t) = y - v(t) - m(x - u(t)).$$

Hence

$$\frac{\partial f(x, y, t)}{\partial t} = -\dot{v}(t) - \dot{m}\cdot(x - u(t)) + m\dot{u}(t).$$

To find the envelope of the family of all normals to C, we must now solve the last two equations for x and y. From the second of these we get

$$x = u(t) + \rho(t)\cdot(\dot{v}/\dot{s})|_t,$$

by working out the derivatives of m and using the usual formulae (5.4.2), (5.5.1) for $\dot{s}$ and ρ. Substituting in the first equation we obtain

$$y = v(t) - \rho(t)\cdot(\dot{u}/\dot{s})|_t,$$

so each point $(x, y) = A$ on the envelope satisfies the equations (5.5.3) for the evolute. But A lies on the normal at P, so by theorem 5.6.1 this normal is a tangent to the envelope and hence to the evolute. This completes the proof.

Exercises 5.6.2

1. In exercise 5.5.1(6) find parametric equations for the evolute E of the cycloid Γ. Using the transformations:

$$X = x - t\pi, \qquad Y = y + 2r, \qquad t = \pi + \theta$$

show that E is another cycloid. By comparing points on E and Γ with the same parameter t, verify that any tangent to E cuts Γ orthogonally. [SU]

2. In the (I, J)-plane $\mathbb{R}^2$, let Γ be the curve $4I^3 + 27J^2 = 0$. Show that the cubic equation $x^3 + Ix + J = 0$ (with real coefficients I, J) has a multiple root iff $(I, J) \in \Gamma$. Sketch Γ and indicate the region of the (I, J)-plane for which the cubic equation has three simple real roots. Each point $P \in \Gamma$ can be expressed in the form $(-3t^2, 2t^3)$ for some $t \in \mathbb{R}$. If $t \neq 0$ show that the normal at P to Γ has equation $f(I, J, t) = 0$ where:

$$f(I, J, t) = 2t^4 + 3t^2 - tJ + I.$$

If (I, J) is fixed and not on Γ show that $\partial f/\partial t$ has only one real root; and hence that there are either two or no normals from (I, J) to Γ. Show that the envelope of the normals to $\Gamma - \{0\}$ has parametric equations $x = 3t^2(1 + t^2)$, $\quad y = 2t(3 + 4t^2)$. [SU]

3. Let $Q = (\cosh\theta, \sinh\theta)$ be a point on the hyperbola $x^2 - y^2 = 1$. By expressing the equation of the normal to H at Q in the form $y = f(x, \theta)$, show that the evolute E of H is given by the parametric equations: $x = \pm 2\cosh^3\theta$, $y = 2\sinh^3\theta$ so E has the equation $x^{2/3} - y^{2/3} = 2^{2/3}$.

   The vertical line through $(x, 0)$ cuts H, E in y_1, y_2 respectively. Show that

$$y_1^2 - y_2^2 = 3((2x^2)^{1/3} - 1)^2$$

and hence that E never cuts H. Make a rough sketch of E and H, to indicate their mutual relationships, and in each appropriate region S of the plane, indicate the number of normals that can be drawn to H from a typical point of S. [SU]

4. Show that for the ellipse $x^2/a^2 + y^2/b^2 = 1$ the evolute is the astroid $(ax)^{2/3} + (by)^{2/3} = (a^2 + b^2)^{2/3}$. By putting: $ax = \alpha \cos^3 t$, $by = \alpha \sin^3 t$ where $\alpha = (a^2 - b^2)^{1/3}$, investigate the curve. (Calculate $\dot{y}/\dot{x}$ to find singularities in dy/dx: these are cusps.)

   Let $b = 1$ and let a vary. Show that if a is near 1 then the astroid lies inside E, but if $a = 3$ (for example) it has two vertices outside E. Hence show that for some value of a, the astroid has two vertices on E, and find this value.
5. The 'super-ellipses' are given in the Cartesian form by $(x/a)^k + (y/b)^k = 1$. The ellipse has $k = 2$, and the astroid above has $k = 2/3$. Investigate the shape, evolute, and parallels of super-ellipses for different values of k. Why is the parametrization $x = a \cos^{2/k} t$, $y = b \sin^{2/k} t$ not suitable for computation? Show that a suitable substitute is $a \cdot \operatorname{sgn}(\cos t) \cdot |\cos t|^\alpha$, $b \cdot \operatorname{sgn}(\sin t) \cdot |\sin t|^\alpha$, where $\operatorname{sgn}(z) = z/|z|$ if $z \neq 0$, and $\operatorname{sgn}(0) = 0$. Investigate the shapes formed for various a, b, α. [$\alpha = 2$ gives a diamond.]
6. Let Γ denote the graph of the catenary $y = \cosh x$ in the (x, y)-plane, and let F be the family of circles of radius a (fixed) with the centres lying on Γ. Find in parametrized form the envelope E of the family F. Show that if $a < 1$ then E consists of two disjoint regular curves, but that if $a > 1$ then one of these curves has a pair of singularities at $(x, y) = (\pm b, 2\sqrt{a})$ where $b = \cosh^{-1}\sqrt{a} - \sqrt{a^2 - a}$. Find the evolute K of Γ, and verify that the singular points of E both lie on this evolute ($a > 1$). Sketch Γ, E, and K on the same diagram. Give a very brief interpretation in terms of geometric optics. [SU]

5.7 CAUSTICS

As a further example, consider the curve that is observed when sunlight shines on a teacup. The curved surface of the cup reflects the light, and we see a type of envelope on the surface of the tea. The curve which is enveloped is known as a **caustic**. To analyse the curve, we use Fig. 5.7.1, which shows a horizontal section of the cup as a circle, centre **0**, radius a. A light ray travels along a horizontal line, and hits the cup at (say) $P = (a \cos \phi, a \sin \phi)$. At P it is then reflected along a line which we denote by C_ϕ (to indicate the dependence on ϕ, which we allow to run through some interval I containing 0). The envelope of the family of lines C_ϕ, as ϕ runs through I, is the caustic that we see.

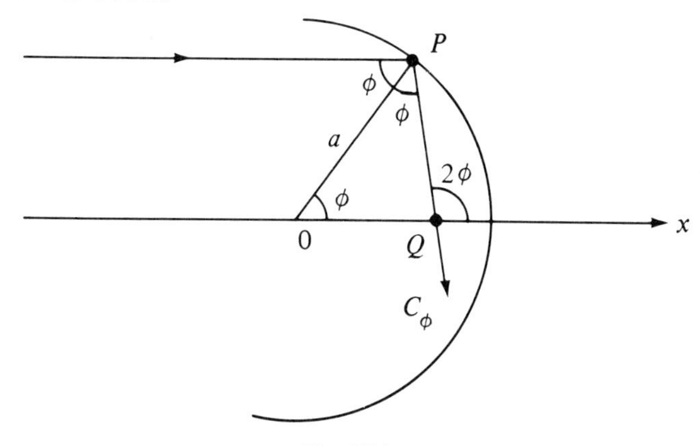

Fig. 5.7.1.

To find the equation of the line C_ϕ, suppose the radius $\mathbf{0}P$ makes an angle ϕ with $\mathbf{0}x$. Then the horizontal ray at P also makes an angle ϕ with $\mathbf{0}P$, so the angle between $\mathbf{0}P$ and C_ϕ is ϕ, because of the law of reflection: 'the angle of incidence equals the angle of reflection' (see section 4.3.1). Therefore C_ϕ makes an angle 2ϕ with $\mathbf{0}x$, because the exterior angle of triangle $\mathbf{0}PQ$ equals the sum of the base angles. Now $P = (a \cos \phi, a \sin \phi)$ so the equation C_ϕ is

$$y - a \sin \phi = \tan(2\phi)(x - a \cos \phi), \tag{5.7.1}$$

and to avoid infinities in $\tan(2\phi)$ we restrict ϕ so that $-\pi/4 < \phi < \pi/4$.

Multiplying through by $\cos(2\phi)$, this equation becomes

$$x \sin(2\phi) - y \cos(2\phi) = a[\sin(2\phi) \cos \phi - \cos(2\phi) \sin \phi],$$

and the square bracket simplifies to $\sin \phi$, by the addition formula for $\sin(A - B)$. Therefore C_ϕ is the contour at level zero of

$$f(x, y, \phi) = x \sin(2\phi) - y \cos(2\phi) - a \sin \phi.$$

Hence

$$\frac{\partial f(x, y, \phi)}{\partial \phi} = 2x \cos(2\phi) + 2y \sin(2\phi) - a \cos \phi,$$

and to find the envelope we must solve the equation $f = \partial f/\partial\phi = 0$. These give

$$x = \tfrac{1}{2}a[2\sin \phi \sin(2\phi) + \cos \phi \cos(2\phi)] = \tfrac{1}{2}a \cos \phi\,(2 - \cos(2\phi))$$

$$y = \tfrac{1}{2}a[\cos \phi \sin(2\phi) - 2\sin \phi \cos(2\phi)] = \tfrac{1}{2}a \sin \phi\,(1 - \cos(2\phi)),$$

where we have used the addition and half-angle formulae to simplify each square bracket.

These are the parametric equations of the envelope of the reflected rays, and they make sense if now ϕ is allowed to run from $-\pi$ to π. If we use a graph plotter, or sketch this planar curve, we find it is a closed kidney-shaped loop as in Fig. 5.7.2; hence it is called a 'nephroid' from the Greek 'nephros' = 'kidney'.

Since $x(-\phi) = x(\phi)$ and $y(-\phi) = -y(\phi)$, the curve is symmetrical in the x-axis; and there is a cusp at $B = (a/2, 0) = (x(0), y(0))$, at which $\dot{x} = 0 = \dot{y}$. Indeed, near $\phi = 0$, x and y have Taylor expansions

$$x = \frac{a}{2}\cdot\left(1 - \frac{\phi^2}{2} + \ldots\right)\left(2 - \left(1 - \frac{4\phi^2}{2} + \ldots\right)\right) = \frac{a}{2} + \frac{3a\phi^2}{4} + \ldots,$$

$$y = \frac{a}{2}\cdot(\phi + \ldots)\left(\frac{4\phi}{2} + \ldots\right) = a\phi^3 + \ldots,$$

so the nephroid near B is approximately the semi-cubical parabola $27ay^2 = 64X^3$, where $X = x - a/2$.

It is this cusp which we see in the teacup as a caustic. However, our analysis has not shown why the caustic looks so bright—this is because we have used a model that regards light rays merely as straight lines, with no mention of brightness. The

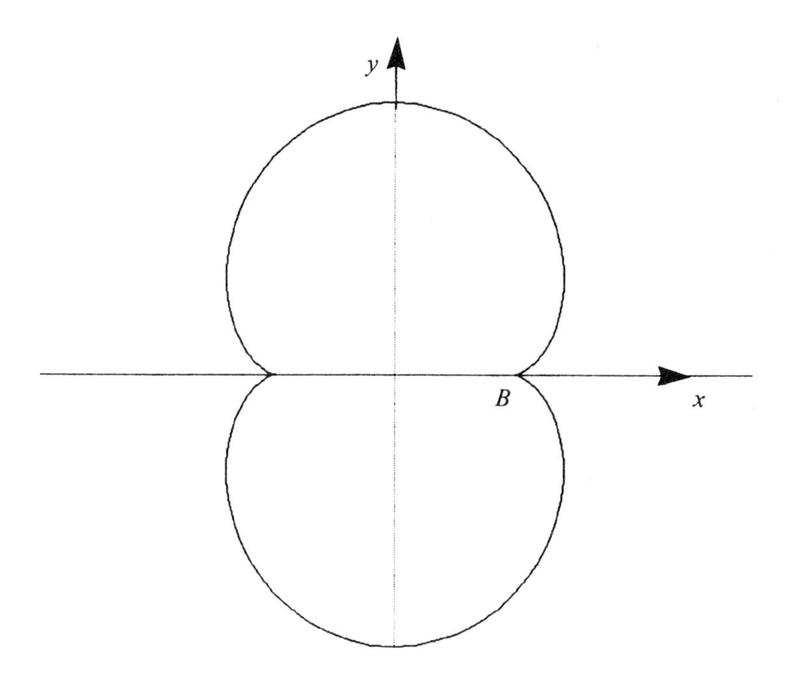

Fig. 5.7.2. The nephroid

wave theory of light gives a model in which brightness plays a role, but it gives an *infinite* light intensity at the cusp; only with a quantum theory of photons can the full story be predicted, and we have to refer readers to the physics books and an arduous study. Meanwhile, the geometry of the nephroid can be explored with the following program.

```
REM Program 5.7.1, Caustic of a circle
READ a, tmin, tmax, dt , xmax, ymax, k
DATA 2, 0   , 6.3 , 0.1, 3   , 2   , 4
FUNCTION fx(t) =  a*COS(t)
FUNCTION fy(t) =  a*SIN(t)
FUNCTION dx(t) = -a*SIN(t)
FUNCTION dy(t) =  a*COS(t)
FUNCTION ft(y) =  ASIN(y/a)
SET MODE "graphics"
SET WINDOW -xmax,xmax , -ymax,ymax
FOR t = tmin TO tmax STEP dt
  PLOT fx(t),fy(t);
NEXT t
PLOT
FOR y = a TO -a STEP -a/32
  LET  t = ft(y)
  LET  x = fx(t)
  LET xd = dx(t)
  LET yd = dy(t)
```

Program 5.7.1 continued

```
    LET xr = xd^2 - yd^2
    LET yr = 2*xd*yd
    LET  r = SQR(xr^2 + yr^2)
    LET  m = k/r
    PLOT -xmax,y ; x,y ; x+m*xr,y+m*yr
  NEXT y
  FOR t = tmin TO tmax STEP dt
    LET x = a*COS(t)*(2 - COS(2*t))/2
    LET y = a*SIN(t)*(1 - COS(2*t))/2
    PLOT x,y;
  NEXT t
  END
```

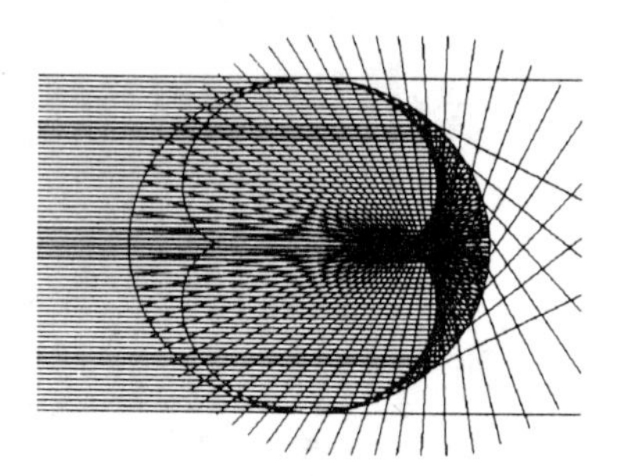

Sample output

The nephroid has other geometrical properties, and is one of a class of plane curves which can be obtained by placing a mark on a circular disc D, and rolling D round the circumference of a fixed disc, F. The path of the mark is a plane curve, which exhibits various properties that depend on the ratio of the radii of D and F, and whether D rolls on the inside of F or the outside. Such curves are treated in classical texts on geometry, which show that they have equations of the form

$$x = a\cos\phi - b\cos\left(\frac{a\phi}{b}\right), \qquad y = a\sin\phi - kb\sin\left(\frac{a\phi}{b}\right) \tag{5.7.2}$$

where a, b are constants, and k is 1 or -1 according as D rolls on the outside or inside of F (and the curve is accordingly an **epicycloid** or a **hypocycloid**—a **hypo**dermic syringe goes *inside* the **epi**dermis, which is the skin on the *outside* of the body). Again, ϕ runs from $-\pi$ to π (or from 0 to 2π).

At one time mathematics students had to learn this mass of detail. Engineers needed to know it to design various gears and drives—indeed, a nephroid-shaped cog is visible on the outside of older sewing machines, and drives the spool of cotton from side to side. We hope we have given readers enough background to read that material themselves, should they have the need or the curiosity. See for example Gibson (1956), Lamb (1907) which contain much detail on calculus generally.

Exercises 5.7.1

1. Use a graphics calculator or graph plotter to display the nephroid and other hypo- and epicycloids from their parametric forms.
2. Use the addition formulae of trigonometry to express the above equations of the nephroid as:

$$x = \tfrac{1}{4}a(3\cos\phi - \cos(3\phi)), \qquad y = \tfrac{1}{4}a(3\sin\phi - \sin(3\phi)).$$

Hence show that the nephroid is an epicycloid.

3. Investigate caustics for elliptical teacups and other curves C as follows. Instead of the radius at P on the circle, use the normal at P which has gradient $\tan(2\phi)$ where now $\tan\phi$ is the gradient $-\dot{x}/\dot{y}$ of the normal. Therefore equation (5.7.1) must have $\tan(2\phi)$ changed to $2\tan\phi/(1-\tan^2\phi)$, and the differentiation of $\partial f/\partial\phi$ becomes rather more complicated. Once done, however, it can be used to modify the envelope program in exercise 5.5.2(7), and then the caustic curves can be investigated especially when the algebra would be otherwise difficult (or impossible) to manage.
4. Investigate the curves (5.7.2) for various integers a, b.
5. The cycloid is defined in exercise 5.2.1(3). Show that, as t runs from 0 to π, an arch-shaped curve (the 'true' cycloid) is swept out, followed by further copies as t increases. When t is an integral multiple of π, we observe a cusp between two neighbouring cycloids. Show that the evolute is another cycloid, and calculate the radius of curvature.
6. Let $Z=(p,q)$, $X=(p,0)$, $Y=(0,q)$, $r=\mathbf{0}Z$ and angle $Y\mathbf{0}X=\theta$. Show that if $r=f(\theta)$ then the envelope of the line XY is given by: $x=a\cos^2\theta\cdot g(\theta)$, $y=a\sin^2\theta\cdot h(\theta)$ where $g(\theta)=\cos\theta\cdot f(\theta)+\sin\theta\cdot f'(\theta)$, $h(\theta)=\sin\theta\cdot f(\theta)-\cos\theta\cdot f'(\theta)$. Verify that in exercise 5.6.1(11), case (ii), $g(\theta)=a\,(2\sin^2\theta-\cos^2\theta)$, $h(\theta)=a\,(2\cos^2\theta-\sin^2\theta)$.

   Use your calculator or computer to display this curve as a 'Maltese cross'with eight cusps, and composed of four 'butterflies'. Now try $f(\theta)=\sin 3\theta, \sin 4\theta$, etc.
7. A way of making attractive patterns is known as 'curve stitching in a circle'. Mark n points at even spaces round a circle (e.g., take $n=36$) and label them with the numbers $0, 1, 2, \ldots, n-1$. Now choose a multiplier m (e.g., take $m=3$). Now, for each point i calculate j obtained as the remainder when the product mi is divided by n (e.g., $i=5$ gives $j=15$, $i=20$ gives $j=60 \bmod 36=24$), and join the points i and j by a line segment. The set of these lines envelopes a curve—in the case above this should be a nephroid. Obtain the equation of the line ij and hence find the envelope. Show that in the general case the envelope is that of a hypocycloid. The following program can be used to explore the pictures.

```
REM Program 5.7.2, Hypocycloid envelopes
READ n  , a, xmin, xmax, ymin, ymax
DATA 128, 2, -3  , 3   , -2  , 2
OPTION BASE 0
DIM x(128),y(128)
SET MODE "graphics"
SET WINDOW xmin,xmax, ymin,ymax
FOR i = 0 TO n
  LET t = i*2*PI/n
  LET x(i) = a*COS(t)
  LET y(i) = a*SIN(t)
  PLOT x(i),y(i);
NEXT i
INPUT PROMPT " m : " : m
FOR i = 0 TO n
  LET j = MOD(m*i,n)
  PLOT x(i),y(i) ; x(j),y(j)
NEXT i
END
```

m:4

Sample output

5.8 POLAR COORDINATES

Another useful way of specifying a plane curve is to use polar coordinates, which we used in (4.7.6) to find the orbit of a planet. With these we can locate a point P by using polar coordinates (r, θ) where r is the distance from the origin $\mathbf{0}$ to P, so $r = \sqrt{(x^2 + y^2)}$, taking the positive square root. We call r the **modulus** of P, and denote it by $|P|$. Here, θ is the polar angle Pol(P), as shown in Fig. 5.8.1; it is measured in radians, anticlockwise, and lies between 0 and 2π. We do not attempt to define θ if $P = 0$, since $r = 0$ locates $\mathbf{0}$ precisely. If $P \neq 0$, then the x- and y-coordinates of P can be written as: $(x, y) = (r \cos \theta, r \sin \theta)$. Probably the reader has tried exercises of the form 'plot the locus given in polar coordinates by $r = f(\theta)$', and here we must emphasize that r has to be positive. Thus, if (for example) $f(\theta)$ is $\cos 2\theta$, then the correct locus is the undotted curve in Fig. 5.8.2. The dotted part consists of points with polar coordinates $(-f(\theta), \theta + \pi)$ for which $f(\theta)$ is negative. We can of course allow such curves provided we state clearly what our convention is.

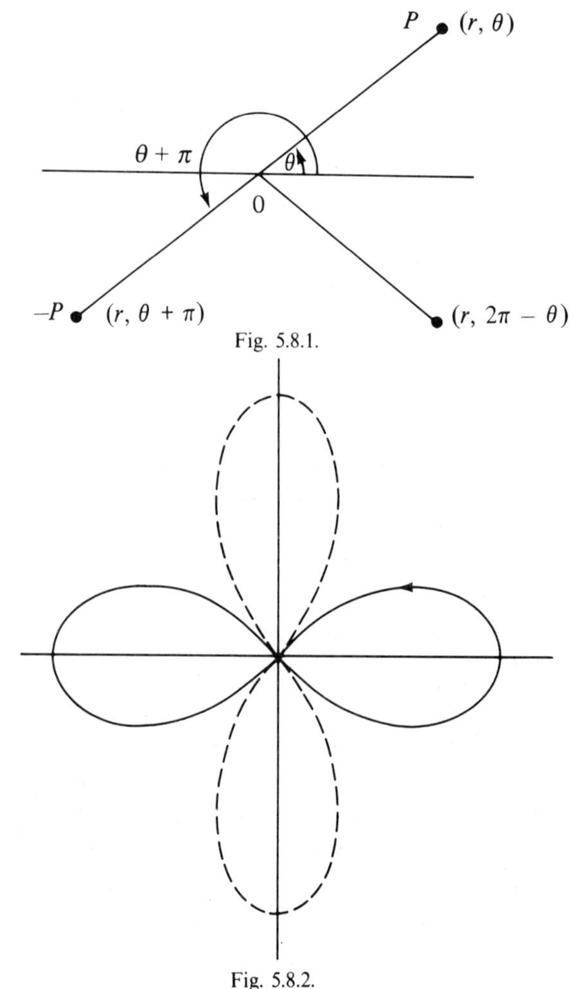

Fig. 5.8.1.

Fig. 5.8.2.

We have used Fig. 5.8.1 to indicate what the polar angle is, but for some problems it is necessary to be able to give an analytic definition. This is more complicated than at first appears: often, people simply write $\theta = \tan^{-1}(y/x)$, but this is very wrong since it can only give a (possibly negative) angle in the range of the function $\tan^{-1}: \mathbb{R} \to (-\pi/2, \pi/2)$, and says nothing when $x = 0$. Instead, we shall need to use the inverse cosine, that is, the function $\cos^{-1}: J \to [0, \pi]$, where J is the interval $[-1, 1]$. One reason for being especially careful is that we wish θ to have partial derivatives. Thus we shall construct a function

$$\mathrm{Pol}: \mathbb{R}^2 - \{0\} \to [0, 2\pi), \qquad \theta = \mathrm{Pol}(x, y), \tag{5.8.1}$$

with partial derivatives which turn out to be:

$$\frac{\partial}{\partial x}\mathrm{Pol}(P) = \frac{y}{r^2}, \qquad \frac{\partial}{\partial y}\mathrm{Pol}(P) = \frac{x}{r^2}, \tag{5.8.2}$$

where $r = |P|$, provided P does not lie on the positive real axis, $\mathbb{R}_+$. On $\mathbb{R}_+$, $\mathrm{Pol}(x, y)$ jumps from 0 to 2π as $\mathbf{z}$ crosses from y positive to y negative (just as in Fig. 5.8.1). These are the properties which the reader needs to remember, since they are the ones that are used in practice; the *construction* need not be remembered, once it is over.

The details are as follows. Using the inverse cosine, mentioned above, we define the polar angle by

$$\mathrm{Pol}(x, y) = \begin{cases} \cos^{-1}\left(\dfrac{x}{r}\right), & \text{if } y \geq 0 \\ \pi + \mathrm{Pol}(-x, -y), & \text{if } y < 0. \end{cases} \tag{5.8.3}$$

Let us test this definition, to see if it gives the correct angles for known points. From the first line, we have

$$\mathrm{Pol}(1, 0) = \cos^{-1}(1/1) = 0, \qquad \mathrm{Pol}(-1, 0) = \cos^{-1}(-1/1) = \pi;$$

and from the second,

$$\mathrm{Pol}(-1, -1) = \pi + \mathrm{Pol}(1, 1) = \pi + \cos^{-1}\left(\frac{1}{\sqrt{2}}\right) = \frac{3\pi}{4},$$

and

$$\mathrm{Pol}(1, -1) = \pi + \mathrm{Pol}(-1, 1) = \pi + \cos^{-1}\left(-\frac{1}{\sqrt{2}}\right) = \frac{7\pi}{4},$$

and these are the answers suggested by Fig. 5.8.1.

It remains to calculate the derivatives, and these follow from the chain rule if $y \neq 0$, because we know the derivatives of $\cos^{-1}$. It is clear from the definition that Pol is not differentiable on the positive x-axis, because of its jump in value as we cross; on the negative x-axis, we could establish differentiability from first principles. A shorter way, however, is to observe that

$$\mathrm{Pol}(x, y) = \pi + \tan^{-1}(y/x), \qquad \textit{if } x < 0, \tag{5.8.4}$$

so, since $\tan^{-1}(y/x)$ is always differentiable if $x \neq 0$, then the derivatives of Pol (with $x < 0$) are those of $\tan^{-1}$.

Exercises 5.8.1

1. Sketch graphs of the following curves specified in terms of plane polar coordinates (r, θ), where a is a positive constant.

   (a) $r = a \sin \theta$ (b) $r = a \sin 3\theta$
   (c) $r = a \sec 2\theta$ (d) $r = a \operatorname{cosec} \theta$

   What are the differences if r is allowed to take negative values?
2. Find out how to use your graphical calculator or graph-plotting software to plot polar graphs such as those above. Which convention about the sign of r does it follow?
3. What sort of curve does the polar equation $r^2 - 2r \cos \theta + 1 = 0$ represent?
4. The fixed points F and G are a distance $2a$ apart. The point P moves so that the product of its distance from F and G is equal to a^2. Find the Cartesian equation of P in a suitable coordinate system. Find the polar equation of P. Sketch the curve traced out by P. What type of curve is it?
5. More generally the **Cassini oval** is the locus of a point which moves so that the product of its distance from two fixed foci $(\pm c, 0)$ is a constant (a^2). Prove that its equation in Cartesian coordinates is

   $$(x^2 + y^2)^2 - 2c^2(x^2 - y^2) + c^4 - a^4 = 0$$

   or in polars:

   $$r^4 - 2c^2r^2\cos 2\theta + c^4 - a^4 = 0$$

   whence

   $$r = c\sqrt{[\cos 2\theta \pm \sqrt{(b^4 - \sin^2 2\theta)}]}, \qquad b = a/c.$$

   What are the symmetries of the curve? Draw sketches (or get calculator/computer pictures) that show how the curve changes as a changes from 0 to ∞.
6. Consider the curve $r = b + \cos 2\theta$ where b is a constant. Sketch or draw the curve for each of the following cases:

   (a) $b > 1$ (b) $b = 1$ (c) $0 < b < 1$
   (d) $b = 0$ (e) $-1 < b < 0$ (f) $b = -1$ (g) $b < -1$.

7. A family of ovals C_α is given by the polar equation $C_\alpha : r = a \sin^\alpha\theta$ as α varies. Show that C_α is the locus of a point P such that $\mathbf{0}P = PM^c$ for some constant c, where PM is the distance P from the axis $\theta = 0$. Show also that C_α has parametric equations:

   $$x = a \sin^\alpha\theta \cos \theta \qquad y = a \sin^{\alpha+1}\theta \cos \theta$$

   so that C_α is a closed (oval-shaped) curve. Investigate the curvature at $\mathbf{0}$, and the evolute. How do these change with α or a?
8. Verify that the definition (5.8.3) of Pol(x, y) gives the values π, $5\pi/4$, $3\pi/2$, $7\pi/4$ when (respectively) P lies on the negative x-axis, and $P = (-1, -1), (0, -1), (1, -1)$. Compare the values of Pol(P) when $P = (1, 0.1)$ and $P = (1, -0.1)$, to show that they differ by almost 2π.
9. Use the formulae (5.8.2) to compute the derivatives of Pol(P) when P is $(1, 0)$, $(0, 1)$, $(1, 1)$, $(1, -1)$, $(-1, -1)$.

10. Find a formula like (5.8.4) when $x > 0$, to show that the derivatives of $\mathrm{Pol}(x, y)$ are related very closely to those of $\tan^{-1}(y/x)$.
11. The cardioid C is given in polar coordinates by

$$r = (1 + \cos\theta).$$

Convert to Cartesian coordinates, and parametrize C by θ $(=t)$. Show that the speed s and the curvature κ at a typical point $(x(t), y(t))$ are related by $\kappa s = 3/2$. [SU]

Polar coordinates also occur when we use complex numbers, and a function $\mathrm{Arg}(z)$ is used to measure the polar angle, but now in the interval $(-\pi, \pi]$. We explain this as follows.

5.9 COMPLEX NUMBERS: THE ARGAND DIAGRAM

Readers probably know that complex numbers are of the form $z = x + y\sqrt{(-1)}$, and they may have met the idea of representing z as the point (x, y) in the coordinate plane $\mathbb{R}^2$. In this way, the real numbers form the x-axis, since y is then zero. Thus every real number is a complex number, but not vice versa. It is usual in this context to refer to the plane as the **Argand diagram**, after J. R. Argand (1768–1822, Geneva); but he was only one of several people who introduced the idea around 1800, the first of whom was Caspar Wessel.

In the plane $\mathbb{R}^2$, a typical point P has Cartesian coordinates (x, y) where x, y are real numbers, but when we think of P as a complex number in the Argand diagram we write P as $z = x + \mathrm{j}y$; and we use j as the symbol for $\sqrt{-1}$ instead of the more traditional i to correspond with the vector notation $\mathbf{z} = x\mathbf{i} + y\mathbf{j}$ where $\mathbf{i}$, $\mathbf{j}$ are the unit vectors along the x and y axes respectively. (So the vector $\mathbf{j}$ corresponds to the number $z = 0 + \mathrm{j}\cdot 1 = \mathrm{j}$.)

Thus z is called 'real' if $y = 0$, and then the real numbers fill out the x-axis. Hence the vector $\mathbf{i} = (1, 0)$ is the real number 1. Complex numbers are manipulated algebraically as expected; they are added by the vector rule:

$$(x + \mathrm{j}y) + (u + \mathrm{j}v) = (x + u) + \mathrm{j}(y + v) \tag{5.9.1}$$

and multiplied according to the rule

$$(x + \mathrm{j}y)\cdot(u + \mathrm{j}v) = (xu - yv) + \mathrm{j}(xv + yu). \tag{5.9.2}$$

(As a check, the reader should use this formula to verify that $\mathrm{j}^2 = -1$.)

Thus j is the square root of -1 and has coordinates $(0, 1)$. Once we know this, then (5.9.2) is a very natural consequence of multiplying out the right-hand side by the ordinary laws of algebra. Thus there is nothing mysterious about the 'square root of minus one', provided we widen our concept of the set of numbers from a one-dimensional model into two dimensions.

When we think of the plane as $\mathbb{R}^2$ (for doing coordinate geometry) we denote a typical point by P; but when we work in the Argand diagram, we denote P by z. Also when we write $z = x + \mathrm{j}y$, *it is always intended that* x, y *are real*, unless further explanation is given.

Each complex number z has an associate, called the **conjugate** of z, and this is the number $\bar{z} = x - \mathrm{j}y$. By (5.9.2), we have

$$z\bar{z} = x^2 + y^2 \quad (= r^2, \text{ say}), \tag{5.9.3}$$

which is a real number ≥ 0. It is zero if, and only if, $z = 0$ (so $x = 0 = y$). We write $r = |z|$, the **modulus** of z; it is the length of the vector $(x \quad y)$.

Exercises 5.9.1

1. Show that the locus given by $|z| = 9$ is a circle. (The problem is to find the set of all z in the Argand diagram for which $|z| = 9$.)
2. If k is a real positive constant, show that the locus $|z - a| = k|z - b|$ is a circle unless $k = 1$. (Square each side to get an equation in Cartesian coordinates.) Sketch the family of curves obtained by letting k vary. (The circle is the **circle of Apollonius**.)
3. Let S_r denote the circle $|z| = r$. Show that if z makes one revolution round S_r then w makes n revolutions round S_q, when $w = z^n$ and $q = r^n$. If, instead, $w = z - z^2$ show that w traces out a heart-shaped curve (a **cardioid**) K_r, and use (5.9.3) to show that $|w|$ increases if r increases. Hence show that if $r < s$, then K_r lies inside K_s.

If $z = x + \mathrm{j}y$ is not zero (so $(x, y) \neq \mathbf{0}$), then (x, y) has polar coordinates (r, θ), and r is just $|z|$. Also we can write z as: $z = r(\cos\theta + \mathrm{j}\sin\theta)$. Here, $x = r\cos\theta$ and $y = r\sin\theta$. If now $w = s(\cos\phi + \mathrm{j}\sin\phi)$, then it is proved in Chapter C that

$$zw = rs(\cos(\theta + \phi) + \mathrm{j}\sin(\theta + \phi)), \tag{5.9.4}$$

so in particular the modulus is multiplicative: $|zw| = rs = |z|\cdot|w|$. For brevity, we write cis θ for $\cos\theta + \mathrm{j}\sin\theta$, and then z becomes r cis θ, so that (5.9.4) becomes

$zw = rs\,\mathrm{cis}(\theta + \phi)$, and taking $w = z$ gives **'de Moivre's Theorem'**:

$$z^2 = r^2\,\mathrm{cis}(2\theta),\ z^3 = r^3\,\mathrm{cis}(3\theta), \ldots, z^n = r^n\,\mathrm{cis}(n\theta).$$

Let us use this to find the nth roots of unity (where 'unity' is an old-fashioned name for the number 1). These are the complex numbers

$$1, \quad \mathrm{cis}(2\pi/n), \quad \mathrm{cis}(4\pi/n), \quad \ldots, \quad \mathrm{cis}(2(n-1)\pi/n), \tag{5.9.5}$$

or more briefly $\omega_k = \mathrm{cis}(2k\pi/n)$ $(k = 0, 1, 2, \ldots n - 1)$, for

$$(\omega_k)^n = \mathrm{cis}(n\cdot 2k\pi/n) = \mathrm{cis}(2\pi) = \cos 2\pi + \mathrm{j}\sin 2\pi = 1.$$

Note that the numbers in (5.9.5) are precisely the powers of ω_1, since $\omega_1^k = \mathrm{cis}^k(2\pi/n) = \mathrm{cis}(2k\pi/n) = \omega_k$.

In the Argand diagram, these n numbers lie on the unit circle S which (in ordinary coordinates) has equation

$$S\colon x^2 + y^2 = 1$$

and they form the vertices of a regular polygon of n sides, starting at $\omega_0 = 1$, as shown in Fig. 5.9.1.

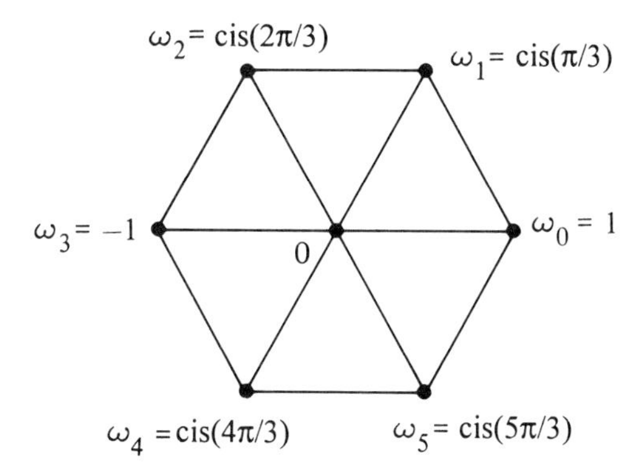

Fig. 5.9.1.

Exercises 5.9.2

1. If $|z| = r$, show that $z\bar{z} = r^2$, and that if $z \neq 0$ then z has the 'multiplicative inverse' $1/z = \bar{z}/r$.
2. Show that $(x^2 + y^2)(u^2 + v^2) = (xu - yv)^2 + (xv + yu)^2$ when all the numbers are real. (Note: you can verify that the equation holds by direct but tedious multiplication; but how did anyone dream up the right-hand side? To see why it is so, recall the general fact that modulus is multiplicative. 'Mathematics is seeing the truth without effort'.) Conclude that if integers m, n are each the sum of two squared integers, then so is mn.
3. If the complex number z is $r \text{ cis } \theta$, show that it has three complex cube roots, given by the points of the circle C (centre $\mathbf{0}$, radius $r^{1/3}$) with polar angles $\theta/3, \theta/3 + 120°, \theta/3 + 240°$.
4. If $z = r \text{ cis } \theta$ prove that its nth roots are the numbers $s \cdot \omega_k \cdot \text{cis}(\theta/n)$, where s is the ordinary (real) nth root of r. Sketch these points in the Argand diagram, and show that they form the vertices of a regular polygon and lie on a circle, of centre $\mathbf{0}$ and radius s. (Note: the equation $z^n = 1$ has no more than n complex roots; see corollary 6.6.1.)
5. Sketch on separate Argand diagrams the regions defined below. Shade the regions defined, indicating by continuous lines the boundaries to be included, and by broken lines, the boundaries to be excluded.

   (a) $|z - 2 \text{ cis}(\pi/3)| \leq 4$ (b) $\pi/4 \leq |\arg z| \leq \pi/2, \quad |z| < 1$
   (c) $|z - 2| - |z - 3\text{j}| \geq 1$ (d) $|z + \text{j}| + |z - \text{j}| \leq 3.$

   Show that all the roots of the equation $(1 + z)^{2n+1} = (1 - z)^{2n+1}$ are given by $\pm \text{j} \tan(k\pi/(2n + 1))$ where $k = 0, 1, \ldots, n$.
6. Express in the form $a + \text{j}b$ with a, b real:

   (a) $(3 + 4\text{j})/(2 - 5\text{j})$, (b) $(1 + \text{j}\sqrt{3})^8$, (c) $\sqrt{(-2\text{j})}$ and
   (d) the roots of the quadratic equation $z^2 - (3 + 5\text{j})z + 8\text{j} - 4 = 0$.

7. Find the set of complex numbers z for which the real part of $(z + 1)/(z - \text{j})$ equals 2.
8. Let $z = (2 + \text{j})/(3 - 4\text{j})$. Express in the form $a + \text{j}b$ with a, b real, the numbers: $z, z^{-1}, z^6, z + z^2 + z^3 + z^4$. Find all the fifth roots of z. Show that they form a regular pentagon and find its area. Find the set of points ω such that $|\omega| = 2|\omega - z^3|$.
9. Express the following in the form $x + \text{j}y$:

$$(2 - 3\text{j})^2(4 + 5\text{j}), \qquad (1 + 2\text{j})^2/(1 - \text{j}), \qquad (1 + 3\text{j})^{-1} - (1 - \text{j})^{-1}.$$

10. Find the roots of the following equations:

$$z^2 = -5 + 2j \qquad z^2 - (3 + j)z + (4 + 3j) = 0.$$

11. Prove that $\overline{(z/w)} = \bar{z}/\bar{w}$.
12. Assuming that $|z + w| \leq |z| + |w|$ for all complex numbers z, w, prove that $|z - w| \geq |\,|z| - |w|\,|$.
13. Prove that $|z + w|^2 + |z - w|^2 = 2|z|^2 + 2|w|^2$.
14. Find the most general solution to the equations

(a) $2z^2 - 2(\sqrt{3} + j)z + (1 + j\sqrt{3}) = 0$

(b) $z = \sinh^{-1}\dfrac{1}{2}$

(c) $z^3 = 2 + 2j$

(You may give your answer in polar form.) [SU]

Next, if $z = x + jy = r \operatorname{cis} \theta$, we have said that (r, θ) are the polar coordinates of z; but we need again to be able to give an analytic formula for θ in terms of z. For technical convenience, it is usual in this branch of the subject to measure the angle as shown in Fig. 5.9.2, so that it lies between $-\pi$ and π, depending on the quadrant Q_i in which z lies. This measure is denoted by Arg(z). Formally, then, we shall be obtaining a function

$$\text{Arg}: \mathbb{R}^2 - \{0\} \to (-\pi, \pi], \tag{5.9.6}$$

such that

$$\text{Arg}(\bar{z}) = -\text{Arg}(z) \tag{5.9.7}$$

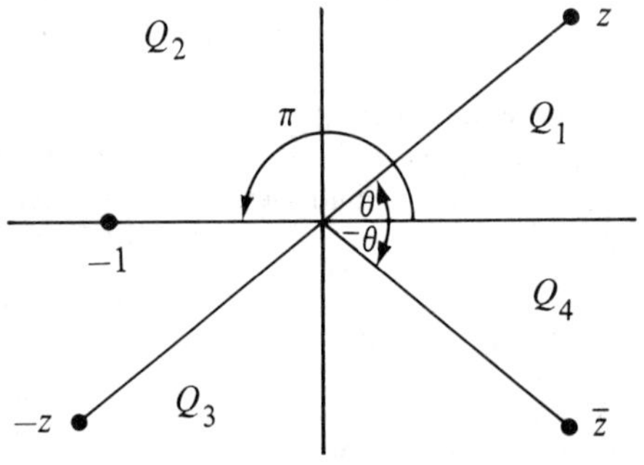

Fig. 5.9.2.

which (as with the function Pol in (5.8.1)) we wish to have partial derivatives; it will turn out that these are given by the same formulae as in (5.8.3), but this time x must not lie on the negative x-axis $\mathbb{R}_<$. Also on $\mathbb{R}_<$, Arg jumps in value by 2π.

To define Arg(z), it suffices by (5.9.7) to assume that z ($= x + jy$) is in Q_1 or Q_2, which means that $y \geq 0$. Then we set

$$\text{Arg}(z) = \text{Pol}(x, y) \qquad (y \geq 0); \tag{5.9.8}$$

and if, instead, $y < 0$, we take $\text{Arg}(\bar{z}) = \text{Arg}(z)$ as given by (5.9.7).

We leave it as an exercise for the reader to test this definition, as we did earlier for the Pol function, to see if it gives the correct angles for known points—for example, that, as suggested by Fig. 5.9.2,

$$\text{Arg}(1 + \text{j}) = \pi/4 = -\text{Arg}(1 - \text{j});$$

$$\text{Arg}(-1 + \text{j}) = 3\pi/4 = -\text{Arg}(-1 - \text{j}).$$

Note that if $y = 0$, then by (5.9.8), Arg(z) is 0 or π, according as x is positive or negative; and here $\bar{z} = z$, so (5.9.7) still holds. However, if z is just below $\mathbb{R}_<$ and hence of the form $-x - \text{j}y$ with x and y positive, then by (5.9.7), $\text{Arg}(z) = \text{Arg}(-x + \text{j}y)$, which is nearly $-\pi$ when y is nearly zero. Thus Arg *is not continuous on the line* $\mathbb{R}_<$.

The partial derivatives of Arg(z) come immediately from those of Pol, except when z lies on $\mathbb{R}^+$. But when $x > 0$ we have $\text{Arg}(z) = \tan^{-1}(y/x)$, which is differentiable; so we have (5.9.3) by the chain rule.

Exercises 5.9.3

1. Find Arg(z) when $z = 2 + 5\text{j}$, $-2 + 5\text{j}$, $2 - 5\text{j}$, $-2 - 5\text{j}$. Compare your answers with the corresponding values of Pol(P) when $z = P$. Compare the values of Arg(z) when $z = -1 + \text{j}(0.1)$ and $z = -1 - \text{j}(0.1)$, to show that they differ by almost 2π.
2. Show that Arg(zw) differs from Arg(z) + Arg(w) by at most an integral multiple of 2π. What happens when $z = w = -1$?
3. Sketch the region of the complex plane specified by

   (a) $1 < |z + 1 - 2\text{j}| < 2$;
   (b) $\pi/3 < \arg(z + 1 - 2\text{j}) < 2\pi/3$. [SU]

4. Use de Moivre's theorem to find

   (a) $\cos 4\theta$ in terms of $\cos \theta$;
   (b) $\tan^3\theta$ in terms of sines and cosines of multiples of θ. [SU]

5. (a) Simplify $(1 + \text{j}\sqrt{3})/(\sqrt{3} + \text{j})^3$.
   (b) Determine the square roots of $1 + \text{j}$.
   (c) If $(qz + 1)/(z - 1) = \text{e}^{\text{j}\alpha}$, where q and α are real constants, show that

   $$z = \frac{(1 - q)(1 + \cos \alpha) - \text{j}(1 + q)\sin \alpha}{1 - 2q\cos \alpha + q^2}.$$

   (d) Solve the equation $\text{e}^z + \text{e}^{-z} = 0$, showing that $z = x + \text{j}y$ with $x = 0$ and $y = (2n + 1)\pi/2$ for any integer n. [SU]

6

Control space and phase space: polynomial functions

In the last chapter we met families of curves defined by functions f_t with one parameter t. Many models involve several parameters, and in this chapter we begin with families of algebraic polynomials. Their zeros move as the parameters change, and give rise to some interesting plane curves.

6.1 QUADRATIC FUNCTIONS: THE CONTROL PLANE

A quadratic function is one of the form: $y = ax^2 + bx + c$, with $a \neq 0$. Therefore we may divide by a, and 'complete the square' to have

$$y = a \cdot \left[\left(x + \frac{b}{2a} \right)^2 + \frac{c}{a} - \left(\frac{b}{2a} \right)^2 \right] = a \cdot \left(u^2 - \frac{d}{4a^2} \right)$$

where

$$u = x + \frac{b}{2a}, \qquad \text{and} \qquad d = b^2 - 4ac.$$

Thus we obtain the famous 'quadratic formula' for the zeros of y (which are the roots of the equation $y = 0$):

$$u = \frac{\sqrt{d}}{2a} \qquad \text{or} \qquad x = \frac{-b \pm \sqrt{(b^2 - 4ac)}}{2a}. \tag{6.1.1}$$

As written, these formulae include the case when all the numbers involved are complex; but for many applications the coefficients a, b, c are real (as we suppose always in the sequel), so it is important to know when the zeros are real. Clearly, this occurs when the **discriminant** d is non-negative, that is,

$$d = b^2 - 4ac \geq 0. \tag{6.1.2}$$

Since it is the function y/a which is of primary interest, we consider now the set of all functions of the form

$$f_{bc}(x) = x^2 + 2bx + c \tag{6.1.3}$$

where the factor 2 is inserted to simplify the algebra to come. In this case d takes the form:

$$d = b^2 - c. \tag{6.1.4}$$

Here, there is one quadratic function f_{bc}, for each pair (b, c) of real numbers (called **control parameters**). For example, $f_{2,-3}$ is $x^2 + 4x - 3$. We call the (b, c)-plane of coordinates the **control plane**, and to be clear as to which coordinates are in question we denote the control plane by $\mathbb{R}^2_{b,c}$. Each point of $\mathbb{R}^2_{b,c}$ corresponds to a quadratic function of the form (6.1.3), of which the graph lies in the usual plane $\mathbb{R}^2_{x,y}$, which is called the **phase space**. As we vary the control parameters in $\mathbb{R}^2_{b,c}$ the graph of f_{bc} will gradually vary in $\mathbb{R}^2_{x,y}$, but some variations are more significant than others, as we now discuss.

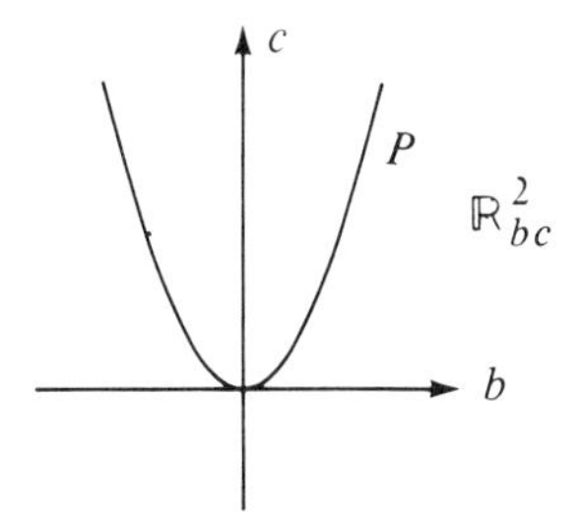

Fig. 6.1.1.

The control plane is divided into two regions as in Fig. 6.1.1, by the parabola P: $c = b^2$, and if a point (b', c') lies above P, then $c' > b'^2$ so by (6.1.3) the function $f_{b'c'}$ has no real zeros. In geometrical language, the graph P' of $f_{b'c'}$ (in $\mathbb{R}^2_{x,y}$) does not cut the x-axis. If we change the control parameters, P' will move, but it will never cut the x-axis as long as (b', c') remains above P. If (b', c') lies on P itself, then $d = 0$ and there is just one zero ($= -b'$, hence real). Geometrically, P' and the x-axis touch at $x = -b'$. But if (b', c') lies below P then $f_{b'c'}$ has two real zeros, because $d > 0$. (Strictly, d in (6.1.4) is a function of b' and c', so here we should have said $d(b', c') = 0$, rather than the abbreviated form.) In this last case, we say that P' cuts the x-axis **transversely**—meaning that the intersections are proper crossings and not contact points. Notice that it is at a contact point that real roots are created or destroyed, depending on the direction of passage through it, as the control is varied.

Therefore, there are only three basically different types of function (6.1.3), and Fig. 6.1.2 displays all the relevant information by showing $\mathbb{R}^2_{x,y}$ in a 'portrait gallery' with a small picture of the xy-graph of a typical function f_{bc}, hung at (b, c) when (b, c) lies either on P or on either side of it. With this figure, and the discriminant formula, we now have all the essential information about real quadratic functions.

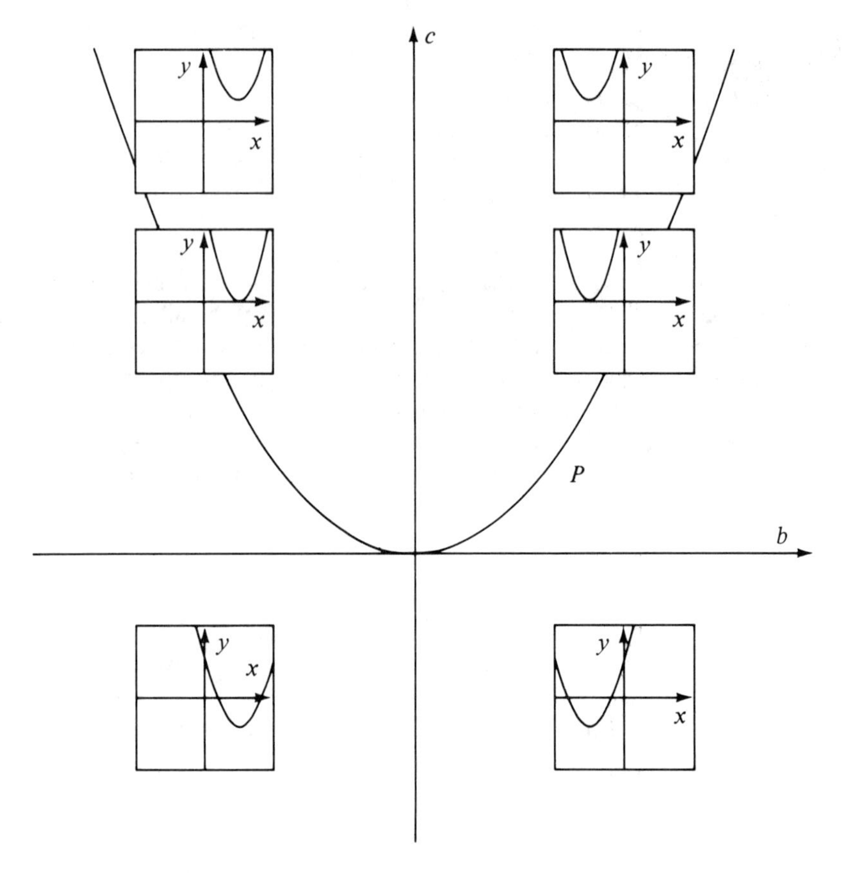

Fig. 6.1.2.

Exercises 6.1.1

1 Show that f_{bc} in (6.1.3) can be written in 'reduced' form as $z^2 + \text{d}$, with $d = b^2 - c$. Then show that the set of all (b', c') for which $f_{b'c'}$ has this same reduced form, lie on the parabola $c = b^2 - d$ in $\mathbb{R}^2_{bc}$.

2. Write a computer program that will show how the graph of f_{bc} changes, as (b, c) runs round the unit circle (centre $\mathbf{0}$, radius 1) in the control plane.

It is interesting to see how the zeros of f_{bc} change with b and c, and we illustrate this in Fig. 6.1.3, which will need some information to 'read'. First, we look at all points (b, c, x) in $\mathbb{R}^3_{bcx}$ for which $f_{bc}(x) = 0$; these fill out the surface Q shown in Fig. 6.1.3 (which is adapted from the excellent set of diagrams in Broecker and Lander (1978). Now, the quadratic formula tells us that the roots of f_{bc}, when the discriminant d in (6.1.4) is positive, are given by two functions of b and c, namely

$$r = -b + \sqrt{(b^2 - c)}, \qquad s = -b - \sqrt{(b^2 - c)}. \tag{6.1.5}$$

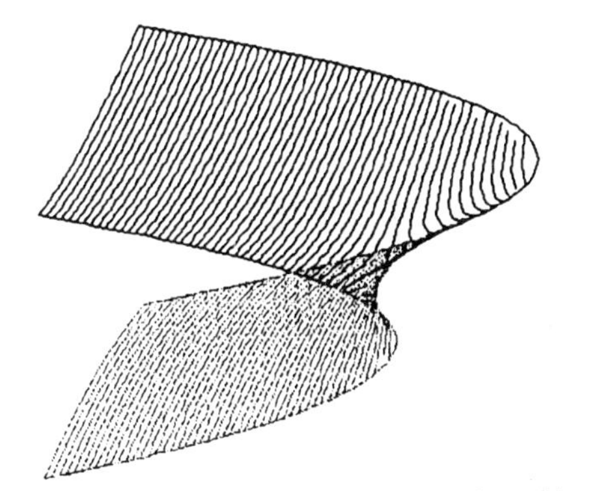

Fig. 6.1.3. The surface Q: the darker portion is that with $x \geq 0$.

If we keep c fixed and plot the graphs of r, s against b in the plane $\mathbb{R}^2_{bx}$, we obtain curves that form the sections of Q in Fig. 6.1.3 by the plane $c =$ constant, in $\mathbb{R}^3_{bcx}$. These curves sweep out Q as c varies. [If no confusion is likely, we often write $\mathbb{R}^3_{bcx}$ etc. instead of $\mathbb{R}^3_{b,c,x}$ etc.]

Readers with access to software which plots surfaces of the form $z = f(x, y)$ can rewrite r in (6.1.5) as $z = -x + \sqrt{(x^2 - y)}$, and similarly for s.

Exercises 6.1.2

1. Write a computer program that will show the curves sweeping out the surface Q as in the text.
2. (For readers familiar with the necessary calculus, or otherwise refer to Chapter D.) Regard r, s in (6.1.5) as functions of b and c, and calculate their partial derivatives. Check that you get the same answers as by using implicit differentiation on the equation $f_{bc}(x) = 0$.

6.2 CUBIC FUNCTIONS

Less familiar than the quadratics are cubic functions of the form

$$y = x^3 + ax^2 + bx + c, \qquad \text{or} \qquad f(x) = (x - u)(x - v)(x - w),$$

where u, v, w are the roots of the equation $y = 0$. We again assume that the coefficients a, b, c are all real, and then we need to find a criterion to tell us when the roots are also real.

Always, at least one of the roots is real, by the intermediate value theorem, as used in exercise 4.4.1(1). Some graphs of standard cubics are shown in Fig. 6.2.1, but the full range is shown in Fig. 6.2.3.

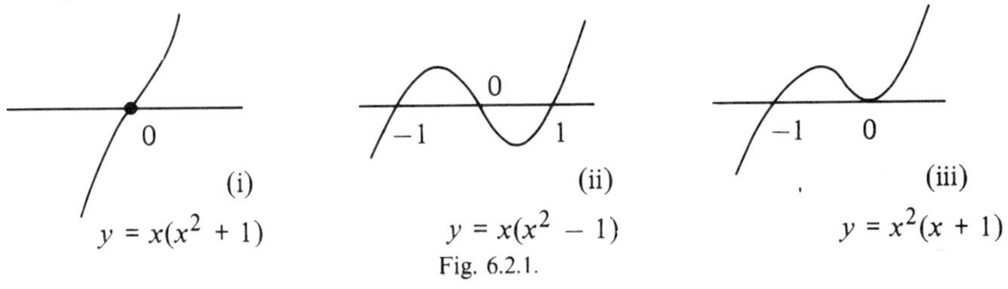

Fig. 6.2.1.

Next, when we multiply out the factorized form of y, and compare coefficients, we find that

$$a = -(u + v + w). \tag{6.2.1}$$

Writing $x = z - a/3$, we find that the coefficient of z in y is zero, and y takes a form that we call **reduced** and write in a standard way:

$$y = z^3 + 3Iz + J. \tag{6.2.2}$$

By (6.2.1), the sum of the zeros is now zero; we have moved the origin to their centroid. Also we now have only two control parameters, I, J, so (as with the quadratics) we have a control plane now denoted by $\mathbb{R}^2_{IJ}$, and we can refer to the function in (6.2.2) as f_{IJ}. The phase space is still $\mathbb{R}^2_{x,y}$. Rewriting (6.2.2) with x now in place of z we can consider examples such as:

(i) $f_{1,0} = x\cdot(x^2 + 3)$; (ii) $f_{1,0} = x\cdot(x^2 - 3)$; (iii) $f_{0,0} = x^3$;

of which (i) has 0 as its only real zero (the other zeros being the complex numbers $\pm\sqrt{(-3)}$). The three zeros of (ii) are all real: viz. $0, \pm\sqrt{3}$; and, since they are all different, we call them **simple roots**, in contrast with (iii) in which we count the number 0 three times over as a zero, to conform to the rule that a cubic equation has three (complex) roots. But $f_{-1,2}$ factorizes as $(x - 1)^2\cdot(x + 2)$, which has zeros 1 (counted twice) and -2 as a simple zero. For brevity we refer to the **multiplicity** of a zero as 1, 2, or 3 according as it is simple, or we count it twice or thrice (and it is then a 'repeated' root).

Exercises 6.2.1

1. Use your calculator to locate the zeros (and their multiplicity) of f_{IJ} when $(I, J) = (-1, 0)$, $(1, 2)$, $(-7, -6)$, $(-5, 2)$, $(3, 2)$.
2. Show that if $(x - u)^2$ is a factor of f_{IJ}, then $(x - u)$ is a factor of $d/dx\, f_{IJ}$, and the graph f_{IJ} touches the x-axis at $x = u$. (Hence, u is real.)
3. It is often convenient to write the general cubic polynomial in the form $f_{abc}(x) = x^3 + 3ax^2 + 3bx + c$. Here, we have a control space, denoted by $\mathbb{R}^3_{abc}$, in which the plane $\{a = 0\}$ is essentially $\mathbb{R}^2_{IJ}$ since the coefficient of x^2 in f_{0bc} is zero.

   Show that then the reduced form (6.2.2) has $I = b - a^2$, $J = 2a^3 - 3ab + c$. Hence show that the set of all points (a, b, c) in $\mathbb{R}^3_{abc}$, for which f_{abc} has this same reduced form, consists of the curve C_{IJ} traced out by $(a, a^2 + I, J + 3aI + a^3)$ as a runs from $-\infty$ to ∞. Sketch C_{IJ}, by erecting it over the curve $b = a^2 + I$. (The latter lies in the plane $\{c = 0\}$ in $\mathbb{R}^3_{abc}$; note that the plane $\{a = 0\}$ cuts C_{IJ} where $b = I, c = J$.)
4. The following problem arises in a study of certain cubic equations. For each fixed s and non-zero x consider the plane curve F_s in the (p, q)-plane, parametrized by y and with $z = y^2$ for brevity, so that $F_s(y) = (p, q)$ where:

   $$p = \frac{(s - z)(s - 3z)}{x^2(1 + z)} \qquad q = \frac{y(s - z)(2 + s - z)}{4x^3(1 + z)}.$$

   Consider first the case $s = -3$. Show that F_{-3} maps each line $L_a\colon x = a > 0$ to a parabola-shaped curve P_a and that these curves P_a fill the whole of the region $D\colon \{(p, q)\,|\,4p^3 - 27q^2 > 0\}$, and they approach the boundary as a increases. Show that on the region $X: \{x > 0\}$ of the (x, y)-plane, F_3 is an injection and maps X onto D. Now change s to -1, and show that F_{-1} maps X onto D but it is not an injection because it maps each line L_a to a curve that crosses itself on the p-axis.

Use computer graphics to find a value s_0 of s between -1 and -3, such that if $s > s_0$ then F_s is an injection on X, but not if $s < s_0$.

Our aim is to have a test for deciding when a given reduced cubic f_{IJ} has three real roots but, as we saw earlier, we must reckon with the possibility of non-simple roots. The test is described in terms of the (cubic) discriminant

$$D = 4I^3 + J^2 \tag{6.2.3}$$

and reads as follows.

Test 6.2.1
The cubic function f_{IJ} has three real roots or one according as the discriminant D is negative or positive. If D is zero, then f_{IJ} also has real roots, one of which is simple and the other of multiplicity two, except at $I = 0 = J$ when it has a single real root of multiplicity three.

Before we prove this, let us see what the test tells us about the control plane, which we now denote by $\mathbb{R}^2_{IJ}$. By (6.2.3), those points (I, J), at which D is zero, all lie on the curve K given by $J = \pm 2(-I)^{3/2}$; here we must have $I < 0$ in order to take the square root before cubing. The sharp point K is a 'cusp', but the curve K itself is often called 'the cusp', rather than the older 'semi-cubical parabola'.

If we now refer to Fig. 6.2.2, we see that $D < 0$ at any (I, J) within the arms of K, for then $J^2 < 4I^3$, as we see with $f_{-1,0}$; and similarly, if (I, J) lies outside those arms, then $D > 0$. Hence $\mathbb{R}^2_{IJ}$ is divided into two regions $\{D < 0\}$, $\{D > 0\}$ which are separated by K; and Fig. 6.2.3 shows $\mathbb{R}^2_{IJ}$ as a portrait gallery in which we have chosen a sample of 'typical' points (I, J) at which to affix a sketch of the xy-graph of f_{IJ} to show how it changes as the control parameters change.

In this, as in similar diagrams throughout the book, there is a technique of construction which readers should use for themselves; if it is understood, we can omit a lengthy explanation as each diagram occurs. The following explanation refers to Fig. 6.2.3 as a case in point, from which the general procedure should become clear with practice.

6.3 CONSTRUCTION OF GALLERIES

Step 1
Find convenient control points for which the corresponding portrait is easy to draw; here, take $J = 0$ and $I = \pm 1/3$, so the functions f_{IJ} are $x(x^2 + 1)$ and $x(x - 1)(x + 1)$, with portraits labelled 1 and 2. The graph G_{IJ} of f_{IJ} cuts $\mathbf{0}x$ in three distinct points when $(I, J) = (-1, 0)$, and continues to do so as it rises when J is increased for a while.

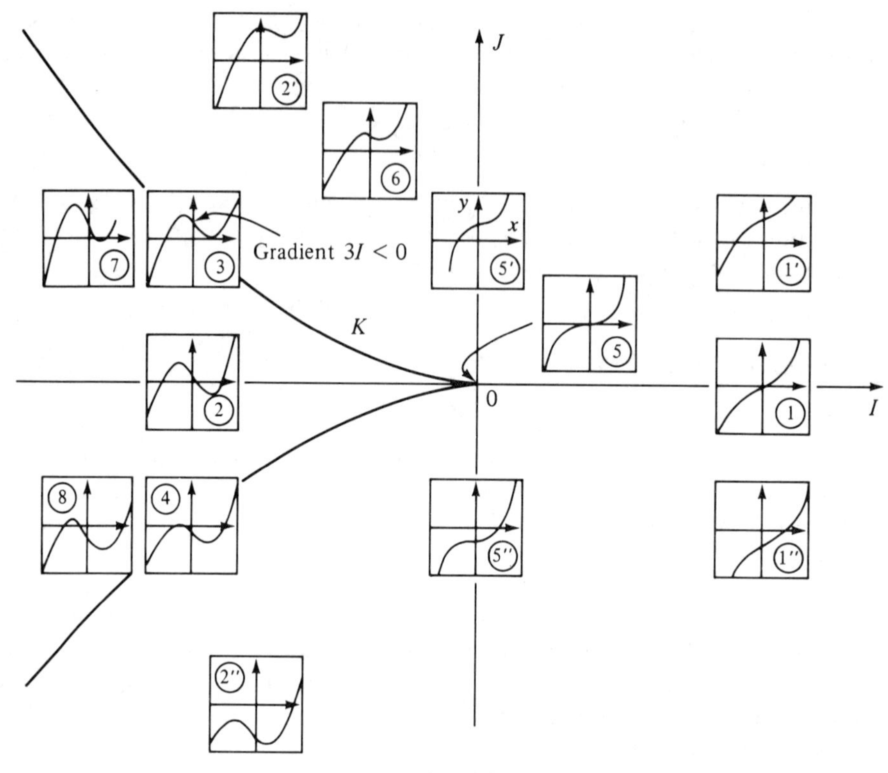

Fig. 6.2.2.

Step 2
Significant change can occur only with the appearance of a contact point, which happens here when the 'trough' becomes tangent to $\mathbf{0}x$ as in portrait 3. As J increases further, the trough clears $\mathbf{0}x$ and thereafter the graph cuts $\mathbf{0}x$ only once. Had we decreased J, then G_{IJ} would move down until the 'crest' touches $\mathbf{0}x$, as in portrait 4.

Step 3
The types of tangency in 3 and 4 are clearly different, yet one must change continuously to the other as we move around the curve K. Thus a transitional stage is needed, and portrait 5 displays it; it is more degenerate than the others, since f_{00} has zero as a 'triple' root.

(*Note*: tangency is hard to demonstrate by computer graphics because it is unstable; small errors in the computer arithmetic will turn the tangent into a chord.) As I increases, portrait 2 changes through 5 to the form 1, from which 1′, 1″ and 5′, 5″ are obtained by increasing or decreasing J. Similarly, portraits 2′ and 2″ are raised and lowered versions of 2, while 3, 4 are intermediate transitions to them from 2. Similarly 6 is a transition from 2′ to 5′, and 7, 8 are obtained from 3 and 4 by decreasing I.

6.3.1 An important surface

As with the quadratics, we can draw a surface to see how the zeros of f_{IJ} change with I and J; the surface is illustrated in Fig. 6.3.1, but we need further information in order to 'read' it. It is not now practicable to express each root as an explicit function of I and J. Instead, we keep I fixed and plot the graph of

$$J = -x(x^2 + 3I) \tag{6.3.1}$$

against x in the plane $\mathbb{R}^2_{xJ}$, to obtain a curve G_I as in Fig. 6.3.1(a–c). Note that in (6.3.1), J is the function $-f_{I,0}$.

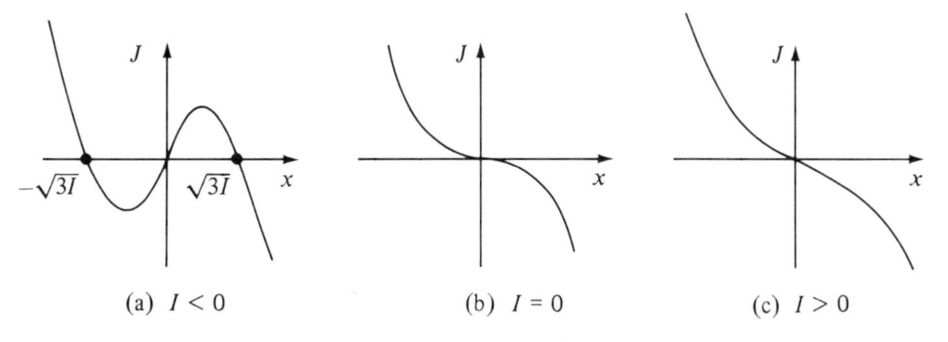

Fig. 6.3.1(a). The graph G_I.

Now imagine G_I as lying in the plane I = constant, in $\mathbb{R}^3_{xIJ}$. Then as I varies, the curve G_I sweeps out a surface C in $\mathbb{R}^3_{x,I,J}$ as shown in Fig. 6.3.1(d). Since each point X on C has coordinates (x, I, J) such that (6.3.1) holds, this means that x is a zero of the function f_{IJ}.

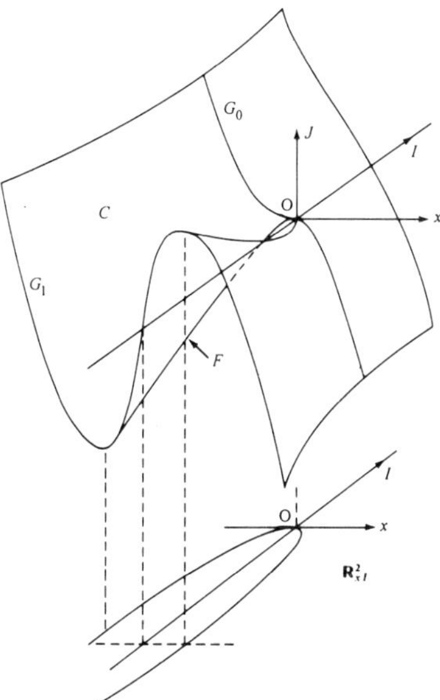

Fig. 6.3.1(d). The graph G_I.

As I varies, the max and min points of G_I sweep out a curve F that runs along the folds in C. The projection of F on the plane $J = 0$ is the parabola shown below C in Fig. 6.3.1(d). Its projection on the control plane $\mathbb{R}^2_{IJ}$ is the cusp curve, which (together with C) forms a structure that has become famous in the literature of catastrophe theory, and we shall return to it in Chapter 11.

Exercises 6.3.1

1. Apply test 6.2.1 to f_{IJ} when (I, J) is: (1, 2), (−1, 2), (−2, 1), (−1, −2), (−3, 11).
2. Use your graphical calculator to decide whether you agree that our sample in Fig. 6.2.2 is 'typical'.
3. Write a computer program that will show the curve G_I sweeping out the surface C as in the text. Show that points on the folds are of the form $(x, -x^2, 2x^3)$, by finding the maxima and minima of J.
4. (Continuation of exercise 6.2.1(3).) Use the following method to show that, just as K separates $\mathbb{R}^2_{IJ}$ in Fig. 6.2.2, there is a surface P that separates $\mathbb{R}^3_{abc}$ into two portions, corresponding to cubics with 1 or 3 real zeros; for each point (a, b, c) on P, f_{abc} has a zero of multiplicity 2 or 3.

Step 1

Factorize f_{abc} in the form $(x - l)^2(x - m)$, multiply out and compare coefficients to show that

$$a = -\tfrac{1}{3}(2l + m), \qquad b = \tfrac{1}{3}l(l + 2m), \qquad c = -l^2m. \tag{6.3.2}$$

These equations give a mapping $F: \mathbb{R}^2_{lm} \to \mathbb{R}^3_{abc}$. Show that F maps distinct points to distinct points. Hence the image P of F, consisting of all points $F(l, m)$ as (l, m) runs through the plane $\mathbb{R}^2_{lm}$, is a surface, sketched in Fig. 6.3.2(a), which looks like a sheet of lined paper folded down the middle, with each line having a cusp at the fold. This leads us to call P a 'page of cusps'.

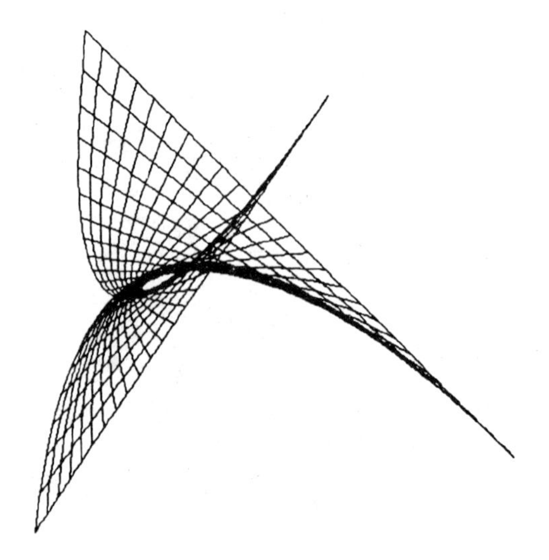

Fig. 6.3.2(a). The page of cusps.

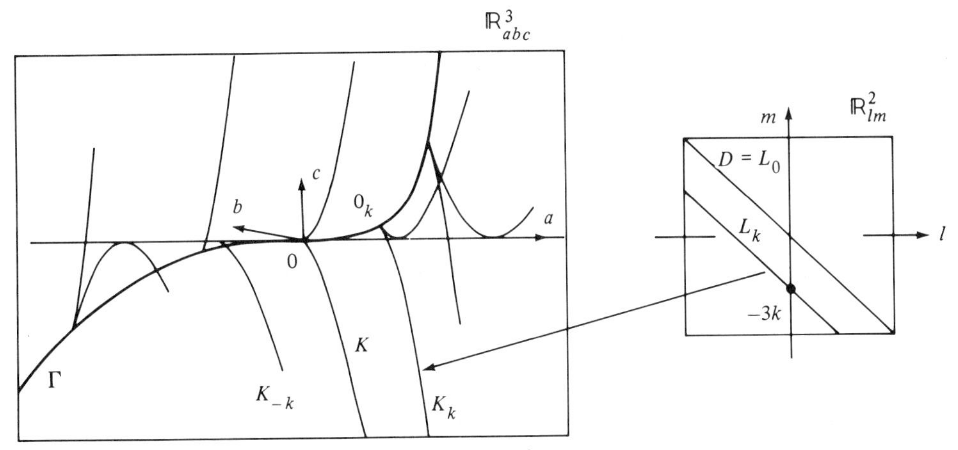

Fig. 6.3.2(b). The mapping F of $\mathbb{R}^2_{lm}$ onto the page of cusps.

Step 2

Show that this surface is P as required by noting first that each point of P is of the form $F(l, m)$ and corresponds to a cubic with root m of multiplicity 2 (if $l \neq m$) or 3 (if $l = m$). Next, show that P cuts the plane A: $\{a = 0\}$ in points of the form $(0, -l^2, 2l^3)$, and these lie on K when we regard A as $\mathbb{R}^2_{IJ}$.

Step 3

Show that each plane X_k: $\{a = k\}$ cuts P in a cuspidal curve K_k like K, by showing that the only points (l, m) mapped by F into X_k are those on the line L_k: $m = -2l - 3k$. In particular $F(-k, -k) = 0_k$ (say) $= (k, k^2, k^3)$ which we take as the origin in X_k. Thus as k runs from $-\infty$ to ∞ points 0_k form a space curve in $\mathbb{R}^3$, called a 'twisted cubic'. We now think of the planes X_k as being strung out along this curve, which pierces X_k at 0_k. Note: in step 2, A is X_0.

Step 4

Each point T of L_k can be written $(-k - t, -k + 2t)$ for some t. Show that $F(T) = 0_k + (0, -t^2, -t^2(3k + 2t)) = M_k(S)$ where $S = (-t^2, 2t^3)$ on K (see step 2) and M_k: $A \to X_k$ is the linear mapping that sends (I, J) to $0_k + (-I, -3kI + J)$. Thus K_k is the image of K by M_k, and corresponds to the 'lines of print' mentioned above.

Step 5

Show that $F(S)$ is also on the curve C_{IJ} in exercise 6.2.1(3), when $(I, J) = S$. Thus the various curves C_{IJ} run along P, roughly parallel to the fold, and crossing each curve K_k just once. By considering curves C_{IJ} that cut X_0 away from K, show that for each (a, b, c) in $\mathbb{R}^3_{abc}$, f_{abc} has three real zeros or one, according as (a, b, c) is enclosed by the fold, or lies outside it.

5. Devise computer sketches to display the various curves in exercise 4 above.
6. (*The eliminant*) Obtain the equation of the page of cusps as follows. First, given a cubic $C(x) = x^3 + 3ax^2 + 3bx + c$, a quadratic $Q(x) = x^2 + 2px + r$, and any number u, let $P(x) = C(x)Q(u) - C(u)Q(x)$. Show that $P(x)$ is of the form $(x - u)(\alpha x^2 + \beta x + \gamma)$ where

$$\alpha = u^2 + 2pu + r$$
$$\beta = 2pu^2 + u(r + 6ap - 3b) + 3ar - c,$$
$$\gamma = ru^2 + u(3ar - c) + 3br - 2pc.$$

Now suppose that u is a common root of C and Q; show that $P(x)$ is identically zero (i.e., zero for every x), so that $\alpha = \beta = \gamma = 0$. Consider these as equations in u^2, u, and 1, and show that

$$\det \begin{pmatrix} 1 & 2p & r \\ 2p & r + 6ap - 3b & 3ar - c \\ r & 3ar - c & 3br - 2pc \end{pmatrix} = 0.$$

When the determinant is expanded, the result is the **eliminant** of C and Q, because x has been eliminated. Now take $Q(x)$ to be $C'(x)$, and thus find the equation of the page of cusps in $\mathbb{R}^3_{abc}$. (See also Milne (1930), p. 470.)

7. Using the same technique, replace $C(x)$ by the quadratic $R(x) = x^2 + 2ax + b$ to obtain the eliminant of R and Q as $4(1 - p)(pb - ar) - (b + r)^2 = 0$.

   (Note: these exercises show that if there is a common root, then the eliminant is zero; the converse is true but harder: see Griffiths (1981b). The theory works with any polynomials, and is referenced under 'Bézout's method' in algebra books.)

8. One means of making a computer sketch of the surface C in Fig. 6.3.1(d) is, for each I, J, to calculate one root of the resulting cubic numerically, using an iterative technique such as interval bisection or Newton–Raphson, and to use this to find the other roots, if they exist. Then a point can be plotted (after projection) for each root (i.e., 1, 2, or 3 points (I, J, x) for each I, J). The following uses a function `root(I, J)` to return one root, found by interval bisection, and a procedure `project(I, J, x, p)` which applies a projection to (I, J, x) to get a screen point (sx, sy), which is plotted if $p = 0$ and draws a line if $p \neq 0$.

```
REM Program 6.3.1, Cubic surface
FUNCTION f(x,I,J) = x*x*x + 3*I*x + J
DECLARE FUNCTION root
DECLARE SUB project
READ Imin , Imax, Istep, Jmin, Jmax, Jstep , eps
DATA -0.8 , 0.6 , 0.1 , -2  , 2   , 0.1   , 0.001
READ xmin, xmax, ymin, ymax, m1   , m2 , floor
DATA -3  , 3   , -2.2, 0.8 , -0.5, 0.3, -8
SET MODE "graphics"
SET WINDOW xmin,xmax, ymin,ymax
FOR I = Imax TO Imin STEP -Istep
   FOR J = Jmin TO Jmax STEP Jstep
      LET x = root(I,J)
      CALL project(I,J,x,0)
      LET k = -12*I - 3*x^2
      IF k >= 0 THEN
         CALL project(I,J,(-x + SQR(k))/2,0)
         CALL project(I,J,(-x - SQR(k))/2,0)
      END IF
   NEXT J
NEXT I
SET COLOR 2
FOR J = Jmin TO Jmax STEP Jstep
    CALL project(Imax,J,root(Imax,J),1)
NEXT J
CALL project(Imax,Jmax,floor,0)
FOR I = Imax TO Imin STEP -Istep
    CALL project(I,Jmax,root(I,Jmax),1)
```

Program 6.3.1 continued

```
NEXT I
CALL project(Imin,Jmax,floor,0)
FOR x = root(Imin,Jmax) TO root(Imin,Jmin) STEP 0.01
    CALL project(Imin,-x*(x^2 + 3*Imin),x,1)
NEXT x
CALL project(Imin,Jmin,floor,0)
FOR I = Imin TO Imax STEP Istep
    CALL project(I,Jmin,root(I,Jmin),1)
NEXT I
PLOT
CALL project(Imin,Jmin,floor,1)
CALL project(Imin,Jmax,floor,1)
CALL project(Imax,Jmax,floor,1)
CALL project(Imax,Jmin,floor,1)
CALL project(Imin,Jmin,floor,1)
FOR t = -SQR(-Imin) TO SQR(-Imin) STEP 0.1
    LET I = -t^2
    LET J = 2*t^3
    CALL project(I,J,floor,1)
NEXT t
PLOT
FOR t = -SQR(-Imin) TO SQR(-Imin) STEP 0.01
    LET I = -t^2
    LET J = 2*t^3
    CALL project(I,J,t,1)
NEXT t
STOP
FUNCTION root(I,J)
      LET xlow  = -10
      LET flow  = f(xlow,I,J)
    LET xhigh =  10
      LET fhigh = f(xhigh,I,J)
      DO
        LET xmid = (xlow + xhigh)/2
        LET fmid = f(xmid,I,J)
       IF fmid*flow > 0 THEN
           LET flow = fmid
           LET xlow = xmid
        ELSE
           LET fhigh = fmid
         LET xhigh = xmid
        END IF
      LOOP UNTIL ABS(fmid) < eps
      LET root = xmid
END FUNCTION

SUB project(I,J,x,p)
    LET sx = -J*0.7 + m1*I*2
    LET sy =  x*0.2 + m2*I*2
    IF p = 0 THEN PLOT sx,sy ELSE PLOT sx,sy;
END SUB

END
```

Proof of test 6.2.1

To prove that the above test 6.2.1 is valid, we first suppose that f_{IJ} has a double root u (i.e., of multiplicity two). Then as seen in exercise 6.2.1(2), df_{IJ}/dx is zero at $x = u$, so

$$f_{IJ} = u(u^2 + 3I) + J = 0 = f'_{IJ} = 3u^2 + 3I.$$

Substituting from $u^2 = -I$ gives $J = -2Iu$, squaring gives $J^2 = -4I^3$, so $D = 0$ at (I, J) which therefore lies on the curve K in $\mathbb{R}^2_{IJ}$.

Next, every point X on K corresponds to a cubic with a double zero. For, since X has a negative I-coordinate, we can write this as $-t^2$ for some real number t. So by (6.2.3), the J-coordinate of X is $2t^3$ and X corresponds to the cubic $x^3 - 3t^2 + t^3$, which factorizes as $(x - t)^2(x + 2t)$, and we have a double zero t, together with a simple zero, $-2t$, as claimed—unless $t = 0$ which gives the special case of multiplicity three.

Now suppose (I, J) lies off the curve K, and for brevity write g for f_{IJ}. We shall show that g cannot have three real roots unless $I < 0$. For, if g has even two real zeros, then the graph must have a horizontal tangent at $x = c$ (say) between them (by Rolle's theorem); so $dg/dx = 0$ at $x = c$, whence dg/dx has a real zero. Since $dg/dx = 3\cdot(x^2 + I)$, this cannot be zero unless $I < 0$. Therefore g has no chance of three real zeros unless $I < 0$; and then dg/dx has two real zeros at $x = \pm c$ where c now is $\sqrt{(-I)}$.

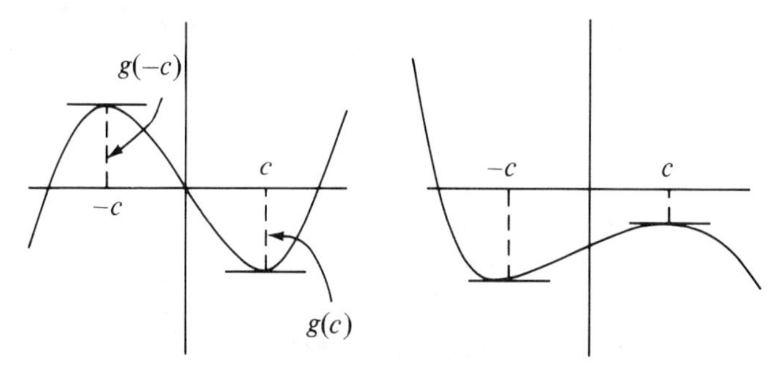

Fig. 6.3.3.

A glance at Fig. 6.3.3 now shows that if g has three real zeros then $g(c)$ and $g(-c)$ are of opposite signs, whereas with only one real zero, these are of the same sign. These correspond, respectively, to the cases:

$$\text{(i) } p(c) < 0, \qquad \text{(ii) } p(c) > 0, \quad \text{where } p(c) = g(c)\cdot g(-c).$$

Since $g(c) = c^3 + 3Ic + J$ and $c^2 = -I$, we leave it to readers to verify that $p(c) = 4I^3 + J^2$. The test now follows from (i) and (ii).

6.4 FORMULAE FOR THE ROOTS OF CUBIC EQUATIONS

We now give the famous formula for the roots of the cubic equation $f_{IJ}(z) = 0$ in

(6.2.2). In exercise 6.4.1(1) below, we lead the reader through the steps of the proof that the zeros of f_{IJ} are:

$$2^{-1/3}((-J + \sqrt{D})^{1/3} + (-J - \sqrt{(D)})^{1/3}). \tag{6.4.1}$$

This formula was first published in Italy by Cardan in 1545, but Tartaglia and Scipio Ferreo also have claims to be the discoverers; see Boyer and Merzbach (1989) for details. The importance of (6.4.1) has led the physicist Richard Feynman (1988) to remark that, with the solution of cubics, for the first time people knew something significant in mathematics, which was unknown to the classical Greeks.

As we know, at least one root of f_{IJ} is real. Thus (6.4.1) *forces* us to calculate this real root by using imaginary numbers—we cannot discard complex numbers as meaningless, in the way mathematicians had done with quadratic equations. The discovery of the formula made it clear that they would have to come to grips with complex numbers, an effort that took some 300 years to set minds at ease.

In fact (6.4.1) is not very practical for specific calculations (although it won Tartaglia a lot of money in public contests). Its theoretical importance first led to the later formula of Ferrari for quartic equations, and then to the search for similar types of formulae for solving equations of nth degree. That search led Japanese mathematicians, for example, to find ever more complicated solutions for specific equations of very high degree, but in the West the search was shown to be fruitless, by the work of Abel and Galois around 1830. This led to questions of what an algorithm ought to mean, and eventually to Turing's notion of computability and the digital computer in the late 1930s, followed by the invention of the necessary hardware in the 1940s.

Exercises 6.4.1

1. Prove (6.4.1) as follows. Write $z = u + v$, so that

   $$z^3 = u^3 + v^3 + 3uv(u + v) = u^3 + v^3 + 3uvz.$$

   Then $f_{IJ}(z) = u^3 + v^3 + 3(uv + I)z + J$. Hence f_{IJ} will be zero if we choose u and v to satisfy the equations

   $$uv = -I, \qquad u^3 + v^3 = -J$$

   so u^3 and v^3 are the roots of the quadratic equation

   $$t^2 + Jt - I^3 = 0.$$

   Solve this equation to obtain (6.4.1).
2. Use (6.4.1) to solve $f_{IJ} = 0$ when $(I, J) = (-5, -126), (11/3, 6), (-27, 54)$.
3. Show that if u, v is one choice of roots in exercise 1 above, then the other roots of the cubic are $\omega u + \omega^2 v$, $\omega^2 u + \omega v$, where $1, \omega, \omega^2$ are the three roots of unity (i.e., cube roots of 1 in the Argand diagram.)
4. Show that the discriminant D in (6.2.3), in terms of a, b, c prior to (6.2.1), is $a^2b^2 - 4d^3 - 4a^3c + 18abc - 27c^2$.
5. Show that $(2 + \sqrt{-121})^{1/3} = 2 + \mathrm{j}$.

Just as the solution of a cubic has been made to depend on that of a quadratic, it is possible to solve quartic equations by solving a cubic. For details see Hall and

Knight (1942). Meanwhile we lead the reader through a sequence of steps, analogous to those of exercise 6.3.1(4), to find a surface that corresponds to quartics with a repeated root.

Exercises 6.4.2

1. We consider only reduced quartics of the form

$$f_{abc}(x) = x^4 + 6ax^2 + 4bx + c \tag{6.4.2}$$

(with no terms in x^3), so that the appropriate control space is again $\mathbb{R}^3_{abc}$. If f_{abc} has a real repeated root l, then we have

$$f_{abc}(x) = (x - l)^2(x^2 + px + m).$$

Step 1
Compare coefficients to show that $p = 2l$ and hence

$$a = \tfrac{1}{6}(m - 3l^2), \qquad b = \tfrac{1}{2}l(l^2 - m), \qquad c = l^2 m. \tag{6.4.3}$$

Step 2
These equations give a mapping $s\colon \mathbb{R}^2_{lm} \to \mathbb{R}^3_{abc}$, with $(a, b, c) = s(l, m)$. Show that s is injective. Then s maps the (l, m) plane into $\mathbb{R}^3_{abc}$, to form a surface S, known as the **swallowtail** and illustrated in Fig. 6.4.1(a) (with a 'whisker', shown dotted, which is not part of S but which is needed in the description). For clarity, the sketch is not to scale.

For later reference, certain parts of S are labelled, and the names are chosen because (with this distortion of scale) the swallowtail resembles more a pair of hydrofoils attached to the hull of a galleon. In Fig. 6.4.2 we display cross-sections of S when it is cut by a plane $X_k\colon \{a = k\}$. Before justifying these sketches, the reader will probably find it helpful to have a little more description of S.

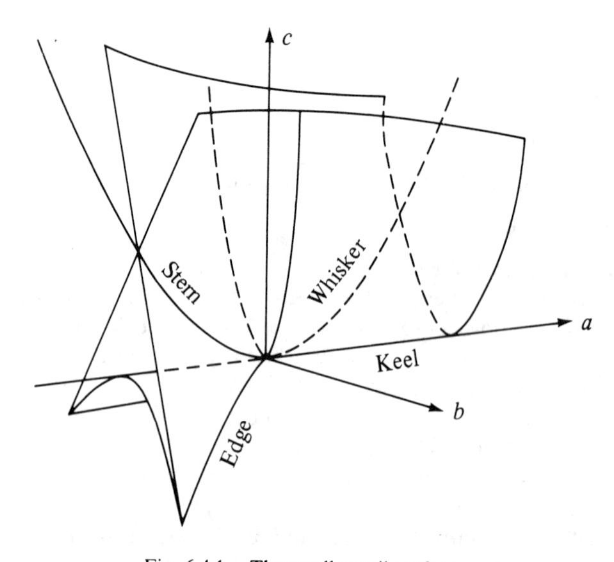

Fig. 6.4.1. The swallowtail surface.

By construction of s, each point $P = (a, b, c)$ on S is such that f_{abc} ($= f_P$ say) has a real root of multiplicity ≥ 2; but if P lies on the 'whisker' in Fig. 6.4.1, then f_P has two distinct non-real roots, each of multiplicity 2. It will turn out that if P lies above the keel, then f_P has no real roots, while if P lies within the conical 'hydrofoils' then f_P has four real roots, and if P lies below the keel, then f_P has just two real roots.

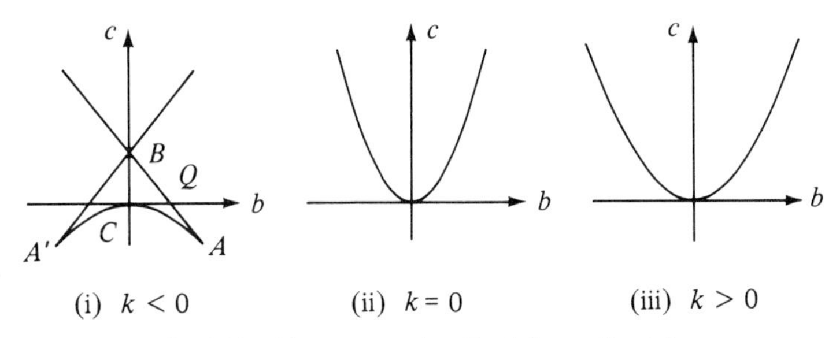

Fig. 6.4.2. The cross-section C_k of the swallowtail.

Step 3

Recall from step 2 the mapping s. Where does s map the axes in $\mathbb{R}^2_{lm}$? Show that:

(a) s sends (l, m) into the section of S given by the plane X_k: $\{a = k\}$, if and only if (l, m) lies on the parabola H_k: $m = 3(l^2 + 2k)$ *in* $\mathbb{R}^2_{lm}$;
(b) the image of H_k is the curve C_k in X_k that has parametric equations $b = -l(l^2 + 3k)$, $c = 3l^2(l^2 + 2k)$;
(c) C_k passes through $\mathbf{0}$ only when $l = 0$;
(d) C_k cuts the c-axis of X_k only when $l = 0$ (if $k = 0$), and also when $l = \pm\sqrt{(-k)}$ if $k < 0$;
(e) the point B in Fig. 6.4.2(i) sweeps out half of the parabola $c = 9a^2$ in the (a, c)-plane (this is the stern in Fig. 6.4.1);
(f) The derivatives db/dl, dc/dl are simultaneously zero only when $l^2 = -k$ (so $k \leq 0$), at points A, A' on C_k, where $A = (k, -2k\sqrt{(-k)}, -3k^2)$.

We have shown A, A' in Fig. 6.4.2, and the theory of Chapter 5 shows that C_k has cusps there. Granted this, A and A' sweep out the edge in Fig. 6.4.1. Show that it projects into the (a, c)-plane as the parabola $c = -3a^2$, and into the (a, b)-plane as the cusp curve $b^2 = -4a^3$.

Step 4

As to the 'whisker' in Fig. 6.4.1, show:

(a) If f_{abc} in (6.4.2) has a repeated complex root z, then the conjugate $\bar{z}$ is also a repeated root;
(b) f_{abc} is of the form $(x^2 + p^2)^2$ with p real, so $(a, b, c) = (0, p^2/2, p^4)$;
(c) the whisker constitutes the missing half of the parabola $c = 9a^2$ in the (a, c)-plane, complementary to the stern.

2. Given the above information, construct Figs 6.4.1 and 6.4.2 by computer. (If a scale is used as in Klein (1939) p. 99, then the 'hydrofoils' look much more like a swallow's tail.)
3. Establish the statements above, concerning the number of real roots of f_{abc}, at least when (b, c) is on one or other of the axes in Fig. 6.4.2. For example, on the c-axis, f_{abc} is a quadratic in x^2, and if its roots are positive, then f_{abc} has four real roots x; on the b-axis,

f_{abc} is $x(x^2 + 6ax + 4b)$, and we can use our knowledge of cubics. This gives a quick memory check to decide which region corresponds to which number of real roots; but more technique is needed to prove the result, away from the axes.

4. Use the method of section 6.3.1 to construct a gallery of portraits of reduced quartics, analogous to that in Fig. 6.2.2, with K there replaced by the cusped curve in Fig. 6.4.2. (Use exercise 3 to get started.)

6.5 DISCRIMINANTS

In (6.1.3) above we saw that the general quadratic function $f_{bc}(x) = x^2 + 2bx + c$ has equal roots iff (= 'if, and only if'), in the control plane $\mathbb{R}^2_{bc}$, the point (b, c) lies on the parabola $P: c = b^2$. (Thus P is the set of points at which the discriminant $d = b^2 - c$ of the polynomial f_{bc} is zero.) We now explain why the curve P can be regarded as an envelope.

If we keep x fixed and let (b, c) vary in $\mathbb{R}^2_{bc}$, then the equation $f_{bc}(x) = 0$ is the equation in $\mathbb{R}^2_{bc}$ of a line L_x, which depends on the parameter x. Thus x is now the control parameter, and the remaining variables b, c constitute the phase point (b, c). This terminology helps to avoid confusion in such shifts of viewpoint. Now, if z is a double root of the equation $f_{bc}(x) = 0$, then the graph of $y = f_{bc}(x)$ touches the x-axis at $(z, 0)$, so $f'_{bc}(z) = 0$.

Therefore, f_{bc} has a double root at z iff the two equations: $f_{bc}(x) = 0 = f'_{bc}(x)$ hold at z. But, for the lines L_x, these are just the equations for finding their envelope! Eliminating x gives $0 = (-b)^2 + 2b(-b) + c$, which is the equation of P.

The cusp

Similarly we can consider the reduced cubic functions $f_{IJ}(x) = x^3 + 3Ix + J$ discussed in section 6.2. Here the control plane is $\mathbb{R}^2_{IJ}$ and the discriminant is

$$D = 4I^3 + J^2. \tag{6.5.1}$$

We saw in (6.2.3) that the equation $f_{IJ}(x) = 0$ has a double root iff, in the control plane $\mathbb{R}^2_{IJ}$, the point (I, J) lies on the cusp $D = 0$. But just as for the quadratics, such a double root z must satisfy both equations $f_{IJ}(x) = 0 = f'_{IJ}(x)$. As before, these are the equations for an envelope, this time of the family of lines L_x: $J + 3xI = -x^3$ in $\mathbb{R}^2_{IJ}$, and eliminating x gives $4I^3 + J^2 = 0$ as the equation of the curve enveloped, which is just $D = 0$.

The swallowtail

Now let us go on to discuss quartic functions in the same way. We consider reduced quartics

$$f_{abc}(x) = x^4 + 6ax^2 + 4bx + c \tag{6.5.2}$$

which lack a term in x^3. In the earlier cases a double root was necesarily real, but it is possible (but relatively rare) for f_{abc} to have a double complex root z, in which case $\bar{z}$ is another. Therefore we confine ourselves for the present to considering double roots which are real. As before, the equation $f_{abc}(x)$ has a real double root t iff t satisfies the equations $f_{abc}(x) = 0 = f'_{abc}(x)$. But we think of x as the control parameter and (a, b, c) as the phase point, then $f_{abc}(x) = 0$ is a linear equation, this time of a plane in the three-dimensional 'phase' space $\mathbb{R}^3_{abc}$. It will therefore simplify matters if we fix a for the moment, and think of the equation as that of a line L_x: $4bx + c + (x^4 + 6ax^2) = 0$ in $\mathbb{R}^2_{bc}$. Thus the two equations (5.6.3) for t give the envelope E_a of the family of lines L_x as:

$$\begin{aligned} b &= -(x^3 + 3ax), \\ c &= -(4bx + x^4 + 6ax^2) = 3x^2(x^2 + 2a). \end{aligned} \tag{6.5.3}$$

Hence E_a is a curve given parametrically by these equations, as x runs from $-\infty$ to $+\infty$; so E_a is the curve C_k (with $a = k$) described in exercise 6.4.2(1), step 3. We shall show how to analyse these curves after the following exercises.

Exercises 6.5.1

1. For each $t = (a, b, c) \in \mathbb{R}^3$, let f_t denote the quartic polynomial $x^4 + ax^2 + bc + c$. The *swallowtail surface* consists of those t for which f_t has a repeated root; let s denote the section of this surface by the plane $a = -1$. Thus, take $t = (-1, b, c)$ throughout.
   (a) Show that if f_t has minima at x_1, x_2 and $f_t(x_1) = f_t(x_2) = v$, then $f_t - v$ has each of x_1, x_2 as a repeated root, so $x_1 + x_2 = 0$, $x_1^2 = 1/2$, and $b = 0$.
   (b) For any t with $b = 0$, regard f_t as a quadratic in x^2 and hence deduce the numbers of its real roots and critical points (as a quartic) when t moves down the c-axis.
   (c) Show that the point Q in Fig. 6.4.1 is at $(-1, 2/(3\sqrt{3}), 0)$, and that if l is the vertical line through Q, then for any $t \in l$, the quartic f_t has three real critical points.
   (d) Given that the quartic g corresponding to Q has no critical point less than its double root, sketch the graph of g, and then a series of sketches to show how the qualitative features of the graphs change, as t moves to $\pm\infty$ on l away from Q. [SU]

2. (a) Show that the cubic function $f_{a,b}$: $\mathbb{R} \to \mathbb{R}$, where

   $$f_{a,b}(x) = x^3 + ax + b$$

   has a non-simple root if, and only if, (a, b) lies on the curve

   $$\Gamma\colon 27b^2 + 4a^3 = 0$$

   in the control plane C.
   (b) Suppose that f_{a_0,b_0} has only one real root, a simple one at x_0. Use the implicit function theorem (8.1.1) to explain why (a_0, b_0) has a neighbourhood U in the plane C for which, if $(a, b) \in U$, then $f_{a,b}$ has exactly one real zero $\theta(a, b)$, where $\theta(a_0, b_0) = x_0$, and $\theta : U \to \mathbb{R}$ is smooth. Explain briefly why this information allows us to assert that (a_0, b_0) must lie to the right of Γ in the plane C. [SU]

A general approach to analysing curves like E_a in (6.5.3) depends on knowing how they rise and fall in the (b, c)-plane. Thus we calculate dc/db, which by the chain rule is c'/b' when $b' \neq 0$. Here, $b' = -3(x^2 + a)$, and $b' = 0$ only when $x^2 + a = 0$. Therefore b' has no zeros if $a > 0$; if $a = 0$ then $x = 0$ is a double zero of b', and if $a < 0$, then b' has the zeros

$$x = \pm v, \qquad v = \sqrt{(-a)}. \tag{6.5.4}$$

Now refer to Fig. 6.4.2 above, wherein C_k is E_a (with $a = k$).

Also $c' = 12x(x^2 + a)$, so except at these zeros of b', we have $dc/db = -4x$. Hence the slope of the curve E_a has sign opposite to that of x. Now, when $|x|$ is large, then $b \cong -x^3$ and $c \cong 3x^4$, so as x runs from $-\infty$ to $+\infty$, E_a starts (with $-x$ large) in the top right and descends as shown, eventually climbing to the top left. If $a = 0$, then the last approximations are equalities, and we can eliminate x to get $c = 3b^{4/3}$. So E_0 is just the graph of this, as in Fig. 6.4.2(ii). If $a > 0$, then $c \geq 0$, so E_a descends, as shown in Fig. 6.4.2(iii), to touch the b-axis at the origin $\mathbf{0}$, then rises steadily as x increases from zero. Finally if $a < 0$, we have the more complicated behaviour shown in Fig. 6.4.2(i). Here, E_a is symmetrical about the c-axis which it crosses when $x = 0$ or $\pm u$, where $u = \sqrt{(-2a)}$ and $u > v$ in (6.5.4). Hence, E_a descends to make a first crossing at B when $x = -u$, then continues to the cusp A shown, approaching tangentially with gradient $-v > 0$. At A, the signs of b' and c' change, so E_a now rises from A to cross the c-axis at C where $x = 0$. By the symmetry, E_a continues to the cusp $A'(x = v)$ then climbs to B and on upwards as shown.

For any individual value of a, of course, the shape of E_a can be found accurately by computer. The main point of our procedure above has been to show that although there is an infinite set of curves E_a, there are only two essentially different types (Figs 6.4.2(i),(iii)), separated by E_0 (which is flatter at the origin). If a is not zero, we can change it slightly without altering the essential shape of E_a, but this is not true for $a = 0$.

Another way of seeing the changes in E_a is to string the curves E_a along the a-axis in three-dimensional space $\mathbb{R}^3_{abc}$ as in Fig. 6.4.1, to form the swallowtail surface there described. In Chapter 11, we shall find its equation in (a, b, c) coordinates, of the form $D = 0$, where again D is the discriminant of f_{abc}.

The butterfly

In principle, the same technique will work for polynomials of any degree, but it becomes less satisfactory because of the dimension of the control space. The strengths and weaknesses are shown when we try the obvious next case, which is to decide when the reduced quintic

$$f_u(x) = x^5 + 10px^3 + 10qx^2 + 5rx + s, \qquad u = (p, q, r, s) \tag{6.5.5}$$

has a double real root t. (The form of the coefficients is chosen to simplify some later arithmetic.) Because of a particular application later, it is appropriate for us to choose to keep p, q fixed, and then as before we find the envelope of a family of lines parametrized by x, in this case, of lines

$$L_x\colon s + 5rx = -(x^5 + 10px^3 + 10qx^2)$$

in (r, s)-space $\mathbb{R}^2_{rs}$. Applying the usual technique we find that the envelope in $\mathbb{R}^2_{rs}$ is the curve B_{pq} given parametrically by the equations

$$r = -(x^4 + 6px^2 + 4qx), \qquad s = 4x^5 + 20px^3 + 10qx^2, \tag{6.5.6}$$

with x running from $-\infty$ to ∞, and p, q kept fixed.

As before, for sketching the curve, we look at ds/dr which is

$$\frac{20x^4 + 60px^2 + 20qx}{-(4x^3 + 12px + 4q)} = -5x,$$

provided the denominator is non-zero, that is, provided

$$x^3 + 3px + q \neq 0. \tag{6.5.7}$$

Thus B_{pq} is smooth except at points corresponding to real roots of this cubic, which has three real roots or one according as its discriminant D is negative or positive, respectively. By (6.5.1), $D = 4p^3 + q^2$, so if (p, q) lies outside the cusp $D = 0$ in the control plane $\mathbb{R}^2_{pq}$, then B_{pq} has just one point at which it, too, has a cusp; and this occurs when the parameter x is the single real root x^* of (6.5.7). The general features of B_{pq} are shown in Fig. 6.5.1. Before we give further explanations, the reader may find the following program helpful in exploring B_{pq} for various values of p, q.

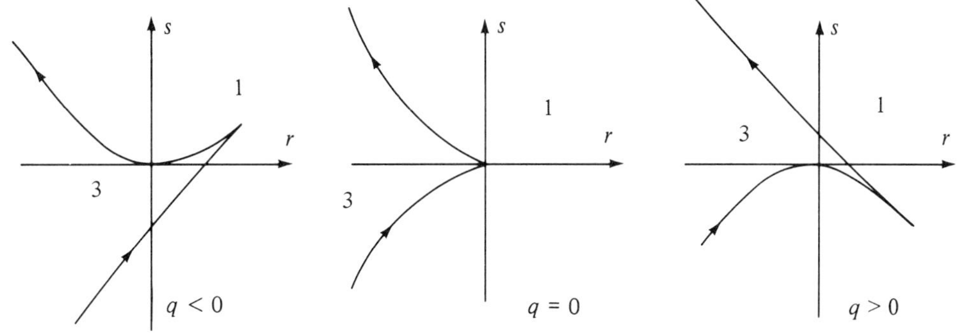

Fig. 6.5.1. B_{pq} when $p \geq 0$.

```
REM Program 6.5.1, Butterfly Bpq
READ xmin, xmax, xstep
DATA -1  , 1  , 0.02
FUNCTION r(x) = -(x^4 + 6*p*x^2 + 4*q*x)
FUNCTION s(x) = 4*x^5 + 20*p*x^3 + 10*q*x^2
SET MODE "graphics"
SET WINDOW -6,6 , -4,4
DO
  INPUT PROMPT " p, q : " : p,q
  IF p = 99 AND q = 99 THEN EXIT DO
  CLEAR
  FOR x = xmin TO xmax STEP xstep
    PLOT r(x),s(x);
  NEXT x
  PLOT
LOOP
END
```

In Fig. 6.5.1 the numerals give the number of real roots possessed by the quintic f_u in (6.5.5) when (r, s) is in the appropriate region of the appropriate figure. These can be checked when (r, s) lies on the axes, and then we use the result of exercise 6.7.1(5). We leave it as an exercise for the reader to check that the above figures do describe the state of affairs.

Next, we must look at what happens when $p < 0$. First we keep q at 0, to see that the cubic (6.5.6) has roots $x = 0$, and $\pm l$ where $l = \sqrt{(-3p)}$; also $s = 0$ when $x = \pm m$

and $m = \sqrt{(-5p)}$, and $r = 0$ when $x = \pm n$ where $n = \sqrt{(-6p)}$. Thus, as x runs from $-\infty$ to ∞, it passes through the points $-n, \ldots, n$ as shown:

$$\xrightarrow[\quad -n \quad -m \quad -1 \quad 0 \quad 1 \quad m \quad n \quad]{} x$$

and then $B_{p,0}$ is the 'butterfly' curve sketched in Fig. 6.5.2 with the corresponding points $N, M, \ldots, N'$ (and $M = M'$).

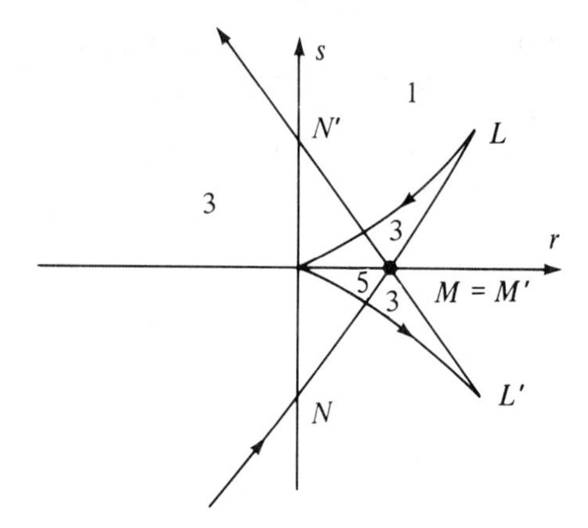

Fig. 6.5.2. The butterfly curve $B_{p,0}$ $(p < 0)$.

The diamond-shaped region from **0** to M corresponds to quintics f with five real roots because, on the r-axis, f is the quartic $x^4 + 10px^2 + 5r$ multiplied by x, and this has real positive roots x^2 if $p^2 \geq r/5$ (i.e., if $0 < r < M$). In this way we can check that the numbers of real roots are as indicated.

Finally, the shapes when $p < 0$ and $|q|$ is non-zero and 'large' are shown in Fig. 6.5.3. When $|q|$ is small, B_{pq} is just a slight perturbation of Fig. 6.5.1; note that by continuity there must then be a value q_0 for which the arm marked A must pass through the upper cusp. The cusped 'triangle' eventually decreases and disappears as $|q|$ increases.

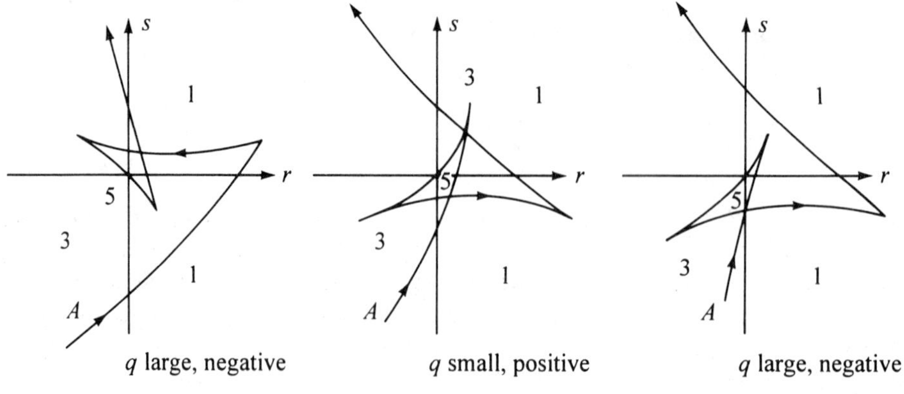

Fig. 6.5.3. B_{pq} when $p < 0$.

Exercises 6.5.2

1. Fix a value of $p < 0$, and use computer graphics to locate the value of q_0 mentioned in the text, for which A passes through the upper cusp. (Make pictures with $q < q_0$ and $q > q_0$.)
2. The arm A meets the r-axis in a zero z of f in (6.5.6). Show that $\partial z/\partial q = -z/(4pz + pq)$ so z increases as q decreases from zero.

Again we leave the reader the exercise of filling in the detailed arguments, and to use computer graphics to assemble the curves B_{pq} into surfaces like that shown in Fig. 6.5.4 (first drawn by E. C. Zeeman). Another associated surface will be considered in Chapter 11. As we have seen, families of curves lead in a natural way to surfaces in three dimensions. Techniques for working with three variables are developed in Chapter D.

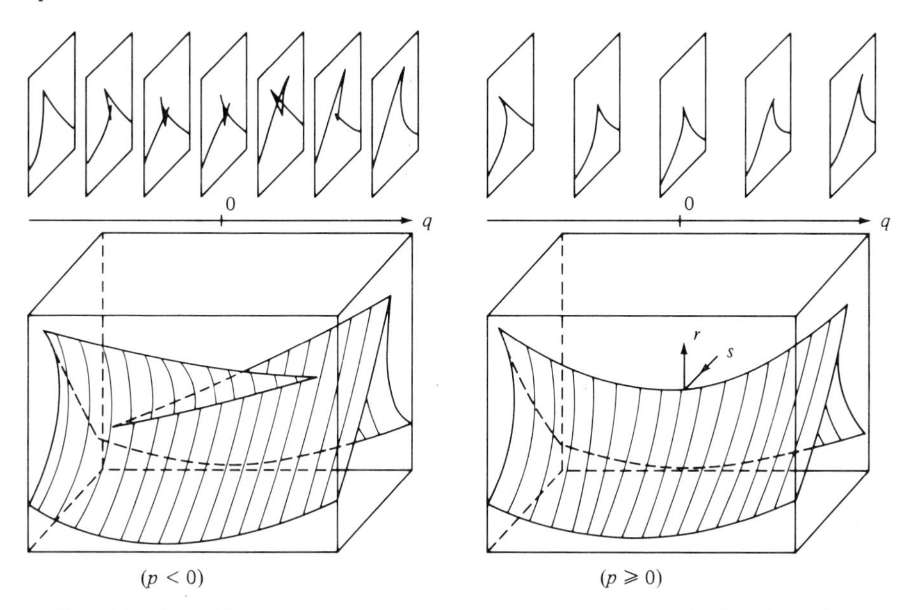

Fig. 6.5.4. Assembling the curves B_{pq} into a surface. The value of p is fixed, and q varies.

6.6 POLYNOMIAL FUNCTIONS

Now we shall consider the general polynomial function

$$f(x) = a_n x^n + a_{n-1} x^{n-1} + \ldots + a_1 x + a_0, \qquad (a_n \neq 0), \tag{6.6.1}$$

where we allow each of the coefficients a_r to be a complex number; but if all the a's are real then f is a 'real polynomial'. For brevity, $f(x)$ is often written as $\sum_{r=0}^{n} a_r x^r$. Since x^n is the highest power of x that occurs in f, we say that f has **degree** n; if the degree is 0, 1, 2, or 3, then accordingly f is constant (a_0), linear, quadratic, or cubic. One of the simplest examples is the polynomial $p_n(x) = x^n - u^n$, which is zero when $x = u$. Now, $(x - u)$ is a factor of $p_n(x)$, for $p_n(x) = (x - u)g_n(x)$ where $g_n(x) = \sum_{r=1}^{n} x^{n-r}u^{r-1}$, as can be seen by multiplying up and collecting terms. Readers have probably used this process on polynomials of low degree by finding a zero (say u) and then saying that $x - u$ is a factor. In fact, this is the correct procedure for a general polynomial.

Theorem 6.6.1
Suppose that in (6.6.1), $n > 1$ and the complex number u is a zero of f. Then there is a polynomial $h(x)$, of degree $n - 1$, such that $f(x) = (x - u)h(x)$.

Proof
Since $f(u) = 0$, then $f(x) = f(x) - f(u) = \sum_{r=0}^{n} a_r p_r(x)$ with $p_r(x)$ as above. There we saw that $x - u$ is a factor of each term, so it is a factor of f; and the remaining factors $(a_r g_r(x))$ add to make the required polynomial $h(x)$.

This result is a form of the **remainder theorem**. Now it may happen that $x - u$ divides $h(x)$, and so by factorizing as long as we can, we may write

$$f(x) = (x - u)^m g(x) \tag{6.6.2}$$

with m as large as possible and $g(u) \neq 0$. We then call m the **multiplicity** of the root u. For example, to say that f has 'coincident' (or 'repeated') roots means the same as saying: 'f has a root of multiplicity > 1'. A root of multiplicity exactly 1 is called a simple root. By the fundamental theorem of algebra (exercise E.9.2(4), every polynomial of degree >0, with coefficients which are complex numbers, has a (possibly complex) root. The following exercise should now allow the reader to complete a proof of the next corollary.

Corollary 6.6.1
If f has degree $n > 0$, then f has exactly n roots, if we count each with its multiplicity.

Exercises 6.6.1

1. Use computer graphics to sketch the graphs of various polynomials $f(x)$ of degree 7, say. In each case chose a polynomial with the full number of real roots, and then investigate how these change as you alter a_0, and then a_1, and so on. (In order to estimate a suitable range for $f(x)$, observe that if $|x| \leq X$, then f will be largest when all the a's are positive. Hence prove that $|f(x)| \leq \Sigma\, |a_r| X^r \leq (n + 1)Y$, where $Y = X^n$ if $X > 1$, and $Y = 1$ otherwise.)
2. Show that if u is a simple root of $f(x)$ in (6.6.1), then the derivative $\partial f/\partial x \neq 0$ when $x = u$. (Use (6.6.2) to see what happens to $\partial f/\partial x$ when u is not simple.) Show that when u is real and not simple, the graph of f touches the x-axis at $x = u$.
3. Prove corollary 6.6.1 as follows. First show that f has at most n complex zeros. (Use the method of induction: clearly if $\deg(f) = 1$ then f has exactly one zero. Assume the result holds for any polynomial of degree $< n$, and suppose $1 < \deg(f) = n$, with u a zero of f. By theorem 6.6.1, $f(x) = (x - u)h(x)$. Show that if $v(\neq u)$ is another zero of f, then $g(v) = 0$, so v is one of the k zeros of g. By the inductive assumption, $k = \deg(g) < n$. Now show that the zeros of f consist of u plus those of g—at most n in all.) Note that this proof does not use the fundamental theorem of algebra.
4. Now refine the proof to take into account the multiplicity of each root. (By the fundamental theorem, f has a root u. Now start with $f(x) = (x - u)^m g(x)$.)

5. Reduce the solution of the quartic equation $x^4 + 2px^3 + qx^2 + 2rx + s = 0$ to that of a cubic as follows. The idea is to add $(ax + b)^2$ to each side and choose a, b to get an equality between two perfect squares: $(x^2 + px + k)^2 = (ax + b)^2$. By comparing coefficients, show that a, b, k must satisfy

$$a^2 = p^2 + 2k - q, \qquad ab = pk - r, \qquad b^2 = k^2 - s,$$

and hence that the cubic $r(k)$ is zero, where

$$r(k) = 2k^3 - qk^2 + 2(pr - s)k - p^2s + qs - r^2.$$

If now we take a root k of this cubic, we can find a, b; and then x satisfies either of the quadratics

$$x^2 + px + k = \pm(ax + b).$$

Try the quadratic with p, q, r, s equal to $-1, -5, 5, -3$, to get a value -1 for k, with $a^2 = 4 = b^2 = -ab$. Hence find the roots $x = (3 \pm \sqrt{5})/2$ and $(-1 \pm \sqrt{13})/2$.
6. [Hard] Show that the cubic $r(k)$ in exercise 5 above has a repeated root if, and only if, the original quartic has one.
7. An alternative method is due to Descartes: first change x to $y + c$, and find c so that the coefficient of y^3 is zero. By equating coefficients, now find k, l, m so that the left side factorizes as $(y^2 + ky + l)(y^2 - ky + m)$. Hence obtain a cubic equation to find k, then l and m, and hence all four roots y of the reduced quartic. Try this on a numerical example.

6.7 VARIATION OF ROOTS WITH COEFFICIENTS

As we have seen, the formulae for the roots of equations of degree four, are complicated; and when $n > 4$ analogues cannot exist in general. Now, by sketching graphs of polynomials (as in exercise 6.6.1(1)), we guess that the roots of $f(x)$ in (6.6.1) will vary continuously as the coefficients a_j vary. This guess turns out to be correct, at least when the roots are simple. In that case, each simple root is a function ϕ of the a_j's, and then it is surprisingly easy to use calculus for getting information about the derivatives of ϕ. The following explanation is for readers who already know the necessary calculus given in Chapter D.

For definiteness, let $n = 3$, and suppose x is a simple root of the reduced cubic, $x^3 + 3Ix + J = 0$. This is an equation of the form $F(x, I, J) = 0$ so it follows from the general theory (see Chapter 8, theorem 8.2.1) that this equation is also satisfied for all x of the form $x = \phi(I, J)$, when (I, J) varies in some region D of $\mathbb{R}^2_{IJ}$, so we write it as $F(\phi(I, J), I, J) = 0$, valid for all (I, J) in D. Hence by the chain rule of differentiation, we have (with $x = \phi(I, J)$):

$$\begin{aligned} \frac{\partial \phi}{\partial I} &= -\frac{\partial F}{\partial I}\bigg/\frac{\partial F}{\partial x} = \frac{-3x}{3x^2 + 3I} = \frac{-x}{x^2 + I}, \\ \frac{\partial \phi}{\partial J} &= -\frac{\partial F}{\partial J}\bigg/\frac{\partial F}{\partial x} = \frac{-1}{3(x^2 + I)}; \end{aligned} \qquad (6.7.1)$$

since x is simple the denominator is non-zero, by exercise 6.6.1(2).

Consider the case when $I > 0$, so (by test 6.2.1) there is just one root x and x is simple. Then $\partial\phi/\partial I > 0$ if $x < 0$, so x increases as I increases; whereas $\partial\phi/\partial J < 0$ so

x decreases as J increases. But if $I < 0$, and x is such that $x^2 < -I$, then x increases as J increases; and if further $x < 0$, then $\partial\phi/\partial I < 0$ so x decreases when I increases. This confirms what we see when we draw graphs.

Exercises 6.7.1

1. Draw graphs to test the last assertion. Check also the case when $I < 0$, $x^2 < -I$, and $x > 0$.
2. Find formulae analogous to (6.7.1) for the roots of the quadratic $x^2 + 2bx + c$, and compare them with the result of directly differentiating the formulae (6.1.1).
3. Let $q(x) = x^4 - 2a^2x^2$, where a is real and positive. Show that its graph is symmetrical about the y-axis, with a maximum and two minima when x is $0, -a, a$, respectively; and that the two minima are on the same horizontal level $(-a^2)$. To $q(x)$ we now add bx to form the function $q_b(x)$, which, if b is small enough, still has a maximum and two minima. Find which of these will be higher than the other, and nearer or further from $\mathbf{0}$, by studying how the zeros of dg/dx vary with b. (It is probably hopeless to try to find formulae for these zeros themselves. For details see Griffiths (1984).)
4. (a) Let $g_{ab}(x) = x^3 + ax + b$. Show that the equation $g_{ab}(x) = 0$ cannot have a non-simple zero unless $4p^3 + 27q = 0$.
   (b) Now suppose that a, b are strictly positive. Sketch the graph of g_{ab} and show that the equation $g_{ab}(x) = 0$ has just one root $x = \theta(a, b)$ which is simple, and $< -2a/3$.
   (c) By implicit differentiation, show that at $x = \theta(a, b)$,
   $$\frac{\partial\theta}{\partial a} = -\frac{x}{3x^2 + a}, \qquad \frac{\partial\theta}{\partial b} = -\frac{1}{(3x^2 + a)x}.$$
   Hence show that if a or b is increased, then x will decrease. [SU]
5. Following (6.5.5) we asserted that if X is a region of Fig. 6.5.1, and $P \in X$ lies on an axis of $\mathbb{R}^2_{rs}$, then for any $Q \in R$, the quintics f_P, f_Q have the same numbers of zeros (all simple). To prove this let $\gamma: I \to X$ be a smooth path in X such that $\gamma(0) = P$, $\gamma(1) = Q$. For each $t \in I$ let $F(t, x) = f_{\gamma(t)}(x)$. Then F is smooth and
   $$\frac{\partial}{\partial x}F(t, x) = \frac{d}{dx}f_{\gamma(t)}(x) = f'_{\gamma(t)}(x).$$
   Show that if $f_{\gamma(t)}(x) = 0$ then $\partial/\partial x\, F(t, x) \neq 0$ since γ does not meet the boundary of X. Now apply exercise 8.1.2(6) to complete the proof. (Clearly, this proof applies to polynomials of all degrees, not just to quintics.)

7

Quadratics, conics, and quadrics

7.1 INTRODUCTION

In $\mathbb{R}^2$ and $\mathbb{R}^3$, the next most complicated shapes after the linear ones are quadratic—the curves and surfaces known as 'conics' and 'quadrics'. Their algebraic descriptions are simplified when we can find suitable coordinate transformations, so this chapter uses freely the linear mathematics of matrices and vectors (details of which may be found in Chapter B). We begin by working in the plane, and survey the conics; these are useful in several parts of mathematics, as we shall see throughout the book. Each has an equation which is a special case of the general equation

$$ax^2 + 2hxy + by^2 + 2gx + 2fy = c, \tag{7.1.1}$$

and when a, b, c, f, g, h are specified numbers, we shall give an algorithm which will tell us the main features of the curve that (7.1.1) then represents. This study will give practice in handling 2×2 matrices, their eigenvalues and eigenvectors; and we conclude by showing that there are only four essentially different types of 2×2 matrix. Afterwards we follow a similar program in $\mathbb{R}^3$, but we leave more details for the, by then practised, readers to fill in themselves.

7.2 CONICS

The reader has met the standard equations of certain plane curves of second degree (for example, in Chapter 5), such as the parabola

$$y^2 = 4ax \tag{7.2.1}$$

and the ellipse

$$\frac{x^2}{a^2} + \frac{y^2}{b^2} = 1. \tag{7.2.2}$$

As with lines, each curve is the set of points that satisfy the corresponding equation. If $a = b$ in (7.2.2), then the curve is a circle of centre $\mathbf{0}$ (the origin) and radius a. Such

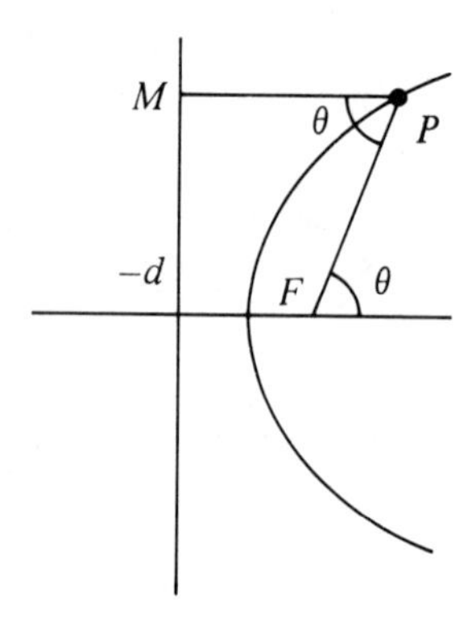

Fig. 7.2.1.

curves have been studied from ancient times; they are among those which the classical Greek geometer Apollonius (262–190 BC) recognized as sections of a cone. Such a 'conic' was first defined in terms of a constant $e > 0$ called the **eccentricity**, a fixed point F called the **focus**, and a fixed line called the **directrix**. The conic was then defined to be the set of points P satisfying the law:

$$\text{distance } PF = e \cdot \text{distance } PM$$

where M is the point on the directrix that is nearest to P. Translating this into the language of coordinate geometry, we suppose $P = (x, y)$, $F = \mathbf{0}$, and the directrix is the line $x = -d$ (see Fig. 7.2.1).Then if $PF = r$ and angle$(MPF) = \theta$, we have

$$r = e \cdot (r \cos \theta + d) = e \cdot (x + d). \tag{7.2.3}$$

Recalling that $r^2 = x^2 + y^2$, we can square and add to get

$$x^2 + y^2 = e^2(x^2 + 2xd + d^2)$$

that is

$$x^2(1 - e^2) + y^2 = e^2(2xd + d^2). \tag{7.2.4}$$

which is of the form (7.1.1). Also (7.2.3) can be rearranged in polar form to give:

$$r = ed/(1 - e \cos \theta). \tag{7.2.5}$$

The case when $e = 1$ is therefore quite special; if we write $a = d/2$, and $x' = (x + a)$ then the equation becomes $y^2 = 4ax'$, again as in (7.1.1). Here F is now moved to $x' = a$, and the curve is a parabola.

If $e \neq 1$, then we can divide (7.2.4) by $1 - e^2$ ($= k$, say) and complete a square to put (7.2.4) into the form

$$\left(x - \frac{de^2}{k}\right)^2 + \frac{y^2}{k} = \frac{(ed)^2 + e^4/k}{k} = \frac{e^2 d^2}{k^2},$$

which looks simpler if written in the form

$$X^2 + \frac{y^2}{k} = a^2, \qquad a^2 = \frac{e^2 d^2}{k^2}, \tag{7.2.6}$$

where $X = (x - de^2/k)$ and we have changed the old origin to $x = de^2/k$, $y = 0$.

If $e < 1$ then $k > 0$ so k has a (real) square root. We can therefore define $b = a\sqrt{k}$ so $b^2 = a^2(1 - e^2)$; hence $d = b\sqrt{k}/e$ and d is called the **latus rectum**. Now divide (7.2.6) by a^2 to obtain (7.2.2) since $b^2 = ka^2$. The curve is an ellipse.

If $e > 1$ then $-k$ has a real square root and we write $b^2 = a^2(e^2 - 1) = -ka^2$; so (7.2.6) simplifies to

$$\frac{X^2}{a^2} - \frac{y^2}{b^2} = c, \tag{7.2.7}$$

with $c = 1$. However, (7.2.7) makes sense for all values of c, and the curve it represents is called a hyperbola. When $c = 0$, the curve then consists of the pair of straight lines

$$\frac{X}{a} + \frac{y}{b} = 0, \qquad \frac{X}{a} - \frac{y}{b} = 0$$

since

$$\frac{X^2}{a_2} - \frac{y^2}{b^2} = \left(\frac{X}{a} + \frac{y}{b}\right)\left(\frac{X}{a} - \frac{y}{b}\right).$$

These lines are called the **asymptotes** of the hyperbola with $c = 1$. The various curves are sketched in Fig. 7.2.2. Their curious names arise from the Greek: with the ellipse, k 'falls short' of 1, and with the hyperbola k 'overshoots' 1. (In English 'ellipsis' is the omission of words, and 'hyperbole' is exaggeration.)

Exercises 7.2.1

1. Program 7.2.1, below, can be used to explore the focus–directrix definitions of the conic by varying the value of e. You may need to adjust the screen dimensions for certain values. Adapt it to plot the locus P, rather than the lines FP, PM.

```
REM Program 7.2.1, Focus/directrix
LET d = 1
SET MODE "graphics"
SET WINDOW -3,3, -2,2
PLOT -3,0 ; 3,0
PLOT -d,-2 ; -d,2
INPUT e
FOR theta = -PI+0.01 TO PI STEP PI/16
   LET r = e*d/(1-e*COS(theta))
   LET x = r*COS(theta)
   LET y = r*SIN(theta)
   PLOT 0,0 ; x,y ; -d,y
NEXT theta
END
```

7 0.7

Sample output

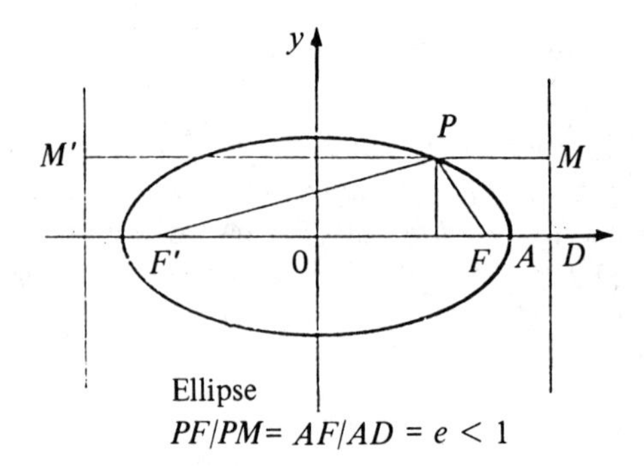
y
P
M'
M
F'
0
F
A
D
Ellipse
PF/PM= AF/AD = e < 1

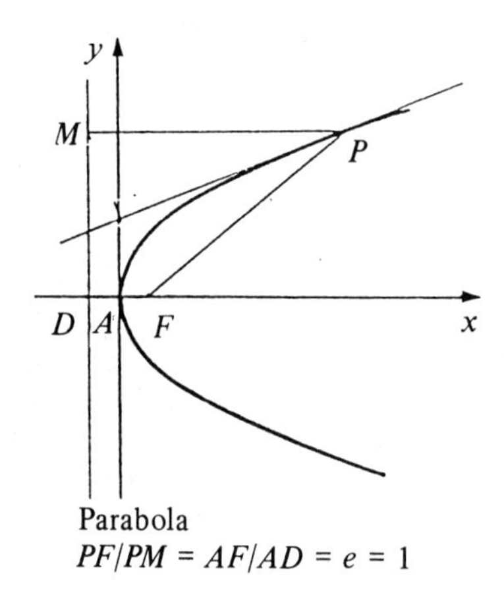
y
M
P
D
A
F
x
Parabola
PF/PM = AF/AD = e = 1

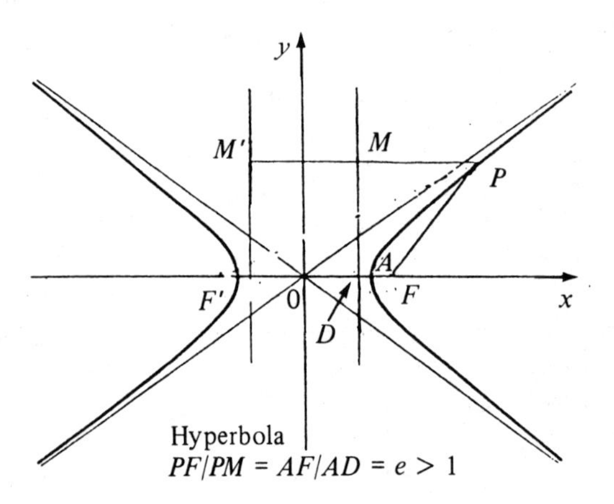
y
M'
M
P
A
F'
0
F
x
D
Hyperbola
PF/PM = AF/AD = e > 1

Fig. 7.2.2.

2. The form in $x, y: f(x, y) = c$ as in (7.1.1) is known as an **implicit equation** for the curve K. In special cases this can be rearranged to give, say, an explicit formula for y in terms of just x and constants. In the next section we shall meet implicit forms which cannot be so rearranged. In these cases a program such as program 7.1.2 below can be very helpful (if slow) in giving an impression of the curve by shading one side of it. The plane is divided into the three sets of points given by $f(x, y) < c$, $f(x, y) = c$, and $f(x, y) > c$. Use the program to explore the possible shapes of the bounding curve K for various values of a, h, b, g, f, c.

```
REM Program 7.2.2, Conic boundary
READ a, h, b , g, f , c
DATA 2, 1, -1, 3, -2, 4
FUNCTION conic(x,y) = a*x^2 + 2*h*x*y + b*y^2 + 2*g*x + 2*f*y
READ xmin, xmax, xstep, ymin, ymax, ystep
DATA -6  , 6   , 0.1  , -4  , 4   , 0.2
SET MODE "graphics"
SET WINDOW xmin, xmax,  ymin, ymax
FOR x = xmin TO xmax STEP xstep
  FOR y = ymin TO ymax STEP ystep
    IF conic(x,y) < c THEN PLOT x,y
  NEXT y
NEXT x
END
```

Sample output

3. (a) What is the equation of the circle which passes through the points $(5, -3)$, $(5, 5)$, and $(1, 5)$?
   (b) Find the polar equation of the ellipse $3x^2 + 4y^2 = 12$ with the pole at a focus and the initial line in the direction of the positive x-axis. Give the polar equation of the ellipse with the pole at the other focus and the initial line in the direction of the positive x-axis.
4. Find the eccentricity and equation of each of the following curves and sketch each curve:
   (a) ellipse, foci at $(0, 2)$ and $(8, 2)$ and the major axis of length 10;
   (b) ellipse, centre $(2, 2)$, one focus at $(-1, 2)$ and major axis of length 10;
   (c) hyperbola, centre $(2, 1)$, one focus at $(-2, 1)$ and distance 6 between the two branches;
   (d) parabola, vertex $(2, 0)$ and directrix $x = 0$;
   (e) parabola with focus $(3, -2)$ and directrix $x = -1$.
5. The orbit of the Moon is an ellipse with the Earth at one of the foci and eccentricity 0.055. Taking the length of the major axis to be 478,000 miles, find the nearest and furthest distances between the Earth and the Moon.

6. Sketch the hyperbolae

$$3x^2 - 2y^2 = 1, \qquad 2x^2 - 3y^2 = -1$$

to show their asymptotes. Hence find their points of intersection. One of these lies in the first quadrant; express it in the parametric form $(a \cosh \theta, b \sinh \theta)$ and show that

$$\theta = \ln(\sqrt{3} + \sqrt{2}).$$

7. (a) What is the eccentricity of a rectangular hyperbola?
(b) Obtain the equation of a rectangular hyperbola referred to its asymptotes as axes.

8. Let $P = (a \cosh \theta, b \sinh \theta)$ be a point on the hyperbola $x^2/a^2 - y^2/b^2 = 1$. Find the equation of the normal at P, and find the point P' where this normal cuts the y-axis. Let $Q = -P$; show that the distance $P'Q$ is $2 \sinh \theta \times (a^2 + b^2)/b$.
9. Find the equations of the tangent and normal to the curves

$$\frac{x^2}{a^2} + \frac{y^2}{b^2} = 1, \qquad \frac{x^2}{a^2} - \frac{y^2}{b^2} = 1,$$

at $(a \cos \theta,\ b \sin \theta)$ and $(a \cosh \theta,\ b \sinh \theta)$, respectively.
10. The ellipse/hyperbola/Cassini oval may be defined as the locus of a point which moves so that the sum/difference/product (respectively) of its distances from two fixed points is constant. Show (either algebraically or geometrically) that, in general, the locus of a point which moves so that the *quotient* of its distances from two fixed points $(\pm c, 0)$ is a constant k $(k \neq 0)$, is a circle (known as the circle of Apollonius). Express the centre and radius of the circle in terms of c and k.
What value of k does *not* give a circle and what curve is obtained in that case?
11. Show that if a ray of light leaves the focus of a parabola, then the curve reflects it parallel to the axis. (Hence the design of certain lamps and light reflectors.) If the curve is replaced by an ellipse, show that a ray from one focus is reflected to pass through the other focus.

The standard equations (7.2.1), (7.2.2), (7.2.6) are all examples of the general equation (7.1.1), which we must now analyse in order to describe its locus, that is, the set K of points in $\mathbb{R}^2$ that it represents. For this purpose, we shall use the phrase 'K is empty' to mean that no point of $\mathbb{R}^2$ satisfies the equation, even though there may be other solutions which involve complex numbers.

We multiply (7.1.1) through by a and complete the square in the quadratic part, to get, with $X = ax + hy$ and $D = ab - h^2$:

$$ac = X^2 + Dy^2 + 2gX + 2y(af - hg), \tag{7.2.8}$$

and we look at three separate cases, as follows.

(a) $D = 0 = (af - hg)$. Then $(X + g)^2 = ac + g^2$ $(= d$, say), which represents the pair of parallel lines $ax + hy + g = \pm\sqrt{d}$ if $d \geq 0$; if $d = 0$ the lines coincide, and if $d < 0$ then K is 'empty', since only non-real numbers can satisfy the equation.
(b) $D = 0 \neq (af - hg)$. Then (7.2.8) becomes $(X + g)^2 = Y$, where $Y = d - 2y(af - gh)$, so K is a parabola.
(c) $D \neq 0$. In this case, we write $x = x' + s$, $y = y' + t$, and leave the reader to check that (7.1.1) becomes

$$ax'^2 + 2hx'y' + by'^2 + 2x'(as + ht + g) + 2y'(ha + bt + f) = c',$$

for some c'. We now get rid of the linear terms by finding s, t to satisfy the linear equations $as + ht + g = 0 = hs + bt + f$; there is a unique solution since $D \neq 0$. Hence, x', y' must then satisfy an equation that occurs frequently, the **general homogeneous equation** of **second degree**:

$$ax^2 + 2hxy + by^2 = c' \tag{7.2.9}$$

which resembles (7.1.1) but lacks the linear (f and g) terms. With the shift of origin above, this form is then known as that of a **central conic**.

To decide what sort of curve K the equation (7.2.9) represents, we prove the following theorem.

Theorem 7.2.1
Let K be the locus represented by equation (7.2.9) and let $D = ab - h^2$. Then K is an ellipse if $D > 0$ and a hyperbola if $D < 0$. If $D = 0$, then K is a pair of (possibly coincident) parallel lines provided $ac' \geq 0$; otherwise K is empty.

Note that, since $f = g = 0$, parabolas are not mentioned here; they appear only in (b) above. In a sense the parabola can be thought of as a limiting case of an ellipse whose centre is 'going to infinity along the x-axis' and so cannot be represented as a central conic in the form (7.2.9).

Before going on to the proof of theorem 7.2.1, the reader should try the following exercises to grasp what the theorem is saying.

Exercises 7.2.2

1. Assuming the validity of theorem 7.2.1, classify the curve K in (7.2.9) for the cases $c' = 1, -1, 0$ and (a, h, b) takes the values $(1, 2, 3)$, $(-1, 2, 3)$, $(1, -2, 3)$ $(1, 3, 2)$, $(1, 3, -2)$, $(1, \sqrt{2}, 2)$. (This gives a quick check on computer output.) Sketch K when $a = 0 = b$, $h = 1$, $c' = 2$.
2. If $a = 1$, but b, h, c' are all zero, show that K is a pair of coincident lines. What do the equations $(2x - 3y)^2 = c'$ represent, when $c' = 0$, 1, or -1?
3. When we write $x = x' + s$, $y = y' + t$ as in (c) above, what is c' in (7.2.9)? What are the x'-, y'-coordinates of the old origin? What are the old coordinates of the new origin $\mathbf{0}'$? This process is called 'translating coordinates' from $\mathbf{0}$ to $\mathbf{0}'$. Clearly, it cannot alter the shape of the curve K.
4. Strictly speaking, the term $(x + d)$ in (7.2.3) should have been taken positive, since it is a distance; but this distinction is lost when we square. Show that the ellipse and hyperbola therefore each have two foci F, F', at $x = \pm ae$ and two directrices (the lines $x = \pm a/e$). Hence use the property (7.2.3) to prove that for any point P on an ellipse, $FP + PF'$ is a constant, namely, $2a/e$. (This gives a practical method for drawing an ellipse with a pencil and string.) Show that for a hyperbola, $FP - PF' = 2a/e$ and see if you can devise a pencil and string method for drawing a hyperbola.
5. For each of the following equations, sketch the corresponding curve:

   (a) $3x^2 - 5y^2 = 15$;
   (b) $3x^2 + 3y^2 - 6x + 18y + 30 = 0$;
   (c) $4x^2 + 8y^2 - 4x - 24y - 13 = 0$;
   (d) $x^2/4 + y^2/16 = 1$.

6. Identify the curves corresponding to each of the following equations.

   (a) $x^2 + y^2 + 5x - 3y + 1 = 0$;
   (b) $x^2 + y^2 - 2x + 4y + 9 = 0$;
   (c) $xy + y + 2x = 0$;
   (d) $9x^2 + y^2 + 6xy - 4 = 0$;
   (e) $x^2 + 4y^2 + 3y - 6x + 4xy = 0$;
   (f) $x^2 + 3y^2 - 4xy + x + y = 0$.

7. Show that

$$ab(x^2 - y^2) + (a^2 - b^2)xy - a^2bx + ab^2y = 0$$

represents a pair of perpendicular lines. Prove that they intersect on the circle $x^2 + y^2 - ax = 0$. Comment on the case $a = 0$.

Proof of theorem 7.2.1 (Part 1)
We first note again that case (b) following (7.2.8) cannot arise here, since $f = 0 = g$. Hence, when $D = ab - h^2$ is zero, the required conclusion is covered by case (a) there, since d is now ac'. So far, this mode of proof might strike the reader as somewhat pernicketty in style. This often happens when dealing with 'rare' cases, which nevertheless cannot be set aside for various reasons. However, the rest of the proof of theorem 7.2.1 is much more algorithmic, and uses matrix notation. In the next section, we shall first revise some properties of matrices and coordinate changes (which will also have more general usefulness) and then the rest of the proof proper is quite short.

7.3 QUADRATIC FORMS

When using matrices we often need to think of the row vector $(x \quad y)$ as the column vector $\begin{pmatrix} x \\ y \end{pmatrix}$, written, alternatively, as $(x \quad y)^{\mathrm{T}}$. If A is a 2×2 matrix, we can form the matrix product $A(x \quad y)^{\mathrm{T}}$ as a column vector, and its transpose is the row vector $(x \quad y)A^{\mathrm{T}}$. The transpose A^{T} of A has as its rows the columns of A (and thus is the 'reflection' of A in the diagonal from top-left to bottom-right, which is known as the 'leading diagonal'). A matrix is called **symmetric** if it is its own tranpose. So we can write the quadratic expression (7.2.9) as a product of a symmetric matrix and vectors with a scalar result:

$$ax^2 + 2hxy + by^2 = (x \quad y)A(x \quad y)^{\mathrm{T}}, \qquad A = \begin{pmatrix} a & h \\ h & b \end{pmatrix} \tag{7.3.1}$$

For example,

$$3x^3 - 4xy + 5y^2 = (x \quad y)\begin{pmatrix} 3 & -2 \\ -2 & 5 \end{pmatrix}(x \quad y)^{\mathrm{T}}.$$

With $\mathbf{z} = (x \quad y)^{\mathrm{T}}$ this becomes just $(\mathbf{z}^{\mathrm{T}})A\mathbf{z}$, and is called a **quadratic form**. But if B is a second matrix and $\mathbf{z} = B\mathbf{w}$, with $\mathbf{w} = (u \quad v)^{\mathrm{T}}$ then, since transposition reverses order,

$$\mathbf{z}^{\mathrm{T}} = (x \quad y) = (\mathbf{w}^{\mathrm{T}})(B^{\mathrm{T}})$$

so

$$(\mathbf{z}^{\mathrm{T}})A\mathbf{z} = (\mathbf{w}^{\mathrm{T}})(B^{\mathrm{T}})AB\mathbf{w} = (\mathbf{w}^{\mathrm{T}})C\mathbf{w} \text{ (say).} \tag{7.3.2}$$

Thus the change from $\mathbf{z}$ to $\mathbf{w}$ changes the quadratic form (7.3.1) into one of the same kind, but this may be simpler if we choose B carefully. The **determinant** $\det(B)$ of a 2×2 matrix

$$B = \begin{pmatrix} b_{11} & b_{12} \\ b_{21} & b_{22} \end{pmatrix}$$

is given by $\det(B) = b_{11}b_{22} - b_{12}b_{21}$, and thus $\det(B^{\mathrm{T}}) = \det(B)$. Now, since $C = B^{\mathrm{T}}AB$ and det is multiplicative (see (B.4.2)) then

$$\det(C) = \det^2(B)\det(A). \tag{7.3.3}$$

The matrix R_θ which represents an anticlockwise rotation through an angle θ about the origin $\mathbf{0}$ is given by:

$$R_\theta = \begin{pmatrix} \cos\theta & -\sin\theta \\ \sin\theta & \cos\theta \end{pmatrix}.$$

Taking this matrix as B, and writing $c = \cos\theta$, $s = \sin\theta$ (for simplicity), then we leave it for the reader to show that the resulting matrix $C = B^{\mathrm{T}}AB$ is of the form:

$$C = \begin{pmatrix} p & q \\ q & r \end{pmatrix} \qquad \text{where } q = h(c^2 - s^2) - (a - b)sc.$$

By the double angle formulae, we have $2q = 2h\cos 2\theta - (a - b)\sin 2\theta$ and then we choose $\theta = \phi$ (say) to make $q = 0$; thus we are rotating axes through an angle ϕ where $\tan 2\phi = 2h/(a - b)$ (so $\phi = \pi/4$ if $a = b$ and $h \neq 0$; if $h = 0$ we take $\phi = 0$.) Hence we have now ensured that C in (7.3.2) is a **diagonal matrix**, that is,

$$C = \begin{pmatrix} p & 0 \\ 0 & r \end{pmatrix}.$$

We now use the self-explanatory notation $\operatorname{diag}(p, r)$ for C, so

$$(R_\phi)^{\mathrm{T}}AR_\phi = C = \operatorname{diag}(p, r). \tag{7.3.4}$$

Proof of theorem 7.2.1 (Part 2, the conclusion)
We have now found a rotation of the axes, so that in the new coordinates (7.2.9) becomes:

$$(\mathbf{w}^{\mathrm{T}})C\mathbf{w} = pu^2 + rv^2 = c'. \tag{7.3.5}$$

By (7.2.2) and (7.2.6) this equation represents an ellipse if p and r have the same sign, and a hyperbola if the signs are opposite, that is, its locus K is an ellipse or hyperbola according as pr is positive or negative. But $pr = \det(C)$ so by (7.3.3) we have

$$pr = \det(C) = \det(A) \tag{7.3.6}$$

since $\det(B) = \det(R_\phi) = \det(R_\phi)^{\mathrm{T}} = \cos^2\phi + \sin^2\phi = 1$. Therefore K is an ellipse or hyperbola according as $D = \det(A)$ is positive or negative. In our case $D = ab - h^2$ so we have proved theorem 7.2.1 in the case that $D \neq 0$.

The case when $D = 0$ was dealt with in section 7.2, so the proof of theorem 7.2.1 is now complete. Since the construction of C did not depend on whether or not $D = 0$, we have also shown, in the 2×2 case, the following proposition.

Proposition 7.3.1

If A is a symmetric matrix, there is a rotation matrix R_ϕ such that the matrix $C = R_\phi^{\mathrm{T}} A R_\phi$ is in diagonal form.

With a general invertible matrix B in (7.3.3) we have, instead of (7.3.6), that $\det(C)$ and $\det(A)$ have the same sign; for $\det^2(B)$ is positive since $\det(B) \neq 0$. Hence by theorem 7.2.1 the locus K' of the equation $(\mathbf{w}^{\mathrm{T}})C\mathbf{w} = c$ has the same shape as K itself—even though B will not usually keep the coordinate axes at right angles.

Exercises 7.3.1

1. Experiment with the following program to display the effect of a particular B on specific shapes (ellipse, line pair xy, etc.). Note that if a curve is a closed loop, or in two pieces, then so is its image. Note also that we are here thinking of a coordinate transformation as a linear mapping.

```
REM Program 7.3.1, Transform of curve
READ b11, b12, b21, b22, a, b, tmin, tmax, tstep, xmax, ymax
DATA 1  , -2 , 1  , 3  , 2, 1, 0  , 6.3 , 0.1  , 6   , 4
FUNCTION x(t) = a*COS(T)
FUNCTION y(t) = b*SIN(t)
SET MODE "graphics"
SET WINDOW -xmax,xmax, -ymax,ymax
FOR t = tmin TO tmax STEP tstep
  PLOT x(t),y(t);
NEXT t
PLOT
FOR t = tmin TO tmax STEP tstep
  LET xt = b11*x(t) + b12*y(t)
  LET yt = b21*x(t) + b22*y(t)
  PLOT xt,yt;
NEXT t
END
```

2. For the matrix $A = \begin{pmatrix} a & b \\ b & d \end{pmatrix}$ form the matrix $B = \begin{pmatrix} d & -b \\ -c & a \end{pmatrix}$ and the products AB and BA. If the determinant $\Delta = \det(A) = ad - bc \neq 0$ then the matrix A is called **invertible** and

its inverse A^{-1} is given by: $A^{-1} = \begin{pmatrix} dk & -bk \\ -ck & ak \end{pmatrix}$ where $k = 1/\Delta$. Show that the products AA^{-1} and $A^{-1}A$ are both equal to $\begin{pmatrix} 1 & 0 \\ 0 & 1 \end{pmatrix} = I$, called the **identity matrix**.

3. By rotating and translating the axes change the equation

$$194x^2 + 120xy + 313y^2 + 776x + 240y + 607 = 0$$

into standard form and hence identify the curve which it represents.

4. (a) Find the equation of the curve

$$3x^2 - 2xy + 3y^2 - 8x + 8y - 2 = 0$$

with respect to new coördinate axes given by the translation

$$x = X + 1, \qquad y = Y - 1.$$

(b) Obtain a further simplification of the equation of the original curve by rotating the new axes through an angle $\alpha = \pi/4$.

(c) What type of curve does the original equation represent? Could you have identified the type of curve immediately from the original equation? (Give a reason for your answer.) [SU]

7.4 EIGENVALUES AND EIGENVECTORS

Since $R_\phi^{\mathrm{T}} = R_{-\phi} = R_\phi^{-1}$, proposition 7.3.1 says that the symmetric matrix A is an example of a matrix B, for which there is an invertible matrix S such that $S^{-1}BS$ is a diagonal matrix. We now ask when this is possible in general, without the stronger requirement that S be a rotation matrix like R_ϕ above. For precision, a definition is required.

Definition 7.4.1
Two matrices P, Q are said to be **similar** if we can find an invertible matrix R such that $P = R^{-1}QP$.

Thus we wish for an answer to the following question.

Question 7.4.1
When is a given matrix B (which need not be symmetric) similar to a diagonal matrix?

Before giving a reply, we indicate the point of introducing similar matrices. Firstly, when we study a transformation $\mathbf{w} = A\mathbf{z}$, then an invertible matrix R allows us to assign new coordinates $\mathbf{z}'$ to $\mathbf{z}$ by saying that $\mathbf{z}' = R\mathbf{z}$, so $\mathbf{z} = S\mathbf{z}'$ where $S = R^{-1}$; similarly $\mathbf{w}$ has new coordinates $\mathbf{w}'$, so $\mathbf{w} = S\mathbf{w}'$, whence

$$S\mathbf{w}' = \mathbf{w} = A\mathbf{z} = AS\mathbf{z}'.$$

Now $S\mathbf{w}'$ and $AS\mathbf{z}'$ are both (equal) vectors, so operating on each of them with the matrix R gives $R(S\mathbf{w}') = R(AS\mathbf{z}')$, that is,

$$RS\mathbf{w}' = RR^{-1}\mathbf{w}' = I\mathbf{w}' = \mathbf{w}' = RAS\mathbf{z}' = S^{-1}AS\mathbf{z}' = B\mathbf{z}' \text{ (say)}$$

and, so $S^{-1}AS = B$, or, similarly, $A = R^{-1}BR$. In the new coordinates, therefore, it is the matrix B that is doing the transforming, and it is similar to A in the above sense. Hence we try to find the simplest such B and study it, and then work back to the properties of A.

For example, to work out the powers $A^2, A^3, \ldots, A^n$ of A, we have

$$A^2 = (R^{-1}BR)^2 = (R^{-1}BR)(R^{-1}BR) = R^{-1}(BIB)R = R^{-1}(B^2)R,$$

$$A^3 = A(A^2) = R^{-1}BR(R^{-1}(B^2)R) = R^{-1}BI(B^2)R = R^{-1}(B^3)R,$$

and so on.

If, as in (7.3.4), B is a diagonal matrix: $B = \operatorname{diag}(\lambda, \mu)$, then A is similar to the diagonal matrix B, and

$$A^n = R^{-1}(B^n)R = R^{-1}\begin{pmatrix} \lambda^n & 0 \\ 0 & \mu^n \end{pmatrix} R. \tag{7.4.1}$$

Other examples will be seen later. Meanwhile, in order to answer question 7.4.1, we need to use **eigenvalues** and **eigenvectors**; and we now explain the two-dimensional versions in some detail.

When we have a transformation of the form $\mathbf{w} = A\mathbf{z}$ as in (7.3.2), it is often important to know whether there is some point $\mathbf{z}$ that is merely stretched (or even unchanged) rather than rotated. Of course, the origin $\mathbf{0}$ is always left unchanged. Thus we really want to know whether there is some number λ and some *non-zero* vector $\mathbf{v}$, such that

$$A\mathbf{v} = \lambda\mathbf{v}. \tag{7.2.4}$$

Here λ is the stretching factor, and if $\lambda = 1$ then $\mathbf{v}$ is left fixed by the transformation.

Example 7.4.1

Suppose $A = \begin{pmatrix} 2 & 1 \\ 1 & 2 \end{pmatrix}$. Then (7.4.2) requires us to solve the equations $2x + y = \lambda x$, $x + 2y = \lambda y$. Eliminating y gives $x(1 - (2 - \lambda)^2) = 0$, so either $x = 0$ or $1 - (2 - \lambda)^2 = 0$. If $x = 0$ then $y = 0$, and we have not found a non-zero solution $\mathbf{v}$. Hence we must choose λ to satisfy $1 - (2 - \lambda)^2 = 0$, and this quadratic has solutions $\lambda = 1$, and $\lambda = 3$. Substituting $\lambda = 1$ in the above equations for x, y gives $\mathbf{v} = (x \quad y)^{\mathrm{T}} = k(1 \quad -1)$ for any non-zero $k \in \mathbb{R}$: and we can check that $A\mathbf{v} = \mathbf{v}$: the whole line from $\mathbf{0}$ through $\mathbf{v}$ is left unchanged by A. If, instead, we use the other value 3 for λ, we have the solution $\mathbf{w} = k(1 \quad 1)$ for any non-zero $k \in \mathbb{R}$, so $A\mathbf{w} = 3\mathbf{w}$ and A stretches $\mathbf{w}$ by the factor 3. The reader should sketch $\mathbf{v}, \mathbf{w}$ in the plane $\mathbb{R}^2$, to see what A does to the parallelogram with vertices $\mathbf{0}$, $\mathbf{v}$, $\mathbf{w}$ and $\mathbf{v} + \mathbf{w}$.

This example illustrates some of the following general theory. Let $\mathbf{v}$ be as in 7.4.2. If the matrix A has real eigenvalues λ, μ and if, say $|\lambda| > |\mu|$ then λ is called the **dominant eigenvalue** and the corresponding vector $\mathbf{v}_\lambda$ is the **dominant eigenvector**. In this case the repeated application of A to a set of vectors will quickly reveal the direction of the dominant eigenvector and the magnitude of the dominant eigenvalue, as is demonstrated in the following program.

```
REM Program 7.4.1, Dominant eigenvalue & eigenvector
READ a11, a12 , a21, a22, a, b, n , xmax, ymax
DATA 1  , -0.5, 1  , 2  , 1, 1, 32, 24   , 16
FUNCTION x(t) = a*COS(T)
FUNCTION y(t) = b*SIN(t)
DIM xp(64),yp(64)
SET MODE "graphics"
SET WINDOW -xmax,xmax, -ymax,ymax
FOR i = 1 TO n
  LET t = i*2*PI/n
  LET xp(i) = x(t)
  LET yp(i) = y(t)
  PLOT 0,0 ; xp(i),yp(i)
NEXT i
FOR j = 1 TO 4
  PAUSE 2
  FOR i = 1 TO n
    LET xt    = a11*xp(i) + a12*yp(i)
    LET yp(i) = a21*xp(i) + a22*yp(i)
    LET xp(i) = xt
    PLOT 0,0 ; xp(i),yp(i)
  NEXT i
NEXT j
END
```

Sample output

This program draws a set of equally spaced 'spokes' of a circle from the origin to points $(xp)i), yp(i))$. Then the matrix A is applied to each of these points in turn, giving the spokes of the 'squashed circle'. The process is repeated four times in all (with a delay of two seconds between each), and the resulting picture shows the position of the line containing the dominant eigenvector. We can start with any vector $\mathbf{z}$ and observe the points $\mathbf{z}_n = A^n\mathbf{z}$. The reader should prove that $\mathbf{z}$ is of the form $\alpha\mathbf{v}_\lambda + \beta\mathbf{v}_\mu$ for some real α, β; so $\mathbf{z}_n = \lambda^n(\alpha\mathbf{v}_\lambda + q\mathbf{v}_\mu)$ where $q = \mu/\lambda$ and $|q| < 1$. Hence $\|\mathbf{z}_{n+1}\|/\|\mathbf{z}_n\|$ converges to $|\lambda|$, and $\mathbf{z}_n$ becomes ever more nearly parallel to $\mathbf{v}_\lambda$ as n increases. (Compare Fig. 2.6.1(iii).) The program has taken a specific A and $\mathbf{z}$, and it does not need to find α or β.

Now let us consider finding eigenvalues and eigenvectors in a 'theoretical' context, in order to study their properties. We have to solve (7.4.2) for $(x \quad y)^{\mathrm{T}} = \mathbf{v}$, and unpack the coordinates to get simultaneous equations:

$$\begin{aligned} (a-\lambda)x + \quad by &= 0, \\ cx + (d-\lambda)y &= 0. \end{aligned} \tag{7.4.3}$$

These are homogeneous equations, and we want a non-trivial solution since $\mathbf{v} \neq 0$. Therefore both equations have to represent the same line (passing through $\mathbf{0}$) and thus:

$$\frac{a-\lambda}{c} = \frac{b}{d-\lambda}, \qquad \text{that is } (a-\lambda)(d-\lambda) = bc$$

which, rewritten as

$$(a - \lambda)(d - \lambda) - bc = 0 = \lambda^2 - (a + d)\lambda + (ad - bc),$$

means that the matrix $B = \begin{pmatrix} a - \lambda & b \\ c & d - \lambda \end{pmatrix}$ of the coefficients of equations (7.4.3) must have a determinant of zero, and λ has to be a root of the quadratic **characteristic equation**:

$$t^2 - Tt + \Delta = 0, \tag{7.4.4}$$

where $T = a + d$ is the trace of matrix A, and $\Delta = ad - bc$ is its determinant.

Hence there can be no more than two eigenvalues, and these are given by the quadratic formula $(T \pm \sqrt{D})/2$, where D is the **discriminant**:

$$D = T^2 - 4\Delta = (a - d)^2 + 4bc. \tag{7.4.5}$$

If D is negative, then its square roots are imaginary numbers of the form $\pm \mathrm{j}t$, where $\mathrm{j}^2 = -1$. Hence we can say that if:

(a) $D > 0$ *then* A *has two real (unequal) eigenvalues;*
(b) $D = 0$ *then* A *has just one eigenvalue, and it is real;*
(c) $D < 0$ *then* A *has two unequal eigenvalues, and they are complex conjugates (i.e., of the form* $p \pm \mathrm{j}q$*).*

A special case of (b) occurs when A is a **similitude** $\begin{pmatrix} a & 0 \\ 0 & a \end{pmatrix}$. Here we have $A\mathbf{z} = a\mathbf{z}$ for every $\mathbf{z}$, so the eigenvalues are both equal to a. Here, also, A is similar to a diagonal matrix, namely itself, since $I^{-1}AI = A$.

Suppose then that A is not a similitude. If λ is a root of (7.4.4), then we can solve (7.4.3) to get an eigenvector which we write as a column in two ways as:

$$\mathbf{v}_\lambda = (b \quad \lambda - a)^{\mathrm{T}} = (\lambda - d \quad c)^{\mathrm{T}}. \tag{7.4.6}$$

Clearly, $\mathbf{v}_\lambda$ is real (i.e., in $\mathbb{R}^2$) if, and only if, λ is real. By (7.4.3), any constant multiple of $\mathbf{v}_\lambda$ would solve (7.4.2). However, recall that an eigenvector has to be non-zero; therefore either form in (7.4.6) is allowable if $bc \neq 0$ because then neither b nor c is zero.

If $b \neq 0$ then we shall call $\mathbf{v}_\lambda$ (given by the first form in (7.4.6)) the eigenvector **corresponding** to λ; and similarly when $c \neq 0$. If $bc = 0$, more care is needed, and we must examine special cases. For then (7.4.4) reduces to $(t - a)\cdot(t - d) = 0$ with roots a and d. Hence if both b and c are zero, then a and d would be unequal since A is not a similitude; so if $\lambda = a$, then the second form in (7.4.6) is non-zero, and $\mathbf{v}_a$ lies along the x-axis; for the root $\lambda = d$, $\mathbf{v}_d$ lies along the y-axis.

If only one of b, c is zero, suppose $b = 0 \neq c$. Then by (7.4.6) the two eigenvectors are $(a - d \quad c)^{\mathrm{T}}$ for eigenvalue a, and $(0 \quad c)^{\mathrm{T}}$ for eigenvalue d. Here the two vectors are distinct except when $a = d$, and we shall return to this anomaly later. Similarly when $c = 0 \neq b$.

We pause to give readers some practice before going further.

Exercises 7.4.1

1. Find the eigenvalues and corresponding eigenvectors of the matrices

$$A = \begin{pmatrix} -2 & 3 \\ 5 & 1 \end{pmatrix}, \quad \begin{pmatrix} 4 & 2 \\ 1 & 3 \end{pmatrix}, \quad \begin{pmatrix} 2 & 0 \\ 0 & 2 \end{pmatrix}, \quad \begin{pmatrix} 1 & 2 \\ 0 & 1 \end{pmatrix}.$$

2. With the same A as in exercise 1, change the 5 to the letter c. Show that the discriminant is $D = 9 + 12c$, so $D \geq 0$ if $c \geq -3/4$. Solve the equation (7.4.4) in the cases $c = -3/4$, $c = -2$. In the latter case, express the eigenvectors in the form $\mathbf{u} \pm \mathrm{j}\mathbf{v}$, where $\mathbf{u}, \mathbf{v}$ are real vectors.
3. Show that equation (7.4.2) can be written as $(A - \lambda I)\mathbf{v} = \mathbf{0}$ where I is the identity matrix and that λ must satisfy the equation

$$\det(A - \lambda I) = 0 \tag{7.4.7}$$

(where matrices are subtracted entry by entry). Verify that (7.4.7) works out as the quadratic equation (7.4.4).
4. Let P, Q be the similar matrices in definition 7.4.1. By multiplying the equation $P = R^{-1}QR$ in front by R, and behind by R^{-1}, show that $RPR^{-1} = Q$, so the relation of being similar is *symmetrical*. Show that it is also *transitive*, in that if Q is similar to S then P is also similar to S. Putting $R = I$ verify also that P is similar to itself, and hence that the relation is *reflexive*. Thus matrix similarity is an **equivalence relation**.
5. Show that similar matrices have the same eigenvalues. (If in (7.4.7) we have $A = R^{-1}BR$ show that $(A - \lambda I) = R^{-1}(B - \lambda I)R$ and use the multiplicative property of det.)
6. Suppose the matrix A has equal eigenvalues. Show that we can always alter A slightly, so as to make the eigenvalues unequal. (Replace a, b, respectively, by $a + s$, $b + t$ and call the resulting matrix $A_{s,t}$. Show that if $bc \neq 0 \neq t$, then $A_{0,t}$ has distinct eigenvalues, which are real if $t < 0$. If $bc = 0$, work with $A_{s,0}$, except when $a = d$.)
7. If A has eigenvalues λ, μ, use equation (7.4.4) to show that

$$\lambda + \mu = \mathrm{Trace}(A), \qquad \lambda\mu = \det(A).$$

8. Show that A satisfies its own characteristic equation, that is,

$$A^2 - TA + \Delta I = 0.$$

(This is the theorem of Cayley and Hamilton.) Use definition 7.4.1 when the eigenvalues are distinct, and when they are equal use exercise 5 above.
9. Suppose A has unequal eigenvalues λ, μ. Define new matrices

$$Q_\lambda = \frac{1}{\mu - \lambda}(A - \lambda I), \qquad Q_\mu = \frac{1}{\lambda - \mu}(A - \mu I);$$

and use the Cayley–Hamilton theorem to show that $Q_\lambda \cdot Q_\mu = 0$ and

$$A^n = (\lambda Q_\lambda + \mu Q_\mu)^n = \lambda^n Q_\lambda + \mu^n Q_\mu, \qquad (n = 1, 2, \ldots).$$

If, instead, $\lambda = \mu$, let $Q = A - \lambda I$. Show that $Q^2 = 0$ and

$$A^n = (\lambda I + Q)^n = \lambda^n I + n\lambda^{n-1}Q, \qquad (n = 2, 3, \ldots).$$

10. Let f be a polynomial function

$$f(x) = a_0 + a_1 x + a_2 x^2 + \cdots + a_n x^n.$$

If we substitute A for the number x (and $a_0 I$ for a_0), we get a new matrix $f(A)$. Show that

$$f(A) = f(\lambda)Q_\lambda + f(\mu)Q_\mu, \qquad \text{if } \lambda \neq \mu,$$
$$f(A) = f(\lambda)I + f'(\lambda)Q, \qquad \text{if } \lambda = \mu,$$

where f' is the (calculus) derivative of f. Hence in each case, express $f(A)$ as a linear function of A.

11. Let λ be an eigenvalue of A. Show that the set of all vectors $\mathbf{z}$ in $\mathbb{R}^2$, that satisfy the equation $A\mathbf{z} = \lambda\mathbf{z}$, is either a line through $\mathbf{0}$, or the whole plane $\mathbb{R}^2$, or just the point $\mathbf{0}$. (These are called the **eigenspaces** of A.) Show that the plane occurs as an eigenspace, only when A is a similitude, and that $\mathbf{0}$ occurs when A is a rotation. When else does $\mathbf{0}$ occur?
    Use (7.4.2) to show that if λ is an eigenvalue of A, then λ^2 is one for A^2, and λ^n is an eigenvalue of A^n $(n = \pm 1, \pm 2, \cdots)$. Show also that λ is an eigenvalue of A^{T}.

We are now ready to formulate and prove an answer to question 7.4.1, in the form of a theorem.

Theorem 7.4.1

If the eigenvalues of the 2×2 *matrix* $A = \begin{pmatrix} a & b \\ c & d \end{pmatrix}$ *are distinct, then* A *is similar to a diagonal matrix.*

Note that we can verify the hypothesis when A is given, by calculating the discriminant D; the eigenvalues are distinct if $D \neq 0$, by condition (b) after (7.4.5). Note, too, that the theorem does imply that the *only* matrices which are similar to diagonal matrices are ones with distinct eigenvalues.

Proof

There are two distinct eigenvalues λ, μ (which may be complex). If $b \neq 0$ then from the corresponding eigenvectors $\mathbf{v}_\lambda$, $\mathbf{v}_\mu$ in (7.4.6) we can form a matrix E with these vectors as its columns, so

$$E = (\mathbf{v}_\lambda, \mathbf{v}_\mu) = \begin{pmatrix} b & b \\ \lambda - \alpha & \mu - a \end{pmatrix};$$

and

$$\det(E) = b(\lambda - \mu) \neq 0 \tag{7.4.8}$$

since $\lambda \neq \mu$ (given) and $b \neq 0$. Therefore E is an invertible matrix.

If $b = 0$ we use the second forms of the eigenvectors in (7.4.6) with $c \neq 0$ and argue similarly, and if $b = c = 0$ we saw above that the eigenspaces are the axes, so we can use $\mathbf{i}, \mathbf{j}$ as $\mathbf{v}_\lambda$, $\mathbf{v}_\mu$ to give $E = I$, the identity matrix. Thus in each of these cases we still obtain an invertible matrix E, which we use to diagonalize A, as follows.

By (7.4.2) and the row-by-column multiplication of matrices, we have $AE = (A\mathbf{v}_\lambda, A\mathbf{v}_\mu) = (\lambda\mathbf{v}_\lambda, \mu\mathbf{v}_\mu) = ED$, where $D = \operatorname{diag}(\lambda, \mu)$, so we can multiply each side by E^{-1} to get

$$D = E^{-1}AE. \tag{7.4.9}$$

Therefore A is similar to a diagonal matrix, as required.

As an example, let A be as in example 7.4.1. Then $\lambda = 1, \mu = 3$ so $E = \begin{pmatrix} 1 & 1 \\ -1 & 1 \end{pmatrix}$ with determinant 2. By exercise 7.3.1(2), $E^{-1} = \frac{1}{2}\begin{pmatrix} 1 & -1 \\ 1 & 1 \end{pmatrix}$ and it can now be verified that $E^{-1}AE = \frac{1}{2}\begin{pmatrix} 2 & 0 \\ 0 & 6 \end{pmatrix} = \begin{pmatrix} 1 & 0 \\ 0 & 3 \end{pmatrix} = \mathrm{diag}(\lambda, \mu)$.

Now, for some applications, it is necessary to use only real matrices; thus theorem 7.4.1 has a weakness because the matrix E in (7.4.8) may not be real. In that case what could we use as a simple form to which A would be similar? An answer is given in the next section.

Exercises 7.4.2

1. For each of the following matrices, compute the corresponding matrix E and its inverse E^{-1}. Then verify equation (7.4.9). (Use exercise 7.3.1(2).)

$$\begin{pmatrix} 1 & 2 \\ -2 & 4 \end{pmatrix} \quad \begin{pmatrix} 2 & 1 \\ 2 & 4 \end{pmatrix} \quad \begin{pmatrix} 2 & 1 \\ -2 & 4 \end{pmatrix}$$

2. Let A, B be 2×2 matrices. Show that $(AB)^{\mathrm{T}} = B^{\mathrm{T}}A^{\mathrm{T}}$ where T denotes transposition; and that if A is invertible, then so is A^{T}, and $(A^{\mathrm{T}})^{-1} = (A^{-1})^{\mathrm{T}}$. Also show that if E in (7.4.9) is orthogonal, then A must be symmetric ($A = A^{\mathrm{T}}$).

7.5 THE STANDARD FORM OF A MATRIX

We can now give the promised improvement of theorem 7.4.1.

Theorem 7.5.1

Suppose $A = \begin{pmatrix} a & b \\ c & d \end{pmatrix}$ *is a real* 2×2 *matrix with eigenvalues* λ, μ. *If* A *is not a similitude then* A *is similar to exactly one of the following three types of real matrix*:

$$\text{(i)} \begin{pmatrix} \lambda & 0 \\ 0 & \mu \end{pmatrix}, \qquad \text{(ii)} \begin{pmatrix} \lambda & 1 \\ 0 & \lambda \end{pmatrix}, \qquad \text{(iii)} \begin{pmatrix} p & -q \\ q & p \end{pmatrix}$$

according as λ, μ *are* (i) *real and distinct*, (ii) *real and equal, or* (iii) *complex conjugates of the form* $p \pm \mathrm{j}q$, ($q \neq 0$).

Proof

The case when λ, μ are real and distinct has already been dealt with in theorem 7.4.1. If $\lambda = \mu$ then $\lambda = T/2 = (a + d)/2$ and $2\lambda = T$ (where, as usual, T is the trace of A). If b and c are both zero then A is already in the form (i), so we assume that $b \neq 0$. We can then write the corresponding eigenvector in the form

$$\mathbf{v}_\lambda = \begin{pmatrix} b \\ \lambda - a \end{pmatrix} = \begin{pmatrix} b \\ s \end{pmatrix}, \qquad s = \frac{d - a}{2}.$$

If $\mathbf{j}$ is the standard unit vector $\begin{pmatrix}0\\1\end{pmatrix}$, then $A\mathbf{j} = \begin{pmatrix}b\\d\end{pmatrix}$ so

$$\mathbf{v}_\lambda + \lambda\mathbf{j} = \begin{pmatrix}b\\s\end{pmatrix} + \begin{pmatrix}0\\\lambda\end{pmatrix} = \begin{pmatrix}b\\d\end{pmatrix} = A\mathbf{j}.$$

We now construct a new matrix $R = (\mathbf{v}_\lambda \quad \mathbf{j})$. Then $\det(R) = b \neq 0$, so R is invertible and

$$AR = (A\mathbf{v}_\lambda \quad A\mathbf{j}) = (\lambda\mathbf{v}_\lambda \quad \mathbf{v}_\lambda + \lambda\mathbf{j}) = RB$$

where B is the matrix in (ii) above, that is, $B = R^{-1}AR$ and A is similar to B. If b had been zero, then $c \neq 0$ and we would have worked analogously, but with $\mathbf{v}_\lambda = \begin{pmatrix}-s\\c\end{pmatrix}$ and with $\mathbf{i}$ replacing $\mathbf{j}$.

There remains the case when λ, μ are complex conjugates. Let $\lambda = p + \mathrm{j}q$ ($q \neq 0, \mathrm{j}^2 = -1$). Then $\mu = \bar{\lambda} = p - \mathrm{j}q$ and for the corresponding eigenvectors we have similarly $\mathbf{v}_\mu = \bar{\mathbf{v}}_\lambda$. As above we cannot have both b and c simultaneously zero. Suppose $b \neq 0$. Then $\mathbf{v}_\lambda = \begin{pmatrix}b\\p - a + \mathrm{j}q\end{pmatrix}$ and $\mathbf{v}_\mu = \begin{pmatrix}b\\p - a - \mathrm{j}q\end{pmatrix}$. Define the real vectors $\mathbf{u}, \mathbf{w}$ by

$$\mathbf{u} = \frac{\mathbf{v}_\lambda + \mathbf{v}_\mu}{2} = \begin{pmatrix}b\\p - a\end{pmatrix}, \qquad \mathbf{w} = \frac{\mathbf{v}_\lambda - \mathbf{v}_\mu}{2\mathrm{j}} = \begin{pmatrix}0\\q\end{pmatrix}.$$

Then $\mathbf{v}_\lambda = \mathbf{u} + \mathrm{j}\mathbf{w}$, $\mathbf{v}_\mu = \mathbf{u} - \mathrm{j}\mathbf{w}$. Hence $A\mathbf{u} = (\lambda\mathbf{v}_\lambda + \bar{\lambda}\bar{\mathbf{v}}_\lambda)/2 = p\mathbf{u} - q\mathbf{w}$, since $p = (\lambda + \bar{\lambda})/2$, $q = (\lambda - \bar{\lambda})/2\mathrm{j} = -\mathrm{j}(\lambda - \bar{\lambda})/2$, and similarly $A\mathbf{w} = q\mathbf{u} + p\mathbf{w}$.

Hence if R is the matrix $(\mathbf{w} \quad \mathbf{u})$ we have

$$AR = A(\mathbf{w} \quad \mathbf{u}) = (A\mathbf{w} \quad A\mathbf{u}) = (p\mathbf{w} + q\mathbf{u} \quad -q\mathbf{w} + p\mathbf{u}) = RC$$

where C is the matrix in (iii) above. Also $R = \begin{pmatrix}0 & b\\q & p - a\end{pmatrix}$ is invertible since $\det(R) = -bq$, and by assumption $b \neq 0$ and $q \neq 0$. (If $b = 0$ then $c \neq 0$ and we would have worked with $\mathbf{v}_\lambda = \begin{pmatrix}p - d + \mathrm{j}q\\c\end{pmatrix}$ instead, to find $R = \begin{pmatrix}q & p - d\\0 & c\end{pmatrix}$ with $\det(R) = cq \neq 0$.

This completes the proof of theorem 7.5.1.

Exercises 7.5.1

1. Show that the matrix (iii) in the statement of theorem 7.5.1 can be written as $r \cdot R_\theta$, where R_θ is a rotation matrix and $\theta = \cos^{-1} p/r$, $r^2 = p^2 + q^2$. Show that $A^n = r^n \cdot R_{n\theta}$, $n = 1, 2, \ldots$.
2. Suppose A is of the form (ii) in theorem 7.5.1. Use mathematical induction to show that $A^n = \begin{pmatrix}\lambda^n & n\lambda^{n-1}\\0 & \lambda^n\end{pmatrix}$, $n = 1, 2, \ldots$.

3. For each of the following matrices A find a real matrix R such that $R^{-1}AR$ is in standard form:

$$\begin{pmatrix} 1 & 1 \\ -2 & 3 \end{pmatrix}, \quad \begin{pmatrix} 2, & -4 \\ 3, & 7 \end{pmatrix}, \quad \begin{pmatrix} 1 & -1 \\ 1 & 3 \end{pmatrix}.$$

Readers may wonder how much of the above theory applies to matrices of order higher than two. The answer is that almost all of the results apply, but the proofs become more elaborate. Thus the definitions of eigenvector and eigenvalue are as above, but the characteristic equation becomes an equation of higher degree, and this naturally complicates the algebra. It is still possible to define $\det(A)$ and it still equals the product of the eigenvalues; and det is still multiplicative. Also the Cayley–Hamilton theorem still holds, as does theorem 7.4.1. But the natural analogue of theorem 7.5.1 is more complicated, and involves the 'canonical form' of C. Jordan. For details, see advanced textbooks of algebra, such as Blyth and Robertson (1986), Chapter 4.

7.6 LAGRANGE'S REDUCTION PROCESS

A useful illustration of 'higher-dimensional' methods occurs with three variables, when we look for an analogue of proposition 7.3.1. There, the essential feature is that we were able to reduce the quadratic form $\mathbf{x}^{\mathrm{T}}A\mathbf{x}$, where A is a symmetric matrix, to the diagonal form $\mathbf{w}^{\mathrm{T}}C\mathbf{w}$. More precisely, we found an invertible matrix R such that

$$\mathbf{x} = R\mathbf{w}, \qquad C = R^{\mathrm{T}}AR, \qquad \mathbf{w}^{\mathrm{T}}C\mathbf{w} = pu^2 + rv^2, \tag{7.6.1}$$

where p, r are the eigenvalues of A; also the latter are real, and form the diagonal entries of C. A similar result and proof holds for real $n \times n$ symmetric matrices: in the following, all matrices are assumed real. Thus if we follow that proof to find C in a practical case, we need to begin by finding the eigenvalues of A, in order to construct R; but this requires the solution of the characteristic equation analogous to (7.4.4) which is now of degree n and therefore difficult in general. A more practical algorithm was found by Lagrange, as we now explain.

With $n = 2$, we drew conclusions about the behaviour of the diagonal form $\mathbf{x}^{\mathrm{T}}A\mathbf{x}$ from that of the diagonal form in (7.6.1), which is determined by only the signs of p and r, and not their actual sizes. Thus we only need the diagonal elements to have the same signs as the eigenvalues of A. For brevity, any matrix of the form $P^{\mathrm{T}}AP$ is said to be **equivalent** to A. It turns out that, if A is equivalent to some diagonal matrix E, then the number s of (strictly) negative diagonal elements in E is always the same, independently of which matrix P is used. (This is 'Sylvester's law of inertia': see Mirsky (1955), Chapter 12.) Therefore any convenient way of diagonalizing A will do; and we call the 'invariant' number s the **index** of A, and denote it by $\mathrm{ind}(A)$. The **rank** of A is a number defined by

$$\mathrm{rank}(A) = \mathrm{ind}(A) + \mathrm{ind}(-A). \tag{7.6.2}$$

A useful algorithm for diagonalizing A was found by Lagrange, and we now explain it with $\mathbf{n} = 3$ to bring out its essence. In it, we use a small amount of calculus for ease of phrasing.

For convenience we write out the form $\mathbf{x}^{\mathrm{T}}A\mathbf{x}$ in full as

$$F = ax^2 + by^2 + cz^2 + 2fyz + 2gzx + 2hxy \tag{7.6.3}$$

where the matrix

$$A = \begin{pmatrix} a & h & g \\ h & b & f \\ g & f & c \end{pmatrix}$$

is symmetric. Suppose first that $a \neq 0$. Then we regard F as a function of x alone, keeping all other variables constant, and differentiate to obtain the 'partial derivative' $\partial F/\partial x = 2(ax + gz + hy)$. Then the reader can check, by multiplying out, that $F - (\partial F/\partial x)^2/4a$ ($= G$, say) is independent of x, and it is a quadratic form in y, z only. Therefore, by (7.6.1), we can diagonalize G, to obtain $G = p\mu^2 + q\nu^2$, whence

$$F = (1/a)\lambda^2 + p\mu^2 + q\nu^2; \qquad \lambda = \tfrac{1}{2}\left(\frac{\partial F}{\partial x}\right). \tag{7.6.4}$$

So we have found a linear transformation $(x, y, z) \to (\lambda, \mu, \nu)$, with matrix P, say, such that $P^{\mathrm{T}}AP$ is equivalent to the diagonal matrix that we write for brevity as $\mathrm{diag}(1/a, p, q)$. Since $a \neq 0$, this transformation is invertible.

Example 7.6.1

Suppose $F = x^2 + by^2 + cz^2 + 6yz + 2zx - 4xy$. Then $\frac{1}{2}(\partial F/\partial x) = x + z - 2y$ ($= \lambda$, say), and $F = \lambda^2 + G$, where $G = (b - 4)y^2 + (c - 1)z^2 + 10yz$, and

$$A = \begin{pmatrix} 1 & -2 & 1 \\ -2 & b & 3 \\ 1 & 3 & c \end{pmatrix}.$$

Hence if (for example) $b = 5$, then $F = \lambda^2 + (y + 5z)^2 + (c - 26)z^2$, which is a sum of squares if $c \geq 26$. If $c = 26$ (then $\det(A) = 0$), the index is 0 and the rank is 2; if $c \neq 26$ the rank is 3, but the index is 0 or 1 according as $c > 26$ or not. If $c < 26$, the linear transformation is given by the equations

$$\lambda = x - 2y + z, \qquad \mu = y + 5z, \qquad \nu = z$$

which are easily seen to be invertible. Note that if $c > 26$, the last equation could be changed (if desired) to $\nu = z\sqrt{(c - 26)}$, without altering the index.

The method, as described above, assumed that $a \neq 0$ in (7.6.3). If instead $b \neq 0$ and we had used $\partial F/\partial y$, we would get a different reduction to diagonal form, but with the same index and rank. But if a, b, c were all zero we need a small preliminary trick: supposing $f \neq 0$, we use the linear transformation $y = u - v$, $z = u + v$, so that F becomes $2f(u^2 - v^2) + 2(g + h)xu + 2(g - h)vx$, to which we can now apply the previous method.

It should now be clear to the reader how to use Lagrange's method in any number of variables.

Exercises 7.6.1

1. Reduce the following quadratic forms to diagonal form, and calculate the rank and index of each.

   (a) $x_1^2 + x_2^2 + x_3^2 + x_1x_2 + x_2x_3 + x_3x_1$
   (b) $x_1^2 + x_2^2 + x_3^2 - x_1x_2$
   (c) $x_1^2 + x_2^2 - 4x_1x_2 - 4x_2x_3$
   (d) $3x_1^2 + x_2^2 + x_3^2 - 2x_2x_2 + 2x_3x_1 - 2x_1x_2$ [SU]

2. Explain Lagrange's method without resorting to calculus.
3. Using the assertion before (7.6.2) that ind(A) is independent of P, show that the number p of strictly positive diagonal terms of D is also independent of P and $p = \text{ind}(-A)$. Hence show that rank(A) is the number of non-zero diagonal terms of D.
4. If A is a symmetric $n \times n$ matrix, show that each eigenvalue λ is real. (Use the defining equation $A\mathbf{x} = \lambda\mathbf{x}$ for an eigenvector to show that $\lambda^2\mathbf{x}^\mathrm{T}\mathbf{x} = (\mathbf{x}^\mathrm{T}A^\mathrm{T})(A\mathbf{x})$; since $\mathbf{x} \neq \mathbf{0}$ then λ^2 is a ratio of real squares.)
5. (Practice in using notation) The usual notation for the n-variable version of (7.6.3) is

$$F = \sum_{i=1}^{n} \sum_{j=1}^{n} a_{ij}x_ix_j.$$

   If $a_{11} \neq 0$ check that $F - 1/4a\,(\partial F/\partial x_1)^2$ has no terms in x_1.
6. Suppose $n = 3$. Show that each eigenvalue λ of A is a root of the cubic equation $\det(A - tI) = 0$. If now A is symmetric, with distinct eigenvalues λ, μ, ν, use the corresponding eigenvectors as columns of a matrix R to show that $R^\mathrm{T}AR = \text{diag}(\lambda, \mu, \nu)$.
7. If you have a graphical calculator which manipulates matrices you may be able to plot the graph of $y = \det(A - xI)$ directly as a function of x and thus locate its roots. TrueBASIC includes a set of matrix commands and functions, and the following program can also be used to locate roots of the characteristic polynomial, by checking for a change of sign in consecutive values of y.

```
REM Program 7.6.1, Characteristic equation
DIM A(3,3),I(3,3),T(3,3)
MAT READ A
DATA 1, -2, 1, -2, 5, 3, 1, 3, 6
READ kmin, kmax, kstep, ymin, ymax
DATA -3, 10, 0.25, -40, 40
MAT I = IDN(3,3)
SET MODE "graphics"
SET WINDOW kmin,kmax, ymin,ymax
PLOT kmin,0 ; kmax,0
PLOT 0,ymin ; 0,ymax
LET last = 0
FOR k = kmin TO kmax STEP kstep
  MAT T = k*I
  MAT T = A - T
  LET y = det(T)
  PLOT k,y;
  IF y*last<0 THEN PRINT k - kstep/2
  LET last = y
NEXT k
END
```

8. Use the power method of iteration (as in program 7.4.1) to determine an eigenvalue of the matrix

$$A = \begin{pmatrix} 2 & -1 & 0 & 0 \\ -1 & 2 & -1 & 0 \\ 0 & -1 & 2 & -1 \\ 0 & 0 & -1 & 2 \end{pmatrix}$$

starting from the column vector $X_0 = (1\ 1\ 1\ 1)^T$, and applying A six times.

Show that the eigenvalues, λ, of A satisfy the equation

$$f(\lambda) = \lambda^4 - 8\lambda^3 + 21\lambda^2 - 20\lambda + 5 = 0,$$

and that there exists an eigenvalue of A in each of the four intervals (0, 1), (1, 2), (2, 3) and (3, 4).

By expressing $f(\lambda)$ as a function of $\mu = \lambda - 2$ and solving the corresponding quartic equation, or otherwise, determine all four eigenvalues of A. Comment on the particular value you obtained by the power method of iteration in comparison with the exact values. [SU]

9. Given that $(a\quad a\quad a\quad a)^T$, $(b\quad 0\quad -b\quad 0)^T$, $(0\quad c\quad 0\quad -c)^T$, where a, b and c are arbitrary constants, are eigenvectors of the symmetric matrix

$$A = \begin{pmatrix} 2 & 1 & 0 & 1 \\ 1 & 2 & 1 & 0 \\ 0 & 1 & 2 & 1 \\ 1 & 0 & 1 & 2 \end{pmatrix}$$

find a fourth eigenvector and determine the eigenvalues of A.

Use the eigenvectors to construct an orthogonal matrix R which is such that the transformation

$$\mathbf{x} = R\mathbf{x}'$$

changes the quadratic form $V = \mathbf{x}^T A\mathbf{x}$ into $V = \mathbf{x}'^T D x'$ with D diagonal. For your choice of R write down D explicitly. [SU]

There is a neat iterative algorithm for finding the coefficients of the characteristic polynomial of a matrix A due to D. K. Fadeev (Renton 1983) and based on recursive formulae for the sums of products of the roots of polynomial equations known as the Newton–Girard formulae. Suppose A is an $n \times n$ matrix whose characteristic polynomial is:

$$k_1\lambda^n + k_2\lambda^{n-1} + k_3\lambda^{n-2} + \cdots + k_n\lambda + k_{n+1} = 0, \qquad k_1 = 1.$$

Let $B_1 = A$. Then k_2 is $-(\lambda_1 + \lambda_2 + \cdots + \lambda_n)$, that is, minus the sum of the eigenvalues of A, and this sum is given by the trace of A (or B_1). Hence $k_2 = -\text{trace}(B_1)$. Now form $B_2 = A(B_1 + k_2 I)$ where I is the $n \times n$ identity matrix. The iteration is now given by:

$$B_i = A(B_{i-1} + k_i I) \qquad k_{i+1} = \frac{-\text{trace}(B_i)}{i} \qquad i = 2, \ldots, n.$$

Also $k_{n+1} = (-1)^n \det(A)$ and, if $\det(A) \neq 0$, then $A^{-1} = -(1/k_{n+1})(B_{n-1} + k_n I)$. Program 7.6.2 implements this algorithm in TrueBASIC.

```
REM Program 7.6.2, Fadeev's Iteration
DIM A(3,3),B(3,3),C(3,3),I(3,3),T(3,3),k(4)
MAT READ A
DATA 1, -2, 1, -2, 5, 3, 1, 3, 6
LET n = 3
MAT I = IDN(n,n)
LET k(1) = 1
MAT B = A
FOR r = 1 TO n
    LET tr = 0
    FOR j = 1 TO n
        LET tr = tr + B(j,j)
    NEXT j
    LET k(r+1) = - tr/r
    IF r = n-1 THEN MAT C = B
    MAT T = k(r+1)*I
    MAT T = B + T
    MAT B = A*T
NEXT r
MAT PRINT k
MAT T = k(n)*I
MAT C = C + T
MAT C = (-1/k(n+1))*C
MAT PRINT C
END
```

7.7 QUADRICS

Corresponding to the conics in $\mathbb{R}^2$, the function F in (7.6.3) gives rise to surfaces in $\mathbb{R}^3$ of the form

$$S_k = \{(x, y, z) \mid F(x, y, z) = k\}, \tag{7.7.1}$$

and called **quadrics**. These are (in a sense) 'models' of F, because they give us additional insight into the behaviour of F. To examine their structure, we suppose that we have changed coordinates in $\mathbb{R}^3$ so that F takes the form (7.6.4). (Note: this change need not preserve right angles.) Using a more harmonious notation, we write (7.6.4) as

$$F = u\lambda^2 + v\mu^2 + w\nu^2, \tag{7.7.2}$$

with *real* coefficients u, v, w. If none of these is zero, F is called **non-degenerate**, and we consider this case first. We want to know what the surfaces $T_k = \{F = k\}$ look like in (λ, μ, ν) space. By changing back to (x, y, z) coordinates, we shall then know the shape of S_k in (7.7.1). The pictures will depend on the index, that is, the number of negative coefficients, and we must look at several separate cases.

Case 1; index 0
Here, all coefficients are >0, so no (real) point (λ, μ, ν) can satisfy (7.7.1) unless $k \geq 0$. If $k = 0$, then only $\mathbf{0}$ satisfies (7.7.2), and in that case, T_k is the single point $\mathbf{0}$. If $k > 0$,

we look at the horizontal slices of T_k in each plane $\{v = v_0\}$; so λ, μ are then constrained to satisfy the equation

$$u\lambda^2 + v\mu^2 = k - wv_0^2. \tag{7.7.3}$$

This cannot be satisfied by any (real) λ, μ unless $k - wv_0^2 \geq 0$; so we get only an empty slice of T_k unless v_0 lies between $\pm\sqrt{(k/w)}$. Between these extremes, the slice is an ellipse, which degenerates to a point at the extremes. Thus T_k looks as in Fig. 7.7.1, and is called an **ellipsoid**. A similar argument shows that the other principal sections of T_k (i.e., slices of T_k in planes perpendicular to the λ- and μ-axes) are ellipses also.

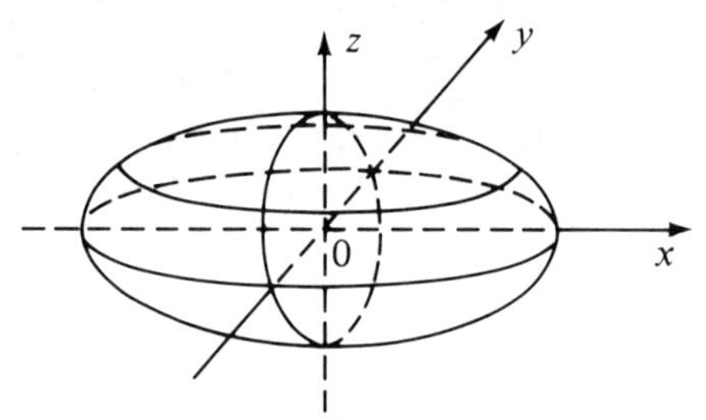

Fig. 7.7.1. An ellipsoid.

Case 2; index 3

Here, all coefficients are negative, so we have the same geometry as in case 1, but with k negative.

Case 3; index 1

Just one coefficient is negative, say $w < 0$. Again we look at the horizontal slices of T_k in (7.7.3), and again these slices are ellipses if $k \geq 0$ (since $-wv_0^2$ is now positive). If $k = 0$, then the slice is just $\mathbf{0}$, and then T_k is a double cone as in Fig. 7.7.2(a). If

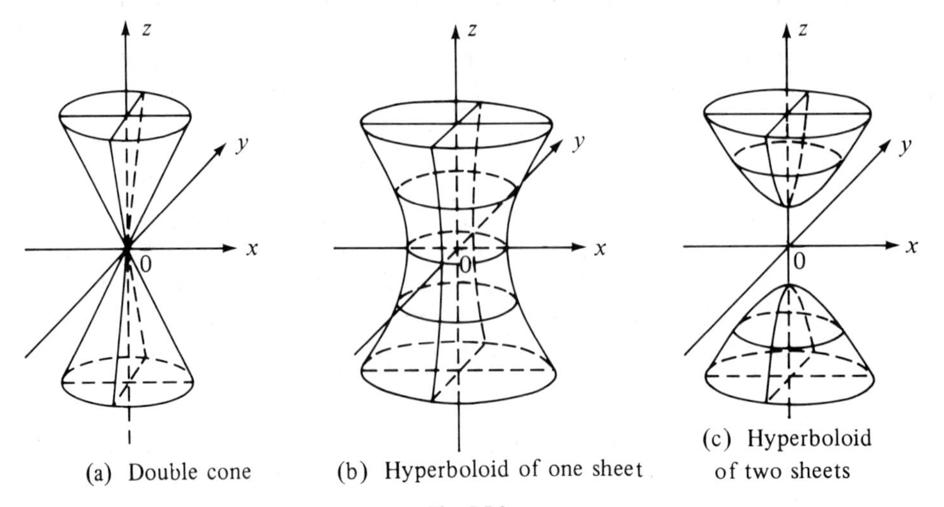

Fig. 7.7.2.

$k > 0$, no slice is degenerate, and we have the 'hyperboloid of one sheet' as in Fig. 7.7.2(b). Its section by a plane $\{\lambda = \lambda_0\}$ perpendicular to the λ-axis has equation

$$v\mu^2 + w\nu^2 = k - u\lambda_0^2$$

which is a hyperbola for any value of k, since $w < 0 < v$. Similarly for sections perpendicular to the μ-axis.

Finally if $k < 0$, then the horizontal slice of T_k in (7.7.3) has no points unless ν_0 lies outside the extreme $\pm\sqrt{(w/k)}$, when we obtain ellipses again. Hence T_k looks as in Fig. 7.7.2(c), and consists of two separate pieces. As before, we see that its other principal sections are hyperbolae, so T_k is called a 'hyperboloid of two sheets'. In each of these three cases, the hyperboloid has an axis, along the ν-axis. If u instead of w had been negative, then T_k would have had the same shape as before but with axis along the λ-axis.

Case 4; index 2

Since now two coefficients are negative, (7.7.3) takes the same form as in case 3 after multiplication by -1. By the previous arguments, therefore, T_k is a hyperboloid of two sheets if $k > 0$ and of one sheet if $k < 0$; and T_0 is a double cone.

We need not dwell on the cases when F is degenerate. For example, if in (7.7.3) w is zero, then when $k \geq 0$ each horizontal section of T_k lies vertically above the ellipse $u\lambda^2 + v\mu^2 = k$ in the (μ, ν)-plane $\{\lambda = 0\}$ so, as sketched in Fig. 7.7.3, T_k is a cylinder of elliptic cross-section (and just the ν-axis if $k = 0$). If $k < 0$, no real points satisfy the equation and T_k is empty.

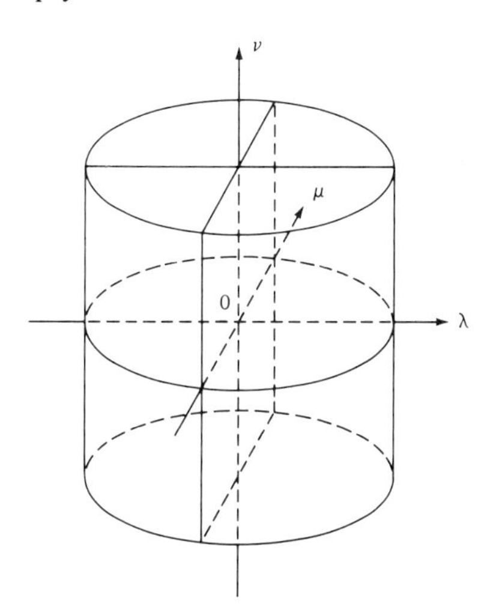

Fig. 7.7.3. Elliptic cylinder.

Exercises 7.7.1

1. Show that if $u = v = 0 \neq w$, then T_k is a pair of horizontal planes if $k \neq 0$, which merge into one plane if $k = 0$.

2. For each quadratic form F in exercise 7.6.1(1), sketch a sample of typical surfaces S_k.
3. Sketch the level surfaces $\{Q = c\}$ for the quadratic form

$$Q(x_1, x_2, x_3) \equiv x_1^2 + x_2^2 + 2\sqrt{2}x_1 x_2 + 2x_2 x_3$$

on $\mathbb{R}^3$, indicating the sign of c for each surface that you draw. [SU]

4. Let $Q(x) \equiv x_1^2 - x_2^2 - x_3^2 + 2x_1x_2 + 2x_2x_3 + 2x_3x_1$ where $x = \{x_1, x_2, x_3) \in \mathbb{R}^3$. By expressing $Q(x)$ in terms of a suitable system of orthogonal coordinates $y = (y_1, y_2, y_3)$, show that the locus $Q(x) = -1$ is a hyperboloid of one sheet. Find the axis of the hyperboloid.

   Denote the hyperboloid by H, and let S_a be the sphere $x_1^2 + x_2^2 + x_3^2 = a^2$. Write down the equation of S_a in the y-coordinates. Use the implicit function theorem 8.1.1 to show that the intersection $H \cap S_a$, if non-empty, is a smooth curve provided $a^2 \neq 1$ or 2. Describe or sketch the form of $H \cap S_a$ for three values $a^2 = \frac{1}{2}, \frac{3}{2}, 3$. [SU]

8

Contours: the implicit function theorem

It is now time to look more closely at the problem of representing a smooth function $f(x, y)$ in a graphical way, in order to gain greater understanding of f. For this the reader will need to know about partial differentiation, which may be found in Chapter D.

8.1 CONTOURS AND THE IMPLICIT FUNCTION THEOREM

Given a function $g(x)$ of a single variable x, a computer can produce a table of values of g, and plot a graph from it. But humans have the additional power of vision and interpretation, by which they can gain extra knowledge of g; they can study the graph of g, and readers will already have had some experience of the value of graphs, as well as of the problem of constructing them. By definition, this graph is the set of all points $(x, g(x))$ in $\mathbb{R}^2$, as x runs through the domain of g.

Similarly, the graph G of a function $f(x, y)$ is the set of all points $(x, y, f(x, y))$ in three-space $\mathbb{R}^3$, as (x, y) runs through the domain of f. In general G will look like a mountain range, with peaks and hollows (local maxima and minima), and other features like mountain passes (saddle points). Such special features occur at the **critical points** of f. For a function of a single variable the critical points are where the derivative is zero. In the case of a function of two variables the critical points are given by the zeros of the 'gradient vector' grad $f = (\partial f/\partial x \quad \partial f/\partial y)$, which is more conveniently given in compact 'slot' notation as grad $f = (f_1 \quad f_2)$. These critical points are linked by contours of f, and we now consider how to draw in these contours at all the remaining points (which we call **ordinary**). The need for sketching contours arises for the reason that geographers study a terrain by constructing a contour map; it is easier both to draw and to interpret than a perspective sketch.

Now, if we want to represent a portion of the Earth's surface on a flat piece of paper in the form of a map, then the surface of the terrain is represented by the graph of the function which gives height above sea-level. Analogously, on the graph G of $f(x, y)$, we regard the height of any point $P = (x, y, f(x, y))$ as its z-coordinate,

$f(x, y)$. We then seek to study our function f by drawing a suitable set of contours. To be precise, if c is any constant, we say that the **contour** of f at height c is the set $\{f = 0\}$ of all points P in the domain of f, such that $f(P) = c$. The set of all the contours is called the **portrait** of f, but we cannot draw the entire infinite set of these: instead, as with a geographical map, we would like to sketch a sample of those contours which are 'typical', in a sense to be defined eventually. To fix ideas, we show in Fig. 8.1.1 a portrait of the function.

$$f(x, y) = x^3 + y^3 - 2x - y, \tag{8.1.1}$$

and we shall refer to it below to illustrate the ideas that follow. After the material of section 8.3 below, we shall be able to explain how it was constructed.

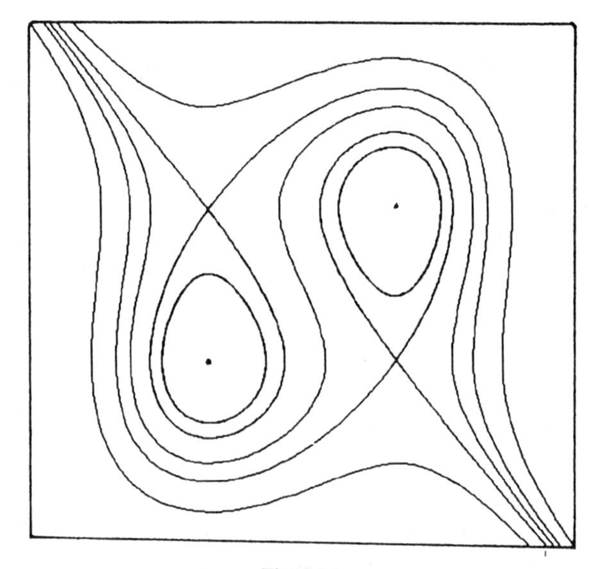

Fig. 8.1.1.

Readers will nowadays expect there to be a computer program that will produce such a portrait, but before we can construct a suitable algorithm, we need to know what to build into it, and how to eliminate (or recognize) false output. As Fig. 8.1.1 indicates, the matter is likely to be fairly complicated (even though f in (8.1.1) is a fairly simple function). Therefore we need some thoughtful theoretical considerations, as follows.

At a first guess, we expect a typical contour to be a smooth curve, like those in Fig. 8.1.1, or on a weather chart (in which the isobars are the contours of the pressure function). But there are some exceptions: for example, the tip of Mount Everest constitutes the contour of the highest points on Earth, and it is a single point, not a curve. Note that some contours in Fig. 8.1.1 have some cross-over points, and are thus not typical near those points; such points are critical, and cannot be regarded as accidental, even though they are rare, as we shall see later.

Exercises 8.1.1

1. Sketch the contours and surfaces corresponding to the following expressions. In each case, calculate the critical points, if any. For example, the contours of (i) are circles, centred at its sole critical point **0**, and the associated surface is the graph, $z = x^2 + y^2$; the latter is a bowl-shaped surface, with the parabolic section $x^2 = z$ in the plane $y = 0$, and similarly in the section $x = 0$.
(i) $x^2 + y^2$, (ii) $x^2 - y^2$, (iii) $x^2 - y + 2$, (iv) $-xy$, (v) $x^2 + 4y$,
(vi) $x + 3y + 9$, (vii) $\sin(x + y)$, (ix) $\ln(x + 2y)$, (viii) $[x + y]$, where $[t]$ means the largest integer $< t$. (This surface is not smooth everywhere!)
2. We shall show later that the equation of the tangent plane at $(a, b, f(a, b))$ to the surface $z = f(x, y)$ is
$$z - z_0 = (x - a)\frac{\partial f}{\partial x} + (y - b)\frac{\partial f}{\partial y}, \qquad z_0 = f(a, b).$$
Find the tangent plane to each surface in exercise 1 when $(a, b) = (1, \frac{1}{2})$.
3. In the last chapter, program 7.2.2 was used to shade a region whose boundary was a given conic. Such a program can easily be adapted to shade in regions bounded by contour lines—and this is much easier than computing the actual contours. Again we scan a grid of points (x, y) in some rectangular region and evaluate $z = f(x, y)$ at each of them. We just need some function $s(z)$ which turns this height into the code number of a colour (which is an integer), and then we plot the point (x, y) in the given colour. The bands of colour on the screen have contours as their boundaries. Use the program to explore different regions and different functions $f(x, y)$. You may need to change `shade(c)`.

```
REM Program 8.1.1,  Contours
FUNCTION f(x,y) = 3*x^2 + 5*x*y - y^2
FUNCTION shade(c) = c/10
READ xmin, xmax, xstep, ymin, ymax, ystep
DATA -6  , 6   , 0.1  , -4  , 4   , 0.1
SET MODE "graphics"
SET WINDOW xmin, xmax,  ymin, ymax
FOR x = xmin TO xmax STEP xstep
  FOR y = ymin TO ymax STEP ystep
    SET COLOR shade(f(x,y))
    PLOT x,y
  NEXT y
NEXT x
END
```

Sample output

4. Let $f(x, y) = 2xy/(x + y)$ if $(x, y) \neq \mathbf{0}$, with $f(\mathbf{0}) = 0$. Sketch the graph of the function $x - f(x, y)$ when (a) $y = 1$, (b) y is large and positive, (c) y is small and positive. Use polar coordinates to sketch the level curves of f.
5. Show that the graph $y = x^4 - x^2 + ax$ has a point of inflexion when $a = (2\sqrt{2})/(3\sqrt{3})\,(= a_0$, say). Sketch the graph when (i) $a = 0$, (ii) $0 < a < a_0$, (iii) $a_0 < a$. Hence or otherwise sketch the contours of the function

$$f_a(x_1, x_2) = x_1^4 + ax_1 - x_1^2 - x_2^2$$

for the same three ranges of a. [SU]

6. Sketch the graph of the function

$$g(x) \equiv -\frac{x^4}{4} + \frac{x^3}{3} + x^2 + 1.$$

Hence or otherwise sketch (a) the graph G and (b) the contours of the function

$$f(x, y) \equiv g(x) - \tfrac{1}{2}y^2.$$

(In (b) take care to include contours that pass through critical points.) [SU]

There is a useful theorem which tells us when a contour is a smooth curve, and some explanation is desirable to help readers understand it. It is called the 'implicit function theorem' because it concerns the solutions of the equation

$$f(x, y) = c, \tag{8.1.2}$$

which—so earlier mathematicians thought—could in principle be solved to define y as a function ϕ of x; but as the function ϕ is not expressly stated, and may be difficult to work out, ϕ is only 'implicit' in equation (8.1.1). Nevertheless, the argument was that since y is a function of x, then we can at least differentiate (8.1.2) straightforwardly with respect to x find $dy/dx(= y')$. For example if

$$f(x, y) = x^3y^2 + 5xy = c,$$

then

$$3x^2y^2 + 2x^3yy' + 5y + 5xy' = 0,$$

which can now be solved to find y'. Some readers may recognize this as the process of 'implicit differentiation', and it is correct once we know the domain of $y = \phi(x)$. But why is it correct?

No proof was offered until the late nineteenth century when theorem 8.1.1 below was formulated (with its hypotheses to exclude cases of breakdown). It tells us that if P is an ordinary point on the contour C, so the vector $\operatorname{grad} f = (\partial f/\partial x \quad \partial f/\partial y) = (f_1 \quad f_2)$ is **non-zero** at P, then (at least near P), C looks like the graph of a function of one variable (and that function is the 'implicit' ϕ): see Fig. 8.1.2. Its proof uses the fact that if P is a solution of (8.1.2), then since f in (8.1.2) is approximately $f(P) + ax + by$, with $(a \quad b) = \operatorname{grad} f|_P$, a first approximation to y in (8.1.2) is $(c - f(P) - ax)/b$ if $b \neq 0$; and this can then be improved.

To be more precise, suppose that f is given on a region D. If $P = (x_0, y_0)$, and P is an ordinary point of f, then $\operatorname{grad} f \neq \mathbf{0}$ at P, so either $f_1(P)$ or $f_2(P)$ is non-zero—

say $f_2(P) \neq 0$. Since (as always) we assume continuity of derivatives, then by the inertia principle (theorem D-2.1), there is some s-neighbourhood N of P on which f_2 remains non-zero. The term 'near P' will then mean 'inside N'. Recall that in $\mathbb{R}$, an r-neighbourhood of x_0 is an interval of the form $\{x_0 - r < x < x_0 + r\}$.

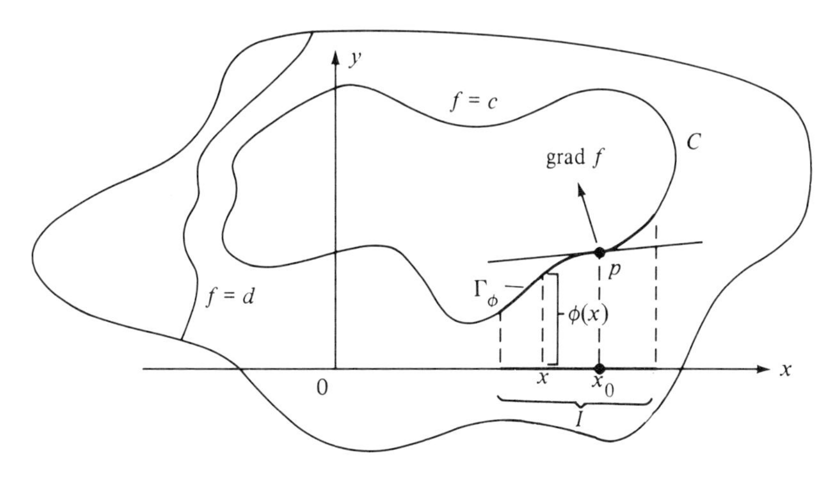

Fig. 8.1.2.

We can now state the following 'implicit function theorem ', known briefly as the IFT.

Theorem 8.1.1 The implicit function theorem (IFT)
Suppose that $P = (x_0, y_0)$ is a non-critical point of f satisfying (8.1.2), and that f_2 is non-zero on the neighbourhood N of P. Then there is an r-neighbourhood I of x_0 in $\mathbb{R}$, and a differentiable function $\phi: I \to \mathbb{R}$, with the properties:

(a) $y_0 = \phi(x_0)$,
(b) *the only solutions (x, y) of* (8.1.2) *in the neighbourhood N are those for which $y = \phi(x)$.*

The set of solutions of (8.1.2) in N is therefore the graph Γ_ϕ of the function, and we say that C is 'locally a graph' at P. Thus, for every x in I we have $f(x, \phi(x)) = c$. Therefore this equation is an identity and we can differentiate by the chain rule to get

$$f_1(x, \phi(x)) \cdot 1 + f_2(x, \phi)) \cdot \phi'(x) = 0,$$

whence, since f_2 is not zero on N, we can divide by f_2 to give

$$\phi'(x) = -\frac{f_1 x, y)}{f_2(x, y)}. \tag{8.1.3}$$

From this, we make the following important deduction.

Corollary 8.1.1

At P, the vector grad f *is orthogonal to the contour through P.*

For, by (a) above, $P = (x_0, y_0) = (x_0, \phi(x_0))$ and the tangent to the contour C through P is the same as the tangent at P to the graph of ϕ; and this has gradient $\phi'(x_0)$. By (8.1.3) it is parallel to the vector $\mathbf{v}_0 = (-f_2(P) \quad f_1(P))^T$. Hence the scalar product of $\mathbf{v}_0$ with grad f is $(f_1(P) \quad f_2(P)) \cdot \mathbf{v}_0$, and this multiplies out to zero; so grad f is perpendicular to $\mathbf{v}_0$ and hence orthogonal to C at P. This proves corollary 8.1.1 which if applied to a contour map, tells us that the most direct way up or down a mountain is to cross each contour orthogonally.

Since grad f is perpendicular to the tangent line L at P, the vector equation of L is

$$(\mathbf{r} - \mathbf{p}) \cdot (f_1(P) \quad f_2(P)) = 0. \tag{8.1.4}$$

The formula (8.1.3) is remarkable, because it tells us the derivative of ϕ, even though we may not be able to find an analytic formula for ϕ itself. We shall not prove the theorem here. In order to state theorem 8.1.1 above, we assumed that $f_2(P)$ was not zero. If, instead, we had assumed that $f_1(P) \neq 0$ then we simply interchange x and y in the statement.

Examples 8.1.1

1. The unit circle S is the contour of height 1, of the function $f(x, y) = x^2 + y^2$; here, grad $f = (2x \quad 2y)$ which is never zero (since $\mathbf{0}$ does not satisfy the equation of S). Also, $f_2 \neq 0$ if $y \neq 0$ (i.e., if P is not one of the two points A, B at which S cuts the x-axis). Of course, without any knowledge of theorem 8.1.1, we can solve the equation $f(x, y) = 1$ to get $y = \pm\sqrt{(1 - x^2)}$, taking the plus sign when P lies above AB and the minus-sign when P lies below. Neither solution has the property of being defined on an interval surrounding A or B, for the geometrical reason that the tangents to S there are vertical. Instead, we can solve at A for x as $\sqrt{(1 - y^2)}$ and again S is locally the graph of that function at A; and similarly for B. Note that even in this simple case, there is no question of being able to express y as one 'globally defined' single function of x.
2. Suppose $f(x, y) = \tan xy + y^3 - x^2$. Here, no algebraic tricks will allow us to solve directly for x or y. We have

$$\text{grad} f = (y \sec^2 xy - 2x \quad x \sec^2 xy + 3y^2)$$

so if P is such that $f_2(P) \neq 0$, for example if $x \geq 0$, then we can apply theorem 8.1.1 and assert that the contour C through P is locally the graph of a function $y = \phi(x)$ with

$$y' = \phi'(x) = -\frac{y \sec^2 xy - 2x}{x \sec^2 xy + 3y^2}.$$

Recall that the height of C at P is $f(P)$. Readers may object that none of this tells us very much about C, but we shall find out more as the theory develops. Have patience!

Exercises 8.1.2

1. With f as in the last example, find the equation of the tangent to the contour of f through P, when P is one of the points $(1, 1)$, $(-2, 3)$. (Use equation (8.1.4)).
2. With the same f, locate its critical points. Verify that there is one at the origin, and that the others lie on the curve $2x^2 + 3y^3 = 0$. Hence decide where theorem 8.1.1 applies to f.
3. For each of the following functions $f:\mathbb{R}^2 \to \mathbb{R}$, which are the points around which it is valid to replace $f(x, y) = 0$ locally by $y = g(x)$ for some smooth function g? What goes wrong geometrically at places where this is *not* valid?
   (a) $f(x, y) = x^2 - 2y^2 - 4$
   (b) $f(x, y) = x^3 - 3x - y^2 + 2$
   (c) $f(x, y) = (x + y)(x^2 + y^2 - 1)$.
4. (a) If $y^3 + 3x^2y = c$ prove that when y is a function of x, $\mathrm{d}^2y/\mathrm{d}x^2 = 2c(x^2 - y^2)/(x^2 + y^2)^3$. Now find a formula for ϕ''in the IFT, theorem 8.1.1.
   (b) Find the critical points of the function $f:\mathbb{R}^2 \to \mathbb{R}$ given by
   $$f(x_1, x_2) \equiv x_1^2 - 2x_2^2 + x_2^3.$$
   Sketch the contours of the function, including in particular those which pass through critical points. Sketch or describe the graph of the function.
   What information does the implicit function theorem give about the nature of the contour $\{f = 0\}$? [SU]
5. In $\mathbb{R}^2$, two smooth curves $f(x, y) = a$, $g(x, y) = b$ intersect at the ordinary point p. Show that they have a common tangent at p iff grad f is parallel to grad g at p, and hence iff the **Jacobian** matrix
   $$J = \begin{bmatrix} \dfrac{\partial f}{\partial x} & \dfrac{\partial f}{\partial y} \\ \dfrac{\partial g}{\partial x} & \dfrac{\partial g}{\partial y} \end{bmatrix}$$
   when evaluated at p, has zero determinant. (For brevity, we sometimes use the classical notation $\partial(f, g)/\partial(x, y)$ for J.)
6. Let $f(t, x)$ be a smooth function, and write $f_t(x)$ for $f(t, x)$. Suppose that the equation $f_t(x) = 0$ has exactly n roots x when $t = 0$, and m when $t = 1$. Show that if $m \neq n$ then there exists some $P = (t_0, x_0)$ such that $\partial f/\partial x = 0$ at P. (Suppose $m < n$ and yet $\partial f/\partial x$ $(= f_t')$ is always non-zero. Then there is a first $s > 0$ such that f_s has $k > m$ zeros $x_1, x_2, \ldots, x_k$. Use the IFT to show that there is an r-neighbourhood I of s, and smooth functions $\theta_i: I \to \mathbb{R}$ $(1 \leqslant i \leqslant k)$, such that $x = \theta_i(s)$ and $f_t(\theta_i(t)) = 0$ if $t \in I$. Now use the uniqueness part of the IFT to obtain a contradiction when $t < s$.)
7. If $z = f(x, y)$ then (x, y, z) represents a point in $\mathbb{R}^3$ on the surface given by $f(x, y)$. A simple projection from points (x, y, z) in $\mathbb{R}^3$ to points (x', y') in $\mathbb{R}^2$ is **oblique projection**, which is the one we most commonly use to represent three-dimensional objects, such as cubes, when drawing on paper. A typical such projection is given by: $x' = y - x/2$, $y' = z - x/2$. If we fix y for the moment at a given value b, then $z = f(x, b)$ gives z as a function of x alone. As x takes values in some interval `[xmin, xmax]` the graph of z against x represents the **section** of the surface $z = f(x, y)$ by the plane $y = b$, and this is much easier to compute than the sections $z = c$ which give the contours. Varying b gives a set of 'salami slices' of the surface parallel to the xz-plane. A similar procedure will give sections by planes $x = a$ which are parallel to the yz-plane. Using the oblique projection the following program represents the surface as a 'wire mesh' of the set of such orthogonal slices: $x = a$, $y = b$. (For more detail on producing such views, see Oldknow (1987) p. 176.)

```
REM Program 8.1.2, Surface slices
FUNCTION f(x,y) = x*x - x*y + y*y
READ xmin, xmax, xstep, ymin, ymax, ystep, zmin, zmax
DATA -4  , 4   , 0.5  , -4  , 4   , 0.5  , -10 , 50
SET MODE "graphics"
SET WINDOW ymin - xmax/2, ymax - xmin/2,  zmin, zmax
FOR x = xmin TO xmax STEP xstep
  FOR y = ymin TO ymax STEP ystep/8
    PLOT y - x/2, f(x,y) - x/2 ;
  NEXT y
  PLOT
NEXT x
FOR y = ymin TO ymax STEP ystep
  FOR x = xmin TO xmax STEP xstep/8
    PLOT y - x/2, f(x,y) - x/2 ;
  NEXT x
  PLOT
NEXT y
END
```

Sample output

8. Use the techniques of the compass needle program 3.3.1, and the contour program 8.1.1, to superimpose small line segments along grad f over the contours of various functions $f(x, y)$ to verify corollary 8.1.1.

8.2 THREE VARIABLES

Next, suppose we have a function $F(x, y, z)$ of three variables, when $P = (x, y, z)$ runs through a domain E in $\mathbb{R}^3$. We cannot now construct a graph of F as before, because in the physical space that we live in, there is no fourth dimension in which to represent the height $F(P)$ of P. We are therefore forced to use the analogue of a contour map, and now the 'contours' will be surfaces rather than curves. These are now usually called **levels** rather than contours (and in physics they are often called 'equipotentials'). Just as with two variables, we can have critical points, and all other points of E are called 'ordinary'. The most familiar examples of three-variable functions are the linear, and the quadratic. Linear functions have no critical points, and the levels are planes; the levels of a quadratic function are quadric surfaces, such as those discussed in the last chapter.

To describe the levels of a general smooth function F on E as above, in the neighbourhood of an ordinary point P, there is an appropriate version of the IFT as follows, which is almost word for word the same as the one above. The level through P is now the set of all points Q satisfying the equation:

$$F(Q) = c, \quad \text{where } c = F(P), \tag{8.2.1}$$

and since P is ordinary, then at least one coordinate of grad F at P is non-zero, say $F_3(P) \neq 0$. By the inertia principle, there is then a neighbourhood N of P in E such that F_3 is never zero on N. With the stage thus set, we have the following theorem.

***Theorem 8.2.1** (IFT for three variables)*
Suppose that $P = (x_0, y_0, z_0)$ is a non-critical point of F such that $F(P) = c$, and F_3 is non-zero on the neighbourhood N of P in the domain E. Then there is an r-neighbourhood U of (x_0, y_0) in $\mathbb{R}^2$, and a differentiable function $\phi: U \to \mathbb{R}$, with the properties:

(a) $z_0 = \phi(x_0, y_0)$,
(b) *the only solutions (x, y, z) of (8.2.1) in N are those for which $z = \phi(x, y)$.*

As with two variables, we do not prove the theorem, because our main aim is to help readers understand and use it. Note that the theorem says that the set S of solutions of (8.2.1) in N is locally the graph of the function ϕ, and therefore a small piece of surface. Thus the level L through P intersects N in S, and L is 'locally a graph' at P, since for every (x, y) in U we have $F(x, y, \phi(x, y)) = c$. Therefore this equation is an identity and by the chain rule we can differentiate with respect to x to get the partial derivative $\phi_1(X)$ where $X = (x_0, y_0)$; thus $F_1(P) + F_3(P)\phi_1(X) = 0$; similarly $F_2(P) + F_3(P)\phi_2(X) = 0$. Hence, by our assumption that $F_3(P) \neq 0$, we obtain

$$\phi_1(X) = -\frac{F_1(P)}{F_3(P)}, \qquad \phi_2(X) = -\frac{F_2(P)}{F_3(P)}. \tag{8.2.2}$$

In order to state theorem 8.2.1 neatly, we assumed that $F_3(P)$ was not zero, if, instead, we had assumed that (say) $F_1(P) \neq 0$ then we simply interchange x and z in the statement of Theorem 8.2.1, with a corresponding change in (8.2.2). These formulae were used in Chapter 6 to investigate the way in which the roots of a cubic vary with the coefficients, and readers may find that the discussion there is now clearer.

Exercises 8.2.1

1. Sketch the level surfaces $F(x, y, z) = c$ corresponding to the expressions:
   (i) $x^2 + y^2 + z$, (ii) $(x^2 + y^2)/z$, (iii) $z^2/(x^2 + y^2)$.
   (It may help to draw some sections $z =$ const. in the (x, y) plane, and similarly $x =$ const., and so on.)
2. Check the formulae (8.2.2) when F is linear of the form

   $$px + qy + rz.$$

3. In $\mathbb{R}^3$ let L be the line with vector equation $\mathbf{r} = \mathbf{p} + t\mathbf{v}$, with $\|\mathbf{v}\| = 1$, and let $f: \mathbb{R}^3 \to \mathbb{R}$ be a smooth function. Regard $f(\mathbf{r})$ as a function $g(t)$. Use the chain rule to show that $g'(t) = \mathbf{v} \cdot \operatorname{grad} f|_\mathbf{r}$ (the **directional derivative** of f along the direction $\mathbf{v}$); and hence that $g'(t)$ is maximum when grad f points along L. (For brevity, grad $f|_\mathbf{r}$ denotes the result of evaluating grad f at the point $\mathbf{r}$.)

4. You are pilot of an overheating space-craft near the sunny side of Mars. The temperature at (x, y, z) is $T(x, y, z) = \exp(-x^2 - 2y^2 - 3z^2)$ where x, y, z are measured in metres. You are currently at $(1, 1, 1)$.
   (a) In what direction should you steer to decrease the temperature most rapidly?
   (b) If the ship travels at 8 metres per second, how fast will the temperature decrease if you go in that direction?
   (c) The skin of the hull will crack if cooled at more than $\sqrt{14}\,e^2$ degrees per second. Describe the set of possible directions in which you may travel while bringing the temperature down safely. [SU]
5. (Tangent planes) In Section 4.2 we explained how the tangent at P on a graph $y = \varphi(x)$ had to be defined by agreement, as the line through P with gradient $\varphi'(P)$. From this, equation (8.1.4) told us the tangent to the curve $f(x, y) = c$ at an ordinary point p; it consists of the set of all vectors $\mathbf{r}$ in $\mathbb{R}^2$ such that $(\mathbf{r} - \mathbf{p})$ is perpendicular to grad f at $\mathbf{p}$. Similarly, then, if $\mathbf{q}$ lies on the level surface $S{:}F(x, y, z) = c$ in $\mathbb{R}^3$, we *define* the tangent plane T to S at $\mathbf{q}$ as follows: T is to consist of all vectors $\mathbf{r}$ in $\mathbb{R}^3$ such that

$$(\mathbf{r} - \mathbf{q})\cdot \text{grad } F|_{\mathbf{q}} = \mathbf{0}. \tag{8.2.3}$$

   Now justify the assertion of exercise 8.1.1 (2) about the equation of the tangent to the surface $z = f(x, y)$. (Take F of the form $z - f(x, y)$.) Test (8.2.3) for 'reasonableness' by showing that if $q = (q_1, q_2, q_3)$ then the plane $\{y = q_2\}$ cuts S and T in the curve $F(x, q_2, z) = F(q)$, and its tangent, respectively. (Use 8.2.2)
6. Find the equation of the tangent plane to the surface

$$x^2 + xy + y^2 + z^2 = 4$$

   at the point $(1, 1, 1)$. Find some more points on this surface.
7. If $F(x, y, z) = (px + qy + rz)(ux + vy + wz)$ and the vector product of $(p, q, r)^{\mathrm{T}}$ and $(u, v, w)^{\mathrm{T}}$ is non-zero, show that F has a *line* of critical points. What if the product is zero? Sketch the level surfaces in each case.
8. Let F be the quadratic function $ax^2 + by^2 + cz^2$ which has a single critical point, namely $\mathbf{0}$. Solve the equation $F = 0$ for z as a function of x and y, and calculate the partial derivatives. Now calculate these by using (8.2.2) to check for agreement.
9. Show that in (8.1.2), y is a smooth function of the pair (x, c). (Apply theorem 8.2.1 to the function $F(c, x, y) = c - f(x, y)$.) This shows that if we smoothly vary the height c of a contour, the contour itself varies smoothly.
10. Make sense of the following problem, expressed in classical language. 'When v is eliminated between equations $y = f(x, v)$ and $z = g(x, v)$, the equation $z = h(x, y)$ is obtained. Prove that

$$\frac{\partial h}{\partial x}\frac{\partial f}{\partial v} = \frac{\partial f}{\partial v}\frac{\partial g}{\partial x} - \frac{\partial f}{\partial x}\frac{\partial g}{\partial v}.$$

   (If $y = f(x, v)$ then $v = \phi(x, y)$ if the IFT applies, and then $z = g(x, \phi(x, y))$; but what are the precise hypotheses required for applying the IFT?) Verify this result when $y = x\cos v - \sin v$, $z = x \sin v + \cos v$.
11. Sketch the surface $z = \sinh(x^2 + y^2)$ and draw the contour lines in the xy-plane. Determine the equation of the tangent plane to the surface at $x = 1$, $y = 2$. [SU]

Because of the way in which (8.2.1) and theorem 8.2.1 are formulated, they immediately make sense when P has n coordinates rather than 3, provided we extend the notions of graph and neighbourhood in the obvious way. (We cannot then picture them accurately in our three-dimensional, physical world.) The resulting statement can be proved, as the IFT for n variables, with corresponding use of the chain rule for finding partial derivatives. The theory can be found in more advanced texts, such as Webb (1991), but before readers attempt it, they will need a good grasp of the practicalities when $n \leq 3$. Even when $n = 3$, it is in general a tricky problem to sketch the level surfaces of a function of three variables, and we now concentrate on the easier (but non-trivial) problem of producing a set of contours when $n = 2$.

8.3 CONSTRUCTING THE PORTRAIT OF A FUNCTION

Given a smooth function $f: D \to \mathbb{R}$, with D a domain in $\mathbb{R}^2$, we shall show how to construct the portrait of f in the sense explained prior to (8.1.1). At a non-critical point P, theorem 8.1.1 tells us that the contours near P look like smooth graphs, because f is locally linear. When P is a critical point, we can summarize the situation as follows (for details see section D. 6).

Everything depends on the sign of $\Delta(P)$, which is the value at P of the **Hessian**

$$\Delta(x, y) = f_{11}(x, y) \cdot f_{22}(x, y) - (f_{12}(x, y))^2. \tag{8.3.1}$$

Recall that P is **non-degenerate** if $\Delta(P) \neq 0$, and such critical points are more typical than degenerate ones. If $\Delta(P) > 0$, then f has a local maximum or minimum at P; if $\Delta(P) < 0$ then P is a saddle point, in the sense of Fig. 3.4.1 (but without the arrows: contours do not have a 'natural' direction). In particular, *there are no spirals among the contours.*

Armed with this information we can construct the portrait of f by going through the following stages.

(A) *Local linearization.* First we draw a sample of the contours in the neighbourhood of each critical point;

(B) *Globalization.* Then we join these fragments of portrait to make a whole.

To do this we need to gather together three principles, which are all consequences of the implicit function theorem.

(B1) *Slope rule.* If z is an ordinary (= non-critical) point, then there has to be (at least) a small piece of a contour C through z, at which the tangent has slope

$$g(z) = -\frac{f_1(z)}{f_2(z)}. \tag{8.3.2}$$

(B2) *No-crossing rule.* No other contour but C passes through z, so contours cannot cross each other at an ordinary point.

(B3) *Exit rule.* This will be explained in detail later, but it will tell us how to extend the small piece of contour through z to its fullest extent.

This summary being granted, we now detail a routine for drawing portraits by hand. As we shall see, this routine will require some human judgement and so is not completely mechanizable into an algorithm, strictly speaking. However, it will allow us to check computer output from a computer algorithm. Also, what we describe is a 'fair weather' routine for simplicity, so we assume we are dealing with a non-degenerate smooth function f with domain $\mathbb{R}^2$; other assumptions will be stated as needed. Once the routine has been practised and absorbed, then we can move on to situations wherein the assumptions can be modified. Already the routine is lengthy to write out, and readers should skim it first to get the gist, and then read the worked examples that follow.

8.3.1 The portrait routine

Step 1
Draw the curves H, V on which the points z satisfy

$$H: f_1(z) = 0; \qquad V: f_2(z) = 0. \tag{8.3.3}$$

(We call H, V, respectively, the **horizontal** and **vertical manifolds** of f.) Find all their points of intersection; these are the critical points of f.

Step 2
Decide whether or not H is itself a contour of f. If it is, then it will appear in the finally constructed portrait; if it is not, then it will be used as a guide for drawing contours. Similarly for V (so our sketches indicate H, V by dashed curves).

Step 3
For each critical point P, calculate $\Delta(P)$ from (8.3.1). By assumption, f is non-degenerate, so either $\Delta(P) > 0$ or $\Delta(P) < 0$. Use theorem D.6.2 to classify P as a max/min or saddle.

Step 4
The curves H, V divide up the plane into a set of regions, each a single connected piece, with boundary composed of pieces of H or V. We assume that the number of these regions is finite. Excepting these boundaries, on each such region X the function $g(z)$ in (8.3.2) is defined (the denominator being non-zero) and has constant sign. According to the sign of g call X 'positive' or 'negative', and mark within X a small segment of positive or negative gradient. In brief, we call the segment a 'marker' and say it is 'of slope = sign X'. This process is called 'signing the regions' (see Fig. 8.3.1).

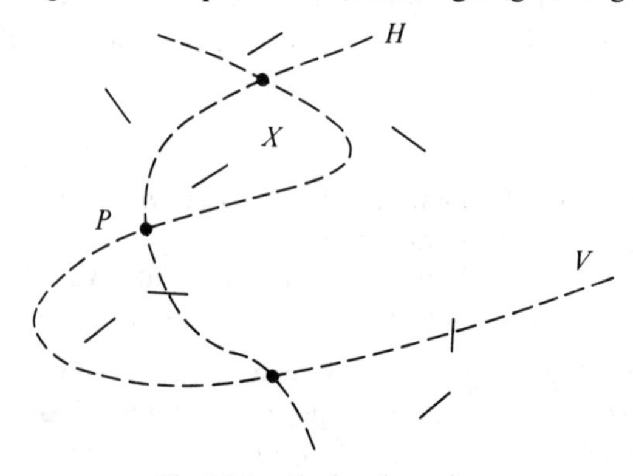

Fig. 8.3.1. Signing the regions.

Comment. By (8.1.3) the tangent to the contour through z in (8.3.3) has gradient $g(z)$. This tangent becomes vertical or horizontal according as z nears V or H, respectively (hence their names).

Step 5

Suppose that there is a saddle point P in step 3. We must now draw the two **separatrices** that cross at P; each of these is a smooth curve, divided into two pieces by P, so we have four 'arms' issuing from P, each lying in one of the four regions that divide up the neighbourhood of P. An arm is part of a contour. Let X be one of these regions. Then (see Fig. 8.3.2 (i)) we draw an arc A from P into X according to the 'slope' rule: slope = sign X, (which summarizes the slope rule (B1) above; we need not be concerned here with the exact value of g.) As A grows longer, Fig. 8.3.2 (ii) shows three possibilities which give us the details of the above exit rule (B3):

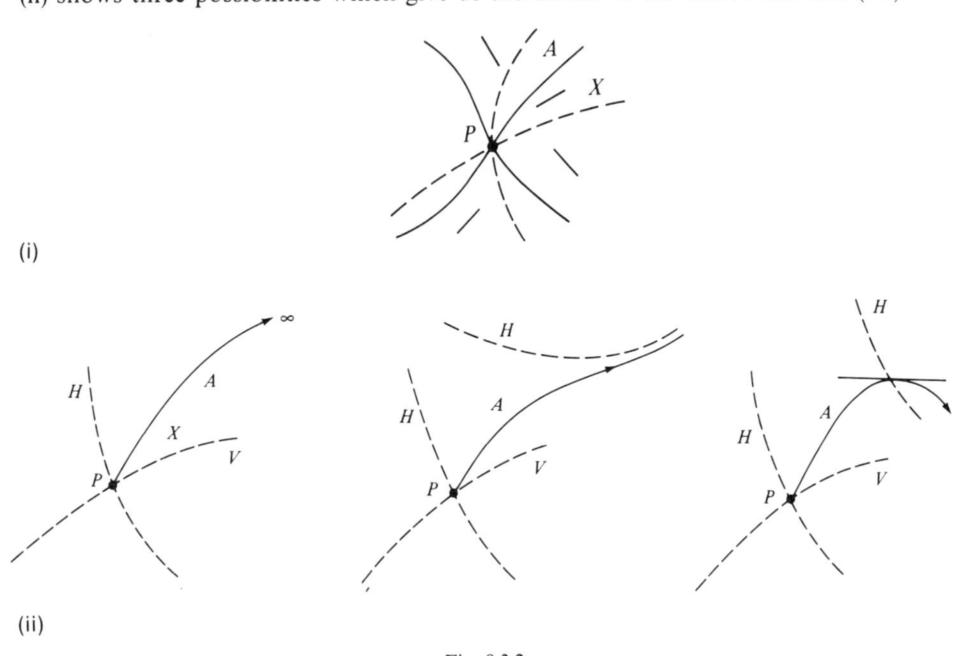

Fig. 8.3.2.

(a) A can head for infinity, without changing the sign of its slope; or A is heading for a portion of H (say) in the boundary of X, and either

(b) H is a contour of f, or

(c) not (these being decided in step 2). Similar remarks apply to V.

If (a), we prolong A as far as convenient.

If (b), we use the 'no-crossing' rule (B2) above: contours cannot cross each other except as separatrices at a saddle point. Therefore we either prolong A to be a separatrix at a second saddle point Q on H, or let it approach H asymptotically, without ever crossing.

If (c), prolong A to cross H with a horizontal tangent there, and into the next region Y, in which A continues with slope = sign Y. (Since H is not a contour, the 'no-crossing' rule is not violated.)

Repeat (a), (b), or (c) as appropriate. (If V is in question, it is crossed at a point on A with vertical tangent.)

Repeat this process with each of the four arms at P. Repeat the entire process for each remaining saddle point.

The plane is now divided by the various separatrices into new regions, and each of the remaining contours of f will lie in one of these, without crossing the boundary.

Step 6

If there are any points P in step 3 with $\Delta(P) > 0$, then as before there are four old regions X that meet at P, but also P lies in one of the new regions Z created in step 5. Draw a simple closed curve C as a contour near P, which surrounds P and lies in Z. As before, when C passes through X, then its slope is that of X.

Step 7

(a) For each new region Z from step 5, draw a curve close to the borders of Z and following the slope rule, and without crossing any contour already drawn.
(b) If any of the resulting contours are felt to be too wide apart, draw another one between them, again subject to the 'slope' and 'no-crossing' rules.

(Here, (b) is a matter of artistic judgement, and is not strictly necessary for drawing a portrait with a representative sample of contours.)

Step 8

The portrait as drawn may suggest that a contour is actually a straight line, and this should be verified analytically. Also a portrait may look roughly symmetrical about this or other lines. Again, these suspicions can be checked analytically, and the drawing adjusted accordingly. (Symmetry about coordinate axes might well be found in step 3, to shorten calculations.)

At this point, we insert an exercise for readers to test their understanding by drawing the portraits of quadratic functions, to get a quick idea of the orientation of the various conics, and so on. One could instead use the algebraic methods of Chapter 6, but it is useful to try out the routine in this very simple case. The routine will not tell us all of what we know from algebra, for example that an asymptote is a straight line, but we are interested only in rough qualitative pictures here: accuracy can be left to the computer.

Exercises 8.3.1

1. Apply the routine to obtain portraits of the following quadratic functions:

    (a) $f(x, y) = 3x^2 + 5xy - y^2$
    (b) $f(x, y) = 2x^2 - 3xy + 5y^2 + x - 7y$.

   Check them against the computer output from programs 8.1.1 and 8.1.2.

8.4 CUBIC POLYNOMIALS

After the quadratics, the next most complicated functions are cubic polynomials, with more interesting portraits as shown by the examples we now discuss. (Active readers may prefer to try their hand at the various examples before reading the explanations.)

Example 8.4.1
Suppose $f(x, y) = x^3 + 6xy + y^2$. Applying the routine of the last Section, we have:

Step 1
H: $0 = f_1(x, y) = 3x^2 + 6y$; V: $0 = f_2(x, y) = 6x + 2y$, so H and V are, respectively, a parabola and a straight line that meet where $3x^2 - 18x = 0$, that is, at the critical points $\mathbf{0}$ and $P = (6, -18)$.

Step 2
Two points on H are $\mathbf{0}$ and $(2, -2)$ at which f takes unequal values, so H is not a contour of f (and neither is V, similarly).

Step 3
The Hessian matrix is $\begin{pmatrix} 6x & 6 \\ 6 & 2 \end{pmatrix}$ with determinant $\Delta(x, y) = 12x - 36$, which is negative at $\mathbf{0}$ and positive at P. Hence $\mathbf{0}$ and P are non-degenerate, and $\mathbf{0}$ is a saddle, while f has a centre at P, which by theorem D. 6.2 is a relative minimum since $f_{11}(P) = 36 > 0$.

Step 4
The two curves H, V divide $\mathbb{R}^2$ into five regions, and to mark them we look at the sign of $g(z)$ in (8.3.2); here $g = -(3x^2 + 6y)/(6x + 2y)$. Taking any convenient point, say $(1, 0)$ in region X (see Fig. 8.4.1, which for clarity is not drawn to scale), we see that $g = -\frac{1}{2} < 0$; and then we note that only one of f_1 and f_2 changes sign as we cross from X to its next-door regions. Therefore we can mark the regions as shown in Fig. 8.4.1.

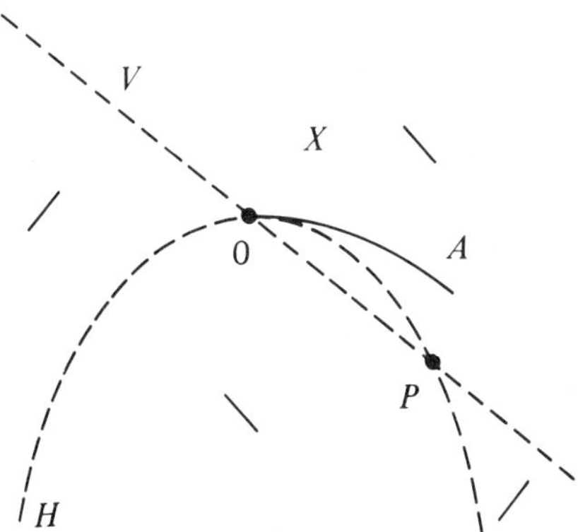

Fig. 8.4.1.

Step 5

We must draw the four arms of the separatrices at **0**. Now, an arm A in the region X of Fig. 8.4.1 must have negative slope, and hence starts off below the x-axis and above the curve H. We shall show that it forms the loop shown in Fig. 8.4.2, and use an argument which applies in many similar situations.

We argue from Fig. 8.4.2, wherein we see that A cannot turn to rise above the x-axis, otherwise it would need to have a horizontal tangent somewhere (at which it would cross H); and in curving round, it would need to have a vertical tangent first. Hence it curves past P to cross V (at R, say) since V is not a contour: see (b) of the routine's step 5. At R, A has a vertical tangent and then turns south-west to have the positive slope of the region Z. Similar argument shows that (as drawn in Fig. 8.4.2)

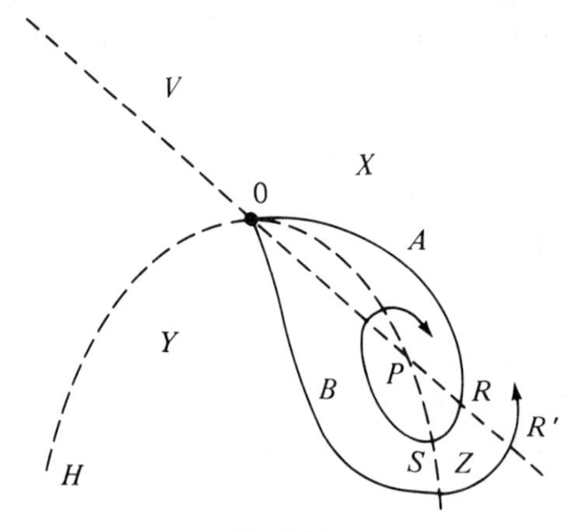

Fig. 8.4.2.

A crosses H with a horizontal tangent at S (say) and then continues north-west. But now consider the arm B in Fig. 8.4.2 which must start in the region Y because of the slope. By arguing as for A we see that B must curve round to meet V at some point R'. This forces a question: is R lower than R? If 'yes' then A is enfolded within B, and would have to continue (by the slope and no-crossing rules (B1) and (B2)) to form an infinitely winding spiral converging to P. But we know that we have a centre at P, in which the contours are all simple closed curves; so this contradiction tells us that R' is not lower than R on V. Reversing the roles of A and B in this argument, we see that R is not lower than R'; so R, R' must be equal, and the arms A, B must close up to form the loop in Fig. 8.4.3.

Similarly the other arms are drawn in, as shown, because of step 5(b) in the general routine. Thus, although the saddle **0** has two separatrices 'locally', the four arms combine to make one curve which has a single cross-over point at **0**; and this curve is the contour of height 0.

Step 6

As P is the sole centre of f, we insert a small simple closed curve C surrounding P.

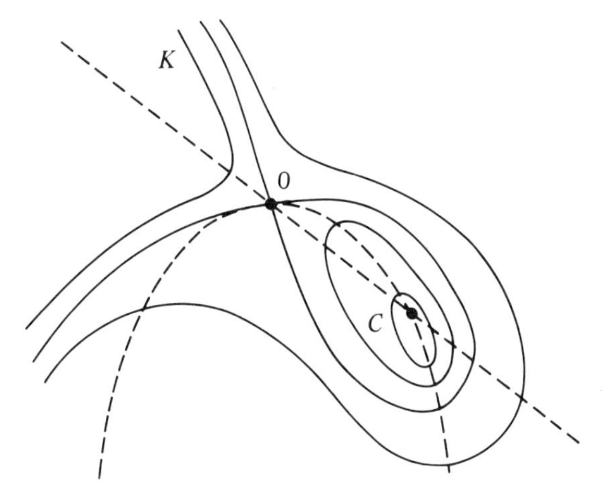

Fig. 8.4.3.

Step 7
The arms we have drawn at **0** define the three new regions of Fig. 8.4.3, and along and near the boundary of each, we draw a smooth curve, of which K is an example in Fig. 8.4.3. This has a roughly hyperbolic shape, and of course is not pointed near **0**. Inside the region formed by the loop, the newly inserted contour is a simple closed curve, but probably more elongated than C in step 6. It is then instructive to insert at least one closed curve between these two, to see how the contours gradually change from one to the other. For clarity here, we have resisted the artistic impulse to insert further typical contours.

Step 8
No symmetry is obvious from the drawing, nor in the function f, except possibly in the line V. (If this guess were correct, then the zero contour of f would cut the line W: $2x - 6y = 0$ [which is perpendicular to V] symmetrically. But on W, $y = x/3$, so f becomes $x^2(x + 19/9)$, which is zero at $x = 0$ and at $x = -19/9$. Hence, W does not cut the zero contour $f = 0$ symmetrically, and the guess is wrong.)

In the next example, H consists of two separate curves, one of which is a contour.

Example 8.4.2
Suppose $f(x, y) = x^2 + 6x^2y + y^2$. Collapsing the enumeration of the steps a little, we have:

$$H: 0 = f_1(x, y) = 2x + 12xy, \qquad V: 0 = f_2(x, y) = 6x^2 + 2y;$$

so V is a parabola, and H consists of the lines $x = 0$ and $y = -1/6$. The first of these is the y-axis, on which $f = y^2$, which is not constant; while on the second, $f = 1/36$, so that line is a contour of f. Clearly f is not constant on V, which is therefore not a contour of f. The critical points are at the intersections of H and V, that is, at the points $\mathbf{0}$, P, Q where $P = (p, -1/6)$, $p = 1/\sqrt{18}$ and $Q = (-p, -1/6)$. The Hessian (8.3.1) is $\Delta(x, y) = 4 - (12x)^2$, which is positive at $\mathbf{0}$ and negative at P and Q. Hence each is non-degenerate, and the latter pair are saddles, while f has a centre at $\mathbf{0}$ which, by the test in theorem D.6.1, is a relative minimum since $f_{11}(0) = 2 > 0$.

Step 4

We must now mark the eight regions defined by H and V. On each of these, we need the sign of $g(x, y) = -(2x + 12y)/(6x^2 + 2y)$, which is <0 at $(1, 0)$. As in example 8.4.1, we note that the numerator of g changes sign as we cross H, and the denominator does so as we cross V. Hence, the signs of the regions alternate as we cross any boundary. Thus we can mark the regions as shown in Fig. 8.4.4.

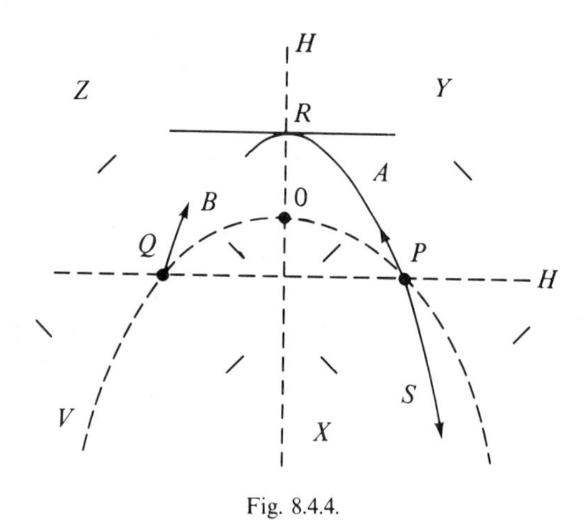

Fig. 8.4.4.

Step 5

Of the three critical points, we must insert the separatrices at P and Q. But since the line $y = -1/6$ already passes through each of them, it must be a separatrix at each. Therefore we have to insert the remaining separatrix S at P, and then at Q. Because of the signs of the regions, the lower arm of S can only go off to infinity in the region X shown in Fig. 8.4.4; and the upper arm A (in region Y) passes between V and the vertical line through P. It then meets the boundary of Y at a point R on H as shown, where it has a horizontal tangent and then turns downwards to have the negative slope of the region Z. Similarly, there is an upper arm B at Q which meets H at some R' on the y-axis; and as in example 8.4.1, we see that $R = R'$ and A and B join to form a loop. (An important alternative argument, that works here, is to anticipate step 8, and check for symmetry across the y-axis. Sure enough, we find that since x appears in f only as a square, then $f(x, y) = f(-x, y)$, so we do have the symmetry we suspected. Therefore $R = R'$ and so A and B join up.)

Step 6

Now insert simple closed curves surrounding the centre $\mathbf{0}$.

Step 7
Finally (step 8 being already in play) we now have seven 'new' regions, bounded by these separatrices, and we insert contours near their boundaries to give Fig. 8.4.5. A few more contours can be added for artistic reasons if required; for these we must use the 'slope' and 'no-crossing' rules mentioned in step 7 of the general Routine.

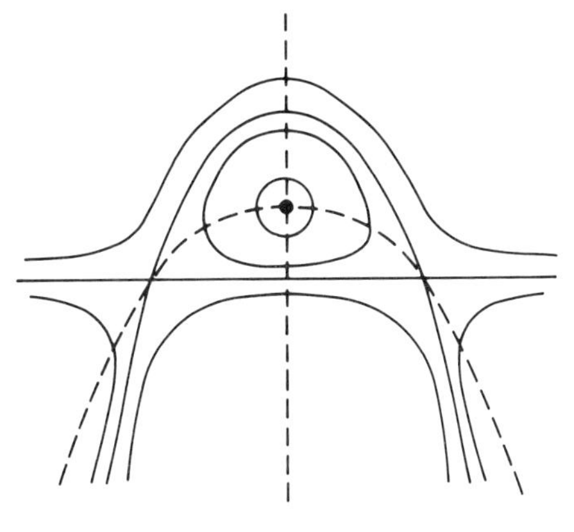

Fig. 8.4.5.

In the next example we work with a degenerate function, to show how symmetry can be used to handle a degenerate critical point.

Example 8.4.3
Suppose $f(x, y) = x^2 + y^3$. Here $f_1(z) = 3x^2$, and $f_2(z) = 3y^2$. Therefore H, V are the lines $x = 0$, $y = 0$ each taken 'twice over', and $\mathbf{0}$ is the sole critical point. On the axes, f is not constant, so neither H nor V is a contour of f. At $\mathbf{0}$ we have $f_{11} = 0 = f_{22} = f_{12}$, so $\Delta(0) = 0$ and $\mathbf{0}$ is a degenerate critical point, but it is simple enough in this case to adapt the routine as follows.

As in Fig. 8.4.6, H and V divide $\mathbb{R}^2$ into the four regions shown. In each of these, $g(x, y) = -x^2/y^2 < 0$, so the slope is always negative. The markers in Fig. 8.4.6 then suggest symmetry in the line $L{:}x = -y$ which leads us to guess that L is a contour of f, and that is easily verified to be true. There are no saddle points and hence steps 5–7 of the routine do not apply, but we treat L as if it were a separatrix and then apply step 7 to obtain Fig. 8.4.7.

It was by applying the routine that Fig. 8.8.1 above was constructed, and we suggest that readers should now attempt this for themselves. They will meet one snag, however; it will not be immediately clear whether the separatrix loops should be horizontal or vertical. The following argument is instructive, and we carry it out for the entire family of functions.

$$f(x, y) = x^3 - 3a^2x + y^3 - 3b^2y \tag{8.4.1}$$

where a, b are strictly positive constants. Here, we have $f(-x, y) = -f(x, y)$, so there

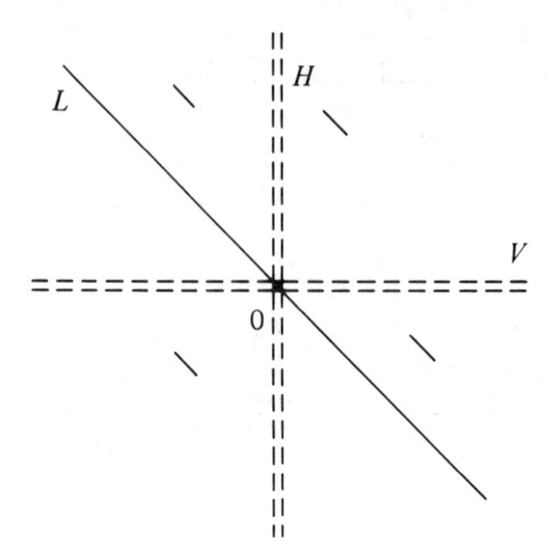

Fig. 8.4.6.

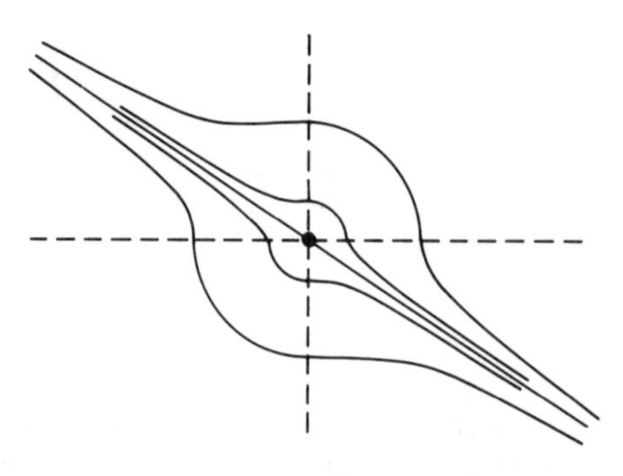

Fig. 8.4.7.

is symmetry by reflection in **0**, when the contour of height c is reflected to that of height $-c$.

The horizontal manifold H consists of the pair of vertical lines $x = \pm a$. (In example 8.4.3, where $a = 0$, these were crushed together into the line $x = 0$ taken 'twice over'.) Similarly, V consists of the horizontal lines $y = \pm b$. Then H and V meet in four points P, Q, R, S as shown in Fig. 8.4.8; as the Hessian is $\Delta(x, y) = 36xy$, it is positive in the first quadrant (for example) so without further calculation we see that f has a centre at P and at R, and a saddle at Q and at S.

At Q the north-west (NW) and NE arms could eventually join to form a vertical loop surrounding P; or instead the NW arm could join the SW arm to form a horizontal loop surrounding R. To decide which, consider the contour K through S at $(a, -b)$, which is of height $f(S)$. The line $x = -a$ cuts K where $f(a, -y) = f(S) = f(a, -b)$, that is where

$$-a^3 + 3a^3 + y^3 - 3b^2y = a^3 - 3a^3 - b^3 + 3b^3$$

so

$$0 = y^3 - 3b^2y + 2(2a^3 - b^3).$$

This last is a cubic equation in y, of the form $y^3 + 3Iy + J = 0$, which, by test 6.2.1 has one real root or three according as

$$D > 0 \quad \text{or} \quad D < 0 \quad \text{where } D = 4I^3 + J^2. \tag{8.4.2}$$

Here, it can be verified that $D = 16a^3(a^3 - b^3)$, and since $a > 0$ we have

(i) $D > 0$ if $a > b$; (ii) $D < 0$ if $a < b$; (iii) $D = 0$ if $a = b$.

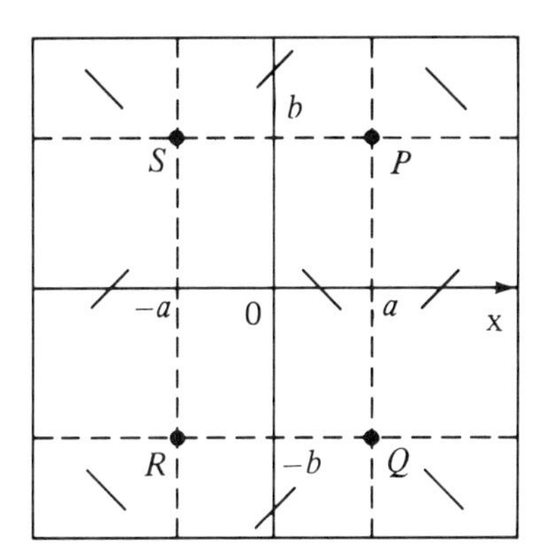

Fig. 8.4.8.

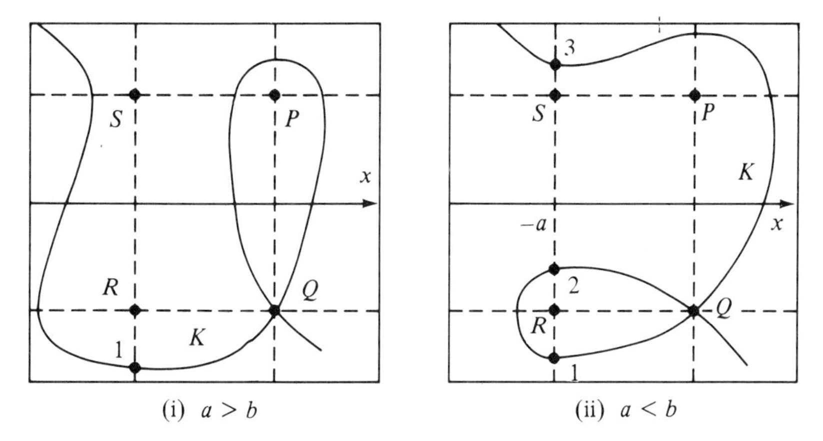

Fig. 8.4.9.

In case (i) therefore, K cuts $x = -a$ only once, so the loop is vertical as in Fig. 8.4.9(i). If (ii) holds, then K cuts the line in three points, of which two are on the loop, which is therefore horizontal as in Fig. 8.4.9(ii).

Curiosity leads us to wonder how, as the control parameters a, b change gradually, Fig. 8.4.9(i) could change gradually into Fig. 8.4.9(ii). It is the 'unstable' case, $a = b$, that allows the transition, as we see in Fig. 8.4.10. For, the NW arm at Q simply continues onwards to become the SE arm at S, and (step 8) this suggests that the line $x = -y$ must be the zero contour; this proves to be the case since $f(x, -x) = 0$

when $a = b$. Also, we then have symmetry about this line because the symmetry in $\mathbf{0}$ (mentioned earlier) carries this line into itself when $a = b$.

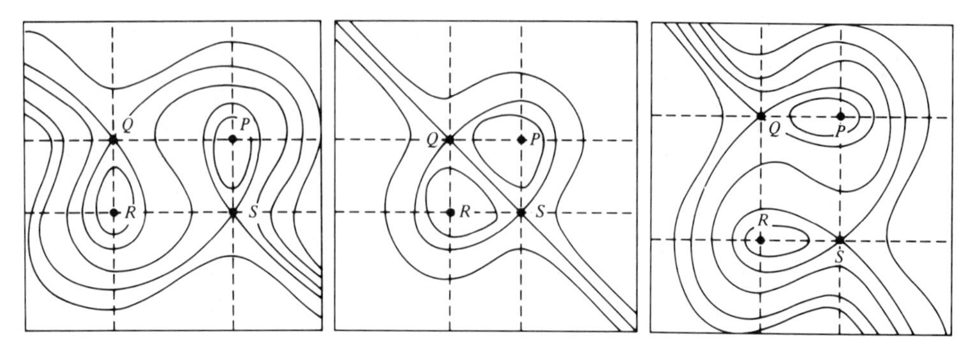

Fig. 8.4.10.

Exercises 8.4.1

1. Verify that D in (8.4.2) simplifies to the form given in the text.
2. Replace the control parameters a^2, b^2 in (8.4.1) by arbitrary numbers p, q and investigate the cases $p = 0 > q$, and p, q both negative to show how the entire family (including also (8.4.1)) evolves from the degenerate function in Example 8.4.3.
3. Show that we can always change the scale by writing $(x, y) = t(X, Y)$, so that in (8.4.1), $f(x, y) = t^3 g(X, Y)$, where

$$g(X, Y) = X^3 - 3pX + Y^3 - 3Y \qquad (p > 0).$$

   (Thus the family of functions (8.4.1) is really governed by the one parameter $p = a/b$.)
4. Sketch a portrait in $\mathbb{R}^2$ of the function f given by

$$f(x, y) = x^2 + y^3 - 3xy. \qquad \text{[SU]}$$

5. If $h(x, y) = y^2 - y\sin x$, show that the critical points of h are all non-degenerate, and consist of saddles at the points $(k\pi, 0)$, and minima at the points $((4k + 1)\pi/2, \frac{1}{2})$, $((4k - 1)\pi/2, -\frac{1}{2})$, $(k \in \mathbb{Z})$. Sketch a portrait of h. (Hint: the separatrices at each saddle are formed by the contour $h = 0$.) [SU]
6. Let $f(x, y) = x^3 - 3xy^2 + ax$, where a is constant. Sketch the portrait of the function $f: \mathbb{R}^2 \to \mathbb{R}$ in the three cases $a = 0, a = -1, a = 1$ indicating clearly the critical points, separatrices (if any), and any symmetry that f may possess. [SU]
7. (a) Show that the critical points of the function

$$f(x, y) = \sin^2 \pi x - \sin^2 \pi y$$

   on $\mathbb{R}^2$ are all the lattice points

$$(\tfrac{1}{2}p, \tfrac{1}{2}q) \mid p, q \in \mathbb{Z}.$$

   (b) Show that each critical point is non-degenerate and that the saddles of f occur at points of the forms

$$(p, q), \qquad (p + \tfrac{1}{2}, q + \tfrac{1}{2}) \qquad (p, q \in \mathbb{Z}),$$

   while maxima and minima occur, respectively, at points of the forms

$$(p + \tfrac{1}{2}, q), \qquad (p, q + \tfrac{1}{2}) \qquad (p, q \in \mathbb{Z}).$$

(c) Show that f takes the values $0, -1, 1$ respectively, at each saddle, minimum or maximum and that the set $\{f = 0\}$ consists of all the lines $y = \pm x + k$ $(k \in \mathbb{Z})$.

(d) Sketch the portrait of f. [SU]

8. Show that the function f given by

$$f(x, y) = x^3 + y^3 + ax^2y + x \qquad (a \in \mathbb{R}),$$

has no critical points unless $a < -3/4^{1/3}$; and then it has just two, on the line $y = \sqrt{-a/3} \cdot x$, and in the first and third quadrants. If (u, v) is such a critical point, show that the Hessian determinant at (u, v) is $12v(3u + 2av)$. For each real k, let C_k denote the level curve $f = k$. Show that the y-axis Y is tangent to C_0 at the origin, by showing that if $k \neq 0$, C_k crosses Y with gradient γ, where $\gamma \to \infty$ as $k \to 0$.

Hence sketch the portraits of f when $a = 1$ and when $a = -1$, indicating the curve C_0 in each case. (Hint: express the equation of the horizontal manifold in the form $y = g(x)$.) [SU]

9. Suppose $g(x, y) = x^3 + 6pxy + y^2$, $p \neq 0$. Show that if $x = \lambda X$, $y = \mu Y$ and $\lambda = p^2$, $\mu = p^3$, then $g(x, y) = \lambda^3 f(X, Y)$ with f as in example 8.4.1. (Thus Fig. 8.4.3 is essentially the portrait of g, for all p.) Show similarly that Fig. 8.4.5 is essentially the portrait of $h(x, y) = x^2 + 6qx^2y + y^2$, for all $q \neq 0$.

10. For each of the following functions $f(x, y)$, construct the portrait and emphasize the curve $C_f = \{f = 0\}$, which is often well known, but better understood when seen within the entire portrait of f. (N.B. For the first six, C_f is a graph of the form $y = \pm g(x)$, with names, respectively, Witch of Agnesi, lemniscate, cissoid, dumb-bell, strophoid and trident.).

$y - a^3/(x^3 + a^2)$
$x(x^2 + y^2) - ay^2$
$y^2(a - x) - x^2(x + a)$
$y^2 - x^4 - x^5$
$x^2 - x^6 - y^6$
$xy(x^2 - y^2) - x^2 + y^2$
$y^4 - x^4 - xy$
$y^2(y^2 - 96) - x^2(x^2 - 100)$
$(x^2 - a^2)(x - a)^2 + (y^2 - a^2)^2$
$(x^2 - 1)^2 - y^2(y - 1)(y - 2)(y + 5)$
$x^4 - x^2y - y^3$
$x^5 + y^5 - 2x^2 + 5xy - 2y^2$
$(y^2 - x^2)(x - 1)(2x - 3) - 4(x^2 + y^2 - 2x)^2$

$x^4 - x^2 + y^2$
$y^2 - x^4 + x^6$
$xy - x^3 + a^3$
$x^2y^2 - x^2 - y^2$
$12x^2 - y^3(4 - y)$
$x^4 + y^4 - a^2xy$
$(x^2 + 2ay - a^2)^2 - y^2(a^2 - x^2)$
$(x^2 - 1)^2 - y^2(3 + 2y)$
$x^4 + y^4 - 2axy^2$
$x^4 + x^2y^2 + y^4 - x(x^2 + y^2)$
$x^4 + x^2y^2 + y^4 - x(x^2 - y^2)$
$(x^2 + y^2 - 3x)^2 - 4x^2(2 - x)$

9

Non-linear models

9.1 TWO-POPULATION EXTENSIONS OF VERHULST'S EQUATION

The linear models we have studied in Chapter 3 have the limitation that if they are robust, there are only two possible fates for the trajectories: these either go off to infinity or converge to the origin. For populations, this means either explosion or death. But there are many systems in biology, mechanics, or electronics, where sustained oscillations occur, precisely to avoid these extreme fates. Consequently they must be modelled by non-linear mathematics. This will require new methods, which nevertheless build on the linear mathematics that was of such excellent service to the mathematical physicists of the nineteenth century.

As a first example, recall that in Chapter 3, we considered a predator–prey model with two populations, which depended on parameters a, b, c, d with appropriate algebraic signs. Just as in Verhulst's equation, where a 'competition' term was added in to the Malthus equation, we can build in terms representing inter-species competition. This model was investigated by the Italian mathematician Vito Volterra, in connection with fish stocks in the Adriatic sea, after the First World War; and he was led to a long study of this and other models in his book (Volterra 1931). Independently, the American biologist A. J. Lotka also considered this model (see his interesting book (Lotka 1956)). Thus we consider the following predator–prey model with two populations N_1, N_2 (often called the **Lotka–Volterra** model):

$$\dot{N}_1 = aN_1 - bN_1N_2, \qquad \dot{N}_2 = -cN_2 + dN_1N_2 \tag{9.1.1}$$

in which a, b, c, d are all positive. Here the signs can be justified as in the linear model in Chapter 3. The equations (9.1.1) form a system which is non-linear because of the product terms; the sum of two solutions will not in general be again a solution. Therefore we need new methods to help us understand such systems, but before we develop these, we need to look at some specific systems (including (9.1.1)) so as to experience some of the things that can happen.

Continuing with (9.1.1), we simplify the notation by putting $N_1 = px$, $N_2 = qy$

with p, q constant; and then, by substitution in the differential equations, we see that these become

$$\dot{x} = ax(1 - y), \qquad \dot{y} = -cy(1 - x) \tag{9.1.2}$$

and we choose $p = c/d$, $q = a/b$. Also, since we clearly need N_1, N_2 to be positive, then $x > 0$ and $y > 0$.

Let us be sure what it means to 'solve' the system (9.1.2). We are supposing that x and y are smooth functions of t, and that, as time goes on, the point $P(t) = (x(t), y(t))$ will move from its initial position $P(0)$ and thereafter trace out a curve in $\mathbb{R}^2$. This curve is called the trajectory through $P(0)$, and for each t its tangent at $P(t)$ has the direction of the vector $(x'(t) \quad y'(t))^T$. Just as in Chapter 3 (program 3.1.1), we can imagine an 'iron filings' diagram, in which at each point (u, v) of $\mathbb{R}^2$ we draw a small arrow parallel to the vector $(au(1 - v) \quad -cv(1 - u))^T$; and then our eyes tend to be drawn along some of the arrows to join them into trajectories. These curves suggest themselves as a solution of our mathematical problem which is: *to find what the functions $x(t)$, $y(t)$ must be, if we know that their derivatives satisfy (9.1.2).*

For any solution $x(t)$, $y(t)$ and for each t we have from (9.1.2):

$$\frac{c\dot{x}(1 - x)}{x} = ac(1 - x)(1 - y) = \frac{-a\dot{y}(1 - y)}{y} \tag{9.1.3}$$

whence, by integrating with respect to t, we get $f(x, y) = \text{constant}$, where

$$f(x, y) = c\ln(x) + a\ln(y) - cx - ay. \tag{9.1.4}$$

We can do this, since x and y are strictly positive. Thus, if the functions x, y satisfy (9.1.2), then, at any time t, the point $(x(t), y(t))$ always lies on a contour of the function f. It will follow from the theory below that conversely, if P is any point on a contour $f = k$, then P lies on a trajectory, so each contour is a trajectory. (Use the contour program 8.1.1 to find the shape of the contours in (9.1.4)—note that (1, 1) is a *fixed point* since both x' and y' are zero there, and that $f(1, 1) = -c - a$, so if both a and c are positive then the contours will need to use negative values for the constant.)

The domain of f is the positive quadrant Q of $\mathbb{R}^2$, excluding the axes. Using the routine of Chapter 8, we see that f has just one critical point, at $M = (1, 1)$, and that the portrait of f is as in Fig. 9.1.1, with M a minimum. In Q, each contour is a closed curve enclosing M. But now, each contour must be viewed as a trajectory, and therefore with a direction of flow along the vector $(\dot{x}(s) \quad \dot{y}(s))^T$ as explained above. Hence the neighbourhood of M looks like that of a centre in a linear system.

As a model of two populations, our portrait shows that, given an initial state (N_1, N_2), the populations will change through values $(N_1(t), N_2(t))$ until, after a finite time T, these return to the initial state; and then the whole cycle repeats. This conclusion is biologically unlikely, if only because we do not observe such exact periodicity in the rise and fall of populations. In spite of this, it is instructive to pursue the mathematics further, so as to reveal some other defects in the model. If the model is simulated by the computer techniques of program 4.7.1 for the numerical solution of coupled systems, then the cycle may not repeat if the algorithm is not sufficiently sophisticated. For example, the Euler method of program 9.1.1 below slips at each step onto another solution curve, and so spirals away from the actual orbit. However the Runge–Kutta method should result in a closed orbit, which the reader should verify.

```
REM Program 9.1.1, Lotka-Volterra
READ a   , c   , x   , y   , t, dt , tmax, xmax, ymax
DATA 0.1, 0.1, 0.5, 0.5, 0, 0.1, 200 , 4.5 , 3
SET MODE "graphics"
SET WINDOW 0,xmax, 0,ymax
FUNCTION xdot(x,y) =  a*x*(1-y)
FUNCTION ydot(x,y) = -c*y*(1-x)
DO
  PLOT x,y;
  LET dx = xdot(x,y)*dt
  LET dy = ydot(x,y)*dt
  LET x  = x + dx
  LET y  = y + dy
  LET t  = t + dt
LOOP UNTIL t>tmax
END
```

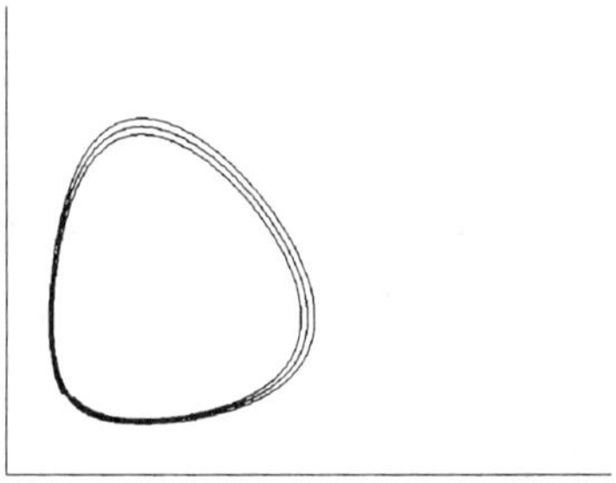

Sample output

Although, for the problem in hand, x and y are positive, the system (9.1.2) nevertheless makes sense at every point (x, y) of $\mathbb{R}^2$. It would therefore be nice not to have to confine our attention to the first quadrant only, but to know what trajectories look like in the plane as a whole. For example, the axes are trajectories, since, if $x = 0$, then by (9.1.2) the vector $(\dot{x} \quad \dot{y})^{\mathrm{T}}$ is $(0 \quad cy)^{\mathrm{T}}$which points towards $\mathbf{0}$ since $c > 0$; and similarly when $y = 0$.

Now if we allow negative x, y then the well-drilled student will change $\ln(x)$ in (9.1.4) to $\ln|x|$, and similarly for $\ln|y|$. The effect is then to change f to a function $g(x, y)$, but with domain D, the whole plane minus the axes. It still has the one critical point M as before (check this). The full portrait of (9.1.2), on $\mathbb{R}^2$ itself, therefore consists of the trajectories along the axes (as described above) plus the trajectories along the contours of g on D (Fig. 9.1.2). It will be useful later to note that everywhere on D we have:

$$\dot{x} = xy \cdot \frac{\partial g}{\partial y}, \qquad \dot{y} = -xy \cdot \frac{\partial g}{\partial x}. \tag{9.1.5}$$

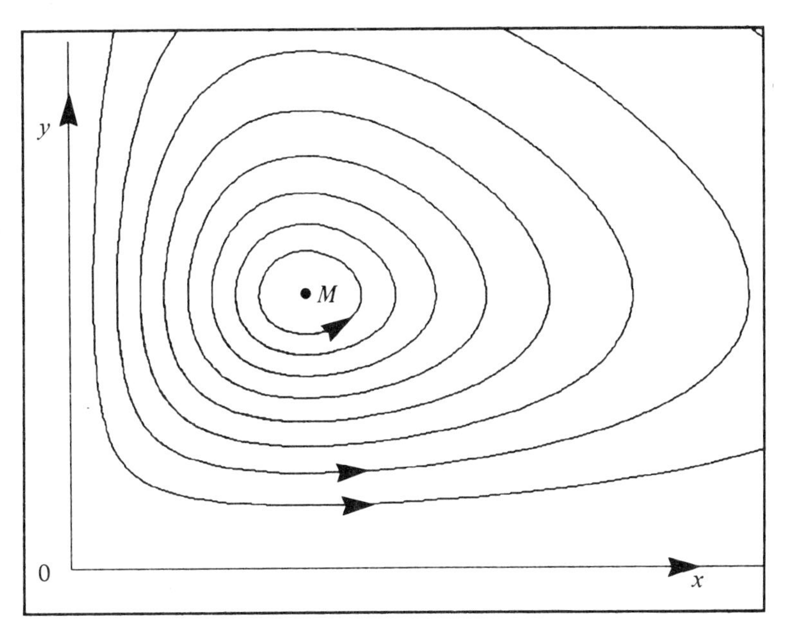

Fig. 9.1.1.

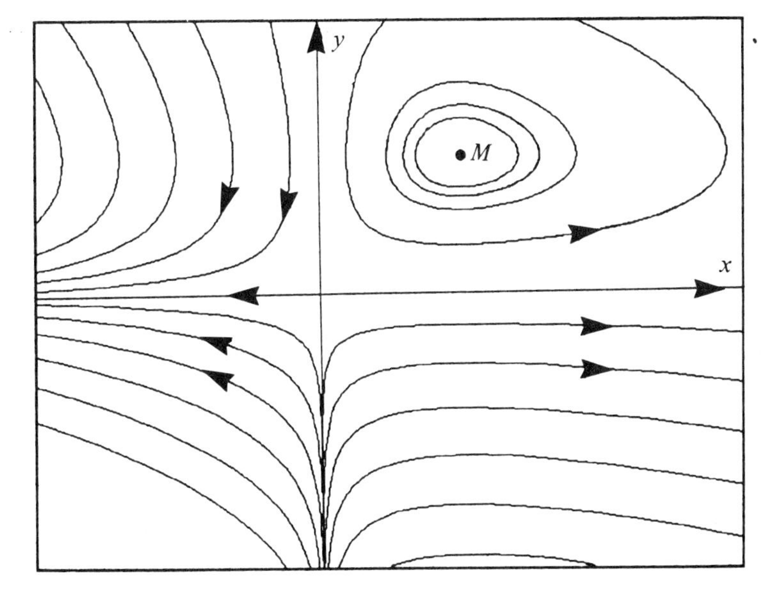

Fig. 9.1.2.

Exercises 9.1.1

1. Use a computer program to produce an 'iron filings' diagram for the system (9.1.2).
2. By writing $Y = y + k$, choose k so that the equations $\dot{x} = ax - pxy$, $\dot{y} = cx + dy - qxy$ take the form (9.1.2). (When $p = q$ these equations were an early model, due to Martini, for studying the way in which a population acquires immunity to an infectious disease:

here x, y are, respectively, the fractions of the population which are vulnerable and immune/affected, and the parameters are constants of proportionality. For details, see Lotka (1956), p. 79.)

3. In the differential equations:

$$\dot{x} = ax - by^2 \qquad \dot{y} = cx^2 - dy$$

a, b, c, d are real constants and $bc \neq 0 \neq a \neq d$. Suppose the equations are rescaled, by writing $x = \lambda u$, $y = \mu v$, $t = ks$ where λ, μ, k are non-zero real constants. Show that the resulting differential equations have the same form as before but with b, c changed to 1, provided we choose

$$k\lambda = (bc^2)^{-1/3}, \qquad k\mu = (b^2c)^{-1/3}$$

and that by suitable choice of k the equations take the form:

$$u' = u - v^2, \qquad v' = u^2 - \sigma v$$

where $u' = du/ds$ and $1 - \sigma = k(a - d)$. (This system will be discussed later.)

The procedure we have followed for solving (9.1.2) will not work in general, because it depends on the fluke that we were able to separate variables in (9.1.3). Why do we call it a fluke? If we go back to the equations (9.1.1), we can write them as

$$\frac{\dot{N}_1}{N_1} = (a - bN_2), \qquad \frac{\dot{N}_2}{N_2} = (-c + dN_1),$$

which say that the average rate of change of N_1 is the linear function $L = a - bN_2$, and similarly for N_2. But a biologist could never measure the parameters so accurately as to be sure that the function L is not really $a - bN_2 + rN_1$ for some small but non-zero constant r. Thus, if the average is a linear function at all, then we should include the term rN_1; the 'fluke' consists in having $r = 0$. With this, the model (9.1.1) acquires the term rN_1^2 which (it might be argued on biological grounds) represents competition within the first species. By rescaling as for (9.1.2), the resulting system is:

$$\dot{x} = ax(1 - y + px), \qquad \dot{y} = -cy(1 - x + qy). \tag{9.1.6}$$

These still retain their general form if small changes (due to errors of measurement) are made in the four parameters a, c, p, q.

The choice of a linear function for the average $\dot{x}/x$ might be justified on the grounds that it is the simplest possible non-constant function. But why is it reasonable to use simplicity as a criterion? One reply could be: for practicality, to keep the mathematics simple but general. Or it could be a philosophical one: that nature is simple. We prefer the first reply in our case, partly because we have to learn to walk before we can ever run. Indeed, each right-hand side in (9.1.6) is a quadratic function, and if it were changed to the most *general* quadratic, then a full solution is not known—after many decades of research by mathematicians in several countries. But we can still handle a rich variety of special cases, and allow certain types of cubic and other functions.

Exercises 9.1.2

1. Given the system:

$$\dot{x} = ax + lxy + ux^2, \qquad \dot{y} = -cy + mxy + vy^2,$$

   show that the rescaling: $x = qz$, $y = pw$ will reduce the system to (9.1.6) if p, q are suitably chosen, and a, c, l, m are not all zero. (Take $p = -a/l$, $q = c/m$.)
2. Investigate the coupled system (9.1.6) with program 9.1.1, and its equivalent with the fourth-order Runge–Kutta technique.

In (9.1.6) we can no longer exploit the fluke that led us to the integral in (9.1.3), so a new technique is needed for solving the resulting system. Its portrait is shown in Fig. 9.2.3 below, but before describing this we give two examples with simpler portraits; these will introduce us to new phenomena that we shall learn to cope with.

We could consider a two-species model of the form

$$\dot{x} = ax + by, \qquad \dot{y} = cx + dy + px^2, \tag{9.1.7}$$

with a small coefficient p for the square term. We shall see in Fig. 9.4.1 below that for a certain range of values of the coefficients, its portrait turns out to be as in Fig. 9.1.3, in which we see two equilibrium points; but since one is a spiral node, we cannot be looking at the portrait of a *function* $f(x, y)$.

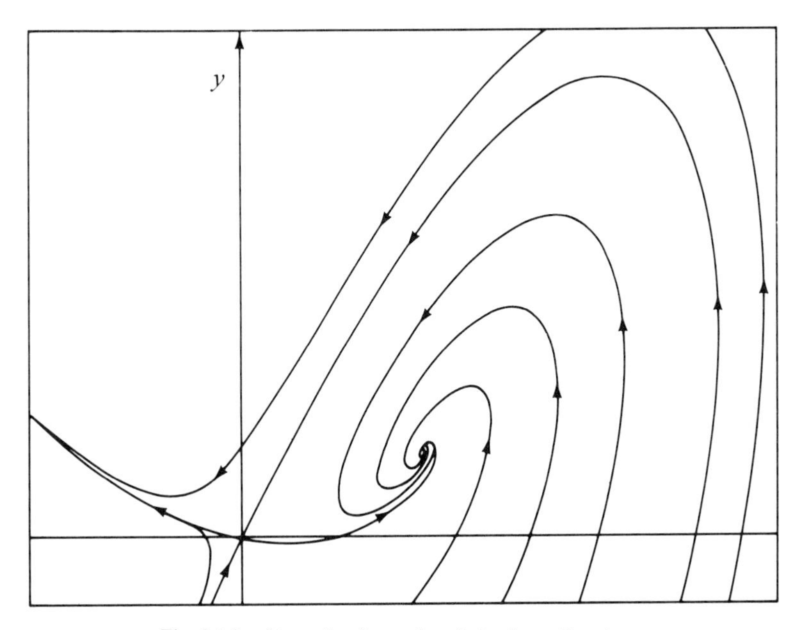

Fig. 9.1.3. Note the changed scale in the y-direction.

Another example was first studied by the Dutch mathematician van der Pol in 1926, to model the behaviour of a thermionic oscillator which at that time was used in radios. The model is a second-order differential equation, and seems to be the earliest non-linear equation needed in circuit theory; the particular circuit has been described by Zeeman as the first to exhibit behaviour of a biological kind (and hence rather late to be discovered). Perhaps this is because the underlying physics is caused by the formation of populations of electrons (although the equation was derived in terms of pre-electronic circuit theory). The equation is

$$\ddot{x} + \mu(x^2 - 1)\dot{x} + x = 0 \tag{9.1.8}$$

where μ is a constant—positive in the physics, but we shall allow it to be negative as well. By introducing the new variable: $y = \dot{x} + \mu(x^3/3 - x)$ we convert the single equation (9.1.8) into the first-order planar system:

$$\dot{x} = y - \mu(x^3/3 - x), \qquad \dot{y} = -x, \tag{9.1.9}$$

and we shall show later than when $\mu = 0$ its portrait is as in Fig. 9.1.4.

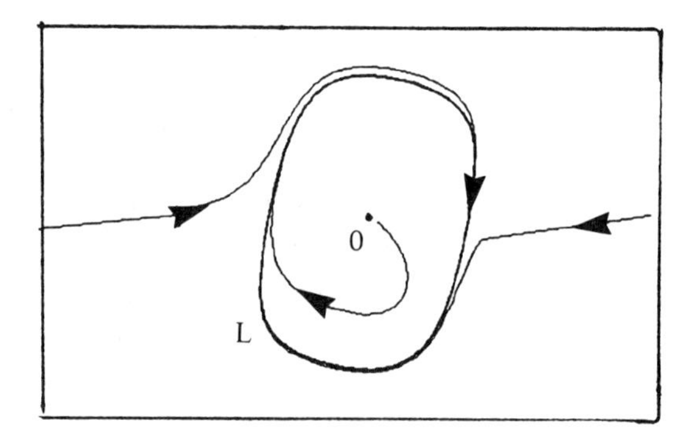

Fig. 9.1.4. The van der Pol equation: $\mu = 1.5$.

There we see a new feature—the closed (but isolated) trajectory L. As we have seen in Chapter 3, the portrait of a linear system can have closed trajectories all surrounding a centre; but then there is an infinity of closed trajectories. The trajectory L in Fig. 9.1.4 is an example of a **limit cycle**, and it has the important feature that all other trajectories spiral onto it; thus any z other than $\mathbf{0}$ is rapidly moved into the neighbourhood of L, and makes oscillations that neither die out nor become infinite. This type of behaviour is regarded as very 'robust', because it persists even if we make small errors in setting the parameter μ or the starting position of z. It is therefore very desirable to design such behaviour into hardware, and reasonable to expect limit cycles to appear in good models of biological systems (such as the heart-beat).

We intend to describe a routine for making a hand drawing of the portrait of any system such as the examples above; and the routine will be very similar to that in Chapter 8 for dealing with a function. With differential equations, a rough check is even more necessary than with functions, because existing computer algorithms for solving differential equations are much more heuristic and the necessary convergence proofs seem to be lacking, in spite of many efforts and partial results.

Before readers begin the theoretical discussion in the next section, they should perform the following exercises, so as to generate intuitive understanding and (we hope) to obtain artistic pleasure if colours are used. This should also suggest that there may not be only one reasonable possibility for drawing the diagrams according to the rules specified up to then. Thus, the need will be seen for the further mathematical guidance that just *had* to be developed, in order to pin down a unique portrait with certainty.

Exercises 9.1.3

1. Consider the system (9.1.7) when a, b, c, d, p are, respectively, $1, -1, -2, -3, 1$. Try to construct Fig. 9.1.3 by using only the principles (A), (B) of local linearization and of globalization that we enunciated in Chapter 8. Recall that with these, we drew contours using the three rules: no-crossing, slope, and exit. For Fig. 9.1.3, we need to substitute 'trajectories' for 'contours', and be prepared to use any of the linear portraits (from Chapter 3) rather than just the saddles and centres.
2. Apply the same rules to fill in Fig. 8.3.1 of Chapter 8, preferably using colours for the various separatrices.
3. Now start with any curves drawn as horizontal and vertical manifolds, sign the regions consistently, and form portraits accordingly (without using any specific equations). Can you arrange matters so as to obtain a 'wave' like those seen in Japanese paintings?
4. Make a drawing to show that in Fig. 9.1.4, the origin could display a node rather than a spiral focus. (The problem is to bend the trajectories so that they approach L suitably: the resulting sketch will, of course, correspond to a system different from (9.1.9); see (9.8.4) below.)
5. Investigate the numerical solution of the van der Pol model (9.1.9) using the Euler method of program 9.1.1 (widening the screen's coordinate window to include (0, 0) as centre). You should find that the Euler and fourth-order Runge–Kutta methods both converge to a limit cycle. Can you explain why this differs from the case with the Lotka–Volterra model?

9.2 EXISTENCE OF TRAJECTORIES: HAMILTONIANS

A century ago, Henri Poincaré taught that it would in general be hopeless to expect to solve such systems as (9.1.6) in 'closed form' with integrals like (9.1.3). All the above systems are examples of a general type, the planar system

$$\dot{x} = P(x, y), \qquad \dot{y} = Q(x, y), \tag{9.2.1}$$

where P, Q are smooth functions with domain $\mathbb{R}^2$. To understand it, we must use qualitative and geometrical methods. Now, just as with the other systems we have met, (9.2.1) gives a vector $\mathbf{w}(X) = (P(X) \quad Q(X))^{\mathrm{T}}$ at each point $X = (x, y)$ in $\mathbb{R}^2$. Our earlier experiences lead us to expect that it will be easier to describe the situation if $\mathbf{w}(X) \neq \mathbf{0}$, when we call X an **ordinary point** of the system. By contrast, if $\mathbf{w}(X) = \mathbf{0}$, we call X an **equilibrium point** (more briefly a **fixed point**). Note that t does not appear explicitly in (9.2.1); for that reason, the system is called **autonomous**. In general, non-autonomous systems are more difficult to handle, because of an increase in dimension.

Exercises 9.2.1

1. Show that the non-autonomous equation $\dot{x} - 3x = t^2$ can be converted to a two-dimensional system (9.2.1), by adding the equation $\dot{t} = 1$. Are there any fixed points?
2. Convert the linear differential equation $\ddot{x} + f(t)\cdot\dot{x} + g(t)\cdot x = 0$ into the form (9.2.1), and into an analogous three-dimensional system when 0 is replaced by $h(t)$.
3. Let R_θ denote the rotation of $\mathbb{R}^2$ about $\mathbf{0}$, and suppose $(x, y) = r(\cos\theta, \sin\theta)$ in polar coordinates. Show that (9.2.1) takes the form $(\dot{r} \quad r\dot{\theta})^{\mathrm{T}} = R_{-\theta}(P \quad Q)^{\mathrm{T}}$.

Our earlier experiences also suggest that there should always be a trajectory $T(t)$ through X, which means: if $T(t)$ has coordinates $(x(t), y(t))$, then $\dot{T}(t) = (\dot{x}(t), \dot{y}(t))$ satisfies (9.2.1). We can think of $\dot{T}(t)$ as the velocity with which the trajectory is described. For example, if X is a fixed point, then (9.2.1) is satisfied by taking $T(t)$ constantly equal to X, so X stays fixed—'in equilibrium'—for all time. The various trajectories will fit together to form the portrait of the system (9.2.1), and as with the linear systems in Chapter 3, our object will be to describe this portrait.

We first consider in detail what happens near X when it is a fixed point. Now, the Taylor expansion of P in (9.2.1) at $X = (u, v)$ is of the form (see D.4.11):

$$P(u + x, y + v) = P(X) + ax + by + O(r^2), \qquad (r = \sqrt{(x^2 + y^2)}),$$

and similarly for Q; and since $P(X) = Q(X) = 0$, then the equations (9.2.1) take the form $\mathbf{z}' = J\mathbf{z} + O(r^2)$, in which $\mathbf{z}$ is the column vector $\begin{pmatrix} x \\ y \end{pmatrix}$. J is the **Jacobian matrix** of (P, Q) evaluated at X and denoted as:

$$J = \frac{\partial(P, Q)}{\partial(x, y)} = \begin{pmatrix} \dfrac{\partial P}{\partial x} & \dfrac{\partial P}{\partial y} \\ \dfrac{\partial Q}{\partial x} & \dfrac{\partial P}{\partial y} \end{pmatrix}. \qquad (9.2.2)$$

Hence it is reasonable to expect that trajectories in a neighbourhood of X will behave like those of the linear system $\mathbf{z}' = J\mathbf{z}$, which is called the **linearization** of (9.2.1) at X. This expectation turns out to be correct, *with an important exception*: the linear system L must not display a centre or a gate (for even *linear* approximations to the latter are likely to exhibit some different form). To be more precise about the meaning of 'behaves like', we have to call on the useful concept of a diffeomorphism (as in Chapter E) to state the following theorem.

Theorem 9.2.1

At the fixed point X of (9.2.1), consider the linearization with matrix the Jacobian J as in (9.2.2). If the determinant and trace of J are both non-zero, then there exists a diffeomorphism $F: U \to V$ between neighbourhoods U of X and V of $\mathbf{0}$, which maps the trajectories of (9.2.1) in U to those in V of the linearized system.

The import of this theorem is indicated in Fig. 9.2.1. We refer readers to Hartman (1964) p. 258 for the formal proof, which is quite complicated. If the system (9.2.1)

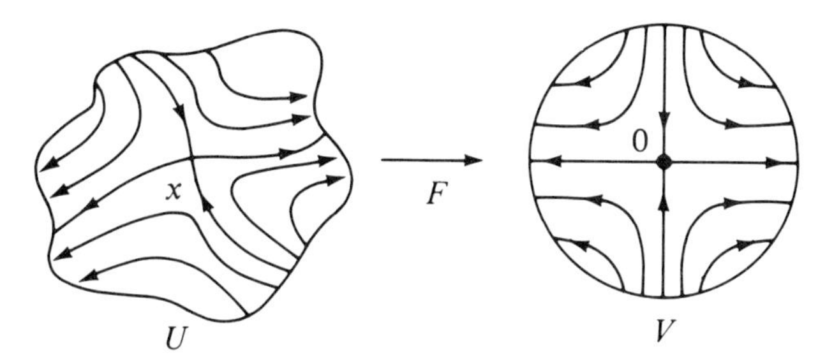

Fig. 9.2.1.

satisfies the conditions of theorem 9.2.1 at each of its fixed points, it is called (rather oddly) **hyperbolic**. We now possess the 'local' step for constructing the portrait S of a hyperbolic system, because (by theorem 9.2.1) our knowledge of the linear portraits gives us a fragment of S at each fixed point. The next step is to show that the short trajectories in each fragment can be hooked up appropriately, as in the portrait routine for a function. Thus, we shall look at what happens near an ordinary point of (9.2.1), after some exercises.

Exercises 9.2.2

1. Show that the following systems are hyperbolic:

   (i) $\dot{x} = x + y - x^2, \quad \dot{y} = 3x;$ (ii) $\dot{x} = x - y^2, \quad \dot{y} = x^2 - cy$

   where c is a constant.
2. Show that the system (9.1.2) is *not* hyperbolic.

From now on, until the end of this section, we shall assume that X is such an ordinary point of (9.2.1). First we need a guarantee that through X there really is a trajectory T, because we cannot give a formula for T, as we could in the linear case. However, there is a precise theorem, the proof of which is discussed in Chapter F.

Theorem 9.2.2
The functions P, Q in (9.2.1) being smooth, suppose X is an ordinary point. Then there is a neighbourhood I of $\mathbf{0}$ *in* $\mathbb{R}$, *and a smooth mapping* $T\colon I \to \mathbb{R}^2$, *such that* $T(0) = X$, *and for each* $t \in I$:

$$\dot{T}(t) = (P(T(t)), Q(T(t))). \tag{9.2.3}$$

Furthermore, T is unique: there is no other mapping with these properties.

Note that by (9.2.3), the gradient along the trajectory is

$$\frac{\mathrm{d}y}{\mathrm{d}x} = \frac{\dot{y}}{\dot{x}} = \frac{Q(x, y)}{P(x, y)}, \tag{9.2.4}$$

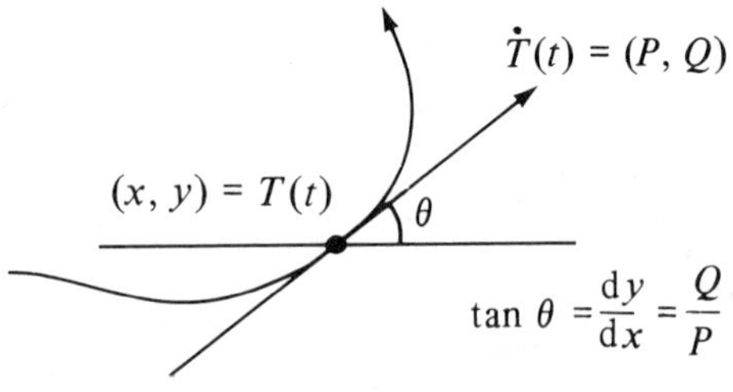

Fig. 9.2.2.

when x and y are evaluated at the time t (see Fig. 9.2.2). This is the same as for the system

$$x' = \phi(x, y)P, \qquad y' = \phi(x, y)Q \tag{9.2.5}$$

(with P, Q as before), since the factor $\phi(x, y)$ cancels from y'/x'. Therefore (9.2.1) and (9.2.5) have the same curves as trajectories, but the velocity in one is a scalar multiple ϕ of the velocity in the other. But new fixed points will appear, at any zeros of ϕ.

The gradient formula (8.3.2) is reminiscent of that in 8.1.3, from which we see that the contours of the function $f(x, y)$ there can be thought of as the trajectories of the system

$$x' = -\frac{\partial f}{\partial y}, \qquad y' = \frac{\partial f}{\partial x}, \tag{9.2.6}$$

(note the switch of x and y!); but now these have a velocity and a *sense of description.* This is why the technique of construction of the portrait of (9.2.1) is just a more complicated version of the routine in Chapter 8.

After the great Irish mathematician W. R. Hamilton, the system (9.2.1) is said to be Hamiltonian if we can find a factor ϕ and a function f, so that

$$\frac{\partial f}{\partial y} = -\phi P, \qquad \frac{\partial f}{\partial x} = \phi Q. \tag{9.2.7}$$

In that case the system (9.2.5) takes the form (9.2.6), and is called **exact**. Thus the problem of drawing the portrait of (9.2.1), when it is Hamiltonian, is reduced to the simpler one of dealing with a function. Since ϕ leads to f, we call ϕ an **integrating factor** for (9.2.1), and f the **integral**. We met a slightly disguised form of this situation with the system (9.1.2), in which

$$P(x, y) = ax(1 - y), \qquad Q(x, y) = -cy(1 - x), \qquad \phi(x, y) = -1/xy,$$

and f is g in (9.1.5). Thus (9.1.2) is a Hamiltonian system. Such systems are common in mechanics, for deep reasons that were clarified by Hamilton; but these reasons are not necessarily valid in biology. (Hamilton is also famous for his algebra of quaternions: see Artmann, Chapter 5.)

Exercises 9.2.3

1. Compare the portraits of the simple harmonic system and that of $\dot{x} = \omega y(2 - y)$, $\dot{y} = -\omega x(2 - y)$, to see how their portraits differ.

2. Show that the equation of the pendulum: $\ddot{x} = -\omega^2 \sin x$ gives rise to a Hamiltonian system. Sketch its portrait and compare with that of the approximation $\ddot{x} = -\omega^2(x - x^3/6)$.
3. Show that the equation $x'' = f(x)$ always gives rise to a Hamiltonian system.
4. Find a Hamiltonian function for the system

$$\dot{x} = 3y^2 + 2x - 1, \qquad \dot{y} = 3x^2 - 2y + 1.$$

   Hence sketch the phase diagram.
5. Explain what it means to say that the system

$$\dot{x} = 1 - 2x^2 - y^2, \qquad \dot{y} = 4x(y + 2)$$

   has Hamiltonian $H(x, y) = y - 2x^2y - y^3/3 - 4x^2$. Sketch the phase portrait of the system, and show how it helps to know that H exists. (Hint: the line $y = 2$ is a contour of H, and H is symmetric about the y-axis.) [SU]
6. In (9.2.1), let $P(x, y) = py + rx^my^{n+1}$, $Q(x, y) = qx + sx^{m+1}y^n$, where all but x, y are constants. Show that x^hy^k is an integrating factor if $(h + 1)/p = (k + 1)/q$ and $h + m + 1/r = k + n + 1/s$, assuming the denominators are non-zero. Hence find an integral when $P = 3y^2 + 10x^2y$, $Q = -2xy - 6x^4$. Sketch the portrait.
7. Suppose that $T(t) = (x(t), y(t))$ is a solution of the system (9.2.7), with t in an interval I and with $X = (x(0), y(0))$. Let $h(t) = f(x(t), y(t))$. Differentiate h, using the chain rule, to show that $\dot{h} = 0$ on I, and hence that the trajectory T lies on the contour $f = f(X)$. If Y is a point on this contour, explain why there is a trajectory of (9.2.7) through Y.
8. The quantity $\partial P/\partial x + \partial Q/\partial y$ is important and is called the **divergence** of P, Q, denoted by $\operatorname{div}(P, Q)$. Show that in (9.2.7) we have $\operatorname{div}(-\phi P, \phi Q) = 0$. Now use theorem E.9.1 to prove the converse, that if in (9.2.1), $\operatorname{div}(P, Q) = 0$, then there exists a function $f(x, y)$ which satisfies (9.2.7) with $\phi = 1$.
9. Find a Hamiltonian system with no critical points.
10. When is a Hamiltonian system hyperbolic?
11. The linear system $\dot{p} = q$, $\dot{q} = -p$ is perturbed by the addition of quadratic terms so that the vector field is given by

$$(q + \alpha pq + 3q^2, -p - 2p^2 - \beta pq).$$

   Find values of α and β such that the vector field is Hamiltonian and find a corresponding Hamiltonian $H(q, p)$. [SU]

9.3 THE THREE RULES

We suggested, in exercise 9.1.3, that a routine for constructing the portrait of a system like (9.2.1) might be obtained from the portrait routine by changing the word 'contour' to 'trajectory' throughout the slope, no-crossing, and exit rules. Also, as we have seen with Fig. 9.1.4, we shall need to reckon with the possibility of limit cycles. It is our aim now to justify the forthcoming routine which will reduce to the earlier one, in the sense that the portrait of a function is that of a Hamiltonian system. To clarify the discussion, we distinguish between the two routines by calling them the 'portrait' and the 'trajectory' routines, respectively.

Let us first establish appropriate analogues of the three rules. In the portrait routine, we have regions and their boundaries. Analogously, we first note that the set of all fixed points of (9.2.1) consists of the common points of the horizontal and vertical manifolds

$$H: \{Y \,|\, Q(Y) = 0\}, \qquad V: \{Y \,|\, P(Y) = 0\}, \tag{9.3.1}$$

which are so-called because of the gradient formula (9.2.4), just as in the case of a function. Indeed, in the system (9.2.7), the equilibria are the same as the critical points of the integral f, and the horizontal and vertical manifolds of the sysem are the same as those of f.

Exercise 9.3.1

Calculate the horizontal and vertical manifolds, and the equilibria, of the systems (9.1.2), (9.1.7), (9.1.9) and (9.1.6).

Just as with the portrait routine, H and V in (9.3.1) divide up the plane $\mathbb{R}^2$ into regions on each of which the function g has constant sign, where at any ordinary point X we have

$$g(X) = \frac{Q(X)}{P(X)}. \tag{9.3.2}$$

Hence by (9.2.4) we have our first rule.

The slope rule

Through each ordinary point X of the region R, there is a unique trajectory T at which the tangent has slope $g(X)$ in (9.3.2), and this slope has constant sign throughout R. Moreover at X, T points in the direction of the vector $(P(X) \quad Q(X))$.

In order to state our next rule, we refer back to theorem 9.2.2, and note that the uniqueness of T implies that we can keep extending the path of T through ordinary points, and hence assume that T is a 'maximal' solution: the interval I cannot be increased further. We then call this maximal T 'the' trajectory through X: there cannot be two maximal trajectories through X. Hence we have our second rule.

The no-crossing rule

Two maximal trajectories cannot cross at an ordinary point, because of the uniqueness of T in (9.2.3).

Computer-generated sketches sometimes appear to break this rule (as with Figs. 9.1.3, 9.1.4); for that reason we often exaggerate hand-drawn sketches for clarification.

Thirdly we need a version of the exit rule in section 8.3 (recall Fig. 8.3.2(ii)), and its import is sketched in Fig. 9.3.1.

The exit rule

Suppose X is an ordinary point in the region R. Then the trajectory through X either leaves R by crossing H or V, or it goes to infinity. In the latter case, it can only become asymptotic to H or V by tending toward the horizontal or vertical.

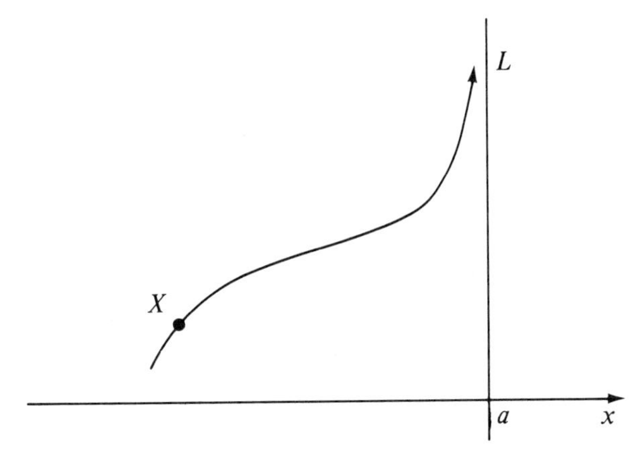

Fig. 9.3.1.

To prove it, we shall need the following extension of theorem 9.2.2, which gives us useful information that we need anyway. It tells us that trajectories which start close together cannot suddenly be wrenched apart. In (9.2.3) the trajectory T clearly depends on X, so we can write T_X to indicate this dependence. We would expect T_X to be a continuous function of X. More precisely we have the following result.

Theorem 9.3.1 (Addendum to theorem 9.2.1)
There is a neighbourhood U of X such that T is a continuous function of (x, y, t) when (x, y) varies in U and t in I.

(Here, if X is (p, q) then $T(p, q, t)$ is just $T_X(t)$. In particular, the same I will work for all points in U, but we cannot assume then that I is always maximal.) Again see Chapter F for a discussion.

Exercise 9.3.2
Use the addendum to justify the above statement about trajectories not being wrenched apart, by proving the following: if T_X is the trajectory through X, with domain I, then for each $t \in I$ and neighbourhood U of $T_X(t)$, there is a neighbourhood V of X such that for each $Y \in V$, $T_Y(t) \in U$. (See Fig. 9.3.2.)

A full proof of the exit rule needs the methods of Mathematical Analysis, so we shall only give a 'skeleton' version of it. Readers may find even this skeleton rather complicated, but we include it to give an idea of the flavour of the more theoretical work that will be met in more advanced texts. It can be skipped at a first reading, until the three rules have been absorbed via the applications in section 9.4 below.

Skeleton proof of the exit rule
Suppose that the trajectory T through X does not cross either H or V; then we must prove that T goes to infinity as stated in the exit rule. Now, g in (9.3.2) is either positive or negative, and accordingly $P(X)$ and $Q(X)$ either have equal or opposite

signs—say equal. Therefore, the coordinates $x(t)$, $y(t)$ of $T(t)$ each strictly increase as t increases in the maximal interval I (and $T(t)$ never leaves R), so the numbers $x(t)$ all have a least upper bound a (say) which could be infinite. Similarly the numbers $y(t)$ have a least upper bound b. If both a and b are infinite then T goes to infinity and stays in R. The next possibility is that $a < \infty = b$. Then T approaches nearer and nearer to the line L: $\{x = a\}$, while its y-coordinate grows ever larger (see Fig. 9.3.2). Thus its gradient tends to infinity, which is the gradient of L. Hence T must be approaching ($=$ asymptotic to) the vertical V. Similarly if $b < \infty = a$ the same argument applies with H instead of V.

Finally in this proof, suppose that both a and b are finite. Then the point $Y = (a, b)$ still lies in R, Y is an ordinary point of (9.2.1), and the trajectory $T(t)$ apparently gets stuck near Y. This will give a contradiction as we now show.

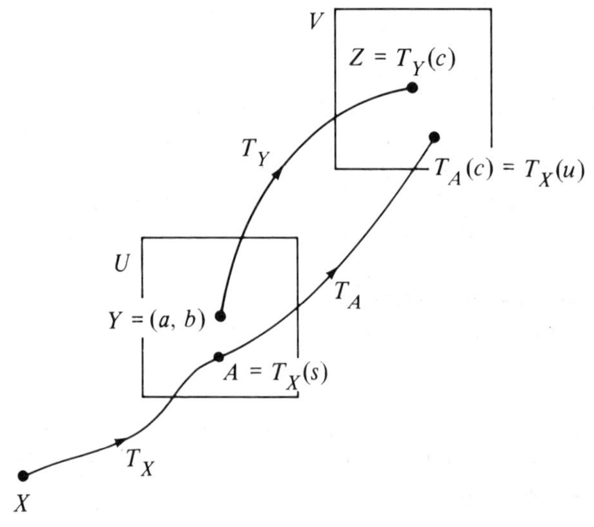

Fig. 9.3.2.

Choose a time $c > 0$ in I and let $Z = T_Y(c)$. Then $Y = T_Y(0)$, and $Z > Y$ (i.e., each coordinate of Z exceeds the corresponding coordinate of Y). So there is a neighbourhood V of Z such that $W > Y$ for every point W of V (see Fig. 9.3.2). By the continuity of T asserted by the addendum in theorem 9.3.1, there is a neighbourhood U of Y such that $T(x, y, c) \in V$ if $(x, y) \in U$. By definition of least upper bound, there exists some time s for which $T_X(s)$ ($= A$, say) lies in the neighbourhood U. But then $T_A(c) \in V$, and so the trajectory through A extends T_X into V; so there is some time u in I with $T_X(u) > Y$. This gives us the contradiction we predicted above, because it contradicts the definition of a and b as upper bounds for the coordinates of T on I. All this assumed that the coordinates of T each strictly increased as t increased in I; but if (say) $x(t)$ had decreased, we would use the greatest lower bound for it, rather than the least upper bound, with obvious corresponding changes. This completes the proof.

9.4 THE TRAJECTORY ROUTINE

Having established the three rules, we shall now incorporate them into a 'fair weather' routine for drawing the portrait of a hyperbolic system such as (9.2.1); then we will try it out on some examples. The steps of the routine are very similar to those of the portrait routine in Chapter 8, so we shall give less explanation. For a reason to appear, we shall have to extend step 2 later. Suppose then that we wish to draw the portrait of a system like (9.2.1).

Step 1
Draw the horizontal and vertical manifolds H, V given by (9.3.1) and find their points of intersection. These are the fixed points of the system. Also mark each region with the sign of $g(X)$ in (9.3.2).

Step 2
Work out $\text{div}(P, Q) = \partial P/\partial x + \partial Q/\partial y$. If this is always zero, then as we saw in exercise 9.2.3(8) above, the system (9.2.1) is Hamiltonian, and the portrait routine applies. (It is not necessary to calculate an integral f, because, as we observed after (9.3.1), H and V here are the same as the manifolds for f.) Therefore we assume that $\text{div}(P, Q)$ is not always zero.

Step 3
For each fixed point X, work out the Jacobian matrix $J(P, Q)$ at X; decide the type of the linear system $\dot{\mathbf{z}} = J\mathbf{z}$. If this is a centre, the system is not hyperbolic, and special methods may be needed (the weather is not fair). Therefore we assume that by checking at each fixed point, we have decided that the system is hyperbolic. In particular, then, theorem 9.2.1 applies. Thus we can insert a diffeomorphic picture of that linear system, in the neighbourhood of X.

Step 4
Suppose that some fixed point is a saddle. Then we insert the separatrices as with the portrait routine, using the three rules established above. It may happen that the exit rule forces a separatrix to go to another equilibrium point. If there are no separatrices, choose significant trajectories (such as the axes for a node) and extend them by using the three rules.

Step 5
Now extend other typical trajectories that stay close to these separatrices or to others. This uses the result of exercise 9.3.2, and can be a check on the drawing, because the sense of the separatrix must be the same as that of its nearby trjectories.

Step 6
Now check your drawing to see that, where you have drawn two diverging flows of trajectories, you have also drawn (correctly) a separatrix that separates them. The absence of such a separatrix in a computer-drawn portrait can indicate that the internal arithmetic has produced unstable rounding errors.

If we try out this routine on the system (9.1.2), we find that there are two equilibrium points, **0** and A, where A lies in the positive quadrant $\mathbb{R}^2_+$. Also **0** is a saddle and A is a centre. We found in (9.1.5) that on the region D away from the axes, the system is Hamiltonian, and the portrait routine of Chapter 8 applies to D. We can then fill in the arrows, by following the appropriate steps of the above routine. The result is Fig. 9.1.3 above.

Next, consider the system (9.1.7). In order not to have distracting algebra for the moment, we consider it with numerical coefficients, say the system

$$x' = x - y = P(x, y), \qquad y' = -2x - 3y + x^2 = Q(x, y). \tag{9.4.1}$$

Here V is the line $y = x$, and H is the parabola shown in Fig. 9.4.1, which we mark as in step 1 of the above routine. The equilibria are at **0** and at $A = (5, 5)$, and each lies in $\mathbb{R}^2_+$ as shown. At any point, the divergence $\partial P/\partial x + \partial Q/\partial y = 1 - 3 = -2 \neq 0$, and the Jacobian matrix J is:

$$\frac{\partial(P, Q)}{\partial(x, y)} = \begin{pmatrix} 1 & -1 \\ -2 + 2x & -3 \end{pmatrix},$$

with determinant -5 at **0**, and 5 at A. Hence **0** is a saddle. At A the trace is -2 and the discriminant is $(-2)^2 - 4(-5) < 0$, so we have there a spiral sink. Hence the system is hyperbolic, and we now carry out step 4 of the routine above.

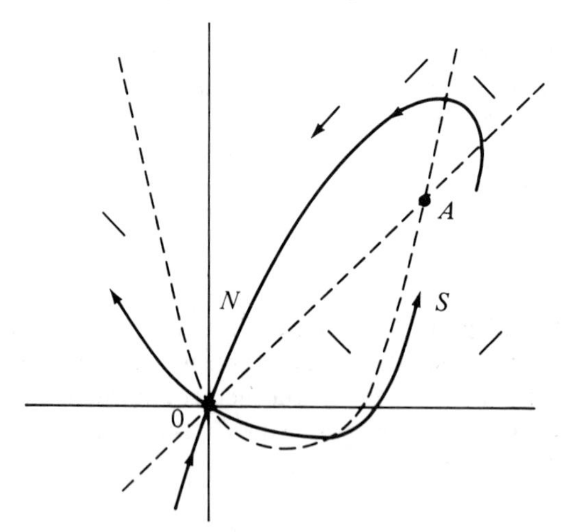

Fig. 9.4.1.

There are four separatrices at the saddle **0**, and the marking of regions in step 1 suggests that three of them will have to look like the ones shown approaching **0** from the north-east and south-west, and the one leaving north-west. On each of these, the sense is worked out as follows. If $x > 0$ and small, while $y > 0$ and large, then both x' and y' are negative, so the velocity points towards **0**, and gives us the direction shown on the north-easterly separatrix N; the senses on all the other separatrices are then automatically determined since **0** is a saddle. Now N cannot have come *from* the sink A, so when we trace it backwards, N curves round A as shown in Fig. 9.4.1. On the other hand, the south-easterly separatrix S must therefore leave $\mathbf{0}$ sloping downwards, and then we apply the exit rule to S, after noting that neither H nor V has a horizontal or vertical asymptote. The rule then forces S to cross H horizontally, then to rise and cross V, and thus to spiral round A.

Finally, we follow steps 5 and 6 to complete the portrait as in Fig. 9.1.3. Thus the effect of the added square term has been to capture certain trajectories of the linear system at $\mathbf{0}$, and divert them into the sink at A: the instability of a saddle has been (partially) tamed by the presence of the sink at A.

One question remains, however: why could the portrait not look like Fig. 9.4.2, in which we have inserted a limit cycle L, round which the separatrix S spirals, with the trajectories within L as shown? It turns out that the system (9.4.1) has no limit cycle, but we now need a way of verifying this fact; so in the next section we introduce a criterion that will allow us to rule out limit cycles from some systems. (At the present time, no general procedure is known that will automatically tell us whether or not a given system has a limit cycle, even when the functions P, Q in (9.2.1) are quadratics or cubics. This is part of Hilbert's 'sixteenth problem', which the great German mathematician first announced in 1901.)

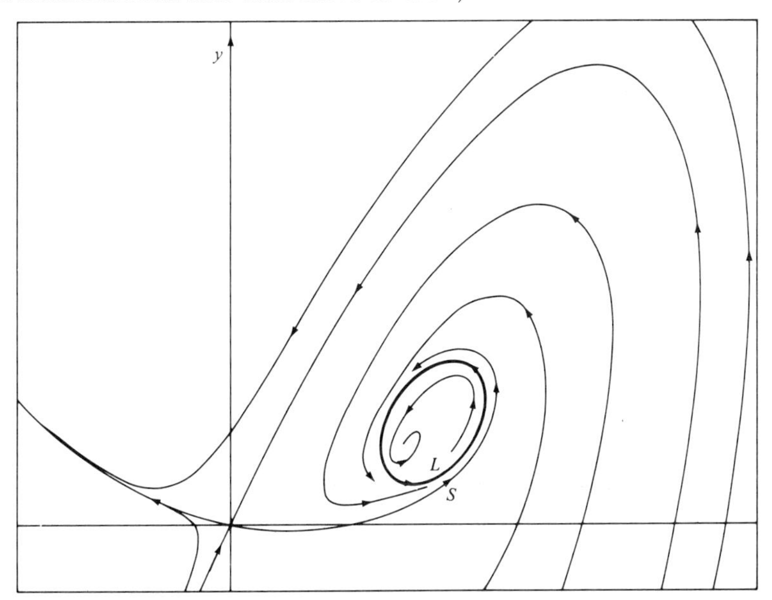

Fig. 9.4.2.

Exercises 9.4.1

1. Assuming for the present that the following systems have no limit cycles, practice the trajectory routine by drawing their portraits:

(a) $\dot{x} = x + 2y^2, \qquad \dot{y} = 3x + y;$

(b) $\dot{x} = x + y^3, \qquad \dot{y} = x^3 - y;$

(c) $\dot{x} = x^3 - y^2, \qquad \dot{y} = x + y^3.$

2. (*Gradient systems*) In $\mathbb{R}^2$, let $\mathbf{v}(t) = (x(t), y(t))$ be a smooth path and $f: \mathbb{R}^2 \to \mathbb{R}$ a smooth function. Suppose that at each of its points $p = \mathbf{v}(t)$, the path is orthogonal to the level of f at p. By corollary 8.1.1, the velocity vector is parallel to grad $f|_p$. Hence show that $\mathbf{v}$ has the same track as a trajectory w of the 'gradient system' of f:

$$x' = \frac{\partial f}{\partial x}, \qquad y' = \frac{\partial f}{\partial y}. \tag{9.4.2}$$

Use the chain rule to show that along w, $d(f \circ w)/dt \geq 0$, so f increases along trajectories; and hence that w cannot be a closed trajectory unless it is a single point. Show that the saddle points of the system are those of f itself, and that a max of f is a source for (9.4.2). Now sketch the gradient system of each function mentioned in Chapter 8.

9.5 CLOSED TRAJECTORIES

In (9.2.3) we saw that the uniqueness of $T: I \to \mathbb{R}^2$ has two important consequences. Firstly, we can extend the trajectory through X to allow a 'maximal' solution, where we may assume that the interval I cannot be increased further. Secondly the 'no-crossing' rule holds: *two distinct trajectories cannot cross at the ordinary point X.*

Now a maximal trajectory T can have two forms, because as a mapping, T is either injective or it is not. If T is injective, then for any distinct times s, t in I, we have $T(t) \neq T(s)$. If T is not injective, then there are two distinct times s, t such that $T(t) = T(s)$, and we call T a **closed** trajectory; in that case, the image of T is a simple closed curve. Indeed, we shall show in exercise 9.5.2 below that there then exists a number $\omega > 0$ such that T is 'periodic with period ω', which means that, for all $t \in I$, we have

$$T(t + \omega) = T(t). \tag{9.5.1}$$

We can see just one such closed trajectory L in Fig. 9.1.4, but several in Fig. 9.1.1. To distinguish these two cases, we call a closed trajectory a **limit cycle** provided it has a neighbourhood that contains no other closed trajectory. In any neighbourhood (however narrow) of a closed trajectory in Fig. 9.1.1, there are clearly other closed trajectories (even though we cannot draw them all). But with Fig. 9.1.4, we cannot fit another closed trajectory in the neighbourhood of L without crossing a spiral, and the 'no-crossing' rule forbids this. Thus, L is a limit cycle.

There is a useful criterion which tells us when there are no closed trajectories: it is called Bendixson's theorem, and we shall use it a few times, before we give a proof below. To state the theorem, we need to know that a function $f(X)$ is said to be **identically zero** when it is zero for all X. (We then write $f \equiv 0$.) Also, recall that for the pair of functions P, Q in (9.2.1), we define the function div(P, Q), the **divergence**, by

$$\operatorname{div}(P, Q) = \frac{\partial P}{\partial x} + \frac{\partial Q}{\partial y}. \tag{9.5.2}$$

Thus the conclusion of exercise 9.2.3(8) can be expressed as follows.

Proposition 9.5.1

If $\text{div}(P, Q) \equiv 0$ *then the system is Hamiltonian.*

Note that, for the two-species model in (9.1.7), $\text{div}(P, Q) = a + d$, which is constant. But in the van der Pol model (9.1.9) the divergence is $-\mu(x^2 - 1)$. These different behaviours are partly responsible for the differences in the two portraits, and with proposition 9.5.1 they indicate that the divergence has a powerful influence on the behaviour of a system.

Theorem 9.5.1 (Bendixson's criterion)

Suppose that the system (9.2.1) has a closed trajectory T, which surrounds a region D. Then the function $\text{div}(P, Q)$ *either changes sign somewhere on D, or else* $\text{div}(P, Q) \equiv 0$ *on D.*

As we have just seen, in (9.1.7) the divergence is $a + d$, which never changes sign; if it is non-zero, then by the criterion, there can be no closed trajectories in (9.1.7). (This is a typical use of the criterion to *exclude* the possibility of a closed trajectory.) In particular, this gives the final justification for Fig. 9.4.1 as the portrait of the system (9.4.1), since $a + d = -2 \neq 0$.

If, for the system (9.2.1), the divergence is identically zero, then proposition 9.5.1 tells us that (9.2.1) is Hamiltonian. If the resulting integral f in (9.2.7) has a maximum or minimum, then there will be an infinity of closed trajectories, which is why the criterion allows for this case.

We saw also that in (9.1.9) the divergence is $-\mu(x^2 - 1) = g$ (say); if $\mu = 0$ then $g = 0$ and the system (9.1.8) becomes simple harmonic with all trajectories closed. But if $\mu \neq 0$, then g is not identically zero, and it is of constant sign on each of the regions $\{x < -1\}$, $\{x > 1\}$, and $\{-1 < x < 1\}$ in $\mathbb{R}^2$. Hence there can be no closed trajectory that lies wholly in one or other of these regions—and L in Fig. 9.1.4 meets all three of them. Note that the *existence* of L does not follow from the fact that g changes sign; other methods are needed, as we shall see.

Exercises 9.5.1

1. An interesting coupled system is one in which a hunter, with speed k, pursues a prey which moves with unit speed on a closed curve, such as a fixed circle. Show that the corresponding model in the case of the circle is:

$$\dot{r} = r(a \cdot \sin \phi - ka), \qquad \dot{\phi} = a \cdot \cos \phi - r,$$

where r is the distance between the prey and the pursuer, while ϕ is a certain angle. Show that there is no closed trajectory, and sketch the portrait. (See Davis (1962), p. 332.)

2. Explore the pursuit problem in exercise 1 by means of the following program.

```
REM Program 9.5.1, Pursuit curve
FUNCTION x(t) = a*COS(t)
FUNCTION y(t) = a*SIN(t)
READ hx, hy, hs , ps, a, t, dt , eps , xmax, ymax
DATA 1 , 0 , 0.6, 1 , 2, 0, 0.1, 0.01, 3   , 2
FUNCTION dist(x1,y1,x2,y2) = SQR((x1-x2)^2 + (y1-y2)^2)
SET MODE "graphics"
SET WINDOW -xmax,xmax, -ymax,ymax
LET k = hs/ps
DO
    LET px = x(t)
    LET py = y(t)
    LET t = t + dt
    LET px1 = x(t)
    LET py1 = y(t)
    LET pd = dist(px,py,px1,py1)
        REM for a circle we know pd accurately
        REM and so we could use  LET pd = a*dt
    PLOT px,py ; px1,py1
    LET hd = k*pd
    LET d = dist(hx,hy,px,py)
    LET hx1 = hx + (px-hx)*hd/d
    LET hy1 = hy + (py-hy)*hd/d
    PLOT hx,hy ; hx1,hy1
    LET hx = hx1
    LET hy = hy1
LOOP UNTIL d < eps
PRINT "gotcha"
END
```

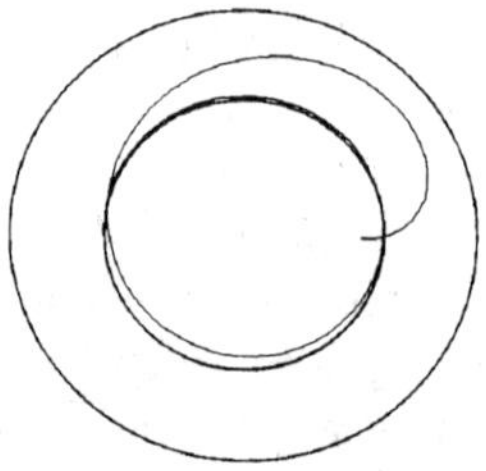

Sample output

3. Show that the system $\dot{x} = f(y)$, $\dot{y} = g(x)$, where f and g are smooth, is Hamiltonian. Change the equation $\ddot{x} = f(x)$ into this form, and investigate the cases when $f(x) = ax + b/(c - x)^n$, when $n = 1, 2, 3$.

   State a criterion to show that the system

$$\dot{x} = -x - y + y^2, \qquad \dot{y} = -2x$$

has no limit cycles, and sketch its portrait. [SU]

$$\dot{x} = ax + y, \qquad \dot{y} = 2x - 8x^3$$

has no closed trajectory if $a \neq 0$. If $a = 0$ show that the system has Hamiltonian function

$$H(x, y) = -\frac{y^2}{2} + x^2 - 2x^4$$

and sketch the portrait.

Show that, if a is fixed and $0 < a < 4$, then the system has three equilibria consisting of a saddle and two spiral sources. Sketch the phase portrait when $a = 3$. [SU]

5. For each $p \in \mathbb{R}$ let S_p denote the system of differential equations

$$\dot{x} = 2x + pxy^2, \qquad \dot{y} = x^3 + y.$$

Sketch phase diagrams for the system S_p, when $p = 0, -1$. In each case, indicate the behaviour of trajectories as y increases along a line $x = c > 0$ when c is small. [SU]

6. Recalling exercise 9.1.1(3), state a criterion that ensures that the rescaled equations have no closed trajectory. Sketch phase portraits for the cases $\sigma = 8$, $\sigma = -8$. [SU]
7. Sketch a possible phase portrait for the system

$$\dot{x} = -(x + y^2 - 2)$$
$$\dot{y} = -y(x^2 + y^2 - 3x + 1).$$

8. Prove Dulac's extension of theorem 9.5.1: suppose there exists a smooth function $\phi(x, y)$ such that $\operatorname{div}(\phi P, \phi Q) \not\equiv 0$ and does not change sign in D. Then the system (9.2.1) has no closed trajectory in D.
9. Investigate the system $\dot{z} = az + bu$, $\dot{u} = cz + du + (kz/(pz + q))$. This occurs in a model due to Starbuck (1973) for deciding the optimal ratio of managers to employees in a large firm: see Griffiths and Rand (1977).

Another (but more subtle) use of Bendixson's criterion tells us when the Volterra system (9.1.5) has closed trajectories, and that it never possesses a limit cycle. This argument is due to Coppel (1966), and we may as well give it here, although readers may prefer to postpone reading it until after they have seen more examples. It is convenient to write (9.1.5) with the more symmetrical notation

$$\dot{x} = x(a + bx + py), \qquad \dot{y} = y(c + dx + qy). \tag{9.5.3}$$

To ensure that the lines $a + bx + py = 0$, $c + dx + qy = 0$ intersect, we assume that

$$E = bq - dp \neq 0. \tag{9.5.4}$$

Since this is modelling a population problem, we are primarily concerned with trajectories in the positive quadrant $\mathbb{R}^2_+$.

Theorem 9.5.2

The system (9.5.3) has no closed trajectory in $\mathbb{R}^2_+$ unless $q(ad - bc) + b(cp - aq)$ ($= F$, say) is zero. If $F = 0$, then the system is Hamiltonian.

Proof

The strategy is to construct an integrating factor of the form $\phi(x, y) = x^{k-1}y^{h-1}$. To calculate h, k, we first find that

$$\operatorname{div}(\phi P, \phi Q) = (A + Bx + Cy)\phi,$$

where $A = ak + ch$, $B = b(k + 1) + dh$, $C = pk + q(h + 1)$. Now choose h, k so as to satisfy the equations $B = 0 = C$; the determinant is E, which is non-zero by (9.5.4), whence

$$k = q(d - b)/E, \qquad h = b(p - q)/E.$$

With this choice of h, k and with F as given above, we find that $A = F/E$. Also, the divergence is then $A\phi$, which can only be identically zero if $A = 0$, that is, if $F = 0$. In that case, we know that the system is Hamiltonian, by proposition 9.5.1. On the other hand, suppose $F \neq 0$. Then, since ϕ does not change its sign on the positive quadrant $\mathbb{R}^2_+$, the Bendixson criterion tells us that there can be no closed contour in $\mathbb{R}^2_+$. This completes the proof.

Proof of Bendixson's criterion
If there is a closed trajectory T surrounding the domain D, then by the divergence theorem (exercise E.7.1(b)) we have

$$J = \int_D \operatorname{div}(P, Q)\, \mathrm{d}x\mathrm{d}y = \int_T Qx' - Py'.$$

But the last integrand is $QP - PQ = 0$, so $J = 0$. Now, the integral of a positive function is strictly positive, and similarly if we replace 'positive' by 'negative'. Hence $\operatorname{div}(P, Q)$ cannot be everywhere of the same sign. This establishes the criterion.

Exercises 9.5.2

1. Show that in the notation of (9.1.6), the system there is Hamiltonian if and only if $ac[ap(1 + q) - cq(1 + p)] = 0$. Hence show that if a, c are fixed and non-zero, and the system is Hamiltonian, then (p, q) lies on a hyperbola unless $a = c$.
2. Prove (9.5.1) as follows. Suppose that $s \neq t$ and $T(s) = T(t)$, with $s < t$. Keeping this s fixed, let u be the smallest such t. Then, for any other such $t > 0$, there is a whole number n such that $t = s + n\omega + r$, where $\omega = u - s > 0$, and the remainder r satisfies $0 \leq r < \omega$. Show that $T(t) = T(s + (n - 1)\omega + r) = \cdots = T(s + r)$, and hence that $r = 0$. Therefore ω is the period required in (9.5.1).
3. Construct a set of portraits for the system (9.1.7), corresponding to various choices of parameters in the control space. (By rescaling time t to $\tau = t/p$, we can assume that $p = 1$. Then consider the cases when V has negative slope, H is inverted, $\mathbf{0}$ is not a saddle, etc., to obtain a finite set of 'typical' portraits.)

Finally in this section, we prove an important property of limit cycles.

Theorem 9.5.3

Suppose the system (9.2.1) has a limit cycle L. Then L encircles some fixed point P, which is not a saddle.

Proof
Consider the vector field v, given by (P, Q) in (9.2.1). Then a point X is a zero of this field if (and only if) X is a fixed point of (9.2.1). By the index theorem E.9.2, the index $I(L)$ is the sum of the indices of all the zeros that it encloses. Also, at each point of L, v is tangent and non-zero. Hence $I(L) = 1$ (see exercise 9.5.1(7)). Therefore not all the enclosed zeros can have index 0, so not all are saddles (by theorem E.9.2). This proves the theorem.

9.6 THE POINCARÉ–BENDIXSON THEOREM

The Bendixson theorem 9.5.1 is useful for telling us conditions under which closed trajectories do *not* exist. But Bendixson also improved a result of Poincaré's, that gives conditions for a limit cycle to exist. We state it here in a version most suitable for our purpose.

Theorem 9.6.1 (Poincaré and Bendixson)
Let D be the region of $\mathbb{R}^2$ that is bounded by two Jordan curves J_1, J_2 with J_1 completely surrounding J_2. Suppose that every trajectory of the system (9.2.1), which starts in D, remains in D for all later times. Then D either contains a fixed point, or a closed trajectory.

The theorem is illustrated in Figs 9.6.1 and 9.6.2. It was first noticed by Poincaré, and later Bendixson gave a proof.

The proof requires some acquaintance with topological notions, so we omit it and refer interested readers to Hartman (1964), p. 151. In Figs 9.6.2(i) and (ii), J_2 encircles a spiral source, so at each of its points, the trajectory enters D. Also, J_1 consists of the portion XP of a trajectory emerging from a saddle point X, together with the segment PX of a vertical manifold. Thus, in Fig. 9.6.2(ii), the trajectory does not leave D; but the Poincaré–Bendixson theory does not allow us to assert the existence of a closed trajectory, because there already exists the fixed point X. However, let D' be the portion of D to the left of the vertical line l shown in Fig. 9.6.2(ii). Then since PX is part of a vertical manifold, trajectories cross l *into* D'. Hence, if there are no fixed points in D', then there is a closed trajectory, by the Poincaré–Bendixson theorem.

The practical need to be able to draw portraits has given rise to the various theoretical results we have discussed. We now use them to deal with the Lotka–Volterra system (9.1.2) and its later modifications.

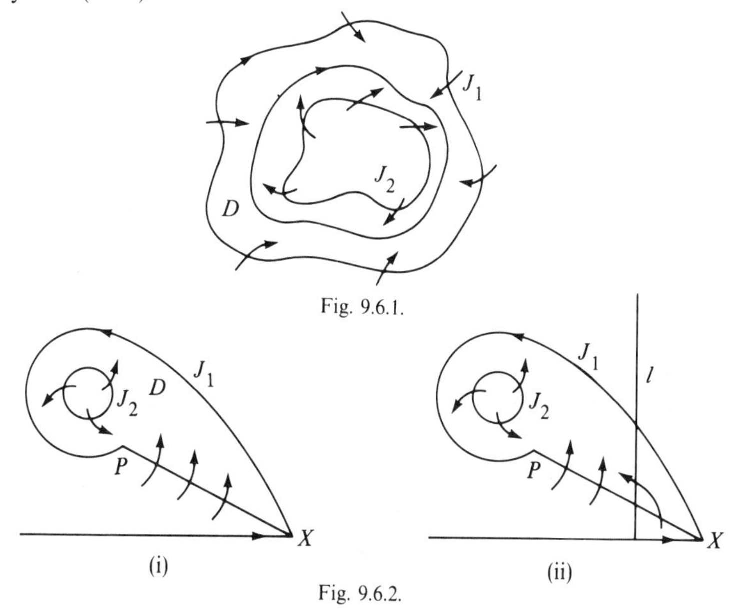

Fig. 9.6.1.

Fig. 9.6.2.

9.7 SKETCHING THE LOTKA–VOLTERRA SYSTEM

Now we know theorems 9.5.3 and 9.6.1, we can apply the trajectory routine to sketch the modified Lotka–Volterra system (9.1.6), which, for ease of reference, we repeat:

$$\dot{x} = ax(1 - y + px), \qquad \dot{y} = -cy(1 - x + qy), \qquad (a, c > 0). \tag{9.7.1}$$

To keep the algebra simpler at first, we shall suppose that p and q are small compared with the other parameters; and we shall look at the case when the system is not Hamiltonian (so (p, q) does not lie in the locus discussed in exercise 9.5.2(1)). Suppose also that p and q are negative. We then go through the portrait routine, knowing that there are no closed trajectories, and hence no limit cycles.

Thus H is a pair of lines, consisting of the x-axis $\{y = 0\}$, and the line L: $y = (x - 1)/q$; while V consists of the y-axis and the line M: $y = px + 1$. Hence we have four fixed points, which all lie in the positive quadrant $\mathbb{R}^2_+$ (see Fig. 9.7.1) since p, q are negative. The intersection B of L and M is approximately at the fixed point A of the original system (9.1.2) that we are now perturbing. The other three fixed points are $\mathbf{0}$ and, on the axes, $X = (-1/p, 0)$ and $Y = (0, -1/q)$. If x, y are large and equal, then since p, q are small:

$$\frac{y'}{x'} \cong -\frac{c}{a} \cdot \frac{q-1}{p-1} \cong -\frac{c}{a} < 0$$

because a and c are assumed positive. Therefore we mark the regions as in Fig. 9.7.1.

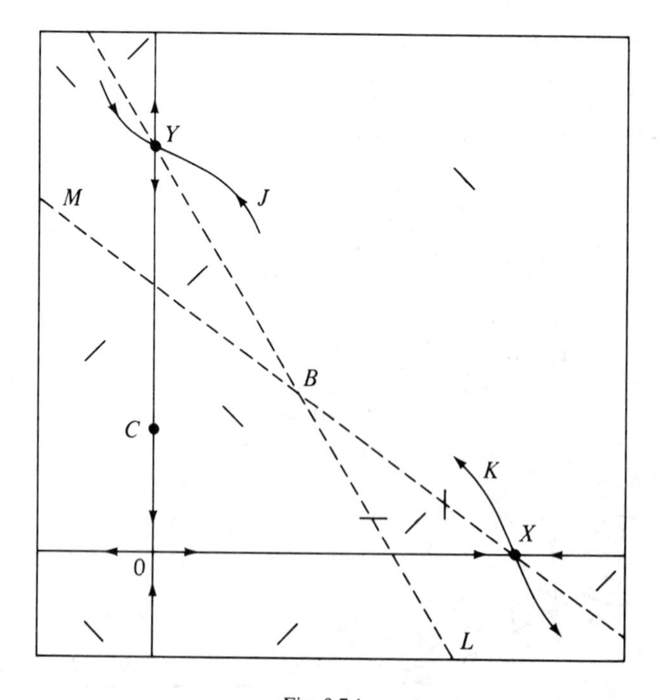

Fig. 9.7.1.

Next we consider the Jacobian matrix of the system:

$$J = \begin{pmatrix} a(1-y) + 2apx & -ax \\ cy & c(x-1) - 2qcy \end{pmatrix}.$$

At $\mathbf{0}$, $\det(J) = -ac < 0$, so $\mathbf{0}$ is a saddle point. At X, $\det(J) = ac(1 + 1/p)$, and this is negative if $-1 < p < 0$; so X is a saddle, and similarly so is Y. Since $B \cong (1, 1)$, we have the following approximations at B.

$$\begin{aligned} &\det(J) \cong ac(1 - 4pq), \qquad \text{Trace}(J) \cong 2(ap - qc), \\ &D \cong (ap - qc)^2 - 4ac(1 - pq) \cong -4ac < 0, \end{aligned} \tag{9.7.2}$$

so B will be a spiral source or sink according as $ap - qc$ is positive or negative. Suppose it is a sink.

Next, we claim that at the saddle $\mathbf{0}$, the four separatrix arms are on the same axes as shown. For, if $C = (0, k)$ is an initial point on the positive x-axis, the equations (9.7.1) give $\dot{x} = 0 > \dot{y}$ if $k < -1/q$; so if C starts below Y, then its trajectory stays on this axis and approaches $\mathbf{0}$. It is therefore one of the arms at $\mathbf{0}$. Similarly for the other arms, as claimed. Similarly also, if C had started on the axis above Y, then it would stay on the axis and flow away from Y. Thus two arms at Y lie on the axis, and similarly two arms at X lie on the x-axis as shown.

But now we know the directions of flow at the saddles X, Y, since one separatrix arm at Y flows down *to* $\mathbf{0}$, and one at X flows *from* $\mathbf{0}$. Hence, another arm J at Y flows from the south-east into Y, and therefore cannot have originated at the sink B. Therefore by the exit principle in reverse, we can produce J backwards, as shown in Fig. 9.7.2 either (i) to become asymptotic to the x-axis, or (ii) to flow out of X. Next, one arm K at X must flow outwards and north-westerly; and by the no-crossing principle K cannot cross J; therefore either $K = J$ as in (ii), or $K \neq J$ and then K is forced to spiral into the sink B. The remaining arms at these saddles all behave straightforwardly as shown. Finally we insert a few typical trajectories as in Fig. 9.7.3.

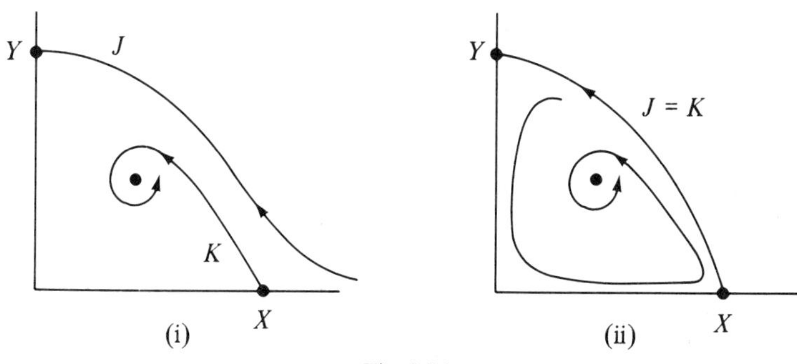

Fig. 9.7.2.

If, instead, B is a source, then we need an addition to the argument: for, the situation of (i) above could conceivably arise whether or not B is a sink or a source.

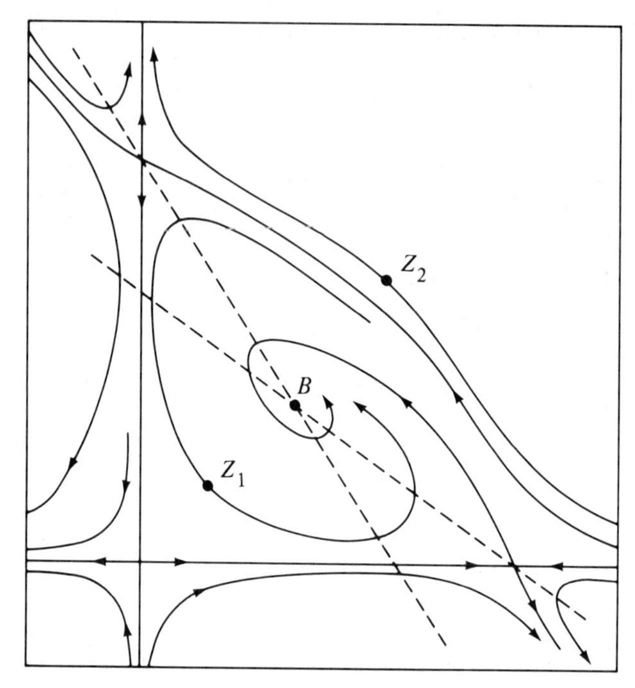

Fig. 9.7.3.

However, we can then surround B by a small circle and use the Poincaré–Bendixson theorem (see Fig. 9.6.3) to show that there would then be a limit cycle. This contradicts theorem 9.5.2. The rest of the construction of the portrait is now left to the reader in the next exercise.

Exercises 9.7.1

1. Draw the portrait of (9.7.1) for the case that B is a spiral source. (Explain why the arm H will then spiral out from the source B, and the arm K will pass it and climb up past Y.)
2. Draw the portrait when (9.7.1) is Hamiltonian and the fixed points are still in $\mathbb{R}^2_+$.
3. Use a computer program to alter the parameters p, q so that the arms H and K coincide in Fig. 9.7.1. (Such a coincidence must happen, to allow the 'sink' configuration, described above, to evolve gradually as B changes through a centre into a source. It probably occurs only when the system is Hamiltonian, but we do not know a precise answer.)
4. The equations (9.7.1) are of the general form $\dot{x} = xP(x, y)$, $\dot{y} = yQ(x, y)$. If X is a fixed point with $xy \neq 0$, find the Jacobian of the system at X and show that the value of its determinant is $xy \det(\partial(P, Q)/\partial(x, y))$.
5. Calculate accurate algebraic formulae instead of the approximations in (9.7.2). Hence find regions of the (a, c, p, q)-space $\mathbb{R}^4$ corresponding to which the system (9.7.1) has at B a spiral sink, or a nodal sink or source, etc. If the parameters are such that there is a node at $\mathbf{0}$ (rather than B), sketch appropriate portraits.

Our motivation for studying the perturbation (9.7.1) of the system (9.1.2) was to show that the original centre, with all the closed trajectories, could not be robust relative to small changes in the parameters. It is Fig. 9.7.3 that is more typical than either Fig. 9.1.1 or Fig. 9.7.2(ii). Nevertheless, from the point of view of modelling two populations, Fig. 9.7.3 shows the defect that an initial position like Z_1 spirals into B, and the two populations rapidly reach a position of making decreasing oscillations about B. An initial point like Z_2 rises up its trajectory, and one population dies out while the other becomes infinite. These do not seem very realistic accounts of the evolution of most real populations. It would be more realistic if, for example, the trjectory of Z_1 approached a limit cycle S surrounding B, but with spirals coming outwards from A to approach S. This happens with the following model, due to Hollings and Tanner (we follow the treatment of Arrowsmith and Pace (1982));

$$\dot{x} = ax\cdot\left[1 - \frac{x}{K} - \frac{wy}{L+x}\right], \qquad \dot{y} = cy\cdot\left[1 - \frac{My}{x}\right], \tag{9.7.3}$$

with all parameters $a, c, w, K, L, M > 0$.

The terms $wy/(L+x)$ and My/x in (9.7.3) are introduced for the following 'ecological' reasons: it is assumed that a predator can capture prey at a rate (r) per unit time, which is small if prey (x) are scarce but which steadily increases to be almost constant ($= w$) when prey are plentiful. Hence y predators will reduce the growth rate $\dot{x}/x$ of prey by the amount yr, and the term $wy/(L+x)$ models these requirements. For the other term, assume that each predator needs at least M prey per unit time for survival. Then the maximum sustainable number of predators is $m = x/M$, so the factor $[1 - My/x]$ ensures that $\dot{y}$ cannot be positive if y increases beyond M.

We restrict ourselves to the subset Q consisting of $\mathbb{R}^2_+$ minus $\{x = 0\}$. Here, H and V are the portions (in Q) of the line $y = x/M$, and the parabola $y = (x+L)(x-K)/wK$, as shown in Fig. 9.7.4.

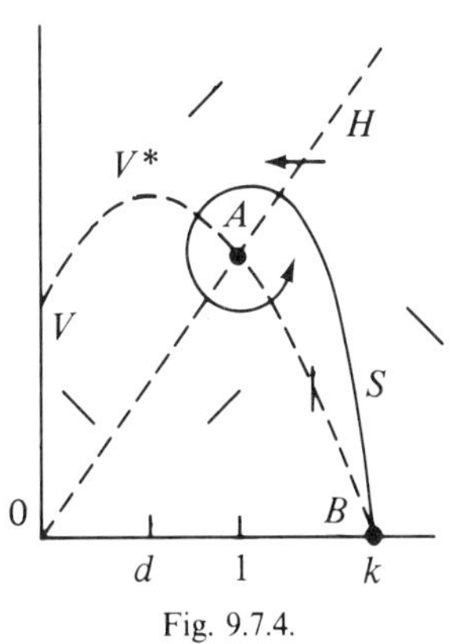

Fig. 9.7.4.

These intersect within Q at just the one fixed point A. Hence, if A has coordinates (u, v), then these are positive. Now rescale by writing $x = uX$, $y = uY$; then (9.7.3) is replaced by the same equations as before (now in the XY-plane) but with L and K changed to $l = uL$, $k = uK$. Also, A is now at $(1, 1/M)$, and since A lies on V, we have

$$wk = (1+l)(1-k)M. \tag{9.7.4}$$

Further, V cuts $\mathbf{0}x$ in $B = (k, 0)$ and (outside Q) at $(-l, 0)$, so its vertex V^* is at $(d, -(k+l)^2/4wk)$, where $d = (k-l)/2$. Hence A lies to the left or right of V^* according as $1 - d$ is negative or positive.

The fixed points are at A and B, and using exercise 9.7.1(4), we see that the Jacobian matrix at B has determinant $-ac < 0$; thus B is a saddle. By that exercise, also, the Jacobian at A is

$$J = \begin{pmatrix} \dfrac{a}{M}\left(-\dfrac{M}{k} + \dfrac{w}{(1+l)^2}\right) & -\dfrac{aw}{1+l} \\ \dfrac{c}{M} & -c \end{pmatrix}.$$

Then

$$\det(J) = \frac{ac}{M}\cdot\left[\frac{M}{k} + \frac{w}{1+l}\cdot\left(1 - \frac{1}{1+l}\right)\right],$$

which is positive since $1 + l > 1$, so A is not a saddle. Moreover, by (9.7.4) the trace T is

$$T = a\cdot\frac{l-k}{k(1+l)} - c = \frac{-2ad}{k(1+l)} - c,$$

so A is either a source or sink, according as $T > 0$ or $T < 0$. If $d < 1$, then (by the remarks after (9.7.4)) V^* is to the left of A, and $T < 0$ since $c > 0$, so A is a sink. If $d > 1$ then T can still be < 0, depending on the size of c (for example); but if $T > 0$ then $d > 1$ and A is a source to the left of V^*, and it can still be either a spiral or node.

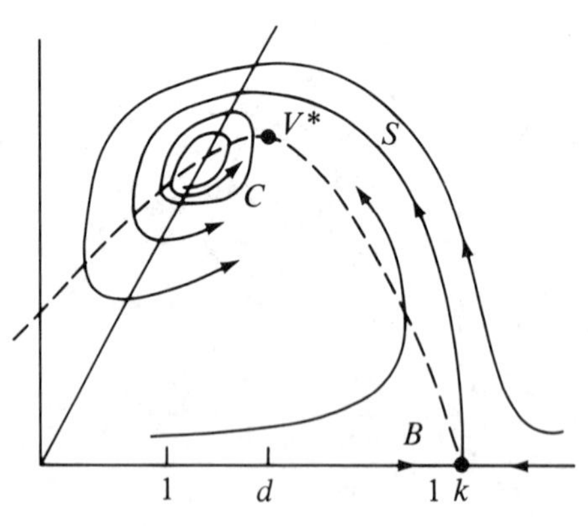

Fig. 9.7.5.

We now construct the portrait (Fig. 9.7.5) by applying the trajectory routine. At $(k, k/2M)$, $\dot{x} < 0 < \dot{y}$, so we sign the regions as in Fig. 9.7.5 and, just as for the system (9.7.1), we see that a separatrix arm S must rise from B and curl round A as shown. But then, if we assume that A is a source, we have the situation of Fig. 9.6.2; so S must wind asymptotically around a limit cycle C that encircles A. Thus we have a model that predicts a stable ecology, in which any initial pair of populations moves and eventually oscillates without damping out, on a limit cycle round A.

A limit cycle like C is said to be **stable**, whereas if nearby trajectories had spiralled away from it, then C is **unstable**. It is conceivable that the limit cycle C may not be unique. If there is another, it must also enclose a fixed point and hence either enclose, or be enclosed by, C. Just how many such cycles can exist does not seem to be known. However, we shall next consider a situation in which it is possible to assert that a limit cycle both exists and is unique. This will cover the analysis of the van der Pol equation in (9.1.8); note that Fig. 9.1.4 indicates just how stable its limit cycle is—the other trajectories home rapidly onto it.

9.8 THE VAN DER POL AND LIÉNARD EQUATIONS

Physical intuition about electrical circuits led van der Pol to formulate the differential equation (9.1.8) and then to guess that there must be a limit cycle. His guess was justified rigorously by the French mathematician Liénard who studied a general class of equations of the following type:

$$\ddot{x} + f(x)\cdot\dot{x} + x = 0, \tag{9.8.1}$$

with f continuous. The equation of van der Pol belongs to this class, since it has $f(x) = \mu(x^2 - 1)$. We now analyse (9.8.1), and first convert it into a system of first order on $\mathbb{R}^2$. Then we shall show that—at least with certain assumptions—the portrait has a unique limit cycle.

With f in (9.8.1), let $F(x) = \displaystyle\int_0^x f(t)\,\mathrm{d}t$

so $F(0) = 0$; and since f is continuous, then F is differentiable and $\mathrm{d}/\mathrm{d}t\, F(x(t)) = f(x(t))\cdot\dot{x}$. Hence if we write $y = \dot{x} + F(x)$ then by (9.8.1), $\dot{y} = -x$. Thus, (9.8.1) is equivalent to the system.

$$\dot{x} = y - F(x), \qquad \dot{y} = -x. \tag{9.8.2}$$

Recall that this was how we converted van der Pol's equation to the system (9.1.9) on $\mathbb{R}^2$:

$$\dot{x} = y - \mu\left(\frac{x^3}{3} - x\right), \qquad \dot{y} = -x \tag{9.8.3}$$

where μ is constant. Readers should keep this in mind as a concrete example, to check the following theory.

Since $F(0) = 0$, the system (9.8.2) has just one fixed point, namely at $\mathbf{0}$; and there the Jacobian matrix is

$$J = \begin{pmatrix} -F'(0) & 1 \\ -1 & 0 \end{pmatrix}, \qquad \det(J) = 1, \qquad \operatorname{trace}(J) = -F'(0). \tag{9.8.4}$$

Thus, since the discriminant D is $F'(0)^2 - 4$, $\mathbf{0}$ will be a node if $|F'(0)| \geq 2$, and otherwise a spiral, provided $F'(0) \neq 0$.

Liénard's theory tells us about the existence of a limit cycle, and applies to (9.8.2) when F is an odd function, that is, $F(-x) = -F(x)$ for all x. In (9.8.3) we have an example with $F(x) = \mu(x^3/3 - x)$ since $f(x) = \mu(x^2 - 1)$; here f is an even function and F is odd, with $F'(0) = -\mu$.

We shall prove the following theorem.

Theorem 9.8.1

Suppose that F in (9.8.2) is an odd function, and that either (i) *F is zero only when* $x = 0$, *or* (ii) *F has just one positive zero, at* $x > 0$. *If* (i) *holds, then (9.8.2) has no closed trajectory; and with* (ii), *(9.8.1) has exactly one limit cycle.*

Before we give the proof (based on a treatment by S. Lynch and colleagues at Aberystwyth), consider the following exercises.

Exercises 9.8.1

1. Show that if (9.8.2) has a closed trajectory, then (9.8.1) has a periodic solution.
2. Show that (9.8.2) has no limit cycle if $F(x) = x^3$ or $F(x) = \sinh x$ or $F(x) = \sin x$.
3. Show that we need only consider the case when $F'(0) < 0$ in (9.8.2) as follows. Write $s = -t$ (thus reversing time), and change y to $-y$ (thus reflecting the portrait in the x-axis), to obtain (9.8.2) again but with $-F$ instead of F. What is the effect on the second-order differential equation (9.8.1)? Show that when $F'(0) < 0$, then the limit cycle predicted by theorem 9.8.1 must be stable, whereas if $F'(0) > 0$, it is unstable.
4. Long before van der Pol, Lord Rayleigh in 1883 had met the equation $\ddot{u} + (-b + c\dot{u}^2)\dot{u} + n^2 u = 0$ while studying mechanical vibrations. Transform it by writing $y = q\dot{u}$, then $x = pt$ (with p, q constant) to get $py' + (-bq^2 + cy^2)y/q^2 + qn^2 u = 0$. Now differentiate and choose $p = n$, $q = \sqrt{(3c/b)}$, $\mu = b/p$, to get van der Pol's equations (9.1.8). (Here, b and c are positive.)
5. Sketch the portrait of the van der Pol system (9.8.3) through the following steps. Here, V is the graph of the function $y = \mu(x^3/3 - x)$. Show that the Jacobian matrix of the system is

$$J = \begin{pmatrix} -\mu(x^2 - 1) & 1 \\ -1 & 0 \end{pmatrix},$$

with determinant 1, and $T = \text{trace}(J) = -\mu(x^2 - 1)$. Therefore $T = \mu$ at $\mathbf{0}$, so the discriminant D is $\mu^2 - 4$. By exercise 3 above, we need only consider the case when $0 < \mu$. (When $\mu = 0$ we have simple harmonic motion).

Show that if $0 < \mu < 2$, then the system exhibits a spiral source at $\mathbf{0}$; and that when $y = 0$, $\dot{y}/\dot{x}$ is positive for large x^2. Hence, mark the regions as in Fig. 9.8.1, and use theorem 9.8.1 to complete the portrait as in Fig. 9.1.4. Thus, for all μ in the interval $I = \{-2 < \mu < 2\}$, we have only *three* qualitatively different portraits of (9.8.3) as shown in Fig. 9.8.2.

How many more portraits are there when μ lies outside I?
6. The limit cycles shown in Fig. 9.8.1 are qualitatively just Jordan curves, but they differ quantitatively as μ varies. Use computer graphics to show that the computed form of the limit cycle takes different curvatures, depending on the size of μ.
7. Fill in the details of the following alternative proof (due to David Rand) that (9.8.3) has a

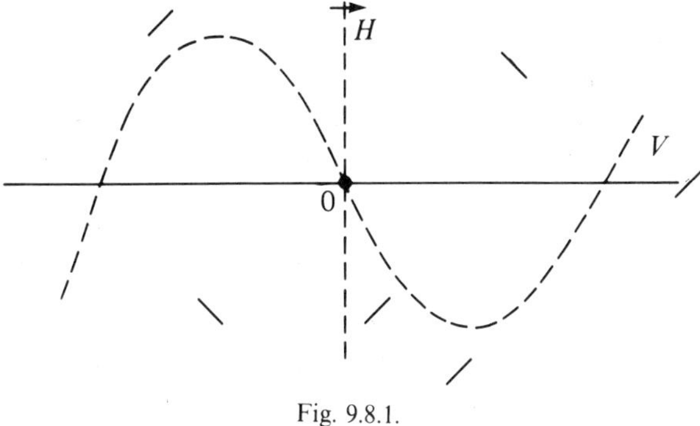

Fig. 9.8.1.

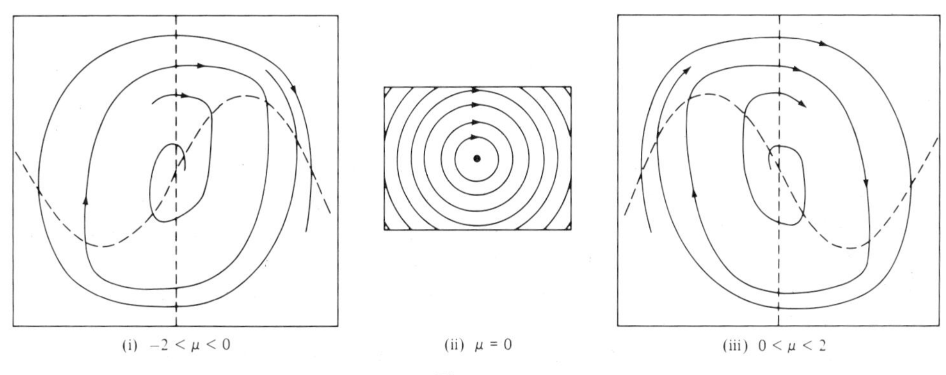

Fig. 9.8.2.

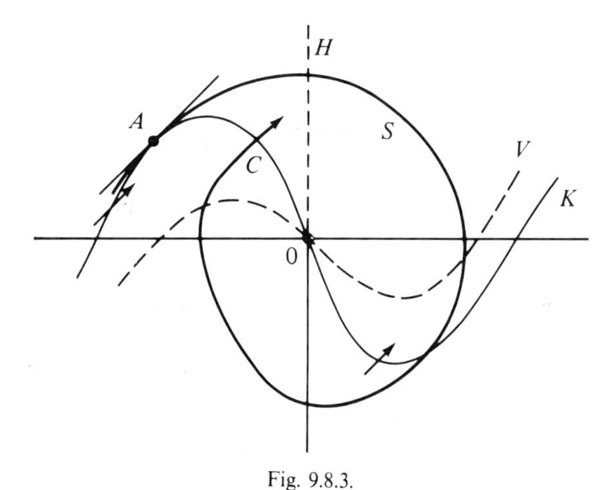

Fig. 9.8.3.

limit cycle if $0 < \mu < 2$. Consider the curve K on which $\dot{x} = \dot{y}$; this is the graph of the cubic function $f(x) = \mu x^3 - \mu x - x$ (see Fig. 9.8.3).
Verify that on K there is just one point $A = (a, f(a))$ with $a < 0$ at which the gradient of K itself is 1. This is also the gradient of the trajectory S through A, by definition of K.

Show that when S has made one clockwise turn about $\mathbf{0}$, it cannot cross K at any point $B = (b, f(b))$ with $b < a$, by the uniqueness of a. Hence S can only cross K again on the arc $A\mathbf{0}$ of K, and this at some point C (say).

If $S = C$ we have a closed trajectory as required, so suppose $S \neq C$. (By Fig. 9.6.3 we now have a limit cycle, but the remaining part of this proof does not use the Poincaré–Bendixson theorem.)

Now consider any point X on $A\mathbf{0}$. The trajectory through it spirals round $\mathbf{0}$ and must meet $A\mathbf{0}$ again (say at Y) since it cannot cross S; so the correspondence $X \to Y$ is a mapping h of $A\mathbf{0}$ into itself. Use exercise 9.3.2 to show that h is continuous. Also, h moves points near $\mathbf{0}$ outwards, and near A inwards, along $A\mathbf{0}$. Explain why there is a fixed point F somewhere on $A\mathbf{0}$, for which $h(F) = F$. Thus the trajectory through F returns to F and hence forms a closed cycle, as required. (The mapping h is an example of a **Poincaré mapping**, which is often used to investigate limit cycles and their stability.)

Proof of theorem 9.8.1

Since F in (9.8.2) is odd, we note that if x, y are solutions, so are $-x$, $-y$; in brief, *the system (9.8.3) is invariant under reflection in* $\mathbf{0}$. Any closed trajectory must encircle $\mathbf{0}$ (by theorem 9.5.3), and it will therefore cut $\mathbf{0}y$ in P say. By the invariance it must cut also at $-P$ (see Fig. 9.8.4). Thus, if we can find a trajectory Z that cuts $\mathbf{0}y$ in two non-zero points P, $-Q$, then the reflection Z' of Z also passes through P and $-P$; and Z' joins up with Z to form a closed trajectory S. (More work will then be needed to decide whether or not S is a limit cycle.) Hence, to find such a Z we can confine activities to the right-hand half R of $\mathbb{R}^2$; moreover, by exercise 9.8.1(3), we may suppose that $F'(0) < 0$, so by (9.8.4), $\mathbf{0}$ is a source.

Next, we note that the vertical manifold V of the system is the graph of $y = F(x)$. Let V_+ denote the portion of this graph in R. At any point $P = (0, p)$ on $\mathbf{0}y$, we have

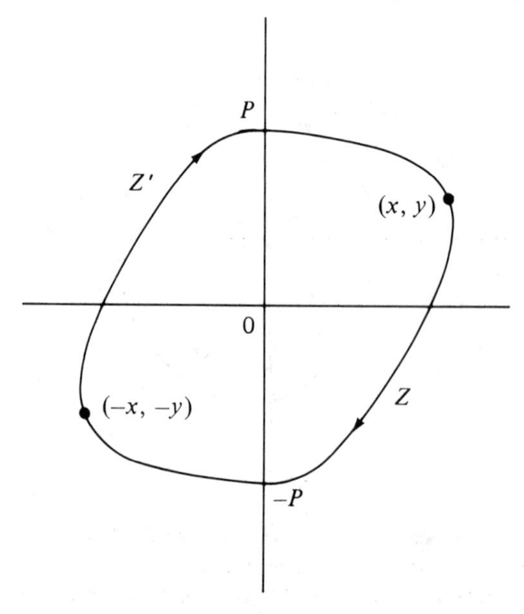

Fig. 9.8.4.

$(x', y') = (p, 0)$ by (9.8.3), so trajectories cross $\mathbf{0}y$ with a clockwise sense. Hence, if M is any point on V_+, then by the exit principle, the trajectory Z through M extends forwards and backwards to meet $\mathbf{0}y$ in C, B, respectively (see Fig. 9.8.5), with B higher than C.

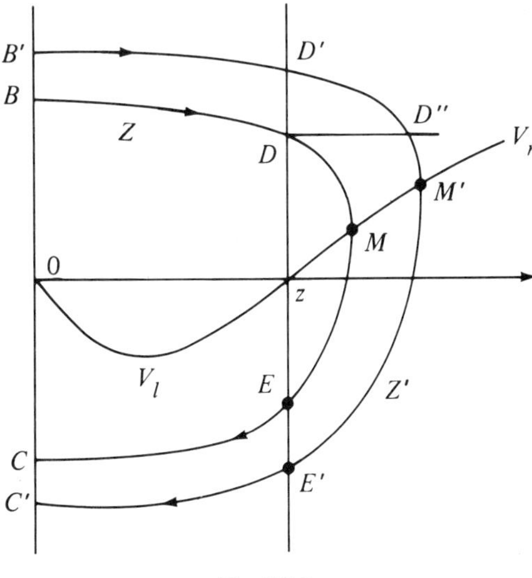

Fig. 9.8.5.

For any point $P = (x, y)$ on Z we have $OP^2 = x^2 + y^2$, and since $y = F(x)$, then by (9.8.2)

$$OP \cdot \frac{\mathrm{d}(OP)}{\mathrm{d}t} = x\dot{x} + y\dot{y} = -x \cdot F(x).$$

Now integrate along Z from B to C; if these points correspond to $t = b, c$ we get

$$\int_b^c F(x)(-x)\,\mathrm{d}t = \int_b^c F(x)\dot{y}\,\mathrm{d}t = \int_B^C F(x)\,\mathrm{d}y = \tfrac{1}{2}(OC^2 - OB^2). \tag{9.8.5}$$

This is a function that depends on M, so we denote it by $G(M)$. By the continuous dependence of solutions on initial conditions (see theorem 9.3.1), G is a *continuous* function of M. Also by the remarks above, Z will be part of a closed trajectory if, and only if, $G(M) = 0$. To prove case (i) of the theorem, it therefore suffices to show that $G(M)$ is never zero; and for case (ii) that $G(M)$ has only one zero. (If there is only one closed trajectory, it must be a limit cycle [Why?].)

To carry out this plan, let I denote the first integral in (9.8.5). Then with case (i) of the theorem, *$F(x)$ does not change sign for positive x*, otherwise $F(x)$ would have different signs at (say) $x = s$ and $x = t$, so F (being continuous) would have a zero somewhere between. This is contrary to the hypothesis (i). Hence, I cannot be zero, and as we saw above, there can then be no closed trajectory.

With case (ii) we have $F < 0$ between 0 and z and $F(x) > 0$ when $x > z$. Then $(z, 0)$ divides V_+ into two portions V_l, V_r as in Fig. 9.8.5 (which is sketched with the above assumption that $F'(0) < 0$). When M is on V_l, the integral I in (9.8.5) has a positive integrand, so $0 < I = G(M)$. Next, we show that if M is on V_r, and M' is higher up that curve than M, then $G(M) > G(M')$; this will show that G is strictly decreasing as M rises up V_r, and hence can have at most one zero. If we can also prove that G is negative somewhere on V_r, then since it is continuous (and positive on V_l), then G actually *has* a zero.

To see that G is decreasing, we refer to Fig. 9.8.5, which shows the two trajectories Z, Z' through M, M'; the vertical through $(z, 0)$ meets Z in D, E, and Z' in D', E'; the horizontals through D, E cut Z' in D'', E''. We break up $G(M)$ in (9.8.5) into a sum of integrals along the arcs of Z; for brevity write I_{PQ} for $\int_P^Q F(x)(-x)\dot{y}\,dt$ (taken along the arc from P to Q). Then by (9.8.5) we have:

$$\begin{aligned} G(M') &= I_{B'D'} + I_{D'D''} + I_{D''E''} + I_{E''E'} + I_{E'C'} \\ G(M) &= I_{BD} + I_{DE} + I_{EC} \\ \Delta G &= G(M') - G(M). \end{aligned} \tag{9.8.6}$$

Since $F > 0$ and $x > 0$ then $I_{D'D'}, I_{E''E'} < 0$; while if the arcs $B'D'$, BD are given by $y = y_2(x)$, $y = y_1(x)$, then

$$\begin{aligned} I_{B'D'} - I_{BD} &= \int_0^z -\frac{xF(x)\,dx}{y_2(x) - F(x)} - \int_0^z -\frac{xF(x)\,dx}{y_1(x) - F(x)} \\ &= \int_0^z -\frac{xF(x)\,[y_2(x) - y_1(x)]\,dx}{[y_2(x) - F(x)]\,[y_1(x) - F(x)]} \end{aligned}$$

which is negative since $0 < z$ and in the integrand every factor other than $F(x)$ is positive. We leave the reader to prove similarly that both $I_{E'C'} - I_{EC}$ and $I_{D''E''} - I_{DE}$ are negative. This accounts for all terms in (9.8.6), so $\Delta G < 0$; and so $G(M') < G(M)$, as required.

Next we show that $G(M')$ can be negative. With the above notation, $G(M') = I_{B'D'} + I_{D'E'} + I_{E'C'}$, and the above calculation of $I_{B'D'} - I_{BD}$ shows that the two outer terms remain bounded as M' moves up V_r.

On the other hand, we have from (9.8.5) that $I_{D'E'} = \int_p^q F(x)\,dy$, where D', E' have y-coordinates p, q. Since $p > q$, and $F(x)$ tends to ∞ as M' moves up V_r, then $I_{D'E'}$ tends to $-\infty$; so $G(M')$ eventually becomes negative. As explained earlier, this completes the proof of theorem 9.8.1.

Exercises 9.8.2

1. Use (9.8.5) to show that if $G(M) < 0$, then we can use the Poincaré–Bendixson theorem to produce a limit cycle, without using the continuity of G (see Fig. 9.6.3).
2. Extend the calculations following (9.8.6), to show that if F has just two positive zeros, then the system has at most two limit cycles. (No general formula is known at present, which relates the number of zeros of F to the maximum number of limit cycles.)

9.9 THE HOPF BIFURCATION

The van der Pol system (9.8.3) depends on the control variable μ, and as μ changes we have the sequence of portraits in Fig. 9.8.1. As we saw in exercise 9.8.1(3), it so happens that the behaviour when $\mu < 0$ is a copy of that when $\mu > 0$; and the interesting point here is the way in which, as μ passes through the value zero, the concentric circles of the centre change to spirals and one limit cycle. We call 0 a **bifurcation** value of μ, since the behaviour changes when μ moves to either side of zero, and this type of behaviour was first systematized by E. Hopf. There is a useful algorithm for deciding when a Hopf bifurcation occurs, so that we have another sufficient condition for constructing models that possess a limit cycle. We shall explain only the algorithm, and refer the reader to detailed proofs in the literature (see Marsden and McCracken (1976)). It seems hardly possible to give a 'plausible' proof in view of the formulae involved.

Briefly, then, suppose we have on $\mathbb{R}^2$ a system S_κ given by equations $x' = P_\kappa(x, y)$, $y' = Q_\kappa(x, y)$ that depend on a control variable $\kappa \in \mathbb{R}$. Suppose also that S_κ always has a fixed point at $\mathbf{0}$ with linearized system L_κ: $\mathbf{z}' = A_\kappa \mathbf{z}$ (the matrix A_κ depending on κ). Then the eigenvalues λ, μ of A_κ are functions of κ, and we suppose that L_0 exhibits a centre; thus λ, μ are of the form $\pm \mathrm{j}\omega$, with ω real and >0. Therefore there is a neighbourhood U of κ in $\mathbb{R}$ such that for all $\kappa \in U$, λ and μ are complex conjugates of the form $u \pm \mathrm{j}v$. Here, u and v are real (smooth) functions of κ, and $u(0) = 0$, $v(0) = \omega$.

Our first assumption is that

$$u_0' = \frac{\mathrm{d}u}{\mathrm{d}\kappa} \neq 0 \qquad \text{when } \kappa = 0. \tag{9.9.1}$$

Next suppose for simplicity that when $\kappa = 0$ the Taylor expansions of P_0 and Q_0 at $\mathbf{0}$ are of the forms

$$P_0(x, y) = -\omega y + p, \qquad Q_0(x, y) = \omega x + q, \tag{9.9.2}$$

where p, q each consist, respectively, of the terms of second and higher degrees. Thus $P_0(\mathbf{0}) = 0 = Q_0(\mathbf{0})$, and the system L_0 exhibits simple harmonic motion. With (9.9.2) we now associate a number H which has a less ghastly form if we use Laplace's operator ∇^2, where, for any $f(x, y)$, $\nabla^2 f = f_{11} + f_{22}$ and $f_1 = \partial f/\partial x$, and so on. Then

$$H = \omega \operatorname{div}(\nabla^2 p, \nabla^2 q) - p_{12}\nabla^2 p + q_{12}\nabla^2 q + (p_{11}q_{11} - p_{22}q_{22}), \tag{9.9.3}$$

where $p_{12} = \partial^2 p/\partial x \partial y$, $\nabla^2 p$, div, etc. are all evaluated at $\mathbf{0}$. Then we can state Hopf's bifurcation theorem.

Theorem 9.9.1

If $H \neq 0$, then the portrait of the system S_κ exhibits a Hopf bifurcation as κ passes through 0. More precisely, for sufficiently small κ, S_κ has a spiral sink or source at $\mathbf{0}$ according as $\kappa < 0$ or $\kappa > 0$; and also when $\operatorname{sign}(\kappa) = \operatorname{sign}(u_0')$, there is a limit cycle. This encircles $\mathbf{0}$, and its size increases with κ; it is stable or unstable according as H is negative or positive.

This allows us to produce a 'quadratic' system with a limit cycle. For example, let

$$\begin{aligned} P_\kappa(x, y) &= \kappa x + \omega y + x^2 + 2xy - 3y^2 \\ Q_\kappa(x, y) &= -\omega x + \kappa^2 y - 2x^2 + 5xy + 4y^2. \end{aligned} \tag{9.9.4}$$

Here the matrix $A_\kappa = J_0$ has eigenvalues $(T \pm \sqrt{D})/2$, where the trace T is $\kappa + \kappa^2$, D is the discriminant $D = T^2 - 4\Delta$, and the determinant $\Delta = \det(A_\kappa) = \kappa^3 + \omega^2$. Now $\Delta > 0$ if $\kappa > -\omega^2/3$, and $D < 0$ when $\kappa = 0$; so by taking a sufficiently short neighbourhood I of 0, we have $D < 0$ for all $\kappa \in I$. Therefore, the eigenvalues are complex conjugates, with real part $u(\kappa) = T/2 = (\kappa + \kappa^2)/2$, so $u_0' = \mathrm{d}u/\mathrm{d}\kappa = \frac{1}{2} > 0$ when $\kappa = 0$. Also at $\mathbf{0}$ we have $p_{12} = 2, q_{12} = 5$ while $\nabla^2 p = 2(1 - 3), \nabla^2 q = (-2 + 4)$; thus $\operatorname{div}(\nabla^2 p, \nabla^2 q) = 0$. Hence $H = 2(-4) - 5(-2) + 10 = -8 < 0$, so by theorem 9.9.1 we have a stable limit cycle when κ is positive and sufficiently small. Unfortunately we do not know whether or not this cycle is unique—we have brought readers within sight of the frontiers of research, and now invite them to have a go themselves! (See, for example, Blows and Lloyd (1988).) Interesting applications, with more detail to justify the specific models, can be found in Arrowsmith and Pace (1982) and the references given there. See also Coppel (1966), who shows that a limit cycle in a quadratic system must be a convex curve.

Exercises 9.9.1

1. Investigate the system (9.9.4) using computer graphics.
2. Show that for the van der Pol system (9.8.3), $u_0' = 1$ but $H = 0$ when $\mu = 0$; so it does not satisfy the conditions of theorem 9.9.1.
3. Let p, q be smooth functions of μ, and let S_μ denote the system:

   $$x' = p(\mu) + \omega y + ax^2 + 2hxy + by^2,$$

   $$y' = -\omega x + q(\mu) + cx^2 + 2kxy + dy^2.$$

   Show that the conditions for a Hopf bifurcation are that $p' + q' \neq 0$ at $\mu = 0$ and $H \neq 0$, where $H = 2[\kappa(c + d) - h(a + b) + 2(ac - bd)]$. Hence produce some quadratic systems for computer graphic display.
4. Suppose that the system (9.9.2) is replaced by one with linear terms $ax + by$, $cx + dy$ (instead of $-\omega y, \omega x$). Show that if the eigenvalues are purely imaginary, then $d = -a$ and $bc > a^2$; and that there is a linear transformation $\mathbf{w} = M\mathbf{z}$ which maps S_0 to a system of the form (9.9.2).
5. Show that the system $x' = 2x + x^2 - \mu x^3 - xy$, $y' = -2x + xy$ has a fixed point at $(2, 4(\mu - 1))$, at which it exhibits a Hopf bifurcation as μ passes through the value $\mu = 1/4$. [SU]
6. Show that the system

   $$x' = \mu x + y + x^2 - y^2, \qquad y' = -x + \mu y + 2x^2 + 3y^3$$

   has no closed trajectory if $\mu < 0$, but that there is a limit cycle encircling $\mathbf{0}$ if μ is positive and sufficiently small. Produce graphics to illustrate the bifurcation as μ passes through $\mathbf{0}$.
7. Show that the system in exercise 5 can have no closed trajectory if $\mu \leq -\frac{1}{2}$. Use computer graphics to investigate the range $-\frac{1}{2} < \mu < \frac{1}{2}$.

8. Show that the system $x' = y - (x^3 - \mu x)$, $y' = -x$ satisfies the conditions of theorem 9.9.1, and hence has a Hopf bifurcation as μ passes through 0. Show also that if $\mu > 0$, then the system can be rescaled to the form of the van der Pol system (9.8.3).
9. State the Poincaré–Bendixson theorem and the important consequence of this theorem in the absence of critical points. State Dulac's theorem.

   Prove that the following system has a unique limit cycle:

$$\begin{aligned} \dot{x} &= -y + (1 - \lambda x^2 - (\lambda + \mu)y^2)x \\ \dot{y} &= x + (1 - \lambda x^2 - (\lambda + \mu)x^2)y \end{aligned} \qquad \lambda, \mu > 0.$$

   (Change to polar coordinates as in exercise 9.2.1(3).) Does the system

$$\begin{aligned} \dot{x} &= x^2 - y - 1 \\ \dot{y} &= y(x - 2) \end{aligned}$$

   have any limit cycles? Justify your answer. [SU]

The contemporary approach to differential equations stresses Poincaré's 'conceptual', geometric view; for further material see the texts (Arnold 1992) (Hirsch and Smale 1974). But some manipulative technique is still required, and readers wishing to acquire it should practice on the many exercises in Piaggio (1952), where the background to several important equations is described.

10

Discrete non-linear dynamics

10.1 STAIRCASES AND COBWEBS

In this chapter, we extend the considerations of Chapter 2 to the case when the dynamics is given by a non-linear function. Manipulative algebra will no longer give much information, so it needs to be supplemented by the use of geometrical methods. The equations of Malthus and Verhulst in Chapter 2 are special cases of the equation.

$$x_{n+1} = f(x_n) \tag{10.1.1}$$

where f: $\mathbb{R} \to \mathbb{R}$ is a smooth function. Our object is to start with an initial value x_0, and then to deduce the behaviour of the orbit of x_0:

$$x_0, \quad x_1 = f(x_0), \quad x_2 = f(x_1) = f(f(x_0)) = f^2(x_0), \quad \ldots, \quad x_n = f^n(x_0)$$

as n increases. For a specified value of n (say 1000), we might use a computer to calculate the first 1000 terms of the orbit, but that still might not be enough for us to be sure of the long-term behaviour of x_n. That is why we need some geometrical considerations to help us make reliable deductions.

Consider first the simplest case, when f is a linear function—say $f(x) = ax + k$ where a, k are constants. In (2.1.7) we have already found an algebraic formula for x_n, but it is helpful to visualize the iterative process graphically, as in Figs.10.1.1, 10.1.2. In these Figures, L denotes the line $y = x$, and G the graph of the equation $y = ax + k$; we have assumed with Fig. 10.1.1 that $0 < a < 1$ so G is not so steep as L, whereas in Fig. 10.1.2, G rises more steeply because $a > 1$. (In the case of the Malthus Model, $a > 1$ and $k = 0$.) Starting with x_0, we can read off x_1 where the ordinate cuts G; then we convert this ordinate to x_1 on the x-axis by taking the horizontal PQ to cut L at $Q = (x_1, x_1)$. Repetition of this process gives the 'staircase' that clearly converges to A in Fig. 10.1.1, where A has coordinates (z, z) say.

No matter where we start with x_0, the process coverges to A, and x_n converges to z, which we call a **fixed** (or **equilibrium**) **point**. Had we taken x_0 to be z itself, then every x_n would also be z, so z is truly fixed. Similarly in Fig. 10.1.2, we obtain a staircase, but it clearly diverges to infinity because G is steeper than L. In Fig. 10.1.2,

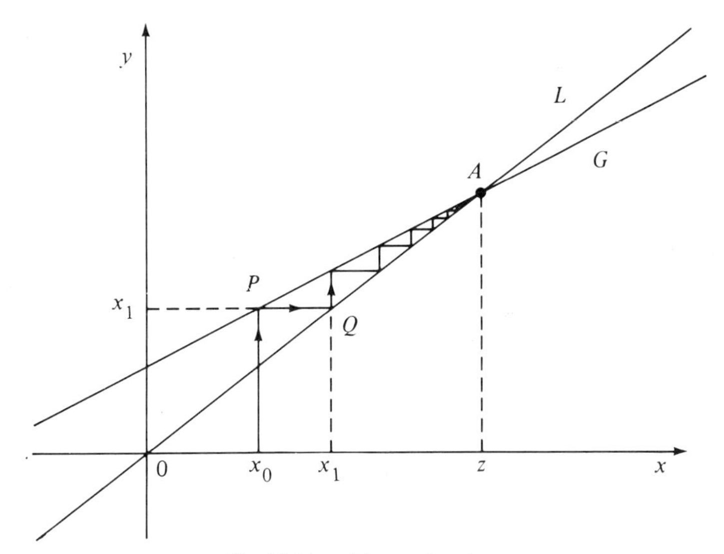

Fig. 10.1.1. A is an attractor.

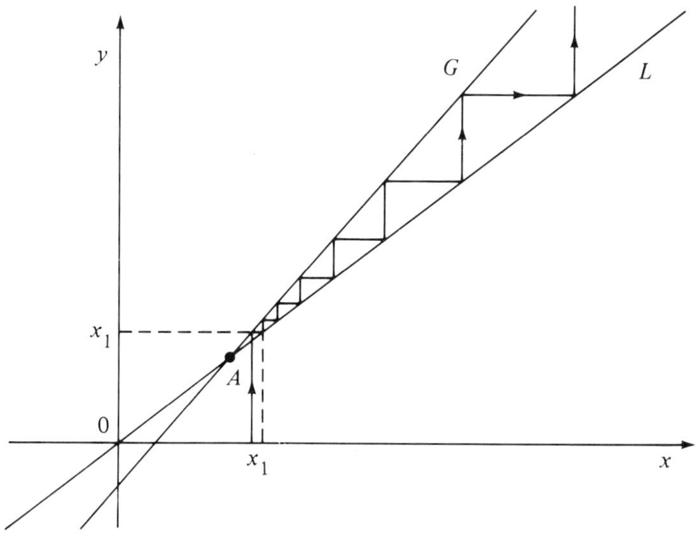

Fig. 10.1.2. A is a repellor.

A is a **repellor**, because the dynamics moves any other point in its neighbourhood out to infinity; the contrasting behaviour near A in Fig. 10.1.1 leads us there to call A an **attractor**. The set S of all initial values x_0 for iterative sequences which converge to the attractor A is called the **basin of attraction** of A.

Exercises 10.1.1

In each of these exercises, express the given sequence in the form $x_{n+1} = f(x_n)$, and sketch the graph of f to find where it is cut by the line $L: y = x$. Then display the behaviour of the sequence by means of a staircase. Check the validity of your sketch by proving algebraically that the sequence behaves as your sketch predicts.

1. $8x_{n+1} = 3x_n + 20$. Show that if $x_0 < 4$, then for each n, $x_n < x_{n+1}$. (Use induction on n to prove that $x_n < 4$, and hence that $x_{n+1} - x_n > 0$.) Your sketch should suggest that x_{n+1} converges to $z = 4$; prove this analytically by showing that if $d_n = 4 - x_n$ then $d_{n+1} = (3/8)d_n$, so that $d_n = (3/8)^n d_0$. Similarly when $x_0 > 4$.
2. If $3x_{n+1} = 8x_n + 20$, find the fixed point and show that it is a repellor.
3. $x_{n+1} = (1 - x_n)/2$. Calculate the first few terms, and on a sketch show that the 'staircase' construction produces a 'cobweb' that encloses the fixed point. Check this algebraically, by showing that one of the sequences $x_0, x_2, x_4, \ldots, x_{2n}, \ldots$ and $x_1, x_3, x_5, \ldots, x_{2n+1}, \ldots$ is increasing, the other decreasing. Show that they have the same limit, $z = 1/3$, so that the fixed point is an attractor.
4. $x_{n+1} = 2(1 - x_n)$. Show that there are cobwebs as before, but that they unwind, so that the fixed point is now a repellor.

The next most complicated type of function is quadratic, as in the Verhulst (or 'logistic') model of the form

$$x_{n+1} = kx_n - bx_n^2 \qquad (k = 1 + a > 0, b > 0). \tag{10.1.2}$$

If, as in (2.2.2), we instead replace x_n^2 by $x_n x_{n+1}$, then

$$x_{n+1} = \frac{kx_n}{(1 + bx_n)} \tag{10.1.3}$$

and f in (10.1.1) is the function $y = kx/(1 + bx)$ with graph shown in Fig. 10.1.3 . This has constantly rising gradient $k/(1 + bx)^2$, which is steeper than that of L at $x = 0$ because $k > 1$. But the curve flattens out (because $b > 0$) toward the asymptote

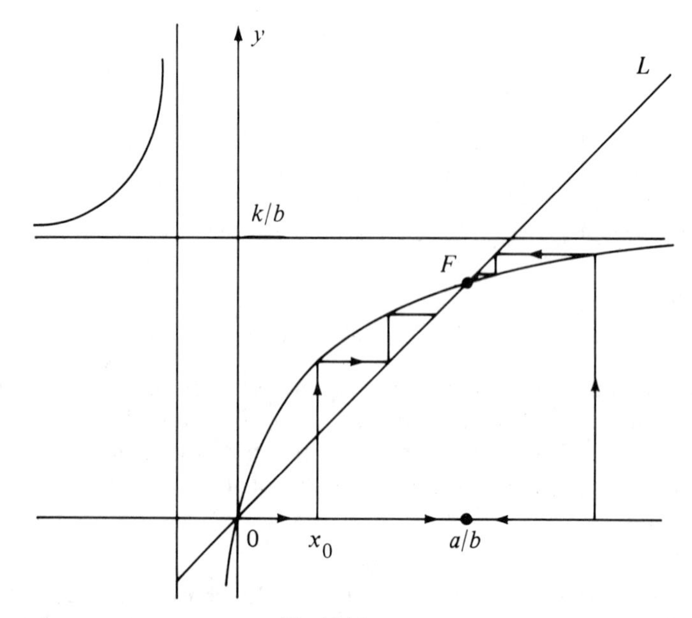

Fig. 10.1.3.

$y = k/b$, so its cuts L at $F = (a/b, a/b)$ as well as at $\mathbf{0}$. Here, F is an attractor, for, if we start with a positive $x_0 < a/b$, we obtain a staircase, as in Fig. 10.1.3, and this climbs to F, while if $x_0 > a/b$, the staircase falls to F. On the other hand, the other fixed point $\mathbf{0}$ is a repellor. This behaviour agrees with that predicted by the differential equation discussed in Chapter 3, in that if the population starts small (or large), it simply climbs (or falls) and levels out at a/b, the equilibrium value. Life is not so simple however; recall that (10.1.3) was obtained from (10.1.2) by the seemingly reasonable approximation of x_n^2 by $x_n x_{n+1}$, but the 'reasonableness' depends dramatically on the size of a, as we shall see. First, the reader should try the following exercises.

Exercises 10.1.2

These are like those of exercises 10.1.1 except that the functions involved are non-linear; the reader should use the same procedures as were required before. The algebra is more tricky, so the essential first step is to see what is happening in a sketch; but care is needed, especially with a cobweb, since a small error in drawing can lead you to believe that it unwinds, when in fact it does the opposite! Later, the arguments required will allow you to check on the validity of computer output.

1. $x_{n+1} = \sqrt{(1 + x_n)}$ and $x_0 = 1$. Prove by induction on n that $x_n \leq x_{n+1}$ and $x_n < 2$ for all n. Show that there are two fixed points, and if z is the positive one, and $d_n = z - x_n$, then $0 < d_{n+1} < d_n/2$. (Note: $z^2 - x_{n+1}^2 = d_{n+1}(z + x_n)$.) To what, then, does x_n converge?
2. If $x_{n+1} = 2/x_n^2$ and $x_0 = 1$, show that x_n does not converge. (Calculate a few terms, and consider what your sketch suggests about the even and odd ones.)
3. $x_{n+1} = (10x_n - 12)/(11 - x_n)$ and $-3 < x_0 < 4$. Prove by induction on n that $-3 < x_n < 4$. (If $-3 < x_n < 4$, show that $1/14 < 1/(11 - x_n) < 1/7$, and use this in $x_{n+1} = -10 - 98/(11 - x_n)$.) Hence prove that $x_{n+1} \geq x_n$, so that x_n is a decreasing sequence, bounded below (in this case by -3). It is a general property of the real number system that we can then say that x_n converges to a limit l. Show that $l = \dot{-}3$.
4. $x_{n+1} = 1/(1 + x_n)$, with various values of x_0.
5. $x_{n+1} = (3 + x_n^2)/2x_n$, $x_1 = 3$. Use the inequality $a^2 + b^2 \geq 2ab$ to prove that $x_n \geq \sqrt{3}$ for all $n \geq 1$, and show that $\langle x_n \rangle$ is a decreasing sequence. Hence find $\lim x_n$.
6. Verify the conclusions of these equations by using program 10.1.1 below to print out the numerical values of x_n. You will need to insert the appropriate function of x, as the given form has been written for a specific application in the next section.

```
REM Program 10.1.1, Sequence tabulator
FUNCTION f(x) = k*x*(1-x)
INPUT PROMPT "Enter values of  k  and  x0  ": k, x0
LET x = x0
LET n = 0
PRINT "n", "x"
DO
   PRINT n, x
   LET x = f(x)
   LET n = n + 1
LOOP UNTIL "cows" = "home"
END
```

7. The following program can be used to plot the graph of the function $y = f(x)$, and the line $y = x$ on the same axes. For a given initial value x0 the program draws a staircase or cobweb. Use this to explore a variety of functions with a variety of initial values.

```
REM Program 10.1.2, Stairs and web
FUNCTION f(x) = k*x*(1-x)
READ xmin, xmax, fmin, fmax, limit
DATA -0.5, 1.5 , 0   , 1   , 100
SET MODE "graphics"
SET WINDOW xmin, xmax, fmin,fmax
DO
  INPUT PROMPT "Enter values for k and x0 : " : k , x0
  CLEAR
  PLOT xmin,0 ; xmax,0
  PLOT 0,fmin ; 0,fmax
  PLOT xmin,xmin ; xmax,xmax
  FOR x = xmin TO xmax STEP (xmax - xmin)/100
    PLOT x, f(x);
  NEXT x
  PLOT
  LET x = x0
  DO
    LET y = f(x)
    PLOT x,x ; x,y ; y,y
    LET x = y
  LOOP UNTIL ABS(x)>limit OR KEY INPUT
LOOP UNTIL "cows" = "home"
END
```

8. The golden section ϕ is the larger root of the equation $x^2 - x - 1 = 0$. Show that the rearrangement: $x_{n+1} = 1 + 1/x_n$ has $x = \phi$ as an attractor. Clearly the process will not converge for *all* x_0, for example, $x_0 = 0$. What, then, is the basin of attraction of ϕ? If $f(x) = 1 + 1/x$ find the inverse function $f^{-1}(x)$ and investigate the attractor and its basin of attraction for the iteration $x_{n+1} = f^{-1}(x_n)$.

10.2 THE VERHULST EQUATION

If with the original Verhulst equation (10.1.2) we rescale by writing $z_n = bx_n/k$, it takes the simpler form

$$z_{n+1} = kz_n(1 - z_n) = f_k(z_n) \tag{10.2.1}$$

corresponding to $y = f_k(x) = kx(1 - x)$. The problem we are modelling requires x_n to be always positive, so we can now restrict attention to the unit interval U: $\{0 \leq x \leq 1\}$. In that case, y takes its maximum value Y(where $f'_k = 0$) at $x = 1/2$, and we must have Y in U as well. Thus $Y = f_k(1/2) = k/4 \leq 1$ so we require that $k \leq 4$. We refer to the set of functions f_k with $0 \leq k \leq 4$ as the **logistic family**.

In the original population problem, we had $k = 1 + a > 1$, but it is instructive to consider what happens as k runs from 0 to 4. We obtain three types of graph as in Figs. 10.2.1–10.2.3, which correspond, respectively, to k in the intervals

$$A: \{0 < k < 1\}, \qquad B: \{1 < k < 2\}, \qquad C: \{2 < k < 4\}. \tag{10.2.2}$$

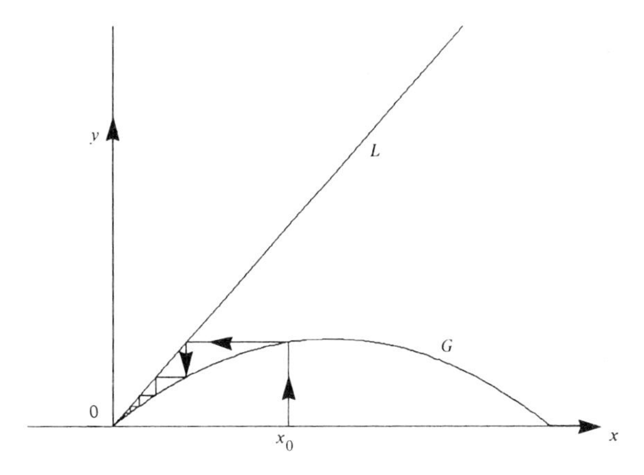

Fig. 10.2.1.

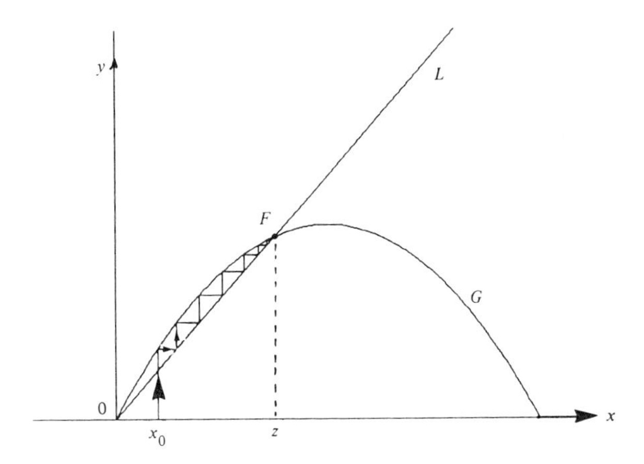

Fig. 10.2.2.

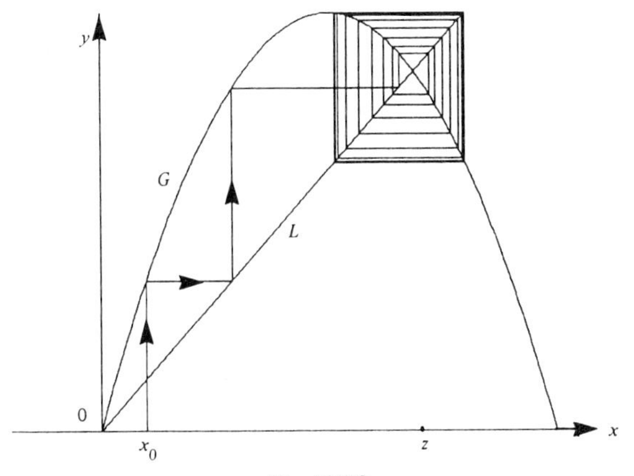

Fig. 10.2.3.

For, we calculate the intersections of the diagonal L with the graph G of $y = f_k(x)$, and find these at $\mathbf{0}$ and $F = (z, z)$, with $z = 1 - 1/k$. If k is in A, then z lies outside the interval U, whereas z lies to the left or right of $1/2$ in U according as k is in B or C. When $k = 1$ or 2, then z is 0 or $1/2$.

In Fig. 10.2.1, the gradient of G at $\mathbf{0}$ is k, and any staircase starting from x_0 in U converges to the fixed point $\mathbf{0}$, which is therefore an attractor: its basin of attraction is all of U. In Fig. 10.2.2, the gradient at F is $k(1 - 2z) = 2 - k$, so G at F is flatter than L, and any staircase starting from $x_0 \leq z$ rises and converges to F; while according as $z < x_0 \leq 1 - z$ or $1 - z \leq x_0 \leq 1$, we obtain a staircase that falls or rises to F. Here we have two fixed points, of which z, $\mathbf{0}$ are an attractor and a repellor; but now $U - \{\mathbf{0}\}$ is the basin of attraction for z. Recall that $k = 1 + a$, where a denoted the survival rate of the population, so that here $0 < a < 1$. This model tells us therefore that regardless of how small b was in (10.1.3), provided $b \neq 0$, the population size approaches the equilibrium value $z = a/b$. Real human populations might be expected to have a in the range $0 < a < 1$, perhaps with $a < 1/2$; but with insects, a can be larger. Before considering Fig. 10.2.3, the reader should try the following exercises.

Exercises 10.2.1

1. By writing $z = sx + 1/2$, show that (10.2.1) takes the form $x_{n+1} = \mu - x_n^2$ if $s = 1/k$ (and $k \neq 0$), where $\mu = k(k-2)/4$. Show that if $k > 4$ then $\mu > 2$; and if $2 < k < 4$, then $0 < \mu < 2$. Find the ranges of μ when k runs through the sets A, B in (10.2.2). Show that for each $k > 0$, when $0 < z < 1$, we have $-k/2 < x < k/2$. (Thus one cost of studying the dynamics of the simpler function $g_\mu\colon x \to \mu - x^2$ is that we have to use an interval for x which varies with μ.)
2. Sketch analogues of Figs 10.2.1–10.2.3 for the transformed equation, noting especially the fixed points and maxima.
3. Do an exercise, similar to exercise 1, that changes (10.2.1) to the form $x_{n+1} = 1 - \mu x_n^2$, with $\mu = 2k/(k-4)$ when $k \neq 4$. Then sketch the analogues of Figs. 10.2.1–10.2.3.
4. Show that if $0 < x_0 < 1$, then the equation $x_{n+1} = 2x_n^2 - 1$ has the explicit solution $x_n = \cos^{2^n}\theta$, where $x_0 = \cos\theta$. Hence show that for each n there is some x_0 for which $x_0 = x_m$ when, and only when, $m = n$. (Thus x_0 'has period n'; see below.) No such explicit solution is known when the coefficient 2 is changed.
5. Let $F_a(x) = ax(1 - x)$. Prove that the discrete dynamical system on $\mathbb{R}$ generated by F_a has the following properties.
   (a) For all $a > 1$ the system has a unique fixed point in the interval $(0, 1)$, with the fixed point losing stability as a increases through 3.
   (b) As a increases through 3 a cycle of period 2 bifurcates from the fixed point.
   (c) The 2-cycle is stable for $3 < a < 1 + \sqrt{6}$ but as a increases through $1 + \sqrt{6}$, a 4-cycle bifurcates from the 2-cycle which itself becomes unstable. [SU]

A new feature arises in Fig. 10.2.3, when k has increased beyond the value 2. We see there how a staircase rises from x_0, and then turns into a 'cobweb' that spirals round F. The picture now suggests three questions as follows.

Questions 10.2.1

(a) Can the spiral wind outwards from F, so that both F and $\mathbf{0}$ are repellors?

(b) If so, how can an orbit settle down between them?
(c) Could the spiral perhaps close up, so as to encircle F without either converging or diverging?

With (c), if the spiral were to close up, then for certain points $x_n, x_{n+1}, \ldots, x_{n+p}$ on it we have $x_n = x_{n+p}$ with $p > 0$, and then the cycle repeats: $x_{n+p+1} = x_{n+1}, \ldots, x_{n+2p} = x_n$, and so on. We then call x_n a **periodic point, of period** p. Real populations often show a rough periodicity, so a good model ought to predict when such periodic points can be present.

The simplest kind of periodic point q would have period 2: here (for any mapping f), $f(q)$ is r (say), and the pair (q, r) satisfies:

$$f(q) = r \neq q, \qquad f(r) = q. \tag{10.2.3}$$

Then q (and r) must each satisfy the equation $x = g(x)$ where $g(x) = f^2(x)$, so

$$x = g(x), \qquad g(x) = f(f(x)). \tag{10.2.4}$$

When f is f_k, then $g(x) = kf(x)(1 - f(x))$, and after substitution of $kx(1 - x)$ for $f(x)$, the equation $x = g(x)$ leads to a quartic equation in x. The fixed points $0, 1 - 1/k$ of f_k also satisfy this equation, which has at most four real roots; hence f_k has either no, or two, points of period 2. We can visualize such points by finding where the graph of f_k^2 cuts the diagonal L (if at all). In Figs 10.2.4, 10.2.5 we show how to construct the graph of f^2 from that of f, for any f like f_k. The second graph will not cut L in other than the fixed points of f_k, until the parameter k increases some way beyond the interval B in (10.2.2). In fact, as we shall see below, once k passes through the value 3, it is large enough for points q, r of period 2 to exist; and then we shall see that F becomes a repellor.

To see the effect of q, r on orbits, it is helpful to look at computer output, first by means of the program 10.2.1 (sequential graph) below, which displays the graph of x_n against n. When (for example) k is set at 3.2, and x_0 is 0.5, the graph displayed alerts us to the possibility that q and r exist, because it looks like a pair of parallel

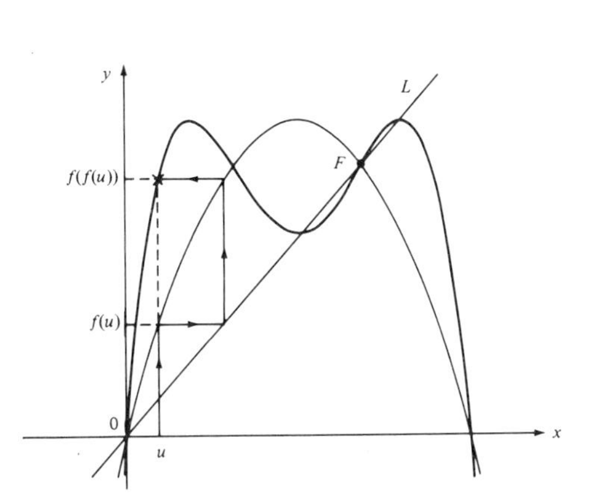

Fig. 10.2.4. Construction of $f^2(u)$.

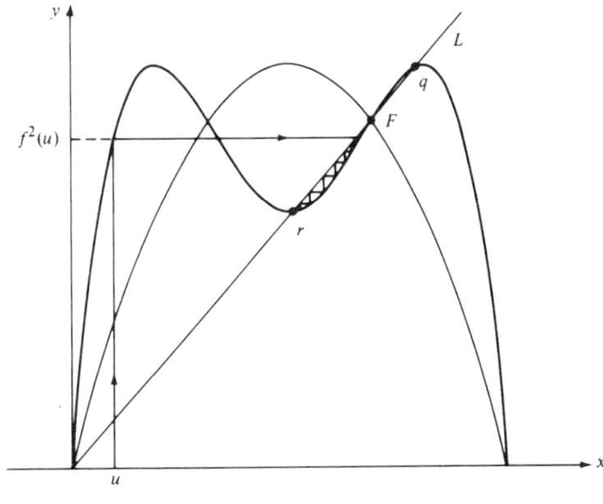

Fig. 10.2.5. F is a repellor.

lines (apart from a few scattered points at the beginning). If the horizontal scale is magnified (say by reducing the variable `nmax` in the program), one sees that these 'parallel' lines consist alternately of an upper point and a lower one, corresponding to the oscillation (10.2.3) between q and r.

The precise location of these points (to 7 decimal places, say) can be read off by using program 10.1.1 (sequential tabulator; see exercise 10.1.2(6)) to print the numerical values of the orbit $\langle x_n \rangle$ of the chosen initial point x_0; for then we soon find that the numbers settle down to two values. If we change the value of x_0, we still obtain the same numbers q, r, possibly after a longer wait for the digits to settle down.

```
REM Program 10.2.1, Sequential graph
FUNCTION f(x) = k*x*(1-x)
INPUT PROMPT "Enter values of  k  and  x0 : " : k, x0
READ nmax, fmin, fmax
DATA 1000, 0   , 1
SET MODE "graphics"
SET WINDOW 0,nmax, fmin,fmax
PLOT 0,0 ; nmax,0
PLOT 0,fmin ; 0,fmax
LET x = x0
LET n = 0
FOR n = 0 TO nmax
  PLOT n, x
  LET x = f(x)
NEXT n
END
```

The reason why the graph program shows us two lines of points is because q and r together form an attracting set, since (unless $x_0 = 1 - 1/k$) the orbit soon settles into an oscillation between q and r in the following precise sense. There are intervals I, J centred round q, r, respectively, such that if $x \in I$ then its orbit $\langle x_n \rangle$ visits I, J alternately, in that $x_{2n} \in I$ and $x_{2n-1} \in J$ for all n. Moreover, the sequences $\langle x_{2n-1} \rangle, \langle x_{2n} \rangle$ converge to r, q, respectively. A similar statement holds if $x \in J$. To allow this to happen without tearing the interval apart, F stays fixed between r and q; and if an orbit comes near to F, it rapidly moves away and into I or J. The pair (q, r) is called an **attracting 2-cycle**. Thus we have provided some (as yet limited) answers to questions 10.2.1 above.

The reader should now experiment with programs 10.1.1, 10.1.2 and 10.2.1, making different choices of k and x_0, to see the variety of possible effects. If k is chosen in an interval centred at 3, we see a great difference in the qualitative behaviour of f_k on either side of $k = 3$, and we call 3 a **bifurcation value**. Also, as k increases further, we see that points x of periods $4, 8, \ldots, 2^n, \ldots$ will appear, as k passes through further bifurcation values $c_2 \cong 3.45, c_3 \cong 3.54, \ldots, c_n, \ldots$. For example, if in program 10.1.2, we set $k = 3.455$, we can see that the orbit oscillates around a clear attracting 4-cycle; but if $k = 3.45$ the clarity apparently depends on our choice of x_0. Similarly when the longer cycles are present. (Unless we know what to expect, it is unlikely that we would recognize a cycle of even such a small period as 2^{12}.) In this way, we first see the phenomenon of period-doubling, until k reaches 3.57 approximately (the precise value is $\lim c_n$, with c_n as above).

After some confused behaviour, there eventually appears a point u of period three, that is, such that $u = f_k^3(u)$. This occurs when k is about 3.824, and then a general theorem given in Li and Yorke (1975) tells us that there must be points of *all* periods! However, the orbit of u consists of u, $f_k(u)$, and $f_k^2(u)$, and is an attracting 3-cycle, whereas the orbits of higher periods do not necessarily form attractors, so we are unlikely to observe them in the graphics of program 10.2.1, unless we hit a value of k for which a cycle has changed from repelling to attracting. As before, once program 10.2.1 leads us to suspect a cycle, it can be located by inspecting the output of program 10.1.1. For example, we find that when $k = 3.84$ and $x_0 = 0.7$, then our 3-cycle consists of the points 0.149408418, 0.488008484, 0.959444782, in that the first, second, and fifth digits do not change after the values 7, 48, 184 of n, and the others settle down. We now suggest that readers experiment in a similar way, and then try the following exercises.

Exercises 10.2.2

1. You can gain a feeling for the iterates of f_k, by viewing their graphs illustrated by program 10.2.2 below (and trying other functions). Note how most points of U are mapped by f^n into narrow bands, the width decreasing as n increases. These correspond to neighbourhoods of an attracting set.

```
REM Program 10.2.2, Graph iterator
OPTION BASE 0
FUNCTION f(t) = k*t*(1-t)
READ k    , n   , fmin, fmax
DATA 3.65, 256, 0    , 1
DIM x(256)
SET MODE "graphics"
SET WINDOW 0,1, fmin,fmax
PLOT 0,0 ; 1,0
PLOT 0,fmin ; 0,fmax
FOR c = 0 TO n
  LET t = c/n
  LET x(c) = f(t)
  PLOT t, x(c);
NEXT c
PLOT
FOR p = 1 TO 10
  GET KEY z
  CLEAR
  PRINT p
  FOR c = 0 TO n
  LET x(c) = f(x(c))
    PLOT c/n, x(c);
  NEXT c
  PLOT
NEXT p
END
```

2. Work out the algebraic formula for $f_k^2(x)$ in (10.2.4), and hence show that the equation for $x = g(x)$ is

$$P(x)(k^2x^2 - k(k+1)x + k + 1) = 0, \qquad P(x) = x(x - (1 - 1/k)).$$

The factor $P(x)$ arises because the fixed points of f are also fixed by g. Show that there will be real solutions when $k > 3$. By substitution in the equation $x = f \circ g(x)$, we see that a point of period 3 must satisfy a (complicated!) equation of degree 8; this heavy algebra can be simplified slightly by using the transformations given in exercise 10.2.1(a): thus f_k becomes $h_\mu(x) = \mu - x^2$ with $\mu = k(k-2)/4$ provided $k \geq 2$. Show that if $v = \mu - \mu^2$, then

$$h_\mu^3(x) - x = (\mu - v^2) - x - 2\mu v x^2 + (2v - 4\mu^2)x^4 + 4\mu x^6 - x^8.$$

This has a factor $(x^2 + x - \mu)$, since the fixed points of h_μ are also fixed by h_μ^3 (but those of h_μ^2 may not be), and can be handled as in Brown (1981). Clearly, higher iterates will become quite unmanageable. It is not surprising that the properties of such an elementary function as f were hard to find until the advent of rapid computers.

3. Use program 10.1.1 to verify the values given above for a cycle of period 3, noting particularly how many iterations are needed to stabilize the first digits, then the second, and so on. If $k = 3.58$, show that the orbit of x_0 seems to be confined within four separate intervals of length 1/10, but with no apparent 4-cycle. (Theory shows that this is not due to rounding errors within the computer.)
4. In Collet and Eckmann (1980) p.36, the following values μ_n are given for which period doubling occurs in the family of functions h_μ: a 2-cycle appears after $\mu_0 = 0.75$, a 4-cycle after 1.25, and then μ_n is:

(2) 1.3680989394	(3) 1.3940461566	(4) 1.3996312389
(5) 1.4008287424	(6) 1.4010852713	(7) 1.401140214699
(8) 1.401151982029	(9) 1.401154502237	(10) 1.401155041989

Find the corresponding values of k for the logistic family. Calculate the numbers $d_n = \mu_n - \mu_{n-1}$ to show that d_n is approximately 4.6 times d_{n+1}. In fact, experimental evidence of this kind led the physicist Feigenbaum to conjecture that d_{n+1}/d_n tends to a limit, which is the same for all families of 'hump' functions (i.e., with a single maximum like the logistic family f_k). His conjecture has since been proved, using some very advanced mathematics, and suggests some intriguing and deep relationships which are not yet understood. The limit is 'Feigenbaum's constant', equal to 4.669196223 approximately, and it turns out to be the largest eigenvalue of a certain linear operator on a space of functions (hence its universality). Many other details are given in Collet and Eckmann (1980). One reason for expecting behaviour to keep repeating on ever smaller scales is seen in Fig. 10.2.6, where the portion inside the box is a miniature (with change of scale) of the whole. We have here a type of 'self-similarity' that we shall study later.

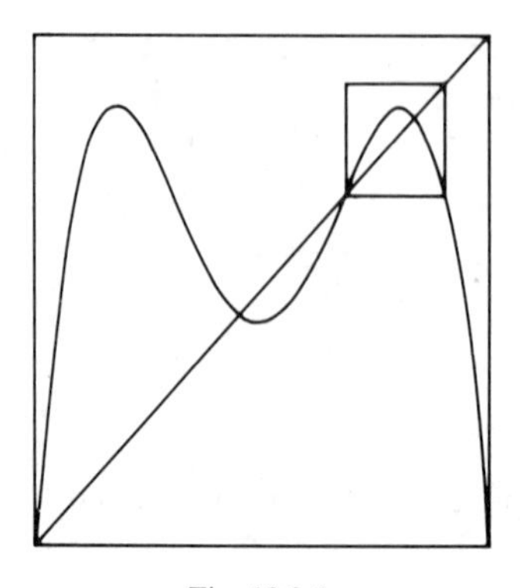

Fig. 10.2.6.

5. Find corresponding bifurcations for the family of functions $q_e(x) = \mu - e \sin x$, when μ is fixed at (say) 0.8, and e runs from -4 to 4. (The fixed points are given by Kepler's equation in section 4.7.) This family is discussed, with others, in Scholl and Thieler (1989).
6. The bifurcations of a function are often shown in a 'pitchfork' diagram. The following program iterates the logistic map for a given number of iterations, from a given starting value x0, and a given parameter value k to obtain a value x. It then plots the successive points (k, f(x)), (k, f²(x)), . . . , (k, fⁿ(x)) for a given n. As k varies the resulting picture shows clearly where the first few bifurcations occur.

```
REM Program 10.2.3, Pitchfork diagram
FUNCTION f(x) = k*x*(1-x)
READ kmin, kmax, sw , x0 , nmax, ymin, ymax
DATA 2.8 , 4   , 319, 0.1, 50  , -0.2, 1.2
SET MODE "graphics"
SET WINDOW kmin,kmax, ymin,ymax
PLOT kmin,0 ; kmax,0
FOR k = kmin TO kmax STEP (kmax - kmin)/sw
  LET x = x0
  FOR n = 1 TO nmax
   LET x = f(x)
  NEXT n
  FOR n = 1 TO 32
    PLOT k,x
    LET x = f(x)
  NEXT n
NEXT k
END
```

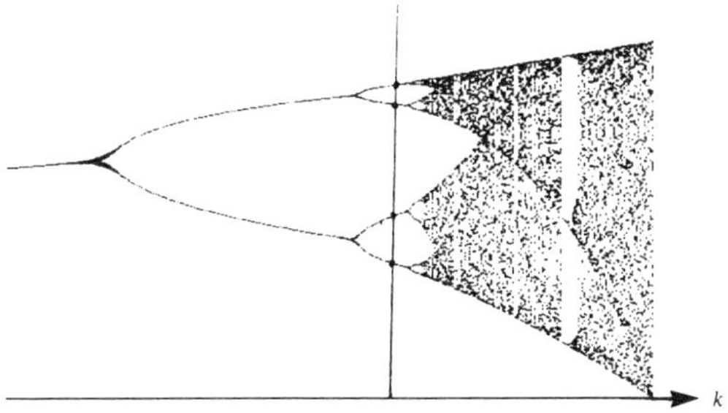

Sample output

Before giving any theory, we shall gain further insight if we consider an example in the plane $\mathbb{R}^2$. Thus we shall have an equation like (10.1.1) but with x_n a vector v_n rather than a number, and we think of f as an instruction that tells us how to move from v_n in $\mathbb{R}^2$ to the point v_{n+1}. The simplest such system is linear as in Chapter 2, where we saw that there is only one fixed point (the origin); and it can be either an *attractor*, a *repellor*, or a *saddle*. However, when f is non-linear, we meet new phenomena, as the following example shows.

10.3 THE MAYNARD SMITH EQUATION

We consider a model, suggested in Maynard Smith (1968) p.51, to describe a population of fruit flies kept under laboratory conditions. The flies feed, and lay eggs, and deaths are compensated by the birth of new individuals from newly hatched eggs after an incubation time of T days (typically $T \cong 30$). If x_n denotes the size of the population after n time intervals, Maynard Smith uses a probabilistic argument to obtain the equation

$$x_{n+1} = ax_n + bx_{n-T} + cx_{n-T}^2. \tag{10.3.1}$$

Here the x's are numbers (not vectors) and a, b represent respectively rates of survival and birth, while $c(<0)$ is a 'braking' action. If $T = 0$, we have essentially equation (10.2.1) again. Of course, this model will not always yield integer values for the x's, but that is a small price to pay. The behaviour at time n is influenced by the delay T, and many social systems are of this type, when an 'elite' (e.g., trainees) is nurtured in order to enrich the society later.

In order to compute x_{T+1} from (10.3.1), we need to know x_0 and x_T; given x_1 we can then compute x_{T+2}, and so on. Thus given the vector $(x_0, x_1, \ldots, x_T)$ in $\mathbb{R}^{T+1}$, we can use (10.3.1) to compute the vectors

$$(x_1, x_2, \ldots, x_T, x_{T+1}), \qquad (x_2, x_3, \ldots, x_{T+1}, x_{T+2}), \ldots$$

Equation (10.3.1) therefore defines for us a dynamical process on $\mathbb{R}^{T+1}$ given by $(x_0, x_1, \ldots, x_T) \to (x_1, x_2, \ldots, x_T, ax_T + bx_0 + cx_0^2)$. This is a mapping F: $\mathbb{R}^{T+1} \to \mathbb{R}^{T+1}$ composed of the two mappings:

$$f\colon (x_0, x_1, \ldots, x_T) \to (bx_0 + cx_0^2, x_1, \ldots, x_T)$$
$$g\colon (x_0, x_1, \ldots, x_T) \to (x_1, x_2, \ldots, x_{T-1}, ax_T + x_0)$$

(so $F = g \circ f$), each of which is very simple; for g is linear, and f is a 'folding' of the x_0-axis. However, the iterations of F are complicated; the biology suggests that F should have orbits that are more or less periodic, but a full analysis of F is not known.

Consider the case when $T = 1$, so our population now has just two generations (e.g., workers and apprentices). Here, F: $\mathbb{R}^2 \to \mathbb{R}^2$ is now a dynamical system in the plane, given by the equations

$$F\colon (x, y) \to (y, ay + bx + cx^2). \tag{10.3.2}$$

If we neglect the non-linear term cx^2, we have a problem like that of the Fibonacci rabbits (Chapter 2), but that term makes a remarkable difference to the dynamics, as we shall see.

It is no longer possible to illustrate the mapping with staircases and cobwebs as with a single dimension. One can try to work with the geometry of f and g above, but it seems difficult (see Griffiths and Rand (1977)), and instead we use analytical methods. We first use the rescaling $(x, y) = -b(x', y')/c$, which magnifies the first quadrant since c is small and negative, so that (10.3.2) becomes $(x', y') \to G(x', y')$ where after dropping the primes,

$$G\colon (x, y) \to (y, h(x, y)), \qquad h(x, y) = ay + bx(1 - x). \tag{10.3.3}$$

As before, we first calculate the fixed points of G by solving the equation $(x, y) = G(x, y)$, to get $(x, y) = (y, h(x, y))$, so $y = x$ and $x = 0$ or $(a + b - 1)/b$. Since the model requires positive coordinates then the fixed points are $\mathbf{0}$ and

$$P = b^{-1}(a + b - 1, a + b - 1), \qquad ((a + b) > 1). \tag{10.3.4}$$

Before making any further theoretical calculations, we display in Figs. 10.3.1–10.3.5 some sketches of the possibilities for the orbits of the mapping G, when $a = 1/2$ and b takes various values. To understand these, recall that the orbit of the point $X = (x_1, x_2)$ starts with X, then visits $G(X) = (x_2, x_3)$, with $x_3 = h(x_1, x_2)$ calculated by the rule (10.3.3); the next point is $G^2(X) = G(x_2, x_3) = (x_3, x_4)$, and so on. The numbers x_n thus generated are the (rescaled) values given by the difference equation (10.3.3), and the nth point in the orbit is (x_n, x_{n+1}).

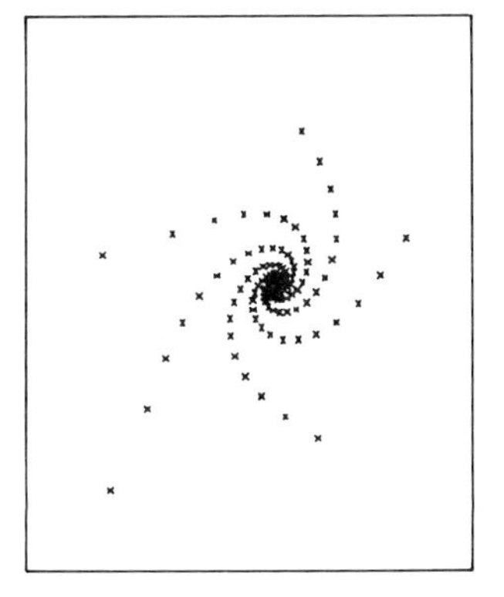

Fig. 10.3.1. $b = 1.95$.

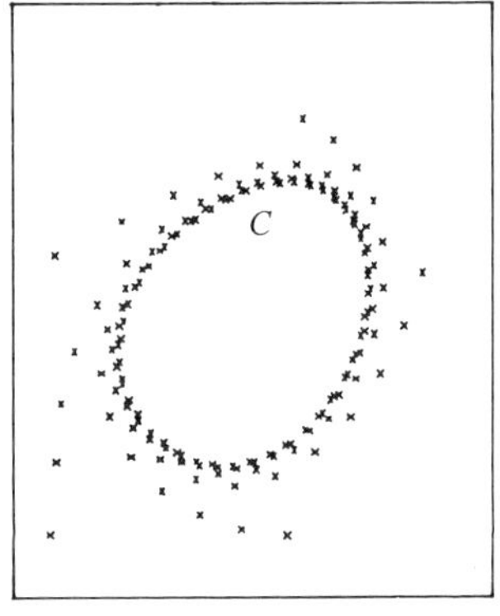

Fig. 10.3.2. $b = 2.03$.

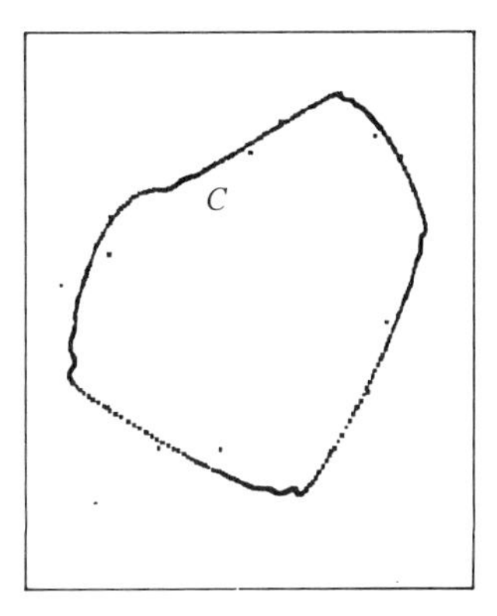

Fig. 10.3.3. $b = 2.3$.

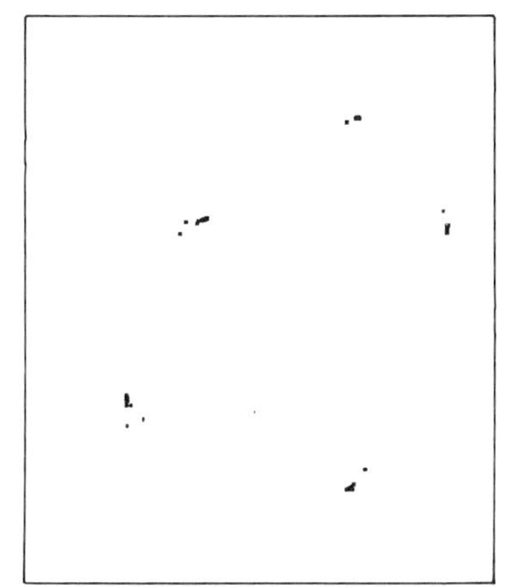

Fig. 10.3.4. $b = 2.36$.

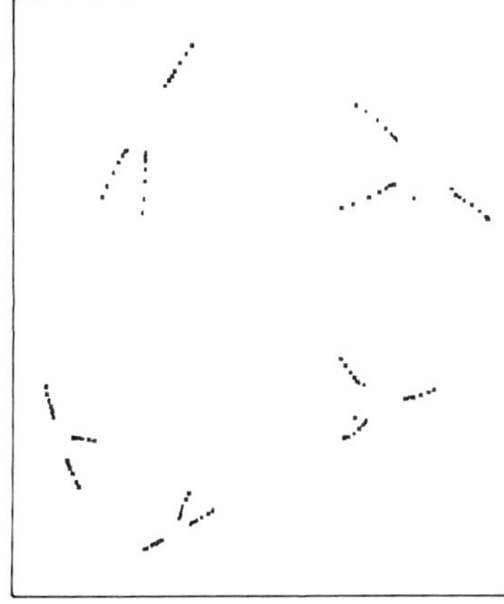

Fig. 10.3.5. $b = 2.46$.

These figures were first produced in the Ph.D. thesis of D. Whitley (1982), who made an extensive study of the dynamics of G. Note how the change from the line to the plane gives a much greater variety of dynamics. Especially, note that an orbit may simply converge to one of the fixed points as in the linear systems, or it may spiral round the oval curve C as in Fig. 10.3.2 (which surrounds P in (10.3.4)). But then notice the way in which, as b increases by quite small amounts, C acquires 'kinks' (due to the folding action of G about the line $x = 1/2$). As b increases, C changes its shape until eventually all that remains looks like a cycle of period 5 in Fig. 10.3.4; but then in Fig. 10.3.5, each of the 5 points acquires a satellite of period 3, and we have a cycle of period 15. This period doubles at $b = 2.47$, and further doublings will follow ($b = 2.472, 2.474$), until we obtain a 'mush' at $b = 2.6$. Although the smooth oval accords well with our intuition of the fluctuations of real populations, the value of b is surprisingly high, and not very stable, in that small changes alter the picture so drastically.

To check these assertions, the reader should investigate these figures further, by means of the two-dimensional plotting program 10.3.1 in the next exercise. (See Whitley (1982) for some analytical proofs.)

Exercises 10.3.1

1. Use the two-dimensional plotting program 10.3.1 to verify Figs 10.3.1–10.3.5 (and create others by varying the parameters). The program allows two choices, the second being a display of the orbit of (0.5, 0.5) under the mapping G; the first choice shows a plot of x_n against n, and produces interesting 'basket-weaves' (e.g., $b = 1.55$), as well as showing periodic cycles clearly. The values of the parameters a, b and (x_1, x_2) can be easily varied. In particular, if it appears that an orbit is closing down on a small version of C (as when $b = 1.49$), you should increase the number of iterations (nmax) to make sure that the orbit is not simply converging to a point.

```
REM Program 10.3.1, 2-D sequence plot
FUNCTION h(x,y) = a*y + b*x*(1-x)
DIM x(1000)
READ nmax, a   , hmin, hmax, x(1), x(2)
DATA 1000, 0.5, 0   , 3    , 0.5 , 0.51
INPUT PROMPT "value of b : " : b
INPUT PROMPT "1 or 2 dimensional display : " : d
SET MODE "graphics"
IF d = 1 THEN
  SET WINDOW 0,nmax, hmin,hmax
  PLOT 0,0 ; nmax,0
  PLOT 0,hhmin ; 0,hmax
  PLOT 1, x(1)
  PLOT 2, x(2)
ELSE
  SET WINDOW hmin,hmax, hmin,hmax
  PLOT hmin,0 ; hmax,0
  PLOT 0,hmin ; 0,hmax
  PLOT x(1), x(2)
END IF
  FOR n = 3 TO nmax
    LET x(n) = h(x(n-1),x(n-2))
    IF d = 1 THEN PLOT n, x(n) ELSE PLOT x(n-1),x(n)
  NEXT n
END
```

(Add semicolons to the end of some of the PLOT statements to draw a joined-up curve.)

Adapt the program to plot the point (x_{n-2}, x_{n-1}, x_n) in $\mathbb{R}^3$ as the point (x^*, y^*) on the computer screen by using an oblique projection: $x^* = y - x/2$, $y^* = z - x/2$, that is,

```
PLOT x(n - 1) - x(n - 2)/2, x(n) - x(n - 2)/2.
```

(You will need to take care over setting the window size.)

2. Given the difference equation $u_n = h(u_{n-1}, u_{n-2})$, with u_1, u_2 specified and $n > 2$, show that if $z_n = (u_n, u_{n+1})$ then $\langle z_n \rangle$ is the orbit of the mapping K: $(x, y) \to (y, h(y, x))$. (What is $K(z_n)$?)
3. Show that, since $c < 0$ in (10.3.2), we can find p, q such that the transformation $x = pz + q$ puts (10.3.2) in the form (after reverting to writing (x, y) for (z, y)):

$$F: (x, y) \to (y, ay + x^2 + b^2/4).$$

4. If z and c are the complex numbers $x + jy, u + jv$, show that the mapping G: $z \mapsto z^2 + c$ can be expressed in real terms as $(x, y) \to (x', y')$, where $x' = x^2 - y^2 + u$, $y' = 2xy + v$.
5. By writing $x_{n-1} = y_n$, convert the difference equation on $\mathbb{R}$:

$$x_{n+1} = ax_n + bx_{n-1} - x_{n-1}^2$$

(where a, b are positive constants) to the form $z_{n+1} = F(z_n)$ on $\mathbb{R}^2$ where $z_n = \begin{pmatrix} x_n \\ y_n \end{pmatrix}$ and F: $\mathbb{R}^2 \to \mathbb{R}^2$ is a smooth mapping.

Show that F has exactly two fixed points, at 0 and at $P = \begin{pmatrix} a + b - 1 \\ a + b - 1 \end{pmatrix}$. Apply the inverse function theorem to show that F is a local diffeomorphism at 0 and P.

Let A denote the Jacobian matrix of F at P. In the positive quadrant of the (a, b)-plane, sketch the set S of points (a, b) for which the eigenvalues of A are not real.

Put $z = P + w$ to approximate at P the given system by the linear system $w_{n+1} = Aw_n$. Assuming that $(a, b) \in S$, write down a standard form for A, and use it to show that the orbits of the linear system spiral in towards P provided that (a, b) lies below the line $2a + b = 3$. [SU]

For any mapping of $\mathbb{R}^2$, such as G in (10.3.3), it is clearly also desirable to know whether there is a region N of $\mathbb{R}^2$, such that if z lies in N, then the orbit of z under G does not wander off to infinity. We would then expect that G maps N into itself. The Julia program 10.3.2 below finds N by taking a fine grid on $\mathbb{R}^2$, and testing each of its points z. Then z, and its successors $G(z)$, $G^2(z), \ldots$, are plotted if its orbit seems to stabilize after a given number of iterations. We call the resulting region of the plane the 'filled' Julia set, its boundary being the 'Julia set' proper. That program will display the filled set N, from which we can 'see' its boundary in enough detail to recognize that it is usually fractal-like. The name Julia honours one of the French pioneers of the subject, who worked (long before the advent of computers) on iterations of mappings of the complex plane, like G_c: $z \to z^2 + c$, with $c =$ constant. Especially, the Julia set J_c of G_c has been studied as c varies, and when $c = -0.12256 + j\,0.74486$ we illustrate in Fig. 10.3.6 the Julia set known as 'Douady's rabbit'. A Julia set can also be seen in the graphics of Newton's method, being the boundary of the set of those points with orbits that do not converge to a root. For details and theory see Devaney and Keene (1989) and Peitgen (1989). We shall return to this topic in section 10.9. We now give the program for a filled Julia set.

Fig. 10.3.6(a). Filled Julia set, as produced by Program 10.3.2.

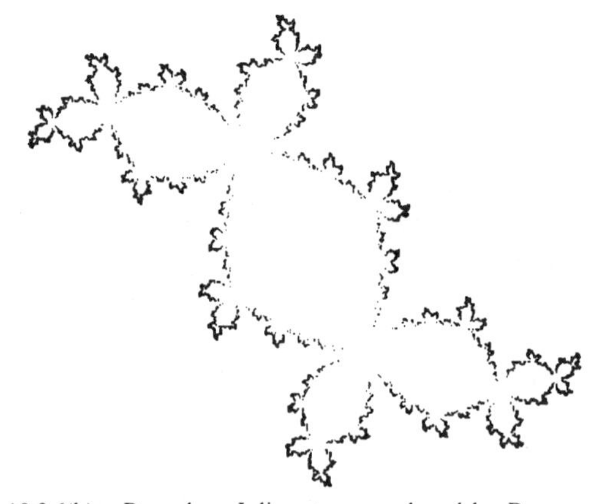

Fig. 10.3.6(b). Boundary Julia set, as produced by Program 10.3.3.

```
REM Program 10.3.2, Filled Julia set
FUNCTION Re(x,y) = x^2 - y*y + cx
FUNCTION Im(x,y) = 2*x*y + cy
READ cx, cy, xmin, xmax, ymin, ymax, sw , sh , nmax, rmax
DATA -1, 0 , -2  , 2   , -2  , 2   , 320, 200, 40  , 4
OPTION BASE 0
DIM x(40) , y(40)
SET MODE "graphics"
SET WINDOW xmin,xmax, ymin,ymax
LET xstep = (xmax-xmin)/sw
LET ystep = (ymax-ymin)/sh
FOR x0 = xmin TO xmax STEP xstep
  FOR y0 = ymin TO ymax STEP ystep
    LET x(0) = x0
    LET y(0) = y0
    LET n = 0
    LET gone = 0
    DO
      LET x(n+1) = Re(x(n),y(n))
      LET y(n+1) = Im(x(n),y(n))
```

Program 10.3.2 continued

```
        LET n = n + 1
        IF x(n)*x(n) + y(n)*y(n) > rmax THEN LET gone = -1
      LOOP UNTIL n>=nmax OR gone = -1
      IF gone = 0 THEN
        FOR n = 0 TO nmax
          PLOT x(n),y(n)
        NEXT n
      END IF
    NEXT y0
  NEXT x0
  END
```

Exercises 10.3.2

1. Show that when $c = 0$, the filled Julia set for G_0 is the interior of the unit disc, and J_0 is the unit circle S^1. Thus when c is small, we can expect J_c to be a small perturbation of S^1; verify this with your computer graphics.
2. If G is the mapping G: $z \to z^2 + c$ of exercise 10.3.1(4), then, for any $z \neq c$ we can try to find a complex number w such that $G(w) = z$ (an 'antecedent' of z). Then $z = w^2 + c$ with solution $w = \sqrt{(z - c)}$. But this is not unique, since the complex number $z - c$ has two square roots whose arguments differ by π. If $|z - c| = r$ and $\arg(z - c) = \theta$ find the two antecedents $w_1(z)$ and $w_2(z)$ of z. Using these we can form an **iterated function system** (IFS) as follows. Choose some probability p and, for each value z_n, generate a random variate r drawn from the uniform distribution $U[0, 1]$. Define the iteration function $f(z)$ by

$$f(z) = \begin{cases} w_1(z) & \text{if } 0 \leq r < p \\ w_2(z) & \text{if } p \leq r < 1. \end{cases}$$

Thus the orbit of any z_0 under the sequence $z_{n+1} = f(z_n)$ is no longer *deterministic*, but it has remarkable properties which the following program illustrates.

```
REM Program 10.3.3, Boundary Julia set by IFS
READ cx, cy  , x , y , xmin, xmax, ymin, ymax, nmax
DATA -1, 0   , 0 , -1, -2  , 2   , -2  , 2   , 5000
FUNCTION arg(x,y)
  IF x>0 THEN LET arg = ATN(y/x)
  IF x=0 THEN LET arg = SGN(y)*PI/2
  IF x<0 AND y>=0 THEN LET arg = PI + ATN(y/x)
  IF x<0 AND y<0 THEN LET arg = -PI + ATN(y/x)
END FUNCTION
SET MODE "graphics"
SET WINDOW xmin,xmax, ymin,ymax
FOR n = 1 TO nmax
    LET dx = x - cx
    LET dy = y - cy
    LET r  = SQR(dx*dx + dy*dy)
    LET theta = arg(dx,dy)
    LET r = SQR(r)
    LET theta = theta/2
    IF RND > 0.5 THEN theta = theta + PI
    LET x = r*COS(theta)
    LET y = r*SIN(theta)
    PLOT x,y
NEXT n
END
```

3. By definition, if z does not lie in the filled Julia set N of G_c, then the iterates of z all converge to infinity. The set W of all such z forms the basin of attraction of ∞. If $z_0 \in W$ define the sequence of antecedents as follows. Choose an antecedent of z_0 and call it z_1; if z_n is defined choose one of its antecedents and call it z_{n+1}. Show that each z_n lies in W, and that if $\langle z_{n(j)} \rangle$ converges to w as j tends to ∞, then w lies in the Julia set of G_c (i.e., on the common boundary of N and W.)

10.4 CHANGE OF STABILITY

It is time to introduce some theory to guide us in further explorations. First, consider again the one-dimensional case of a smooth mapping $f\colon \mathbb{R} \to \mathbb{R}$, and let $P = (p, p)$ be a fixed point of f. If $f'(p) = 1$, then the graph G of f has the diagonal line L: $y = x$ as a tangent at P, but if $f'(p) \neq 1$, then G crosses L transversally (as at $F = (z, z)$ in Fig. 10.2.2). In that case we shall show that the orbit of any point near p converges to p (as at z in Fig. 10.2.2), or moves away from p (as in Fig. 10.2.3, but not necessarily to infinity), according as $|f'(p)|$ is less than or greater than 1. In these two cases, p is an attractor or a repellor for the orbits. (Special cases of this were seen in exercises 10.1.1.) If we imagine a particle P at p which is given a slight push, then it will move back to (or away from) p, according as p is an attractor or repellor; so the two cases are described respectively as stable or unstable. More precisely we have the following theorem.

Theorem 10.4.1
Let p be a fixed point of the smooth mapping $f\colon \mathbb{R} \to \mathbb{R}$, with $f'(p) \neq 1$. Then there is an interval I with centre p such that, for any x in I, the iterates $x_n = f^n(x)$ have the following property:

(a) *if $|f'(p)| < 1$, then the distance $|x_n - p|$ decreases to zero and the entire orbit $x = x_0, x_1, x_2, \ldots, x_n, \ldots$ lies in I and converges to p;*
(b) *if $|f'(p)| > 1$, then the distances $|x_n - p|$ increase, and the orbit moves away from p.*

Proof
We are given that $|f'(p)| \neq 1$. Then, since $|f'|$ is a continuous function, we can use the Inertia Principle (theorem D.2.1) to assert that the interval I exists with centre p such that either $|f'| < 1$ on I, or $|f'| > 1$ on I. Also, if x lies in I, and $n > 0$, then we have

$$\begin{aligned} |x_n - p| &= |f(x_{n-1}) - f(p)| && \text{since } p \text{ is fixed,} \\ &= |f'(c)(x_{n-1} - p)| && \text{for some } c \text{ in } I \end{aligned}$$

by the mean value theorem (Chapter 4). Hence if $|f'| < 1$ on I then $|x_n - p| < |x_{n-1} - p|$, so x_n is closer to p than x_{n-1} and the orbit remains in I and converges to p. If $|f'| > 1$ on I then $|x_n - p| > |x_{n-1} - p|$ so x_n moves further from p as n increases. This completes the proof.

Example 10.4.1
With f as in (10.2.1), $f'(x) = k(1 - 2x)$, so at the fixed point $\mathbf{0}$, we have $f'(0) = k > 0$.

Hence if $k < 1$, the origin is an attractor; but when $k > 1$, it is a repellor. The other fixed point is at $p = (1 - 1/k)$ and $f'(p) = 2 - k$, so p is an attractor when $|2 - k| < 1$, which means $-1 < 2 - k < 1$. Thus p is a repellor when $k < 1$, an attractor when $1 < k < 3$, and then a repellor again. These conclusions justify the statements we made with Figs. 10.2.1 to 10.2.3 above.

With theorem 10.4.1, when p is an attractor, we may ask for the largest possible set we can take as interval I. This set lies within the basin of attraction of p.

Exercises 10.4.1

1. In $\mathbb{R}$, find the fixed point of the mapping $x \to e^{kx}$ and the basin of attraction, for $k < 0$, $k = 0$, $k > 0$. For each of the following mappings $f: \mathbb{R} \to \mathbb{R}$, decide their stability. Find the corresponding basin of attraction.

   $$f(x): \quad x/(1 + x)^3, \quad xe^{k(1-x)}, \quad kx(1 - x^2).$$

   (Each has been proposed as a substitute for (10.2.1) in a population model.)
2. Refine the argument of theorem 10.4.1 for the case when $f'(p) = \pm 1$, by using the Taylor expansion at p to decide whether $f'(c)$ in the proof is < 1.

Example 10.4.2: Newton's Method

There is a well-known method, due to Newton and developed by Raphson, for approximating the roots of an equation $\phi(x) = 0$, and it can be regarded as generating a dynamical system. Suppose $x = a$ is a root of the equation, and $\phi'(a) \neq 0$. Then by the Inertia Principle (theorem D.2.1), $\phi' \neq 0$ on some interval J centred at a. Therefore we can define a new function N_ϕ on J:

$$N_\phi: x \to x - \phi(x)/\phi'(x),$$

and $N_\phi(a) = a$ since $\phi(a) = 0$. Therefore a is a fixed point of N_ϕ, and 'Newton's method' is the assertion that if we start with any x near a, then the iterates $x_n = N_\phi^n(x)$ will converge to a; so a is an attracting fixed point. To verify this, we use the quotient rule to differentiate N_ϕ and get

$$N'_\phi(x) = 1 - \frac{(\phi'(x))^2 - \phi(x)\,\phi''(x)}{(\phi'(x))^2}.$$

Thus, since $\phi(a) = 0$, then $N'_\phi(a) = 1 - 1 = 0$. Hence by theorem 10.4.1, J contains an interval I, centred at a, such that for any x in I, the orbit of x converges to a. (We can find an I without knowing a, by testing that ϕ has opposite signs at the end points.) Then I lies in the basin of attraction of a.

Exercises 10.4.2

1. If $\phi(x) = x(1 - x^2)$, find the appropriate intervals J and I for each of the zeros of ϕ, and then the basins of attraction. Try solving other equations, such as $e^{kx} = x$, by this method, using a calculator.
2. With the same ϕ, let S be the set consisting of the zeros $\pm 1/\sqrt{3}$ of ϕ'. Taking as initial point any x not in S, you can start to compute the orbit $\langle x_n \rangle$ of x under N_ϕ. It may

happen that for some n, x_n lies in S, so the iteration cannot be continued. For example, if $n = 1$, then we say that x is in the set denoted by $N_{\phi}^{-1}(S)$, the 'inverse image' of S. Use this notation to specify a set T in $\mathbb{R}$, such that, if x lies in T, then the orbit of x never hits S. Find intervals K in T such that the orbit of every point in K converges to the same zero of ϕ.

3. Let $f: \mathbb{R} \to \mathbb{R}$ be a continuous mapping, and let p be a fixed integer > 1. Then the f^p-orbit of x is the sequence $f^p(x)$, $f^{2p}(x)(= f^p(x))$, $f^{3p}(x)(= f^p(f^{2p}(x)))$, ... $f^{np}(x)$, ... Show that f maps this orbit to the f^p-orbit of $f(x)$.
4. With f as before, suppose the point a is a point of period 2 and let $b = f(a)$. We have already seen that a and b are fixed points for f^2. Use the chain rule to show that $(f^2)'(a) = f'(b)f'(a) = (f^2)'(b)$. Hence show that if $|(f^2)'(a)| < 1$ then the orbit of a point sufficiently near a will alternately get ever closer to a and b, to justify our earlier description of the nature of an attracting 2-cycle.
5. Suppose that in exercise 4, f depends smoothly on the parameter c in $\mathbb{R}$, say $f = g_c$. Use the implicit function theorem to show that if a_0 is a fixed point of g_0^2 and $|(g_0^2)'(a)| < 1$, then 'a_0 depends smoothly on c', that is, there is a mapping $p: I \to \mathbb{R}$ of some neighbourhood I of 0 such that $p(0) = a_0$, and for each c in I, $p(c)$ is an attracting fixed point of g_c^2.

The process described in exercise 10.4.2(2) applies also in $\mathbb{R}^2$, when we allow all numbers to be complex; and the regions corresponding to the intervals K can be beautifully coloured. The Newton program 10.4.1 will do this for you. As you would expect, the orbit of almost any point 'homes' on some zero of ϕ, but the points of a 'thin' and very complicated set behave differently. For technical details, see Peitgen (1989). The program can be modified to work for any other cubic replacing ϕ.

```
REM Program 10.4.1, Newton
READ xmin, xmax, ymin, ymax, nmax, eps   , sw , sh
DATA -1.5, 1.5 , -1  , 1   , 100 , 0.0001, 319, 199
DIM xp(100), yp(100)
FUNCTION Re(x,y) = (2*x + (x^2 - y^2)/(x^2+y^2)^2)/3
FUNCTION Im(x,y) = 2*y*(1-x/(x^2+y^2)^2)/3
SET MODE "graphics"
SET WINDOW xmin,xmax, ymin,ymax
FOR x = xmin TO xmax STEP (xmax - xmin)/sw
  FOR y = ymin TO ymax STEP (ymax - ymin)/sh
    LET n = 1
    LET xp(n) = x
    LET yp(n) = y
    DO
      LET x1 = xp(n)
      LET y1 = yp(n)
      LET x2 = Re(x1,y1)
      LET y2 = Im(x1,y1)
      LET n = n + 1
      LET xp(n) = x2
      LET yp(n) = y2
    LOOP UNTIL ((x1 - x2)^2 + (y1 - y2)^2) < eps
    IF x2 > 0 THEN
      LET ink = 1
```

Program 10.4.1 continued

```
        ELSE
          IF y2 > 0 THEN LET ink = 2 ELSE LET ink = 3
        END IF
        SET COLOR ink
        FOR i = 1 TO n
          PLOT xp(i),yp(i)
        NEXT i
      NEXT y
    NEXT x
    END
```

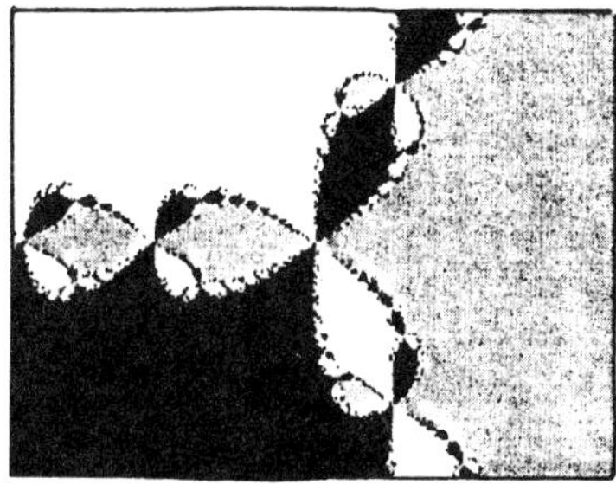

Sample output

10.5 PERIODIC ORBITS

We shall now explain why, in the computer experiments described earlier, we found 'windows' for the parameter k, for which a stable periodic cycle persists. Thus, suppose that f: $\mathbb{R} \to \mathbb{R}$ has a periodic orbit $z_1, z_2, \ldots, z_p$ with period p; here, $p > 0$, $f(z_p) = z_1$, and the p points z_i are all distinct. They form a p-**cycle** Z, and for each point z_i in Z we have

$$f^p(z_i) = z_i, \qquad (1 \leq i \leq p). \tag{10.5.1}$$

Thus z_i is a fixed point of f^p, and we shall now use our knowledge of the behaviour of orbits near z_i to describe their behaviour near Z as a whole. This will make more precise our earlier use of such terms as 'attracting' for a p-cycle, and explain why we could see cycles of period 3, without seeing some of every period when we knew they existed.

The case when $p = 2$ is described in exercise 10.4.2(5), and we now carry out similar calculations for $p > 2$. Our description makes use of a certain number, the 'multiplier' of Z, which arises as follows. Since z_i in (10.5.1) is a fixed point of f^p, we apply theorem 10.4.1 to it; so we calculate the number $(f^p)'(z_0)$, denoted by $g(z_0)$ for brevity. We claim

$$g(z_0) = f'(z_{p-1})\cdot f'(z_{p-2})\cdot f'(z_{p-3})\cdot \ldots \cdot f'(z_0). \tag{10.5.2}$$

To prove this we use the chain rule to begin the following inductive steps:

$$\begin{aligned}(f^{r+1})'(z_0) &= ((f^r)\circ f)'(z_0) = (f^r)'(f(z_0))\cdot f'(z_0)\\ &= (f^r)'(z_1)\cdot f'(z_0)\\ &= (f^{r-1})'(z_2)\cdot f'(z_1)\cdot f'(z_0), \qquad \text{similarly.}\end{aligned}$$

Thus (10.5.2) follows by taking $r = p - 1$ and iterating the last equation. Note that the right-hand side of (10.5.2) depends only on the orbit of z_0 (the p-cycle Z), which is also the orbit of $z_1, \ldots, z_{p-1}$; hence $g(z_i)$ is constant for each point z_i of Z, and this constant is the promised multiplier of Z.

Now suppose that the multiplier g satisfies $|g| < 1$. Then by (10.5.1) and theorem 10.4.1, each z_i in Z is an attracting fixed point of f^p, with a neighbourhood I_i for which, if $x_0 \in I_i$, then the f^p-orbit of x_0 remains in I_i and converges to z_i there. But then f maps z_i into I_{i+1}, and carries the orbit to that of $f(x_0)$, which converges to z_{i+1}. Repeating this argument, we see that the f-orbit of x_0 visits each I_i successively, and gets closer to z_i on each visit. It is in this sense that the original p-cycle Z is 'attracting'. A similar argument describes what happens if $|g| > 1$, and then Z is repelling. (In that case, we do not see repelling cycles on the computer screen, since any test orbit—other than the p-cycle itself—moves rapidly away from the p-cycle.)

We have already noted in computer experiments with the logistic family f_k, that if we see an attracting p-cycle for one value k, then we see one for nearby values of k; we have a 'window' for the integer p. To explain how this happens, we work as in exercise 10.4.2(4). Thus suppose that on some interval I we have a smooth family of mappings $f_c: I \to I$, where c runs through an interval C; and suppose we have a p-cycle $Z: z_1, \ldots, z_p$ when $c = 0$. If the multiplier of Z is $g(Z)$ and $|g(Z)| < 1$, then as we have seen, Z is attracting. We shall apply the implicit function theorem to prove the next theorem.

Theorem 10.5.1

For some neighbourhood J (the 'window') of $\mathbf{0}$ *in C, f_c has an attracting p-cycle Z_c which varies smoothly as c changes in J; and Z_0 is our original Z.*

Proof

Letting $f_c(x) = f(x, c)$, we know that for each i, z_i is a solution of the equation $f^p(x, c) - x = 0$. Abbreviate this to $E(x, c) = 0$, and calculate $\partial E/\partial x$; evaluating at the point $(x, c) = (z_1, 0)$, we get $\partial E/\partial x = \partial/\partial x\,(f^p(x, c) - x) = g - 1$, where g is the multiplier $g(z_1)$ of the p-cycle Z. Hence $\partial E/\partial x \neq 0$ since $|g| < 1$. Therefore by the implicit function theorem, there is a smooth function $v: J \to \mathbb{R}$, with J a neighbourhood of $c = 0$, such that $v(0) = z_1$ and for each c in J, $E(v(c), c) = 0$. Thus $f^p(v(c), c) - v(c) = 0$, so $v(c)$ is a fixed point of f^p. Moreover, if J is small enough, then by the Inertia Principle, $|(f^p)'| < 1$ at $v(c)$, because $|(f^p)'z_1| < 1$. Hence $v(c)$ is an attracting fixed point of f_c^p, so its orbit is the required attracting p-cycle Z_c that varies continuously with c. This completes the proof.

Exercise 10.5.1

Adapt the proof to show that we can change 'attracting' to 'repelling' when $g(Z) > 1$.

10.6 CANTOR SETS AND THE SHIFT MAPPING

The periodic orbits are finite, but it is possible to have an infinite attractor with a special type of structure, called a **Cantor set**. The first such set was constructed for other purposes around 1870 by Georg Cantor, and the instructions are these:

(1) from the unit interval $U = [0, 1]$, remove the 'open' middle third $V = (1/3, 2/3)$, leaving the end points;
(2) the remaining set $U_1 = U - V$ consists of (the points of) the remaining two intervals, from which we remove their open middle thirds, to leave a set U_2 that is split into four intervals;
(3) from each of these four, remove their open middle thirds to leave U_3, split into eight intervals, and so on.

At the nth stage, we are left with a set U_n split into 2^n intervals, from which the next stage removes their middle thirds; and so on. Fig. 10.6.1 shows the first four stages of the construction.

U	————————
U_1	—— ——
U_2	- - - -
U_3	" " " "

Fig. 10.6.1.

Cantor's set C is the set of all those points u of U that remain after all these removals, so u lies in every set U_n. It turns out that C has an uncountable infinity of points, and is totally disconnected; moreover it is 'compact', in that it is bounded and every infinite subset of it contains a limit point in C. Nowadays, any set with these properties is called a Cantor set, and any two Cantor sets can be shown to be homeomorphic.

The original set can also be described as the set of all points of U, for which the digit 1 does not appear in its ternary 'decimal' expansion (using powers of 1/3 instead of 1/10). Thus $u = \sum_{n=1}^{\infty} (a_n/3^n)$ and a_n is either 0 or 2. Therefore the sequence $\langle a_n \rangle$ describes u completely, and we obtain a mapping $s: C \to C$ (called the 'shift') by the rule that $s(u)$ is to correspond to the sequence $a_2, a_3, \ldots$, with a_{n+1} in the nth place. The dynamics of this mapping has been exhaustively studied, and it can be shown that the 'complicated' behaviour of the logistic mappings f_k arises because there is a Cantor set in U, on which f_k behaves like the shift mapping. Also, with the Maynard Smith mapping in section 10.3, a Cantor set may lie on the oval attractor in Fig. 10.3.2, to be the eventual attractor of all orbits.

It is in such cases, where there is an attractor like a Cantor set with the shift dynamics, that people say that the dynamics has become 'chaotic'; but as yet there is no agreed formal definition of 'chaos' that covers all the known experimental examples. For a good introductory discussion, with much more detail, see Devaney's article in Devaney and Keen (1989), which contains a short proof of the 'period 3 implies chaos' result of Li and Yorke, mentioned above.

Exercises 10.6.1

1. Express $x = 0.25$ as a ternary decimal. (Since x lies in the first 'third' U_0 of U, the first digit a_1 must be 0. But $x = 9/36$, and $1/3 = 12/36$, so x lies in the third 'third' U_{02} of U_0. Hence $a_2 = 3$. Now continue.)

2. Show that the length of each set U_n in the Cantor construction is $(2/3)^n$, so the set C has length 0. But is length the appropriate notion of 'content' for C? If C had a 'dimension' d, like 1 for a line, or 2 for a plane, and a non-zero content V, then V would have to satisfy the same scaling law as length and area: thus under a change s of scale, the content would change by the factor s^d. Show that the magnification $x \to 3x$ gives two copies of C, so $2V = (3^d)V$. Hence show that the 'dimension' d of C is $\ln(2)/\ln(3)$. Such a V exists for many very general sets, and is called the **Hausdorff measure**. There is then a corresponding number d, and if d is not an integer, it is now usual to call the set a 'fractal'. See section 10.10 below.
3. Construct a set like C, but using middle fifths. Describe the points in terms of decimals. By removing a variable amount each time, show that we can be left with a non-zero length.
4. The shift mapping s, mentioned above, acts on sequences of 0's and 2's, but it is convenient to replace these by L's and R's. For example, if $q = LRLRLR\ldots$ then $s(q) = RLRLRL\ldots$; verify that $s^2(q) = q$, so q is of period 2. Verify that the sequences $(LRL)(LRL)(LRL)\ldots$, $LLRRLLRR\ldots$ have periods 3 and 4, respectively, and find a sequence of period n. (Such sequences are used for modelling the iterations of a 'hump' function f, symmetrical about $\mathbf{0}$ and with a single maximum there. We associate with each x a sequence q in C as follows: the nth term of q is L or R according as $f^n(x)$ is to the left, or right, of $\mathbf{0}$, and we assume here that no iterate of x is zero. With reasonable hypotheses on f, the mapping $x \to q$ can be shown to represent the attractor of f in C, and relate the dynamics of f to that of s on C. For details, see Bruce *et al* (1990) and Collett and Eckmann (1980).

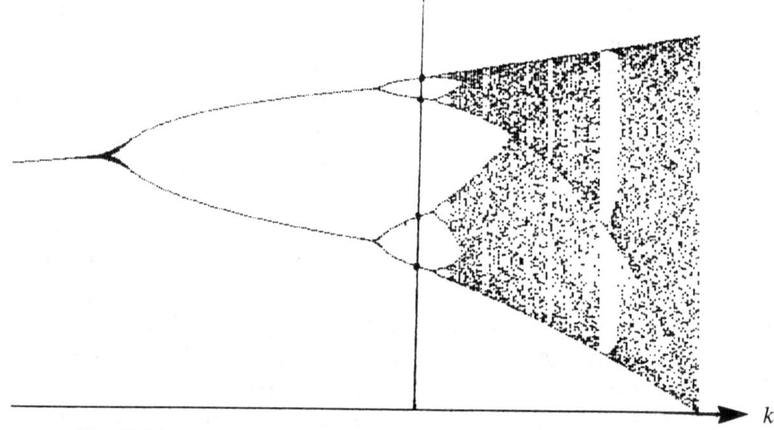

Fig. 10.6.2. Each vertical line $x = k$ meets the set in the attractor of f_k.

We can now understand the famous 'bifurcation diagram', shown in Fig. 10.6.2, for the family of logistic mappings. This seems first to have been constructed experimentally by May (1976), and a full theoretical explanation is not yet complete. The parameter k runs along the horizontal axis; and on the vertical line through each k, the dotted set shows the attractor of the mapping f_k. Thus we see first the periodic attractors of successive periods 2^n, then a mess, then windows with periodic attractors, and so on, to Cantor sets with chaotic dynamics. The amazing fact must be stressed, however, that the theory shows that a similar diagram is 'universal' for any family of hump functions under simple hypotheses. This implies that when making physical models, it is not so much the algebraic formula for f that matters, but rather its general shape.

10.7 DIFFERENTIAL EQUATIONS AND CHAOS

After seeing the possible complexity of the theory of one-dimensional mappings, the reader will not be surprised to learn that the variety is even greater in two dimensions, and with many more gaps in our knowledge. Unfortunately we cannot wash our hands and opt to model with differential equations instead. Firstly, the classical Euler approximation to (= model of) the equation $x' = F(x)$ is the difference equation $x_{n+1} = x_n + hF(x_n)$, and is very useful. Also, one of the earliest examples of 'chaos' arose with the simple-looking system found by the meteorologist Lorenz, as the result of drastically simplifying a more complicated model. His equations are:

$$x' = a(y - x), \qquad y' = bx - y - xz, \qquad z' = cz + xy, \tag{10.7.1}$$

the only non-linearities being the products xy, xz; and the system is autonomous, because the time does not occur explicitly. With such systems, Poincaré had shown that a certain discrete dynamical system could be fruitfully associated. His idea in this case was to find a plane P that is intersected transversally by every trajectory of the system (10.7.1), and then to define a mapping $F\colon P \to P$ as follows. For each point X in P, follow the trajectory through P until it hits P again for the first time at some point Y (as in Fig. 10.7.1). Then $F(X)$ is to be Y. If F exists then it will be smooth;

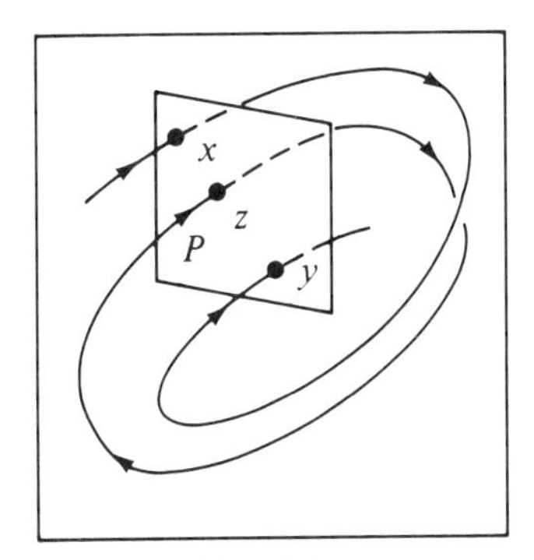

Fig. 10.7.1.

and clearly if it has a fixed point Z, then the trajectory through Z returns to Z and is therefore a closed loop. If (10.7.1) had been a planar system, then we replace P by an arc as in Fig. 9.8.2, and obtain a mapping $f\colon L \to L$ Hence, the search for periodic solutions of the differential equation is related to the fixed points of discrete systems. But because of our access to computers, we know (as Poincaré could not, though he suspected something of the sort) that since F can exhibit 'chaos', then a closed trajectory of the differential equation might therefore be like a very complicated mass of spaghetti, piercing the plane P time after time in the points of a Cantor set. This is illustrated in computer pictures of trajectories of the Lorenz system (see Fig. 10.7.2), which has a 'strange attractor' that has not yet been completely analysed (see the next exercise). In $\mathbb{R}^3$ there is room for curves to wind round each other without crossing, and even to be knotted; whereas the 'no-crossing rule' (see Chapter 9) for autonomous systems in $\mathbb{R}^2$ forces docility.

Fig. 10.7.2. Trajectories for the Lorenz system.

Exercises 10.7.1

1. Show that the Lorenz system (10.7.1) has a fixed point at $\mathbf{0}$ in $\mathbb{R}^3$; and if $c(b-1) < 0$, it has two others, at $(su, su, b-1)$, where $u^2 = c(b-1)$ and $s = \pm 1$. Discuss the linearization at these points by showing first that the Jacobian matrix J is

$$J = \begin{pmatrix} -a & a & 0 \\ b-z & -1 & -x \\ y & x & c \end{pmatrix}.$$

```
REM Program 10.7.1, Lorenz attractor
READ a , b , c          , x, y, z, dt   , xmin, xmax, ymin, ymax
DATA 10, 28, -2.66667, 1, 1, 1, 0.01, -30 , 30  , -30 , 80
FUNCTION xdot(x,y,z) = a*(y - x)
FUNCTION ydot(x,y,z) = b*x - y - x*z
FUNCTION zdot(x,y,z) = c*z + x*y
SET MODE "graphics"
SET WINDOW xmin,xmax, ymin,ymax
DO
    PLOT x,z;
    LET dx = xdot(x,y,z)*dt
    LET dy = ydot(x,y,z)*dt
    LET dz = zdot(x,y,z)*dt
    LET x = x + dx
    LET y = y + dy
    LET z = z + dz
LOOP UNTIL "cows" = "home"
END
```

(Sample output should look like Fig. 10.7.2.)

2. When certain fluids are mixed together (e.g., paint, turpentine and water when we clean a paint brush) interesting patterns are observed as the mixture seethes. This is the **Belousov–Jabotinsky** reaction, discussed in Murray (1978) with equations $\varepsilon\dot{u} = u + v - uv - qu^2$, $\dot{v} = -v + fw - uw$, $\dot{w} = u - w$, where $\varepsilon \cong 2 \times 10^{-4}, q \cong 8 \times 10^{-6}$. Investigate this system by computer.

3. Investigate the following system (due to Rossler), which is even simpler looking than (10.7.1):

$$x' = -(x+y), \qquad y' = x + ay, \qquad z' = ab + xz - cz.$$

In engineering, there are now many systems that have been found to exhibit 'chaos'. For example, consider Duffing's equation for a pendulum:

$$x'' + bx' + ax^3 + cx = f\cos Pt \qquad \left(x' = \frac{\mathrm{d}x}{\mathrm{d}t}, c > 0\right) \tag{10.7.2}$$

in which a, b, c, f, P are constants. If $a = 0$ we have a standard linear equation with the damping term bx' and forcing term $f\cos Pt$; but the cubic term makes a vast difference. The equation can be written as a planar system

$$x' = y, \qquad y' = -(ax^3 + cx + by) + f\cos Pt \tag{10.7.3}$$

which is non-autonomous because of the explicit appearance of the time t. Hence the trajectories no longer need satisfy the no-crossing rule, since a trajectory through (x_0, y_0) is not completely known unless we specify also the time t_0. However, we can write (10.7.3) as the three-dimensional system

$$x' = y, \qquad y' = -(ax^3 + cx + by) + f\cos Pz, \qquad z' = 1,$$

so the third equation integrates as $z = t +$ constant. But now we have an autonomous system, with trajectories that cannot cross; their projections in the (x, y)-plane are those of (10.7.3), so it is then not surprising that these cross.

Exercises 10.7.2

1. By writing $t = us + \pi/2P$, show that Duffing's equation (10.7.2) can take the form $y'' + y + ry^3 = K\sin Qt$ if $b = 0$, $u = 1/\sqrt{c}$, $K = f/c$.
2. Use the method of Chapter 9 to construct a phase portrait for (10.7.3) when $f = 0$.
3. The Duffing program 10.7.2 is adapted from Cook (1991), so that readers can experiment with the system (10.7.3) for themselves. Cook states that interesting effects can be seen when the parameters $(a, b, c\ f)$ take the values: (0, 0.2, 1, 1.6), (0.4, 0.3, 0.1, 2), (0.5, 0.41, 0.5, 6 to 15), (1, 0.3, 0, 10), (0.5, 0.1, −8, −4). Figures 10.7.3–10.7.5 indicate something of the variety of possible trajectories. The first two were found by analogue computers in the 1950s, as explained in Davis (1962) p.394, where the number p is computed as 0.92845.

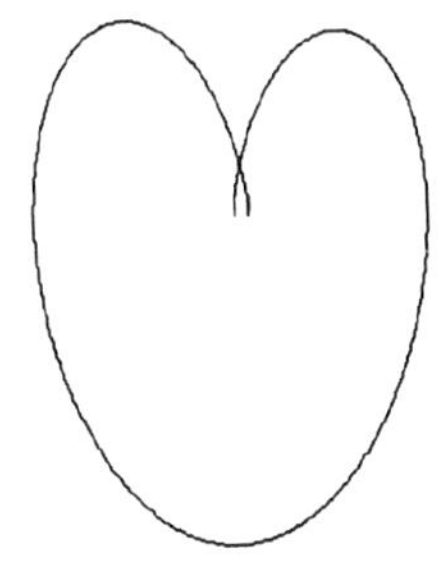

Fig. 10.7.3. $(a, b, c, f, p) = (-\frac{1}{6}, 0, 1, 1, 2p)$.

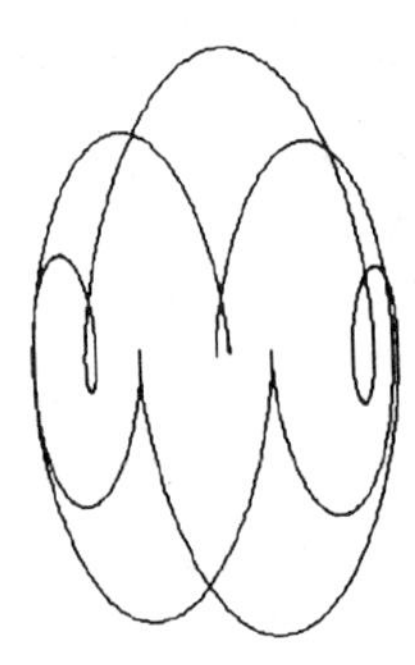

Fig. 10.7.4. $(a, b, c, f, p) = (-\frac{1}{6}, 0, 1, 1, 3p)$.

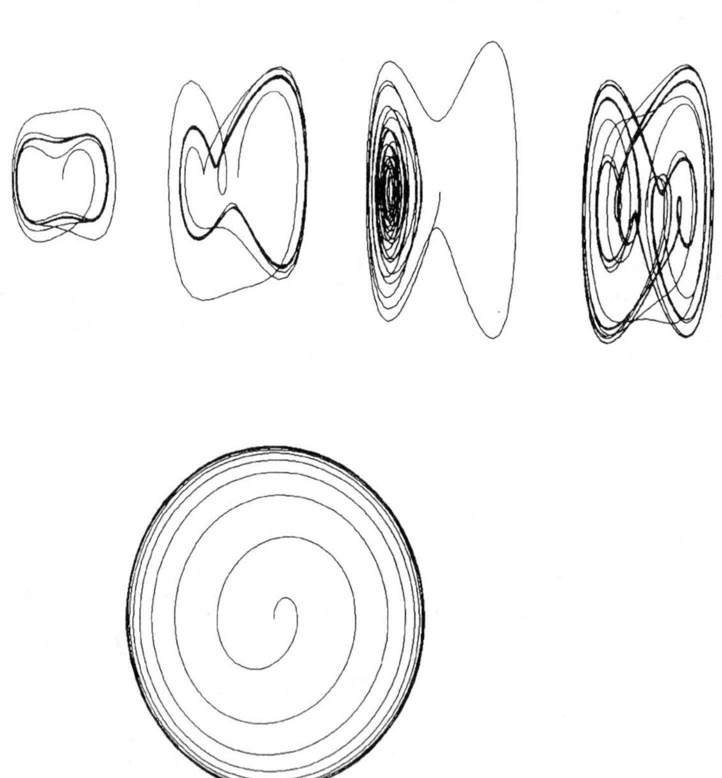

Fig. 10.7.5.

4. Investigate by computer the presence of chaos in the system of exercise 4.8.1(11) when the function f there is cubic rather than linear. (A Josephson junction will give such behaviour.)
5. If in Duffing's equation we have $a = b = 0$, $P = 1$, then we have Mathieu's equation, which models a pendulum for which either the length, or the point of support is oscillating. Investigate Mathieu's equation by computer.

However, computer arithmetic can lead the program to give different results on different computers! This again shows why computer evidence must be treated with

```
REM Program 10.7.2, Duffing's equation
READ a  , b  , c  , f, t, x   , v  , xmax,· vmax, tmax, dt
DATA 0.4, 0.3, 0.1, 2, 0, -0.1, 0.1, 12  , 8   , 100 , 0.1
SET MODE "graphics"
SET WINDOW -xmax,xmax,  -vmax,vmax
FUNCTION f1(t,x,v) = v
FUNCTION f2(t,x,v) = -(a*x^3 + c*x + b*v) + f*COS(t)
LET vv = 0
LET xx = 0
DO
  PLOT x,v;
  CALL RK4(v,x,t,dt,vv,xx)
  LET v = vv
  LET x = xx
  LET t = t + dt
LOOP UNTIL t>tmax
STOP
SUB RK4(v,x,t,dt,vv,xx)
   REM this is as for Program 4-6.1, S.H.M.
END SUB
END
```

suspicion. In fact, there is a fairly full theoretical analysis of the bifurcations of Duffing's equation in Holmes and Rand (1978). More generally, Lorenz has shown how certain algorithms (such as Euler's) for numerical solutions of differential equations can lead to chaos if the step size is too large; yet a smaller size may take too long for a computation that must be done in real time.

We have given no formal definition of 'chaos'. Some writers suggest that a characteristic property of chaos is what lurks in all our examples (including those in $\mathbb{R}$)—the property of 'sensitivity to initial conditions': long-term output is not always a continuous function of input. More precisely, there is some number $a > 0$ such that for all $b > 0$, if the initial points x, y are within b of each other, then distance(x_n, y_n) is eventually $>a$. Also undoubted 'chaos' is present when the dynamics includes an important object called Smale's horseshoe, or even 'homoclinic' orbits; these can be studied in the literature.

10.8 FIXED POINTS: LOCAL THEORY

We cannot begin to give here the supporting theory of iterated mappings, on which several books have been written. We simply hope to make readers aware of what is possible, and content ourselves with a proof of the next theorem, which is a two-dimensional version of theorem 10.4.1 above. It is needed for a first decision on the nature of the fixed points of a mapping F: $\mathbb{R}^2 \to \mathbb{R}^2$, like the Maynard Smith equation above. So, let us suppose we have found a fixed point. Then for simplicity we measure coordinates from there and take it to be the origin $\mathbf{0}$. We give a criterion for $\mathbf{0}$ to be an attractor or repellor, and it resembles that for linear mappings in Chapter 2. As we might expect, to see how F behaves near $\mathbf{0}$, we look first at the behaviour of its linearization, which is given by its Jacobian matrix $A = J_0F$. The simplest situation arises when $\mathbf{0}$ is an attractor or repellor for A; recall that this occurs according as the eigenvalues lie inside or outside the unit circle in the complex plane. If neither case occurs, we saw that the behaviour was more complicated to describe, so it will be even more so for a non-linear F.

But, whereas in the linear case, if $\mathbf{0}$ is a repellor, then orbits flow steadily away from $\mathbf{0}$ to infinity, a non-linear F may make them flow away and then force them back to be repelled again, and so on. Therefore we make a more precise specification.

Definition 10.8.1

$\mathbf{0}$ is a **repellor** of F if there is a neighbourhood U of $\mathbf{0}$, such that for any non-zero z in U, the orbit leaves U. By contrast, $\mathbf{0}$ is an **attractor** for F, if for any U (however small) there exists some V such that every orbit stays in U, if it starts in V.

Note that when $\mathbf{0}$ is a repellor, the definition allows that if $z_m = F^m(z)$ goes out of U, perhaps the orbit hits U again in z_{n+m}; but then its orbit eventually leaves U, so z_{n+m+k} lies outside U, and so on.

Theorem 10.8.1

Suppose that $\mathbf{0}$ is a fixed point of the smooth mapping $F: \mathbb{R}^2 \to \mathbb{R}^2$ at which its linearization has matrix A. If the eigenvalues of A lie inside the unit circle, then $\mathbf{0}$ is an attractor. If they both lie outside, then $\mathbf{0}$ is a repellor.

Before studying the proof, readers may prefer to try exercises 10.8.1.

Proof

Let λ, μ be the larger and smaller of the moduli of the eigenvalues of A. From exercise F.2.2(4), there are positive functions p, q such that for all vectors $\mathbf{z}$, and $n = 1, 2, \ldots,$

$$p(\mu, n)|\mathbf{z}| \leq |\mathrm{A}^n\mathbf{z}| \leq q(\lambda, n)|\mathbf{z}|; \tag{10.8.1}$$

and as n tends to infinity,

$$p(\mu, n) \to \infty \quad \text{if } \mu > 1, \qquad q(\lambda, n) \to 0 \quad \text{if } \lambda < 1. \tag{10.8.2}$$

These two cases occur when the eigenvalues both lie outside, or both inside, the unit circle, respectively.

By the chain rule, a calculation like that of the multiplier in (10.5.2) shows that the Jacobian matrix of F^n at $\mathbf{0}$ is A^n, because $\mathbf{0}$ is fixed. Hence the Taylor expansion at $\mathbf{0}$ gives, when $|\mathbf{z}| < r$ (say),

$$F^n(\mathbf{z}) = (A^n + K_n(\mathbf{z}))\cdot\mathbf{z} \tag{10.8.3}$$

where $|K_n(\mathbf{z})| \leq M_n|\mathbf{z}|$ for some constant M_n, when $\mathbf{z} \in V$. Suppose first that $\lambda < 1$, so we wish to show that $\mathbf{0}$ is an attractor. By (10.8.2), we can choose some n with $q = q(\lambda, n) < 1$, and then we can find some $h > 0$ with $M_n h < 1 - q$ and $h < r$. Let D be a disc, of centre $\mathbf{0}$ and radius $d \leq h$, and let $\mathbf{z} \in D$. Then $|\mathbf{z}| < h < r$ so (10.8.3) applies and gives

$$|F^n(\mathbf{z})| < |A^n\mathbf{z}| + M_n|\mathbf{z}|^2 < (q + M_n h)|\mathbf{z}| = k|\mathbf{z}| \tag{10.8.4}$$

where $k = q + M_n h < 1$. Therefore $\mathbf{z}_n = F^n(\mathbf{z})$ has norm $< h$, and so lies in D. Hence we can replace $\mathbf{z}$ by $\mathbf{z}_n$ in (10.8.4) to see that $\mathbf{z}_{2n} = F^n(\mathbf{z}_n)$ lies in D and by repeating this argument we have:

$$|F^{mn}(\mathbf{z})| < k^m|\mathbf{z}|, \qquad m = 1, 2, \ldots, \tag{10.8.5}$$

so the entire F^n orbit of $\mathbf{z}$ lies in D. But to keep control over the orbit of $\mathbf{z}$ itself, we need to have the pair of neighbourhoods U, V featuring in definition 10.8.1. So, let the neighbourhood U of $\mathbf{0}$ be given. Then we can take d so small that D lies in U; and by continuity there exists in D a neighbourhood V of $\mathbf{0}$ such that if $\mathbf{z} \in V$, then $F^j(\mathbf{z}) \in D$, for each $j < n$ (see Fig. 10.8.1). We assert that V is the neighbourhood required by the definition of attractor. For, if $\mathbf{z} \in V$, then $\mathbf{z} \in D$ and by the above the iterates $\mathbf{z}_n, \mathbf{z}_{2n}, \ldots,$ all lie in D and hence in U. But if $j < n$, then $\mathbf{z}_j$ lies in D, so again its iterates $\mathbf{z}_j, \mathbf{z}_{2j}, \ldots$ all lie in D and hence in U. Therefore the entire orbit of $\mathbf{z}$ lies in U, as required. Moreover, by (10.8.5) the norm of the elements of the orbit converges to zero, and so the orbit itself converges to $\mathbf{0}$.

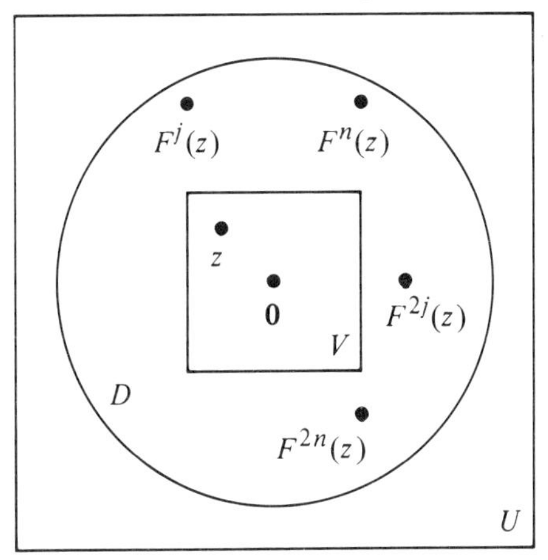

Fig. 10.8.1.

Next suppose that $\mu > 1$. By (10.8.2) there is an n such that $p = p(\mu, n) > 1$ and then by (10.8.3), we can find $h > 0$ such that $M_n h < p - 1$. We assert that the disc, centre $\mathbf{0}$, radius h, will do for the neighbourhood U that is needed to show by definition 10.8.1 that $\mathbf{0}$ is a repellor. For, if $\mathbf{z} \in D$, then from (10.8.3) we obtain, analogously to (10.8.4), that since $|\mathbf{z}| < h$, then

$$|F^n(\mathbf{z})| > |A^n\mathbf{z}| - M_n h > |\mathbf{z}| > h.$$

This shows that the orbit of $\mathbf{z}$ goes outside U, as required, and the theorem is proved.

Using more advanced mathematics, it can be proved that if the eigenvalues of A do not lie on the unit circle, then there is a diffeomorphism between neighbourhoods U, V of $\mathbf{0}$, which carries the dynamics of F onto the dynamics of its linearization. Thus, for example, if the linearization exhibits spirals, so does F. (The proof is intimately related to that for linearizing differential equations: see Hartmann (1964) p. 245.)

Exercises 10.8.1

1. In (10.3.2) let $c = -1$. Show that the Maynard Smith mapping F then has exactly two fixed points, at $\mathbf{0}$ and $P = (a + b - 1, a + b - 1)$. At each of these, calculate the Jacobian matrix A and its eigenvalues, and decide when each is an attractor or repellor. In the control plane $\mathbb{R}^2_{ab}$, sketch the sets of points (a, b) for which the eigenvalues at P are real. Show that the orbits of the linearization spiral in to P provided that (a, b) lies below the line $2a + b = 3$. Use computer graphics to decide whether F also exhibits spirals. [SU]
2. Given the smooth mapping g: $\mathbb{R} \to \mathbb{R}$, and the real number $k \neq 0$, construct a mapping G: $\mathbb{R}^2 \to \mathbb{R}^2$ by the rule: $G(x, y) = (g(x) + y, kx)$. Show:
   (a) G is invertible, and calculate its inverse;
   (b) the Jacobian matrix JG of G has constant determinant;
   (c) each fixed point of G must be of the form (x, kx), where x satisfies $g(x) = x(1 - k)$;
   (d) G corresponds to the difference equation $x_{n+2} = kx_{n+1} + g(x_n)$.
3. Use the computer programs of this chapter to investigate the dynamics of G by taking specific functions for g, such as x^2, $\sin(x)$, and so on. When $g(x) = 1 - bx^2$, G was studied (in a very readable paper) by the physicist Hénon (1976), who found a 'strange attractor' S in the form of a Cantor set lying on a folded ellipse when k, b are near 0.3 and 1.4. When S is magnified in scale, its structure replicates like a fractal. Check that when $b = 1.3$ and $a = 0.29$, S changes to a 7-cycle in a tiny window around $a = 0.299$, then goes fuzzy near $a = 0.3$ and has returned to its former shape S when $a = 0.301$. For high-quality graphics, see Allgood and Yorke (1989).
4. Fix the complex number $c = a + jb$, and let G_c be the mapping that assigns $z^2 + c$ to each complex number z. Show that at z, the Jacobian matrix of G_c has rows (x, y) and $(-x, y)$. Find the fixed points of G_c and decide for which c each is an attractor or repellor.
5. Do the same exercise for the Newton mapping in Example 10.4.2.
6. Let P, F, G: $\mathbb{R}^2 \to \mathbb{R}^2$ be three smooth mappings, with P invertible. We say that F is **conjugate** to G (written $F \approx G$) if there is a diffeomorphism P: $\mathbb{R}^2 \to \mathbb{R}^2$ such that $F = P^{-1}GP$. Show that then $G \approx F$; and if $G \approx H$, then $F \approx H$. Let $w = P(z)$ in $\mathbb{R}^2$. Show that P maps the fixed points of F to those of G, and the F-orbit of z to the G-orbit of w. (Prove $P(F^n(z)) = G^n(W)$.) Thus all the dynamical features of F can be obtained from those of the possibly simpler function G. Does it follow that G is contracting if F is? (Consider linear functions.)
7. The idea of conjugate mappings makes sense in $\mathbb{R}$ (or any metric space, provided we drop the idea of differentiability). Show that in Exercise 10.2.1(1), $f_k \approx g_m$, when $m = k(k - 2)/4$. (Take $P(x) = x/k + 1/2$.)

   Show that in exercise 2 above, if $b \neq 0 \neq k$, then the Hénon mapping G is conjugate to a mapping H that resembles the Maynard Smith mapping, where $H(x, y) = (y, kx + b - x^2)$. (Take $P(x, y) = (y/b, kx/b)$, and verify that $P^{-1}HP(x, y) = G(x, y)$.) Note however, that H is a diffeomorphism but the Maynard Smith mapping is not, because of its fold.
8. Show that in exercise 4 above, G_c is conjugate to the quadratic mapping Q: $z \to z^2 + pz + q$, where $c = q + p/2 + p^2/4$. (Take P to be $z \to z - p/2$.)

10.9 THE MANDELBROT SET

After the linear mappings, the next most simple are those on $\mathbb{R}^2$ of the quadratic type: $z \to z^2 + pz + q$, when we regard $\mathbb{R}^2$ as the plane of complex numbers. By the last exercise, this mapping is conjugate to G_c, where $G_c(z) = z^2 + c$, and c is constant.

We have already seen how complicated the dynamics of G_c can be when z is real, but in $\mathbb{R}^2$ the complexity can be displayed by computer graphics that reveal fantastic beauty. We have earlier seen something of this in the Julia sets of section 10.3, but their computer graphics stem from work of B. Mandelbrot in 1980 from a slightly different point of view. His work was improved by Peitgen and Richter with high-quality computer graphics; once readers have seen their book (Peitgen and Richter 1986), they will find other reproductions very crude. Various other mappings have been similarly examined, and all reveal the presence of a characteristic shape, shown rather crudely in Fig. 10.9.1. That shape is now called the 'Mandelbrot set', and we explain here some of its features.

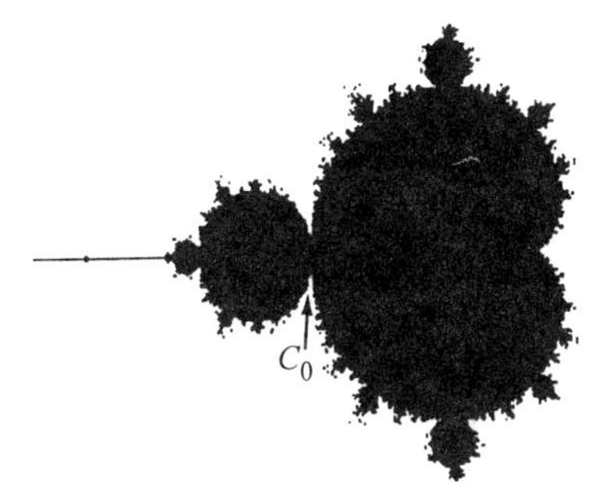

Fig. 10.9.1. The Mandelbrot set.

Whereas there is a Julia set for each G_c, which tests all orbits $G_c(z)$ as z varies, the Mandelbrot set involves testing only the orbit of $\mathbf{0}$, but letting c vary. Specifically M consists of all c such that the G_c-orbit of $\mathbf{0}$ does not diverge to infinity. This orbit consists of the sequence

$$c, c^2 + c, (c^2 + c)^2 + c, \ldots. \tag{10.9.1}$$

Let M' denote the outside of M, so M' can be regarded as the basin of attraction of ∞ in $\mathbb{R}^2$. The beautiful graphics arise when M is painted black (say) and different colours are assigned to points of M', by the rule that if n iterations are necessary to carry z outside some fixed region containing $\mathbf{0}$, then c receives a colour depending on the value of n. (As we shall see below, if $|c| > 2$, then (10.9.1) diverges to infinity; so we know how large to take the region.) If, instead, c lies in the interior of M, then its sequence (10.9.1) may be periodic, or it may converge to a limit, or behave in some other way. But in the periodic case, a colour can also be assigned, depending on the period; and again we observe great beauty and complexity. We offer the Mandelbrot program 10.9.1 below for readers to experiment with; but much depends on the quality of your computer's screen resolution. For example, you may see small blobs apparently disconnected from the parent body, which turn out with higher resolution to be connected to it by thin 'fronds': M can be proved to be in one piece. Also, under magnification, the boundary B between M and M' is seen to have smaller and smaller protuberances, each resembling the parent shape of Fig. 10.9.1. (B is another example of a 'fractal' set, and is often said to be 'self-similar', but this term can mislead: B is not as self-similar as the fractals we shall construct later. After much study, there is a great deal about B that the pictures suggest, but which nobody can explain mathematically (to date).)

```
REM Program 10.9.1, Mandelbrot set
FUNCTION Re(x,y) = x^2 - y*y + cx
FUNCTION Im(x,y) = 2*x*y + cy
READ xmin, xmax, ymin, ymax, sw , sh , nmax, rmax
DATA -2.2, 0.5 , -0.9, 0.9 , 320, 200, 40  , 4
SET MODE "graphics"
SET WINDOW xmin,xmax, ymin,ymax
LET xstep = (xmax-xmin)/sw
LET ystep = (ymax-ymin)/sh
FOR cx = xmin TO xmax STEP xstep
  FOR cy = ymin TO ymax STEP ystep
    LET x = 0
    LET y = 0
    LET n = 0
    LET gone = 0
    DO
      LET xr = Re(x,y)
      LET y  = Im(x,y)
      LET x  = xr
      LET n = n + 1
      IF x*x + y*y > rmax THEN LET gone = -1
    LOOP UNTIL n>nmax OR gone = -1
    IF gone = 0 THEN PLOT cx,cy
  NEXT cy
NEXT cx
END
```

Just from the above information, we might expect repetitions to occur on the boundary B as follows. The G_c-orbit $\langle c_n \rangle$ in (10.9.1) suggests that we define the polynomials $g_n(z) = G_z^n(0)$, and then $c_n = g_n(c)$. So, given a neighbourhood U of c_n, g_n maps some neighbourhood V of c into U. Moreover, unless $g_n'(c)$ is zero, then g_n is a local diffeomorphism at c, so we may assume that g_n maps V diffeomorphically onto some neighbourhood W of c_n. Now B separates U into points of M and M'; so its g_n-image in W separates W similarly. Hence a sharp 'crinkle' of B in U is copied by g_n as a similar item at c_n on B in W; and this is repeated along the orbit for each n, an infinity of times. However, if we magnify U we may find (as at c_0 in Fig. 10.9.1) that the complexities of B near c_0 are approximately repeated within smaller and smaller neighbourhoods of c_0. This is not the exact 'similarity' of the previous repetitions.

To study the process further, we list several provable facts, for each fixed c in $\mathbb{R}^2$.

(a) G_c maps the disc D_k: $\{|z| \le k\}$ into itself, provided $|c| < k(1-k)$; and G_c is also contracting if $k < 1/2$.
(b) If $c \ne 1/4$, $G_c(x)$ always has two fixed points, one of which does not lie inside $D_{1/2}$.
(c) If n is prime, then $G_c(x)$ has an orbit of period n.
(d) If $|c| > 2$ then (10.9.1) diverges to infinity.

(e) The large cardioid shape in M consists of points c for which G_c has an attracting fixed point; it then lies in $D_{1/2}$.
(f) The circular shape within M and to the left contains the disc L of points c for which G_c^2 has an attractive fixed point. L has centre at $c = -1$, and radius $1/4$.

We shall only sketch proofs, and leave details as an exercise for the reader.

Proof of (a)
The fixed points of G_c are the roots u, v of the equation $z^2 - z + c = 0$, so $u + v = -1$. Hence $1 \leq |u| + |v|$, so $|u|$ and $|v|$ cannot both be $<1/2$. (The condition for equal roots is that $c = 1/2$.)

Proof of (b)
If $z \in D_k$, then $|G_c(z)| \leq |z^2| + |c| \leq k^2 + |c|$, whence $G_c(z)$ lies in D_k provided $k^2 + |c| < k$, as required. If also $w \in D_k$, then $|G_c(z) - G_c(w)| = |z - w| \cdot |z + w| \leq 2k \cdot |z - w|$, so G_c is contracting provided $2k < 1$.

Proof of (c)
If G_c has an orbit $z_1, z_2, \ldots, z_n$ of period m, then z_1 is a fixed point of G_c^m, and hence a root of the equation $G_c^m(z) = z$. This is a polynomial equation of degree 2^m, and always has 2^m complex roots. Two of these are fixed points of G_c, and some of the rest may be fixed points of G_c^k, when k is a factor of n. If n is prime, then $k = 1$ or n.

Proof of (d)
For any z, $|z^2 + c|$ is the distance d from z^2 to $-c$, and by the cosine formula for the triangle with vertices $\mathbf{0}$, z^2, $-c$ we have $d^2 = |z|^4 + |c|^2 - 2|c| \cdot |z|^2 \cos\theta$ which is least when $\cos\theta$ is greatest ($= 1$). Thus $d \geq |z|^2 - |c|$, so if $|z|$ and $|c|$ each exceed 2, then $d > 2$. Hence if z and c lie outside D_2, so does $G_c(z)$; and then the whole orbit $\langle c_n \rangle$ in (10.9.1) lies outside. We now leave the reader to show that if $|c| > 2$, then $|c_n|$ increases to infinity.

Proof of (e)
Suppose the root u in (a) is an attractor. Then $|G_c'(u)| \leq 1$ so $|u| \leq 1/2$. Hence $u = sa$, where $0 < s < 1$ and a runs on the circle $D_{1/2}$ of radius $1/2$. Now $c = u - u^2$, since u is fixed by G_c, so c runs on the cardioid K_s: $w = sa - (sa)^2$ as a runs round $D_{1/2}$. But K_s lies within K_1 (see exercise 5.9.1(3)), and (e) is proved.

Proof of (f)
We use the same type of argument as for (e), but replacing G_c by G_c^2 ($= F$). Then a fixed point z of F satisfies the quartic $(z^2 + c)^2 + c - z = 0$. But since two roots are already fixed points of G_c, a factor is $z^2 - z + c$; so the remaining two roots p, q are the roots of $z^2 + z + c + 1 = 0$. Now p is attracting if $|F'(p)| < 1$, and we find $F'(p) = 4p(p^2 + c) = 4pq$ (since $G_c(p) = q$). From the last quadratic, therefore, we require $1 > |4pq| = 4|c + 1|$ so c lies within the circle of centre -1 and radius $1/4$, as described in (f) above.

A conclusion from (a) is that M contains the entire disc of all c with $|c| \leq 1/4$; for then, by the contracting mapping theorem in Chapter F, the orbit (10.9.1) converges. Also, note that the disc L in (f) touches the cardioid in (e) at $c = -3/4$, but this point has a whole neighbourhood in M. A rather difficult theorem tells us that every point c within the cardioid lies in M, because $\mathbf{0}$ is the only critical point of G_c, and this has to lie in the basin of attraction of the fixed point; so its orbit (which includes c) lies in M, by definition of M.

The arguments for (e) and (f) indicate how the other protuberances on M relate similarly to fixed points of G_c^n (and also why the mathematics becomes harder). We have not proved, however, that points within these regions have orbits that do not diverge to infinity. This is a complicated task, because of the several types of possible behaviour.

Mandelbrot's set arose from asking about the behaviour of the orbit (10.9.1) of $\mathbf{0}$ when c varies. Conversely we have the Julia sets: keeping c fixed, we look at the G_c-orbit of z, when z varies. Thus, let M_c be the set ot all z such that the G_c-orbit of z does not diverge to ∞, and let M_c be the basin of attraction of ∞.

Then M_c, M_c' are the analogues (in the phase space) of M, M' above, which lie in the control space. The Julia set J_c is the common boundary of M_c and M_c'. Readers may now like to revisit the Julia program 10.3.2 above, and compare sets M_c with M itself.

Exercise 10.9.1

The basic Mandelbrot program 10.9.1 can be adapted to produce colour in at least two different ways. One way colours the exterior of the set in a colour or tint which depends upon the number of iterations required for the iteration to escape from a given region. The following lines are the additional ones needed, and the `IF` `END IF` structure replaces the `IF gone = 0 THEN PLOT cx, cy`.

```
REM Program 10.9.2, Mandelbrot set, escape colour

FUNCTION shade(n) = 1 + INT(n/5)
   IF gone = -1 THEN
       SET COLOR shade(n)
       PLOT cx, cy
   END IF
```

Try different shading functions and zoom in on interesting portions of the set M. The other adaptation colours the interior differently for different periods of attraction. Again the `IF` `END IF` structure replaces the `IF` line.

```
REM Program 10.9.3, Mandelbrot set, interior colour
LET eps = 0.0001
    IF gone = 0 THEN
        LET xs = x
        LET ys = y
        LET n = 0
        DO
            LET xr = Re(x,y)
            LET y  = Im(x,y)
            LET x  = xr
            LET n = n + 1
            IF x*x + y*y > rmax THEN LET n = nmax + 1
```

Program 10.9.2 continued

```
            LOOP UNTIL (x-xs)^2 + (y-ys)^2 < eps OR n > nmax
            SET COLOR n
            IF n < nmax THEN PLOT cx,cy
      END IF
```

10.10 SELF-SIMILAR FRACTALS

We have used the terms 'fractal' and 'self-similar' in an informal way, partly because there is, as yet, no accepted single meaning for them. One suggested definition of fractal uses the idea of **Hausdorff measure**, and there is an attractively intuitive one due to Keith Wicks (1990a). We shall confine ourselves here to explaining an algorithm for constructing sets X which are obviously self-similar, in the sense that items of structure in one part of X are repeated on a smaller scale elsewhere in X. This technique is due to Barnsley and Demko (1985), and Barnsley's article in Devaney and Reene (1989) mentions its use for economical data compression in computer science. We shall follow the lines of a 'popular' article (Wicks 1990b) and give a version of his fractal program 10.10.1 for generating a class of fractals.

```
REM Program 10.10.1, Fractal
READ theta, r1   , r2   , prob1, prob2, x, y, nmax
DATA 18   , 0.98, 0.15, 30   , 1    , 1, 0, 100000
LET theta = PI*theta/180
READ xmin , xmax, ymin , ymax, sw , sh
DATA -1.53, 1.53, -1.02, 1.02, 320, 200
FUNCTION f1(x,y) = r1*COS(theta)*x - r1*SIN(theta)*y
FUNCTION g1(x,y) = r1*SIN(theta)*x + r1*COS(theta)*y
FUNCTION f2(x,y) = r2*x + 1 - r2
FUNCTION g2(x,y) = r2*y
SET MODE "graphics"
SET WINDOW xmin,xmax, ymin,ymax
LET p1 = prob1/(prob1 + prob2)
PLOT x,y
FOR n = 1 TO nmax
  IF RND < p1 THEN
     LET newx = f1(x,y)
     LET y    = g1(x,y)
  ELSE
     LET newx = f2(x,y)
     LET y    = g2(x,y)
  END IF
  LET x = newx
  PLOT x,y
NEXT n
END
```

The algorithm depends on Hutchinson's theorem (see proposition F.5.1). There we work in $\mathbb{R}$ with two contractions S, T to form a third contraction R, which acts on the space Y of all closed subsets of $\mathbb{R}$. The R-orbit of $\mathbf{0}$ consisted of a sequence $\langle A_n \rangle$ of finite sets that converge to a limit set L (the Cantor set, in that case). Now we work analogously in $\mathbb{R}^2$ with specially simple contracting mappings of the form

$$T(\mathbf{z}) = rA\mathbf{z} + \mathbf{c}, \qquad 0 < r < 1, \tag{10.10.1}$$

where **c** is a constant vector, and A is a rotation matrix. (We call T a 'spiral contraction'; it is invertible.) If we choose two or more of these, we can combine them as before into one contraction C which acts on the space Z of closed subsets of $\mathbb{R}^2$. Then the C-orbit $\langle \mathbf{0}_n \rangle$ of $\mathbf{0}$ will converge to a unique limit set K in Z, which we naturally would like to display by computer graphics. To do this economically needs the following considerations.

Each set $\mathbf{0}_n$ can be represented by computer graphics, as a set of dots on the screen, but $\mathbf{0}_n$ soon becomes too large to compute. The idea of Barnsley and Demko is to use probabilities, to compute a subset $\mathbf{0}'_n$ of only some points of each $\mathbf{0}_n$: and they prove that if $\mathbf{0}$ lies in K then $\langle \mathbf{0}'_n \rangle$ converges to K with probability 1. This proof is given in Wicks (1990b). Our eyes can then extrapolate the shape of the limit K after a relatively small number of computations. More precisely, if we are unable to distinguish visually between elements of Z that lie (in the Hausdorff metric of Z) within d of each other, then by the triangle inequality we cannot distinguish $\mathbf{0}'_n$ from K once $\text{dist}(\mathbf{0}_n, \mathbf{0}'_n)$ and $\text{dist}(\mathbf{0}_n, K)$ are each $< d/2$. To compute $\mathbf{0}'_n$, probabilities p, q are associated with S, T, respectively, and the algorithm chooses (say) S with probability p and plots the point $x_1 = S(0)$. Then it again chooses one of S, T with new probabilities p, q (suppose this is T) and plots the point x_2. Continuing in this way, we compute a sequence $\langle x_n \rangle$, and $\mathbf{0}'_n = \{x_1, x_2, \ldots, x_n\}$, relatively small by comparison with $\mathbf{0}_n$, which could have 2^n members.

In the examples in Wicks (1990b), S is taken to be of the form (10.9.1) with $r = r_1$, $\mathbf{c} = \mathbf{0}$, and A a rotation through an angle θ about $\mathbf{0}$. This ensures that $\mathbf{0}$ lies in K, as required by the algorithm. Then T is taken to be of the form (10.9.1) with the same A but $r = r_2$, and $\mathbf{c} = (1 - r_2, 0)$; thus T contracts about $(1, 0)$ by the factor r_2. Each resulting K is now determined by the three parameters θ, r_1, r_2. If probabilities p, q are associated with S, T, they are chosen to spread the points of the sequence $\langle x_n \rangle$ evenly over the picture; thus the more S contracts the less we choose p, whence we take $p/q = r_1/r_2$.

The program can produce beautiful fern-like patterns which are self-similar in the sense of 'proposition F.5.1 below'. By taking S, T to be of other forms in (10.9.1), upstanding ferns can be produced. These have led to speculation that a similar set of instructions is used in Nature to prescribe types of growth. Clearly, we have here a very intriguing tool for future research.

Fig. 10.10.1. Fractal output from program 10.10.1.

11

Catastrophe sets in modelling

11.1 INTRODUCTION

The subject of catastrophe theory grew up in the 1970s, as a result of work by the great French mathematician René Thom, who had been stimulated to modernize the classical theory of envelopes and to extend the ideas of D'Arcy Thompson's biology text *Growth and form* (Thompson 1961). Thom's book *Structural stability and morphogenesis* (Thom 1975) attracted much attention among scientists of various disciplines as well as pure mathematics, and it had an unusually liberating effect on many, because it allowed speculative and imaginative thinking of unorthodox kinds. The English mathematician E. C. Zeeman introduced several striking interpretations of the theory into the social sciences, which was followed by a spate of popular articles in the Press and elsewhere. These 'catastrophe models' are open to obvious objections, especially when compared with the well-tried models of physics, but they could (and sometimes did) open the eyes of social scientists to the possibility of underlying processes that they had never previously dreamt of. Inevitably there was an eventual reaction that was hostile (for mixed motives) and the subject has fallen out of the limelight. Quietly however, the mathematical theorems and techniques to which it gave rise have been accepted and absorbed, especially in the physical sciences. Indeed, some of the 'catastrophe surfaces' were known to engineers, working on the stability of structures, before they had heard of the catastrophe theory that provided a sounder footing for their work.

In the present chapter, we can only hope to give the slightest introduction, to whet the appetite of readers so that they can then go on to read a formal presentation like that in Poston and Stewart (1978), which establishes the mathematical language needed for proper work and then displays many uses of catastrophe theory in the analysis of such matters as the stability of ships and engineering structures, flows of fluid, refraction of light in droplets, lasers, and certain aspects of social science. Lacking that language, most popular texts are limited and can be misleading; but readers of this book are now in an intermediate position, and we can use some of the techniques they have already developed through using this book, in order to give an idea of the subject's flavour. We begin by giving a detailed analysis of a diagram which is constantly found in popular texts but there in only a partial form. Once this is before us, we can relate it to applications for readers who may not know the physical background.

11.2 THE CUSP CATASTROPHE

Suppose the Nordic god Wotan, known as 'the wanderer', moves along a straight road, subject to only one rule. He is supplied with a graph of a smooth function $f: \mathbb{R} \to \mathbb{R}$, and regards the road as the x-axis. Then Wotan must, at each point x, move forward or backwards according as f is positive or negative at x. If $f(x) = 0$, he must stop and think. The rule can be incorporated into the graph of f as shown in Fig. 11.2.1, by means of the arrows shown on the x-axis; it is usual to refer to these as the 'dynamics' of f.

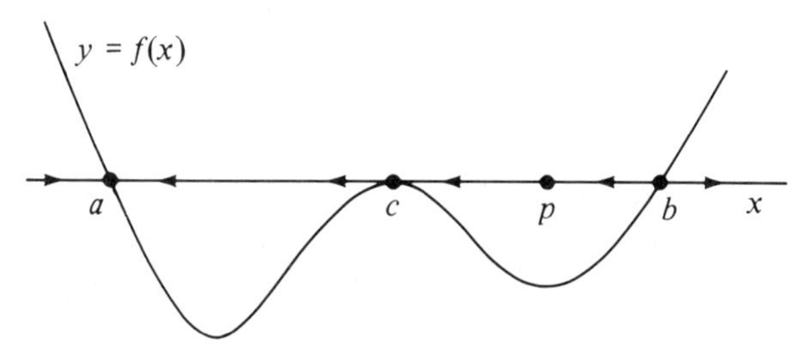

Fig. 11.2.1.

Suppose now that Wotan walks from his home in Valhalla. If this is at the distant left then he walks right (according to the rule) until he meets the point a at which f has a simple zero (and negative gradient). If now he tries to move, either forwards or backwards, then the arrows at a force him to move back there, and he is stuck. If, however, he had started from home at the point b, then the arrows will carry him away if he makes the slightest move. If he moves to the right, he will carry on away from Valhalla; if to the left, he moves through p, and arrives at c. There he can only move to the left, until he arrives at a, where the rule causes him to stick, as before.

At first sight, similar progress would be observed if the road were a wire in a vertical plane shaped like Fig. 11.2.2, on which a heavy bead slides; at b the slightest movement left or right will cause the bead to leave b, and a move leftwards will carry it through p, then through c and on to a. However, being massive, it will carry on

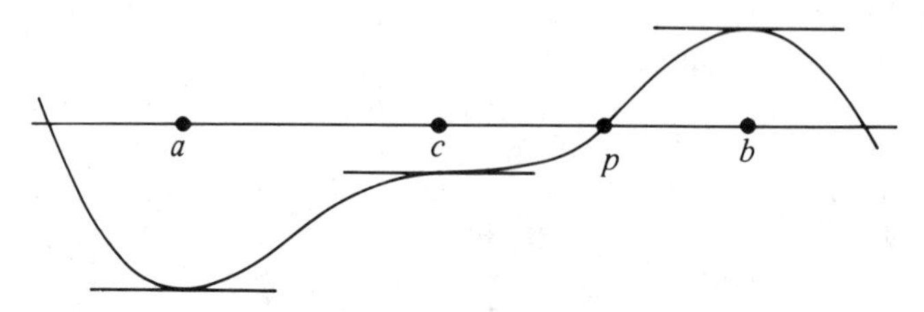

Fig. 11.2.2.

through a, but must slow down because of the braking effect of its governing arrows. It eventually stops and is attracted back to a, goes beyond and returns, to make a series of (decreasing) oscillations. For the moment we are not interested in that kind of dynamics, but only in Wotan's stately 'quasi-static' progress to rest at a. Notice that the points a, b, c are the zeros of f, and for obvious reasons we call a an *attractor* and b a *repellor*. Since f does not change sign as x increases through c, then $f'(c) = 0$, and c is a critical point of f. These zeros are the *equilibrium points* of the dynamics. Similar things would happen if Wotan were governed by an f with other zeros; he will move away from any repellor either to infinity, or to an attractor at which he is caught.

Exercise 11.2.1

Draw diagrams analogous to Fig. 11.2.1 when $f(x)$ takes the forms

(i)	$-3x + 1,$	(ii)	$x^2,$	(iii)	$x^2 - 3x + 2,$
(iv)	$x^2 - 3x + 4,$	(v)	$x^3,$	(vi)	$x^3 - x^2 + x - 1,$
(vii)	$x^3 - 3x^2 + 2x.$				

Now suppose that f arises, as in many physical systems, from the derivation of a 'potential' function V: thus $f = -\mathrm{d}V/\mathrm{d}x$ (the minus sign is included for technical convenience). For example, with Fig. 11.2.1 above, V would have the general form shown in Fig. 11.2.2, with a minimum at a, a maximum at b, and an inflection at c; and we could add any constant to V, since it will vanish from $\mathrm{d}V/\mathrm{d}x$.

Exercises 11.2.2

1. Find V when f is any of the functions in exercise 11.2.1
2. Why is Fig. 11.2.2 qualitatively correct (i.e., of the right general shape even though the slopes may be inaccurate)?
3. Construct similar diagrams when f is replaced by each of the different types of cubic f_{IJ} in Fig. 6.2.3. Check your answers against Fig. 11.5.1 below.

Although Wotan is the chief of the gods, he cannot break a rule, once he is bound by it. So if he is stuck at an attractor, what can he do to gain some freedom? To be specific, suppose the potential function V is a quartic of the form: $V(x) = \frac{1}{4}x^4 + \frac{3}{2}Ix^2 + Jx$. Then $f = -f_{IJ}$, where f_{IJ} is the reduced cubic $F_{IJ}(x) = x^3 + 3Ix + J$ as in (6.2.2). We shall show how Wotan can (by calling for the control point (I, J) in $\mathbb{R}^2_{IJ}$ to be gradually changed) wander back and forth along the road, and eventually get back home to Valhalla. To explain, we assemble our information into the famous form of Fig. 11.2.3, which is based on the familiar surface C of Fig. 6.3.1, but with axes rotated for convenience; C is the set of points $X = (I, J, x)$ in $\mathbb{R}^3_{IJx}$, for which $f_{IJ}(x) = 0$, (so $J = -x(x^2 + 3I)$). Thus, if we now think of Wotan's road as the vertical line through X, we see that he can only be stuck at points of C. In the figure, we have included the associated arrows on X, to indicate the dynamics by which Wotan is governed. Above each point $P = (I_0, J_0)$ of the control plane $\mathbb{R}^2_{IJ}$, there is a vertical line L_P and, like Fig. 11.2.1 (but turned through 90°), it carries the dynamics of $-f_P$, including the equilibria, if any. By the properties of the cusp K (see Fig. 6.3.2) there is one real zero if P lies at A outside the region M enclosed by K, and three if P lies at B within M; but if P lies at Q on K and $Q \neq \mathbf{0}$, then L_Q grazes (i.e., touches) the fold in C at a repeated zero of $-f_Q$. Note that at such a graze point S, the arrows on L_Q point in the same direction, as at c in Fig. 11.2.1. The dynamics on L_A, and the graph of $-f_A$, is shown in Fig. 11.2.3(a); similarly for B, Q in Figs 11.2.3(b),(c).

Now suppose Wotan is at X on the upper portion of C above A in Fig. 11.2.4, with both $I, J < 0$. He is governed by the function $-f_A$, with graph illustrated above. Consider what happens if Wotan moves on C itself. Thus let $R \neq \mathbf{0}$ be a point on K, as shown, and choose a path $p(t)$ in $\mathbb{R}^2_{IJ}$ from A to R with $t \in [0, 1]$, $p(0) = A$, $p(1) = R$, and $p(\frac{1}{2}) = R'$, where the path first crosses K. Now, with his godly powers, let Wotan arrange that through wear and tear, his governing function f changes so that at time t it becomes $f_{p(t)}$. Then, although for each $t < \frac{1}{2}$ he is stuck at the single zero (attractor) $z(t)$ of $f_{p(t)}$, he gradually drifts to the point $Z = z(\frac{1}{2})$ below which $L_{R'}$ grazes C at S. (Thus $z(\frac{1}{2})$ is the one *simple* zero of $f_{R'}$.) Observe that over the region M, C has three sheets; and on the upper one, if $t \geq \frac{1}{2}$, $f_{p(t)}$ has a simple zero (attractor) $z(t)$ and a graph like Fig. 11.2.3(b). The arrows on $L_{p(t)}$ hold Wotan at $z(t)$, but he drifts with time $T = z(1)$, the point at which L_R grazes C. On L_R, $f_{p(t)}$ has a graph like Fig. 11.2.3(c).

Once at T, Wotan is allowed by the arrows on L_R to wander (downwards) on L_R to the point Y at which L_R pierces C again. On L_R, Y is the simple root of f_R, an attractor, and lies on the lower portion of C as shown. Wotan is now stuck at Y, and governed by f_R, but he can then drift down the slope to Valhalla by causing f_R to change: more precisely he chooses a path $u(t)$ in $\mathbb{R}^2_{IJ}$ from R parallel to the J-axis without crossing K again, and $L_{u(t)}$ cuts C in the single attractor $Y(t)$ of $f_{u(t)}$. As $u(t)$ moves to infinity on $\mathbb{R}^2_{IJ}$, so does $Y(t)$ on $L_{u(t)}$, and Wotan is carried along as well. The graph of $f_{u(t)}$ remains similar to that of Fig. 11.2.4(b). He could, of course, have chosen the path $p(t)$ so as to go from X to Y without crossing K at all, but it can be quicker to use the dynamics on L_R than to rely on drift.

But suppose that, at Y, Wotan has second thoughts about going home to his stern wife Fricka, and would like to return to the point X. He can reverse the path $p(t)$ in $\mathbb{R}^2_{IJ}$ and use it to drift on the lowest fold of C to the graze point S on $L_{R'}$. Then, by the rule itself, he can walk (upwards) on $L_{R'}$ from S to its attractor Z on the top fold of C. Should he wish, he can drift from there back to X, along the upper part of C. Here is a case where reversal of the controls produces a return path in the phase space which is different from the outgoing path. Such a circuit is called a **hysteresis loop**.

Note that in none of this movement does Wotan reach the middle fold of C, each point of which is a repellor for the vertical line L on which it lies. Also the movements upwards and dowards *are not movements under gravity*; they are governed by the dynamics in each $L_{p(t)}$, which just happens to be drawn as a vertical line. Also, Wotan is really on his original road, all the time; L_A is a picture of the road when the control is at A.

The possibility of drift is by no means fanciful. In a Ph.D. thesis, Garcia-Cobian (1978), p. 50, has calculated the probability of drifting from a point P outside K to

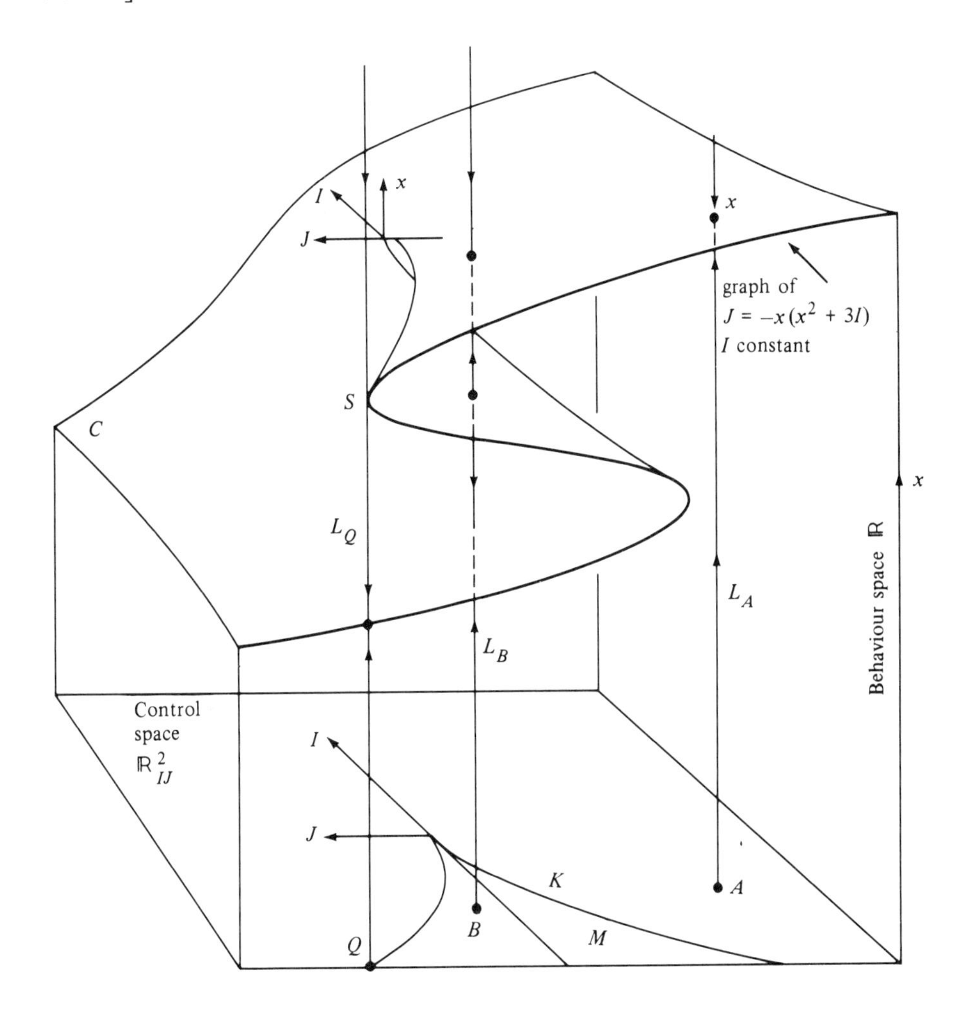

x
I
J
x
graph of
$J = -x(x^2 + 3I)$
I constant
S
C
x
L_Q
Behaviour space $\mathbb{R}$
L_A
L_B
Control
space
$\mathbb{R}^2_{IJ}$
I
J
K
A
B
M
Q

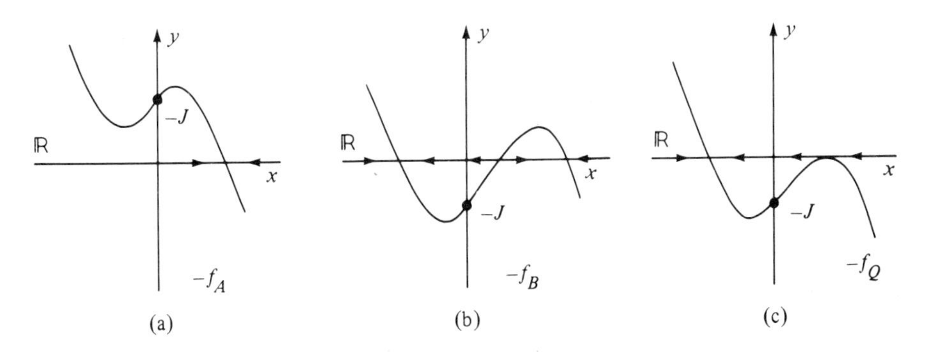

y
$-J$
$\mathbb{R}$
x
$-f_A$
(a)
y
$\mathbb{R}$
x
$-J$
$-f_B$
(b)
y
$\mathbb{R}$
x
$-J$
$-f_Q$
(c)

Fig. 11.2.3.

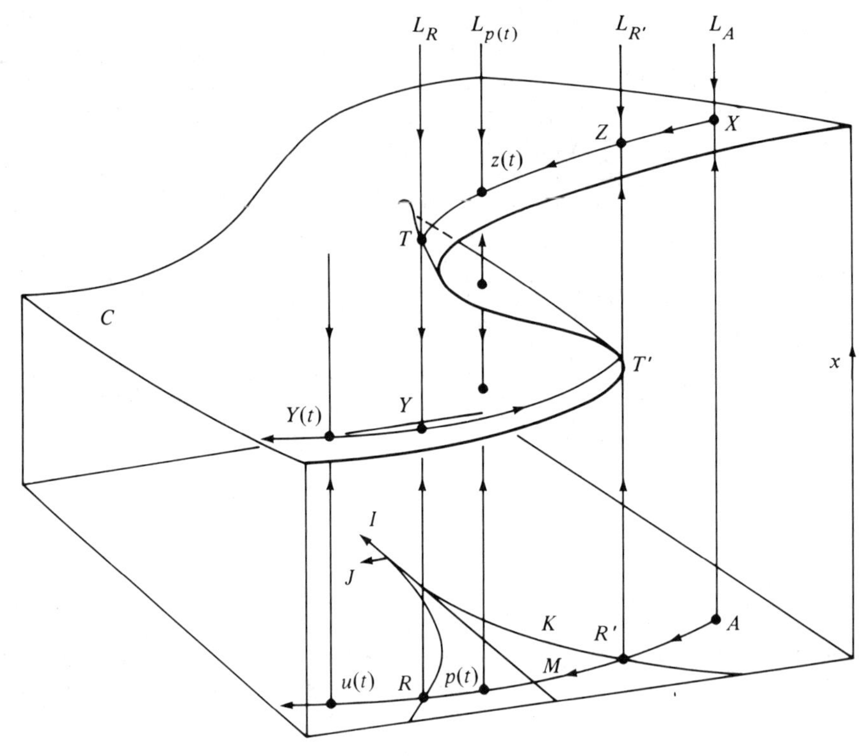

Fig. 11.2.4.

one on K itself, when the drift is by a random walk. Some of the probabilities are surprisingly high, and the expected times of arrival are quite low. (Controllers of nuclear installations please note!)

Exercises 11.2.3

1. Let p, q be two paths in $\mathbb{R}^2_{IJ}$, each starting at $(1, 0)$ and finishing at $(-1, 0)$, but with J always positive on p, and always negative on q, and with $p(t)$ close to $q(t)$ for each t. Show that if Wotan used p, then he ends on the upper fold of C; whereas if he uses q, he ends on the lower fold. This demonstrates the property of **divergence**: nearby paths in the control space may give rise to very different end results in the phase space.
2. Construct a diagram analogous to Fig. 11.2.3, when the cubics f_{IJ} are replaced by the quadratics f_{bc} in (6.1.3). Show that there is no possibility of a hysteresis loop. What about divergence?

We now discuss potentials in some physical systems.

11.3 POTENTIALS

Physical systems can often be described in terms of a scalar function ϕ, usually called a 'potential', and it is now time to look at some of these. In such a system S, ϕ will depend on various numerical variables $x_1, \ldots, x_n$ which together define a point x in $\mathbb{R}^n$; and x might perhaps specify one of the many geometrical positions that S could conceivably occupy. The actual position ($x = p$) that it does occupy can be determined if we know that S is subject to the 'minimum-seeking' rule that ϕ must have a local minimum at p; and then S is said to be in 'stable equilibrium' at p. This was the case with Wotan, in Fig. 11.2.2 above, where $n = 1$; Wotan can occupy positions x on R and the potential is the antiderivative of $-f$ there. The minus sign ensures that an attractor is associated with a minimum.

Minimum-seeking systems S first arose in mechanics, where the sum of the forces acting on S is given by $-\text{grad}\ \phi$. Then by Newton's law, S will be at relative rest if all these forces balance out, which happens at a zero (p) of grad ϕ. Now, if S is any minimum-seeking system with p a local minimum of ϕ, and if S is perturbed to a nearby position x, then $\phi(x) > \phi(p)$, so by the minimum-seeking rule, S will move back to the position p. We shall say that S is **governed** by $-\text{grad}\ \phi$. If $n = 1$, then $-\text{grad}\ \phi$ is just $-\phi'(x)$, and we have dynamics as in the Wotan example, with p an attractor. To fix ideas, let us first consider the following example.

Example 11.3.1

Suppose that a straight spring of natural length l hangs vertically from a fixed point **0**, and a mass m is hung from the lower end. This system is then specified by the distance $l + x$ of m below **0**. When the spring is stretched, it is storing energy known to amount to $\frac{1}{2}kx^2$ units, where k is the stiffness of the spring. Thus the associated potential can be taken as $\phi(x) = mgx + \frac{1}{2}kx^2$, and the system is governed by $-mg - kx$, with a unique minimum at $x_0 = -mg/k$. The mass is in stable equilibrium when it hangs at the distance $l + x_0$ below **0**.

Here, x_0 depends on m, k, and g, which can vary from one experiment to another. Thus m, k, g are 'control variables', and as the point $c = (m, k, g)$ varies in $\mathbb{R}^3$, we see that x_0 varies in the 'phase-space' $\mathbb{R}$. (In practice, if m is too large, the spring will start to 'run' by stretching in a non-linear way, and the system is not then necessarily governed by a potential.) If we pull down the mass to x_1 slightly below x_0, and then release it, it moves back to the position x_0 of equilibrium, goes above it, falls back, and so on, in oscillations subject to the energy equation: $\frac{1}{2}mv^2 + \phi(x) = \phi(x_1)$ where the velocity v is $\dot{x} = \mathrm{d}x/\mathrm{d}t$. Differentiating with respect to time t, and cancelling through by v, we obtain: $m\ddot{x} = mg - kx$, so the oscillations are simple harmonic since $-k < 0$. This is why the equilibrium is called stable, since the system never moves far from x_0, and always returns; and this will always happen because of the minimum property, since then $\mathrm{d}^2\phi/\mathrm{d}x^2 > 0$ (it was k in the present case).

In the general case, similarly, such a minimum p will (in general) depend on other variables that specify the make-up of the system S, sizes of parts, their elasticity, temperatures, and so on. These variables will remain constant in any particular observation of S, and together specify a point c in some space $\mathbb{R}^m$, which is called the 'control' space to distinguish it from the 'phase space' in which x lies. If c is varied slightly (as when we tune an instrument) then p can be expected to move a bit within $\mathbb{R}^n$, and the behaviour of the modified system S_c is not vastly different from that of the original system S. But it may happen that in the space $\mathbb{R}^m$ there is a set E such that if c moves across E, then x moves rapidly or jumps suddenly, and S_c behaves in a drastically different way. A mild example of this was when S_{IJ} is Wotan governed by f_{IJ} in section 11.2, and E is the cusp K; he first drifted from P to R on K and then moved to a very different position along the line L_R.

It is essential that after the jump, the general S_c is still governed by ϕ (which would not hold if we were tuning a violin and a string snapped). In common parlance, we often describe the new behaviour as 'catastrophically' different from the old, which is why the associated mathematics has come to be called *catastrophe theory*, and E the catastrophe *set*. But a 'catastrophe' in this sense need not be a disaster!

Warning

Beware of thinking that each system will have a unique equilibrium point, as in the simplest examples. Economists have in the past made this assumption (with dynamics provided by a mysterious 'hand of the market') which, in view of the complexity (and importance) of contemporary economic systems, is quite unwarranted (and even dangerous). A relatively simple example of a system with two possible stable equilibria is the 'catastrophe machine' of Zeeman, described in the next section.

In example 11.3.1, the behaviour summarized by the stable value x_0, does not change qualitatively with small changes (or errors) in the control variables. It is not sufficiently complicated to bring out a feature, on which catastrophe theory lays great stress, that valid mathematical models should be 'structurally stable'. This important concept was introduced in 1937 by the Russian Mathematicians Andronov and Pontrjagin, and it has had useful effects on the philosophy of modelling even though the definition varies with the type of application. We have already seen examples in Chapters 9 and 10. For our purposes here, it will suffice for the moment to define structural stability of a family of minimum-seeking systems S_c as follows. If S_c has potential function ϕ_c, then for any family of functions, f_c, sufficiently near the family ϕ_c, the two behaviours predicted by either family should be essentially the same.

This 'definition' is still not completely precise, but readers are asked to check the lack of precision when we mention structural stability below, so that they may better appreciate the need for the 'proper' mathematical definition. The mathematics of structurally stable families can work, because it avoids genuinely untypical cases that would not occur in the physical world anyway.

After such generalities, it is best to look at examples, and we shall give two: one a traditional type of model from mechanics, and the other a more unusual model from archaeology.

11.4 THE CATASTROPHE MACHINE

In order to have a simple practical example of a catastrophe in mechanics, Zeeman invented his 'catastrophe machine'. We shall give its mathematical analysis briefly, along the lines of Poston and Stewart (1978), p. 75, and refer the reader to that text for details. (They also give instructions for constructing a practical machine, and a 'gravitational' model that rocks from one position to another.)

Example 11.4.1

The catastrophe machine consists of a disc, pivoted at its centre $\mathbf{0}$, from which it hangs in a vertical plane. Elastic strings are attached to a point on its boundary, and one of these is also tightened and attached to a point A vertically below $\mathbf{0}$. If now the second string is pulled tight at its free end B, then for most positions of B in the plane, the configuration will be in stable equilibrium (see Fig. 11.4.1(a)). But one finds by experiment that there is a region R, shaped like an astroid, with the following property: if C is a point on the right lower boundary arc of R, and B approaches C without crossing the boundary of R, then the disc will have a particular stable position; whereas if B approaches C after crossing the opposite boundary, the disc can have a different stable position. Also near P, if R is towed by hand· up the vertical line below P, the disc is unstable and can flip from one of two positions to the other, with small trembles of the towing hand.

It turns out that we may as well choose $\mathbf{0}A$ and the radius of the disc to have lengths 2 and $\frac{1}{2}$, respectively, and the strings each to have unit unstretched length, and stiffness that we denote by $4k$. Suppose first that B is on the line $\mathbf{0}A$ at distance s above $\mathbf{0}$ (as in Fig. 11.4.1(b)), and $\mathbf{0}E$ makes an angle θ with the downward vertical. Here, s is the control variable, assumed fixed while we investigate the phase variable θ. By symmetry, there is an equilibrium when $\theta = 0$, and we must investigate its stability. Now (neglecting any friction at the pivot), the potential energy $V_s(\theta)$ of the

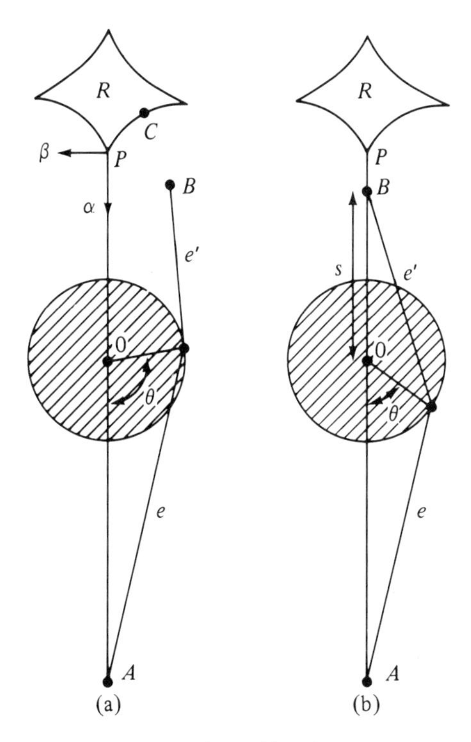

Fig. 11.4.1.

system is the sum of the energies stored in each string; by some trigonometry it can be shown that

$$V_s(\theta) = k\left[\frac{1}{4} + \left(s - \frac{1}{2}\right)^2 + \theta^2\left(\frac{1}{3} - \frac{s(2s-1)}{2(2s+1)}\right)\right] + O(4)$$

after we expand $\cos\theta$ in powers of θ (so $O(4)$ denotes a sum of the form $\theta^3 f(\theta)$ where $f(\theta) \to 0$ as $\theta \to 0$). By the laws of mechanics, the system is minimum-seeking relative to the potential V_s, so we must find whether the candidate $\theta = 0$ is a relative minimum. Clearly $\mathrm{d}V_s/\mathrm{d}\theta = 0$ at $\theta = 0$, and $\mathrm{d}^2V_s/\mathrm{d}\theta^2$ is twice the coefficient (say c) of θ^2 in V_s. We have a minimum or maximum at $\theta = 0$ according as c is negative or positive, so the equilibrium changes from stable to unstable as s increases through the (positive) value h for which:

$$2(2s + 1) = 3s(2s - 1).$$

The positive solution h of this quadratic is 1.40, approximately, and h is the height of the point P at which more complicated behaviour of the system occurs. At P, the coefficient c is zero, and we need more information about the $O(4)$ term in $V_s(\theta)$. If, instead, we consider the equilibrium position with $\theta = \pi$, we can obtain the height of the upper vertex P' of the 'astroid', at 2.46, approximately.

We therefore need to consider the system when B is near P but off the line $\mathbf{0}P$, as in Fig. 11.4.1(a), so we take the coordinates of B to be a, b relative to P as origin. Then a, b are now the control variables, and the potential turns out to be of the form

$$V_{ab}(\theta) = [c_0 + bc_1\theta + ac_2\theta^2 + bc_3\theta^3 + c_4\theta^4] + O(5) \qquad (11.4.1)$$

with $c_0, \ldots, c_4$ being approximately 0.54, 0.24, 0.16, -0.09, 0.045, respectively. When B is at P, $a = 0 = b$, and we have $\mathrm{d}V_{00}/\mathrm{d}\theta = 4c_4\theta^3 + O(4)$; so $\theta = 0$ is a degenerate critical point. It is here that the $O(5)$ term in V_{00} becomes crucial. Suppose, for example, that this term is qx^5, where q is constant, and suppose $c_4 > 0$. If $q > 0$, then V_{00} is always positive near $\theta = 0$, so we have a minimum at $\mathbf{0}$; but if $q < 0$ we have neither a minimum nor a maximum since $\mathrm{d}V_{00}/\mathrm{d}\theta$ does not then change sign at $\mathbf{0}$. This should indicate the unreliability of the traditional way of neglecting higher-order terms simply because they are 'small'.

Now by (C.5.5), a (non-linear, invertible) change of variable from θ to ϕ will carry V_{00} to the form $c_0 + c_4\phi^4$, but this change need not remove the $O(5)$ term from any other V_{ab}. On the other hand, since $c_4 < 0$, we can use the transformation $u = \theta + bc_3/4c_4$ to eliminate the term in θ^3 from all the functions V_{ab} *simultaneously*; and it is desirable to have a similar type of transformation that simultaneously eliminates the $O(5)$ term from all the V_{ab}. The existence of such a transformation is guaranteed by the pure mathematics of catastrophe theory, once we assume that the family of functions V_{ab} is structurally stable, but the proof is very elaborate. Granted this, however, we may say that on neighbourhoods of the origin in each of the control space $\mathbb{R}^2_{ab}$ and the phase space $\mathbb{R}$, there are invertible transformations of coordinates, such that the family of functions $V_{ab}(\theta)$ is transformed to the family

$$W_{IJK}(\phi) = \frac{1}{4}\phi^4 + \frac{3}{2}I\phi^2 + J\phi + K,$$

thus removing the $O(5)$ term altogether. Since our interest is in $\mathrm{d}W/\mathrm{d}\phi$ we can now throw away the constant term K and work with the family

$$W_{IJ}(\phi) = \frac{1}{4}\phi^4 + \frac{3}{2}I\phi^2 + J\phi,$$

in which (I, J) and ϕ lie in neighbourhoods of the origin in $\mathbb{R}^2_{IJ}$ and in $\mathbb{R}$. With the variables thus confined to some region N round $\mathbf{0}$, we now claim that the corresponding portion of Fig. 11.2.3 is sufficient for studying the dynamics of the catastrophe machine.

For, with B at (a, b) in Fig. 11.4.1(a), the physical system is governed by $-\mathrm{d}V_{ab}/\mathrm{d}\theta$ with the associated arrows (dynamics) on a copy of $\mathbb{R}_{ab}$ of $\mathbb{R}$. As $\mathbb{R}_{ab}$ we can take the line in $\mathbb{R}^3_{abx}$, through $(a, b, 0)$ and parallel to the third axis (of x). The attractors and repellors on all these lines $\mathbb{R}_{ab}$ will fill out a surface Z in $\mathbb{R}^3$. Now, the above coordinate transformations carry (a, b) to (I, J), say, and the arrow on R at θ to an arrow on L_{IJ} at $x = \phi$; they carry attractors and repellors on $\mathbb{R}_{ab}$ to similar items on L_{IJ}. Hence everything that happens, in the region N round $\mathbf{0}$ in Fig. 11.2.3 (such as hysteresis or divergence), corresponds to a qualitatively similar happening in $\mathbb{R}^3_{abx}$.

In particular, the cusp K in Fig. 11.2.3 corresponds to a cusp K' at the origin $\mathbf{0}$ in $\mathbb{R}^2_{ab}$ (i.e., at P in Fig. 11.4.1(a)). If q is a path in $\mathbb{R}^2_{ab}$ that crosses K' from left to right, then it corresponds to a path like p in Fig. 11.2.4, with the same possibility of change of attractors as we saw there. This explains the behaviour of the catastrophe machine that we mentioned at the outset. The rest of the 'astroid' in Fig. 11.4.1(b) can be defined and studied in a similar manner.

The general procedure 11.4.1

Many other examples are known in physics, of systems which similarly have a potential of the form (11.4.1). Indeed, Thom refers to this as the **Riemann–Hugoniot catastrophe**, because of examples considered by those authors. Each example needs appropriate physical background to be understood, but catastrophe theory supplies the mathematics that allows us to describe the qualitative behaviour of each system by means of Fig. 11.2.3. In each of them there are two control variables p, q and one phase variable x, and the mathematical procedure has the following steps:

(a) find the potential $V_{pq}(x)$;
(b) find the equilibria by finding the critical points z of V_{pq} (these being the zeros of V'_{pq}, that is, of $\mathrm{d}V_{pq}/\mathrm{d}x$);
(c) classify z as a minimum or maximum if $\mathrm{d}^2V_{pq}/\mathrm{d}x^2$ $(= V''_{pq}) \neq 0$ at z (and thus decide whether z is stable or unstable);
(d) eliminate x to find the 'catastrophe set' C of those control points (p, q) for which V_{pq} has a degenerate critical point z (i.e., such that $V''_{pq} = 0$ at z).

Recall from (5.6.3) that the last step amounts to finding the envelope E of the curves $\{V_{pq} = 0\}$ in the space $\mathbb{R}^2_{pq}$ *if we regard x as the control variable.* In practice the elimination is too difficult, but the qualitative behaviour of the system can be seen by appealing to Fig. 11.2.3, just as we did with the catastrophe machine, provided we are confident that the model is structurally stable; and E corresponds to the cusp

K. Structural stability is usually inherent in well-tested traditional models, and when its lack has been noted in certain newer models, this has led to the insertion of important new variables that had been forgotten (see, e.g., Chillingworth and Furness (1974), concerning the Earth as a dynamo).

11.5 DECAY OF CIVILIZATIONS

The following example, from Renfrew (1978), is one of many from the social sciences, and we choose it because it was developed, not by a mathematician, but by a (scientifically educated) archaeologist. It shows how the mechanism of Fig. 11.2.3 can account for the rapid decay of an ancient civilization, without having to assume (with an earlier generation of archaeologists) that the rapidity indicated an invasion by 'barbarian hordes'. For reference in explaining it we need Fig. 11.5.1, which displays the types of potential V_{IJ} associated with the reduced cubics f_{IJ}, for which $V'_{IJ} = f_{IJ}$. (Read: exercise 11.2(3).) Extracting those labelled 1–4, we relabel them as in Fig. 11.5.2, and observe that 3 acts as a mediator between 1 and 2, which have their peaks skewed to the left and right, respectively.

Renfrew follows the way in which, in several examples, Zeeman has started with Fig. 11.5.2, and used it to get to Fig. 11.2.3. Thus, Renfrew supposes that in a given society, we can measure the degree of central organization by a variable x; further, let u be a measure of the average income minus taxes, and let v measure the investment by the rulers in central control (tax gatherers, police, bureaucrats, priesthood, and so on). Finally, he supposes that the popular support can be measured by a function $S_{uv}(x)$.

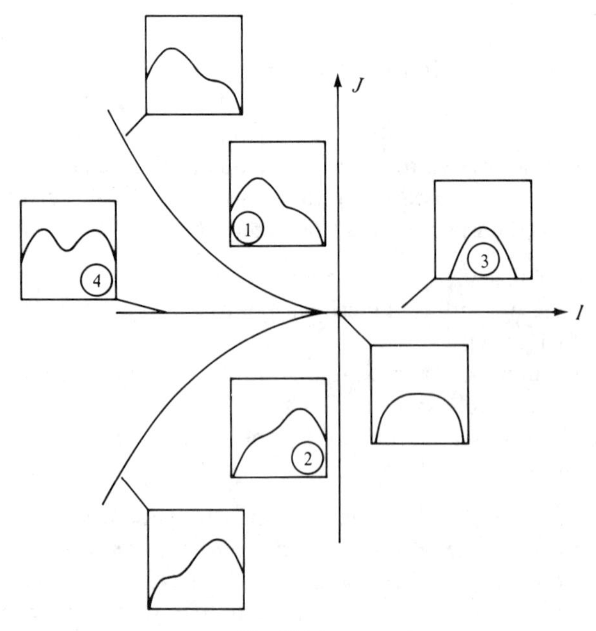

Fig. 11.5.1.

Then if u is high, and v low, the graph of S_{uv} will be skewed as in Fig. 11.5.2(i); for the population is properous and relatively free, and will most likely tolerate a high degree of central organization, at the level x_0 below the peak in Fig. 11.5.2(i). If v is high and u low, then the population is heavily taxed and policed, and x_0 will probably fall as in Fig. 11.5.2(ii). If u and v are both moderate, then S_{uv} will be more symmetrical about x_0. But, if both u and v are high, then S_{uv} may well have two peaks as shown in Fig. 11.5.2(iv), because the population is split between those who, at a time of high prosperity and investment in authority, want only a low, or a high, degree of central organization.

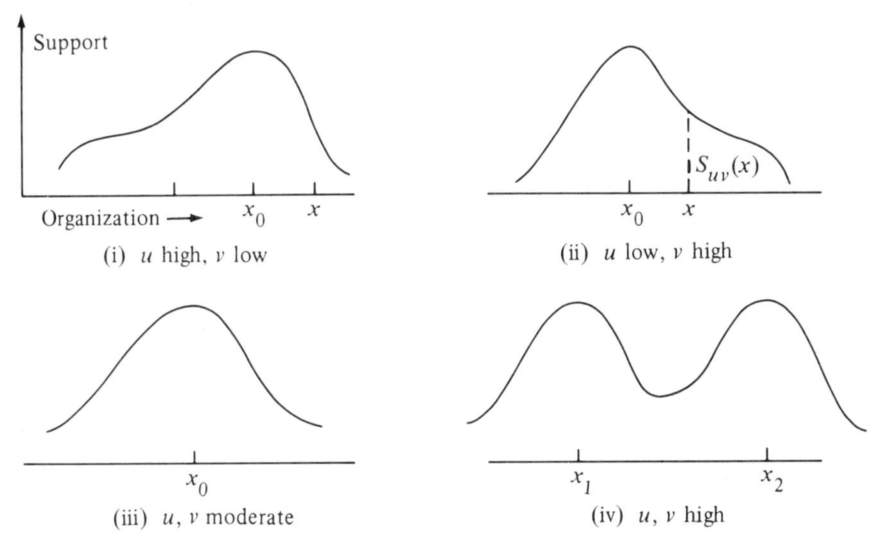

Fig. 11.5.2.

As u, v vary, these figures can change smoothly, one into another, as we move along paths in the control space $\mathbb{R}^2_{IJ}$. Such a path might well arise because of random drift in the variables I, J. Suppose we assume that the functions V_{IJ} form a structurally stable family. Then we may call on catastrophe theory and assert that Fig. 11.2.3 models the dynamics as I and J change. As we have seen earlier, the dynamics can lead to the rapid change from the rich, free, highly organized society of Fig. 11.5.2(i) to the poor, authoritarian, society degenerating into anarchy, represented by Fig. 11.5.2(iii). The first society can leave behind the artefacts and buildings that signify its level of civilization, while the second dissipates its energies and leaves only unrepaired ruins in the historical record. No external 'barbarians' are needed, only careless drift.

Such models give new insights into possible processes of change—even though it is easy to criticize them for being oversimple. But earlier 'social' models have usually been expressed only in literary language, which is difficult if not impossible to use for describing the processes in Fig. 11.2.3. Therefore such processes are less likely to cross the minds of those with only a literary education. Also, it is to be expected that more penetrating mathematical models must at first be simple: as experience grows they can be loaded with further extras. The use of continuous (in place of discrete) variables is common in modelling, but it is more serious to use ill-defined 'potentials'; but then such traditional and useful entities as force and temperature took centuries to come to their present forms, which are still not completely satisfactory in definition.

11.6 COMPROMISE

In various studies, Zeeman (see, e.g., Zeeman (1977)) has considered a catastrophe model in which there are not just two stable equilibria, but where a third allows the possibility of a pause between two extreme states—for example, when a society in a state of war, and at a time of high threat and costly defence, may be split between the hawks (who wish to fight on), the doves (who want to surrender) and the compromisers who opt for negotiation with the enemy. (In a quite different interpretation, this analysis helped in the treatment of anorexia nervosa: see Zeeman (1977), p. 33). To achieve the extra freedom, one needs four control variables, as, for example, with potentials of the form

$$V_{pqrs}(x) = \frac{1}{6}x^6 + \frac{5}{2}px^4 + \frac{10}{3}qx^3 + \frac{5}{2}rx^2 + sx.$$

If, for simplicity of explanation, we return to the tale of Wotan in section 11.2, we now suppose that he is governed by the function $-g_{pqrs}$, where: $g_{pqrs}(x) = V'_{pqrs}(x) = x^5 + 10px^3 + 10qx^2 + 5rx + s$, as in (6.5.5). Here the control and phase spaces are $\mathbb{R}^4_{pqrs}$ and $\mathbb{R}$, respectively, and we now have too many dimensions to display the diagram corresponding to Fig. 11.2.2. But by keeping two variables fixed (in particular p, q) we can display three-dimensional slices S_{pq} of it; thus we obtain the entire set $\{g_{pqrs} = 0\}$ by assembling all these slices. Each slice S_{pq} is constructed like Fig. 11.2.3, except that we replace its cubic graphs by those quintics of the form

$$y = x^5 + 10px^3 + 10qx^2 + 5rx + s, \tag{11.6.1}$$

and our control space is now $\mathbb{R}^2_{rs}$ since we think of p, q as fixed. As before, we obtain a pleated surface, with a fold-curve in $\mathbb{R}^3_{rsx}$ which lies above a bifurcation curve B_{rs} in $\mathbb{R}^2_{rs}$. B_{rs} is the analogue of K in Fig. 11.2.3, but it can now take a greater variety of forms, as in Figs. 6.5.4. When B_{rs} has only one cusp, then S_{pq} resembles Fig. 11.2.3, so the interesting new cases occur when B_{rs} has three cusps.

We consider the case when $p < 0$, $q = 0$, so B_{rs} is the symmetrical 'butterfly' curve; the rest are just skewed perturbations of this. Then Fig. 11.6.1 displays the conventional view given in many books, but it obscures some important detail that we consider in a moment. The important point, for applications, is the existence of the 'pocket', and it may help the reader to make a paper model, as shown in Fig. 11.6.2. The pocket has the property that if Wotan descends a vertical line that grazes the top fold, he can arrive on the middle fold and wander around it before either descending to the lower fold (which consists now entirely of attractors), or returning to the top via a hysteresis loop (as with the earlier cusp). This glimpse should suffice to indicate the extra power of this model, and readers can now turn to Zeeman's articles for the

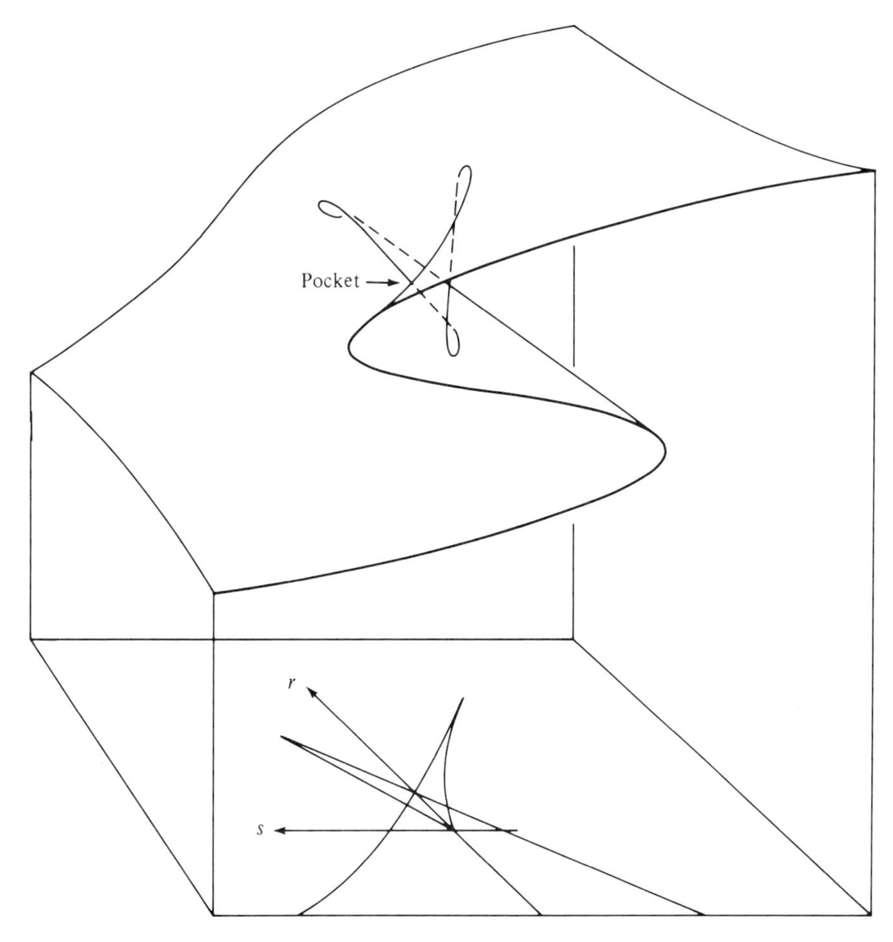

Fig. 11.6.1.

detailed discussion of his applications, which exploit the delay allowed by wandering on the middle fold: delay allows thoughtful negotiation to take place that can avoid hot-headed action. If we use polynomials of odd degree >5, further cascades of folds can be produced, to model a series of delays.

For a better understanding of Fig. 11.6.1, we look at the schematic Fig. 11.6.3, showing the curve B_{rs} which is the projection on the plane $x = 0$ of the curve C that runs along the folds in Fig. 11.6.1. Each of the two cross-over points in B_{rs} lies vertically below two points of C. Typical plane slices of S_{pq} are shown on the left of Fig. 11.6.3, with the graphs of y in Fig. 11.6.1 turned through 90°. In S_{pq}, their critical points (labelled 1–4) lie on C. The middle sections of sketches 2, 3, 4 (between points 2 and 3) are slices of the bottom of the pocket in Fig. 11.6.1. The reader should verify that according to Wotan's original rule for Fig. 11.2.1, a point P on such a section is an attractor for the dynamics on the vertical line through P.

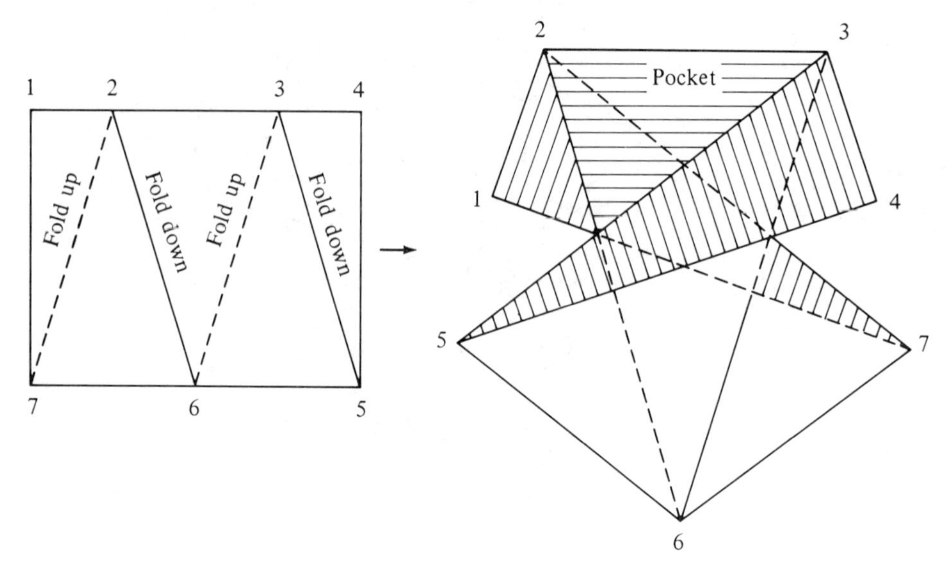

Fig. 11.6.2.

11.7 THE UMBILICS

When we consider families of functions of *two* phase variables, we meet new catastrophe sets, called the **umbilics**; and we now give a brief description of two of these. Thom was interested in these surfaces because they relate to fundamental shapes of natural forms and may be useful in embryology (provided a lot more argument is supplied, as in Zeeman's papers on gastrulation). His speculations on these matters, and the intriguing photographs, may be found in his book. But these catastrophe sets can be met in several other physical contexts, as described in Poston and Stewart (1978), and our approach here will be through a family of differential equations, as we now explain. Readers will need to recall the material on portraits from Chapter 9.

Suppose that for each z in some control space S, we have a smooth dynamical system D_z, with portrait P_z. Then, as we saw in Chapter 9, P_z usually changes only slightly as z changes, but there may be a set T in S, such that if z passes through T, then P_z changes its form in a drastic way. If S is $\mathbb{R}^3$ or $\mathbb{R}^4$, then T is likely to be related to one or other of the standard catastrophe sets, as we now show. Consider a family E_z of the form:

$$E_z: \quad \frac{dx}{dt} = u + rx + y^2, \qquad \frac{dy}{dt} = v + sy + x^2, \tag{11.7.1}$$

so $z = (r, s, u, v)$ runs through $\mathbb{R}^4$. One way in which the portrait P_z will change its form is for the horizontal and vertical manifolds H_z, V_z, to become tangent to each other as z crosses a certain set B; for then two fixed points will coalesce and then

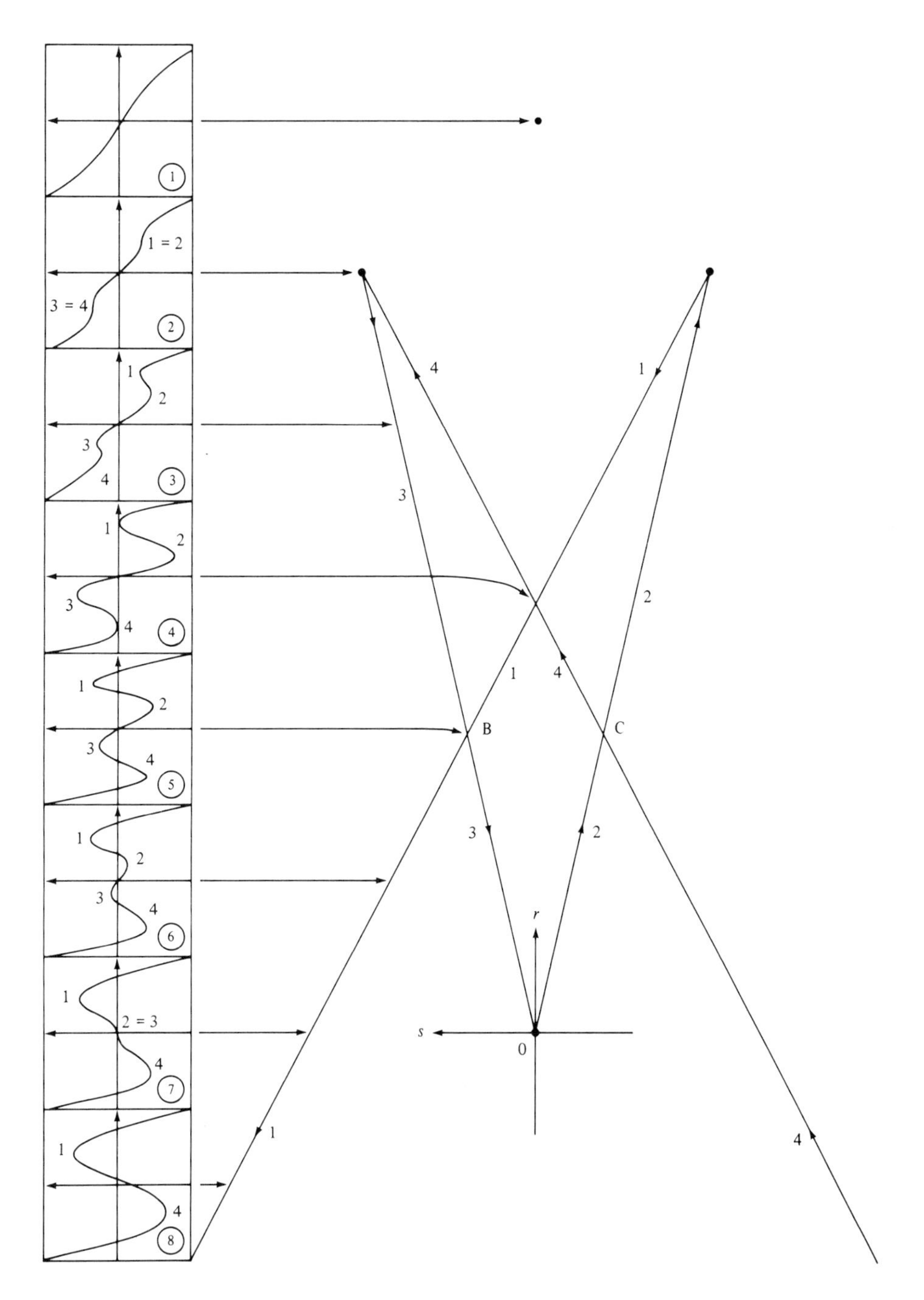

Fig. 11.6.3.

disappear, or a new pair will appear. Another way of changing is for fixed points to change from (say) nodes to spirals. Our interest here is in B, which we call the **contact set**. To compute B, we must find when H_z, V_z touch; their equations are given by: V_z: $u + rx + y^2 = 0$, H_z: $v + sy + x^2 = 0$. To find their intersections, we eliminate y and obtain

$$x^4 + 2vx^2 + rs^2x + (v^2 + s^2u) = 0. \tag{11.7.2}$$

There is a common tangent when the reduced quartic (11.7.2) has a repeated root, and from section 6.5 we know that this occurs when the coefficients in (11.7.2) lie on the swallowtail surface S in $\mathbb{R}^3$. Thus, let θ: $\mathbb{R}^4_{rsuv} \to \mathbb{R}^3_{abc}$ be the mapping given by

$$(a, b, c) = \left(\frac{1}{3}v, \frac{1}{4}rs^2, v^2 + s^2u\right). \tag{11.7.3}$$

Then we have shown: $z = (r, s, u, v)$ *lies on the contact set* B *of the family* E *iff* $\theta(z)$ *lies on the swallowtail* S. It can be verified that for most points p of S, there is a curve of points z such that $\theta(z) = p$, so in $\mathbb{R}^4_{rsuv}$, B is a three-dimensional set that looks mostly like a fattened-up version of S. To get a more accurate picture, we keep a control variable (say d) constant to obtain a three-dimensional slice X of $\mathbb{R}^4$ orthogonal to the d-axis; we then call X a 'd-slice'.

Exercises 11.7.1

1. Find the quartic corresponding to (11.7.2) when x (rather than y) is eliminated.
2. Keeping r or s constant and non-zero in (11.7.3), sketch the corresponding sections of B in an r-slice or s-slice of $\mathbb{R}^4_{rsuv}$.

The family E also leads us to another catastrophe set, the **hyperbolic umbilic**, shown in Fig. 11.7.1(a), with a typical cross-section in Fig. 11.7.1(b). To see this, assume $r, s \neq 0$, and use the rescaling

$$t \to -rt, \qquad x \to -x(rs^2)^{-1/3} \qquad y \to -y(r^2s)^{-1/3}$$

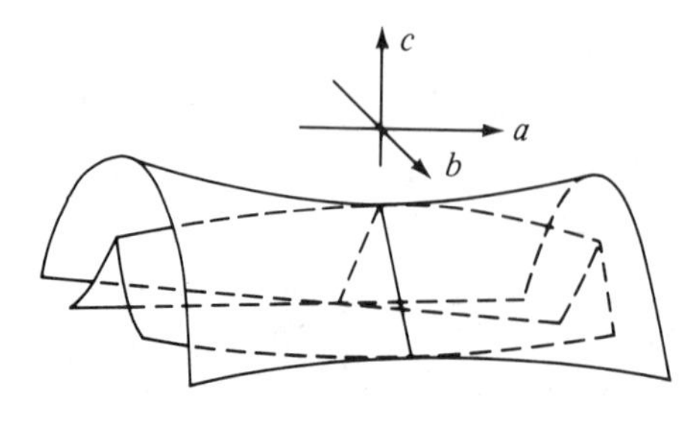

Fig. 11.7.1(a).

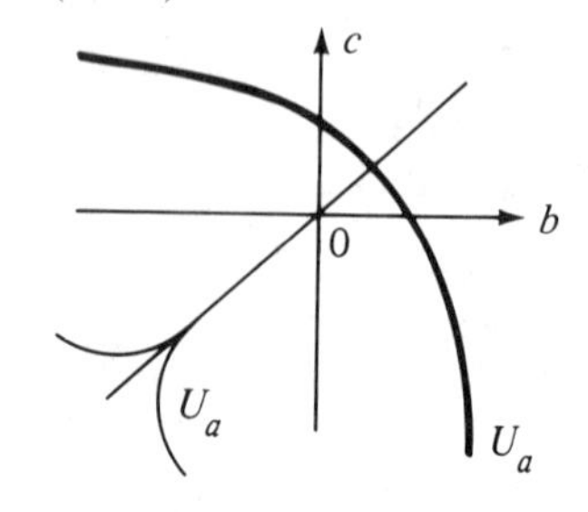

Fig. 11.7.1(b).

to convert the system E_z to

$$D_w: \quad \frac{dx}{dt} = b - x + y^2, \qquad \frac{dy}{dt} = m \cdot (c - y + x^2) \tag{11.7.4}$$

where $w = (m, b, c)$ in the control space $\mathbb{R}^3_{mbc}$. Then it is easily verified that D_w is Hamiltonian (in the sense of section 9.2) iff $m = -1$. In that case, the procedure for finding the contact set B is the same as that for finding the catastrophe set in section 11.4.1 when the Hamiltonian function is used as a potential, except that a critical point must be tested in terms of the Hessian instead of the ordinary second derivative.

This leads us to consider the family of Hamiltonian systems

$$K_{abc}: \quad \frac{dx}{dt} = \frac{\partial H_{abc}}{\partial y}, \qquad \frac{dy}{dt} = -\frac{\partial H_{abc}}{\partial x}, \tag{11.7.5}$$

with Hamiltonian function

$$H_{abc}(x, y) = \frac{1}{3}(x^3 + y^3) - axy + cx + by.$$

Thus when $a = 1$, K_{1bc} is the system D_w in (11.7.4) with $m = -1$. In K_{abc}, the vertical and horizontal manifolds are, respectively, the horizontal and vertical parabolae: $y^2 - ax + b = 0 = x^2 - ay + c$ so the contact set U of the family K is the set of all (a, b, c) for which these parabolas touch at some point $P = (x, y)$. Now $H_{abc}(P) = H_{acb}(y, x)$ so if (a, b, c) lies on U because of P, then (a, c, b) lies on U because of (y, x). Thus the section U_a (with $a \neq 0$) in Fig. 11.7.1(b) is symmetrical about the line $b = c$. We work out its shape as follows.

By exercise 8.1.2(5), such contact at P exists when the Hessian determinant $J = 4xy - a^2$ is zero at P, so since $J = 0$ at P then $x \neq 0$. Hence, taking $a = 2k$ (say) $y = k^2/x$ and then

$$b = 2kx - \frac{k^4}{x^2}, \qquad c = 2\frac{k^3}{x} - x^2, \tag{11.7.6}$$

which are the parametric equations of the section U_a as x runs from $-\infty$ to ∞, excluding zero. Further details are given in exercise 11.7.2(2) below. To find the equation of the surface U, (11.7.6) simplifies to the pair of cubics.

$$x^3 - \frac{b}{a}x^2 - \frac{a^3}{16} = 0 = x^3 + cx - \frac{a^3}{4}, \tag{11.7.7}$$

from which to eliminate x. By exercise 11.7.2(1) below, the eliminant is

$$b^2c^2 + a^2(b^3 + c^3) + \frac{9}{8}a^4bc - \frac{27}{256}a^8 = 0. \tag{11.7.8}$$

In $\mathbb{R}^3_{abc}$, (11.7.8) represents the surface shown in Fig. 11.7.1(a), of which all sections by planes $a = $ constant ($\neq 0$) resemble Fig. 11.7.1(b). Hence the portion $\{a > 0\}$ resembles the portion $\{m > 0\}$ of the contact set of (11.7.3). Finally, Fig. 11.7.2 shows the ways in which the portrait of K_{abc} changes with $w = (a, b, c)$, (when $a = 1$); and

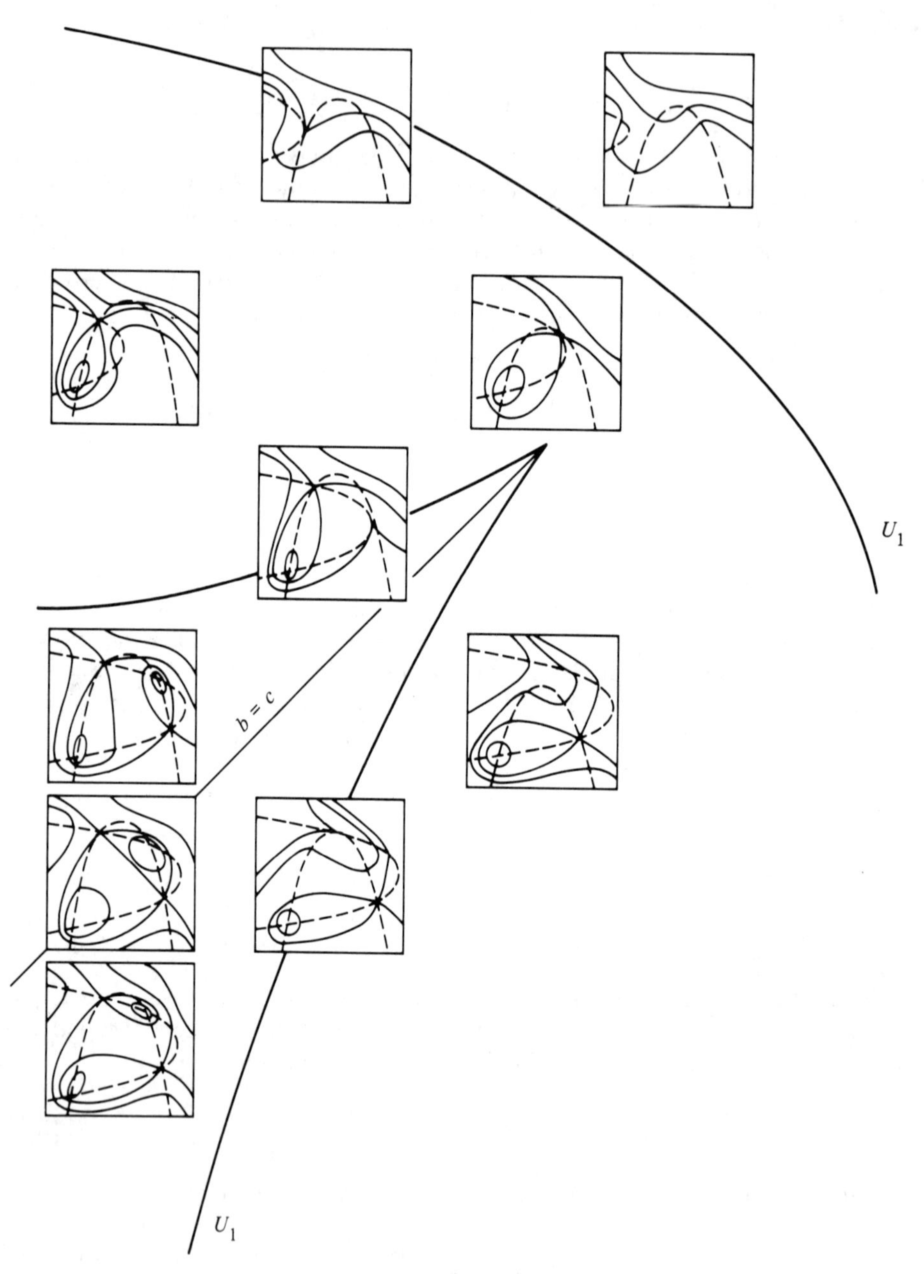

Fig. 11.7.2. The system $K_{1,b,c}$: $\dot{x} = b - x^2 + y^2$, $\dot{y} = -1(c - y + x^2)$. Trajectories are level curves of H_1: $\frac{1}{3}(x^3 + y^3) - xy + cx + by$.

we see there that the significant changes take place when w lies on U_1. These occur because of the changes in the underlying behaviour of the horizontal and vertical manifolds, shown dashed in Fig. 11.7.2. Each of these galleries was constructed by the procedure of section 6.3.

Note: in Fig. 11.7.2, the curves are the contours of the corresponding Hamiltonian function, so we have omitted the usual sensing arrows. The contours have a cusp at degenerate critical points (where H and V touch rather than cross).

Exercises 11.7.2

1. Use exercise 6.3.1(6) to verify (11.7.8). (Take $Q(x)$ there to be the difference of the two cubics in (11.7.7).) Show that if $x = y/a^{1/3}$ then we can take $C(x), D(x)$ to be cubics of the forms $y^3 - uy - t^2$ and $y^3 + vy - t$, with eliminant $f_G(u, v) - g$; where $G = (1 - t)(2t - 1)$, $g = t(1 - t)^2$, and $f_G(u, v)$ is the polynomial $u^3 + v^3 + u^2v^2 + Guv$. Construct portraits of f_G as G varies. (For details see Griffiths (1985).)
2. In (11.7.6), assume $k > 0$. Show that (i) db/dx and dc/dx vanish simultaneously only when $x = -k$; (ii) U_a consists of two pieces A, B corresponding, respectively, to $x > 0$, $x < 0$, and A is smooth, with decreasing gradient; (iii) the curve B is smooth except for just one cusp, at $(b, c) = -(k, k)$, and at its other points, the curve has increasing gradient. By the symmetry under interchange of (b, c), noted above, Fig. 11.7.1 follows.
3. Draw your own version of Fig. 11.7.2 (preferably in glowing colour). In particular, consider just the behaviour of the horizontal and vertical manifolds as (b, c) varies, with $a = 1$.
4. Show that when $a = 0$ in (11.7.5), the section U_a consists of an L-shaped curve. Draw the corresponding gallery of portraits. (See exercise 8.4.1(2).)
5. Consider the system

$$D_w\colon \quad \frac{dx}{dt} = ax + by + y^2, \qquad \frac{dy}{dt} = cx + dy + x^2$$

with control variable $w = (a, b, c, d)$ in $\mathbb{R}^4_{abcd}$. Noting that H_w and V_w here always meet at $\mathbf{0}$ in $\mathbb{R}^2_{xy}$, show:

(a) they will have there a common tangent, iff w lies on the set L: $ad - bc = 0$ in $\mathbb{R}^4_{abcd}$;
(b) the section $d = 0$ of L is a pair of planes;
(c) if k is fixed and $\neq 0$, and T_k is the three-dimensional section $\{d = k\}$ of $\mathbb{R}^4_{abcd}$, then T_k intersects L in a saddle surface.

Suppose next that w is not in L; then any other intersection I of H_w and V_w lies on some line $y = mx$. Show:

(d) H_w and V_w will have a common tangent at I iff the equation $(a + bm) - m^2(c + dm) = 0$ has a repeated root;
(e) if $d \neq 0$, the last equation is cubic of the form $0 = m^3 + 3Hm^2 + 6Im + 3J$, and $(H, I, J) = g(w)$ (say) $= (c/3d, -b/3d, -a/d)$; hence that w is on the intersection of C and T_k iff in $\mathbb{R}^2_{IJ}$, $g(w)$ lies on the 'page of cusps' (see Fig. 6.3.2);
(f) if $d = 0$ but $a \neq 0$, there is a similar mapping h into the page of cusps, where $h(z) = (b/3a, -c/3a, -d/a)$. What happens if $a = 0 = d$?

6. Change (x, y) to $p = x + c/2$, $q = y + b/2$, to transform D_w in exercise 5 to the system E_z in (11.7.1) with $z = (a, d, -(2ac + b^2)/4, -(2bd + c^2)/4)$. This equation gives a mapping $F\colon \mathbb{R}^4 \to \mathbb{R}^4$ with $z = F(w)$. Show that F maps the contact set C of the family D to the

contact set B of the family E. Verify that F is of full rank everywhere except on the set L in (a) above. (Thus F maps the catastrophe sets, L and the cusps, associated with the family D, to the swallowtail and hyperbolic umbilic associated with the family D. Clearly F does not map these homeomorphically, but it often does so on plane sections. Away from L it is a local homeomorphism. For a detailed study, see Griffiths (1985).

11.8 THE ELLIPTIC UMBILIC

We conclude this brief account of the catastrophes with a study of the elliptic umbilic surface $\mathscr{E}$ in $\mathbb{R}^3$, shown in Fig. 11.8.1(a). Thom associates it with the formations of spikes, thorns, and beaks in biology. For applications to fluid flow, see Poston and Stewart (1978), Chapter 11. In this section, we concentrate on the pure mathematics only, because of the new feature—the rotational symmetry of $\mathscr{E}$, shown in Fig. 11.8.1(b).

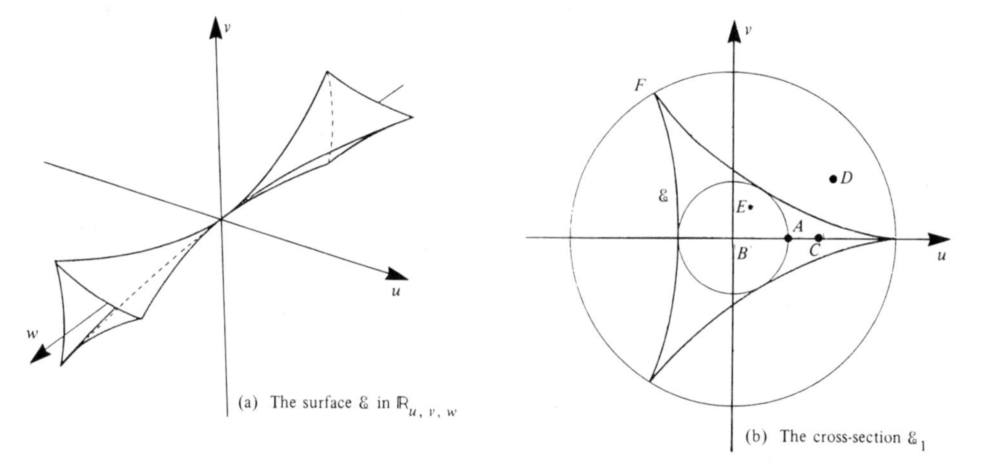

Fig. 11.8.1.

The surface $\mathscr{E}$ is the contact set in the control space $\mathbb{R}^3_{uvw}$ of the Hamiltonian system

$$Q_{uvw}: \quad \frac{\mathrm{d}x}{\mathrm{d}t} = \frac{\partial F_{uvw}}{\partial x}, \quad \frac{\mathrm{d}y}{\mathrm{d}t} = -\frac{\partial F_{uvw}}{\partial y}, \tag{11.8.1}$$

where

$$F_{uvw}(x, y) = \frac{1}{3}x^3 - xy^2 + w(x^2 + y^2) + ux + vy. \tag{11.8.2}$$

Let us form from x, y the complex variable $z = x + jy$. Then the reader can verify that F_{uvw} can be expressed in terms of z as

$$F_{uvw} = \frac{1}{3}\operatorname{Re}(z^3) + w|z|^2 + \operatorname{Re}[(u + jv)z] \tag{11.8.3}$$

where $Re(z)$ denotes the real part of z. Now recall that $\omega = e^{2j\pi/3}$ is a cube root of unity, and ωz is the complex number obtained by rotating z through 120° about $\mathbf{0}$.

Therefore

$$|\omega z| = |z|, \qquad (\omega z)^3 = z^3, \qquad (u + \mathrm{j}v)(\omega z) = (u' + \mathrm{j}v')z$$

where $u' + \mathrm{j}v' = (u + \mathrm{j}v)\mathrm{e}^{2\mathrm{j}\pi/3}$.

The last equation says that a rotation of the phase plane through 120° induces a rotation A of the (u, v)-plane through the same angle. Hence from (11.8.3) we get

$$F_{uvw}(\omega z) = F_{u'v'w}(z). \tag{11.8.4}$$

Hence if $\mathscr{E}_w$ is the slice of $\mathscr{E}$ by the plane of height w, then $\mathscr{E}_w$ is a diagram that is invariant under the rotation A, for if (u, v) lies in $\mathscr{E}_w$, it has a contact point z in $\mathbb{R}^2_{xy}$, and then (u', v') is in $\mathscr{E}_w$ since ωz is a contact point for (u', v'). Next, note that by (11.8.2), $F_{uv(-w)}(x, y) = -F_{uvw}(-x, -y)$, so $\mathscr{E}_w$ looks like $\mathscr{E}_{-w}$, and we need only consider $w \geq 0$. Also if we fix $w > 0$, then

$$F_{uvw}(x, y) = w^3 F_{u'v'1}\left(\frac{x}{w}, \frac{y}{w}\right), \tag{11.8.5}$$

where

$$u' = \frac{u}{w^2}, \qquad v' = \frac{v}{w^2}. \tag{11.8.6}$$

Hence $\mathscr{E}_w$ is obtained from $\mathscr{E}_1$ by the scaling (11.8.6), and we restrict ourselves to $w = 1$. Let then $p = (u, v, 1)$, $G_P = F_{uv,1}$. The horizontal and vertical manifolds H_p, V_p are now hyperbolae:

$$H_p\colon \frac{\partial G_p}{\partial x} = 0 = x^2 - y^2 + 2x + u, \qquad V_p\colon \frac{\partial G_p}{\partial y} = 0 = -2xy + 2y + v,$$

which vary with u, v, but which have fixed centres (at $(-1, 0)$, $(1, 0)$, respectively) and fixed asymptotes. As H_p, V_p vary, so do the critical points of G_p (they are the fixed points of Q_{uv1} in (11.8.1)).

The curves touch when the Hessian determinant $4[(x + 1)(-x + 1) - y^2]$ is zero, that is, on the circle $x^2 + y^2 = 1$, and the contact points must be of the form $x = \cos(t)$, $y = \sin(t)$, for some t such that (x, y) lies simultaneously on both H_p and V_p. Hence $u = -(2\cos(t) + \cos(2t))$, $v = 2\sin(t) - \sin(2t)$, valid for $0 \leq t \leq 2\pi$. These are the parametric equations of $\mathscr{E}_1$, which is therefore the deltoid curve shown in Fig. 11.8.1(b). This is invariant under rotations of 120° about $\mathbf{0}$. By (11.8.6), each $\mathscr{E}_w$ is a scaled version of $\mathscr{E}_1$, so when these curves are strung along the w-axis in $\mathbb{R}^3_{uvw}$, we obtain the surface shown in Fig. 11.8.1(a), and the curved profile arises because of the scale factor $1/w^2$ in (11.8.6).

As $p = (u, v, 1)$ moves on a path in the plane $\mathbb{R}^2_p$, the manifolds H_p, V_p move, with degenerate intersections (= contacts) as the portrait crosses $\mathscr{E}_1$.

The reader should beware that the notions of 'horizontal' and 'vertical' are not invariant under the action of ω in (11.8.4), so Fig. 11.8.2 is not invariant either—but the points of intersection of H and V *are* invariant, because these are the critical points of G_p. Therefore, we start with the degenerate curves of portrait A (at

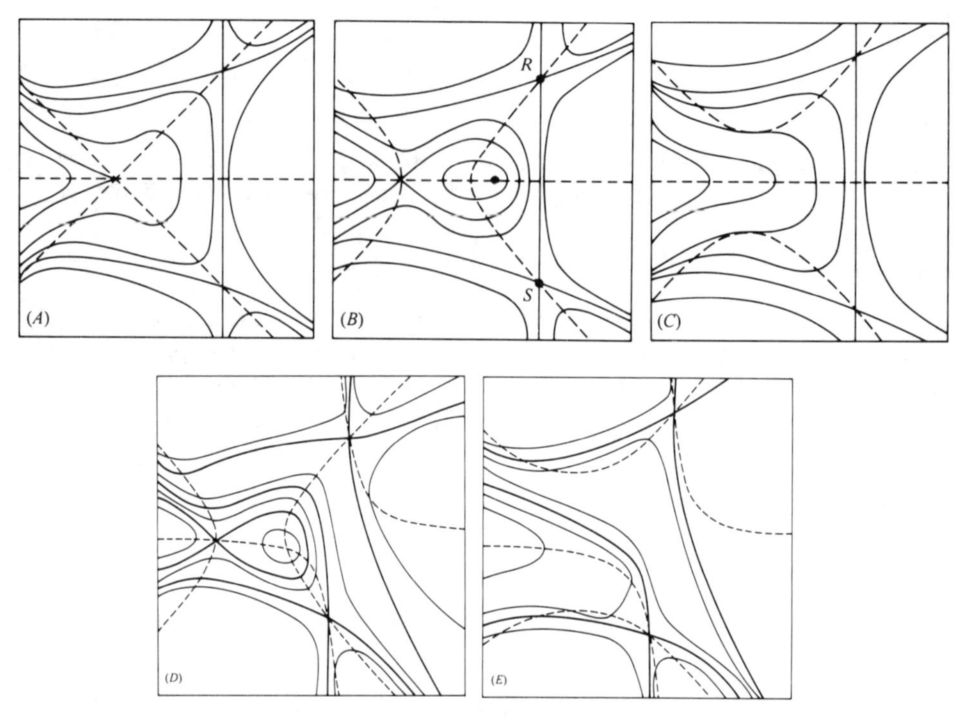

Fig. 11.8.2. Portraits of G_p.

$(u, v) = (1, 0)$); as we then move horizontally or vertically, V_p and H_p alter their shape. The above calculation of the Hessian shows that in all cases, a critical point is a saddle iff it lies outside the unit circle; also if $x = y$ (large) then $\mathrm{d}y/\mathrm{d}x > 0$, and we can sign regions accordingly.

In Fig. 11.8.2 we show portraits of G_p when p is at the points A, B, C, D, E in Fig. 11.8.1(a). With A, B, C, $v = 0$ and u increases through 1 as we move from B through A to C. As we do so, the critical point at P in A disappears, whereas if we move from A to B, P splits into two. Thus we have in B the 'fish-eye' which shrinks to a cusp in A, and then is gone in C. In these three portraits, V_p is degenerate, being a pair of lines; so we now increase v from zero, to obtain V_p as a hyperbola. Thus portraits D, E are the really typical ones, with four and two critical points, respectively. The straight line segment in B, that joins R, S, has now split into two separatrices, one from R, the other from S; for these two continue to be saddles, by continuity of the Hessian determinant.

If now we move from E anywhere outside $\mathscr{E}_1$, the portrait remains qualitatively the same, with just two critical points; only such items as curvatures and gradients of contours will change. Similarly, if we move from D anywhere within $\mathscr{E}_1$, the portrait remains qualitatively the same, with four critical points. But if we cross $\mathscr{E}_1$, from inside to outside, we have to lose two critical points, so two coalesce into a degenerate one as we cross the boundary, and this disappears as we move outside. The fish-eye degenerates to a cusp and then the contour smooths out.

The reader should investigate the detailed changes that occur—in particular when we move across one of the three vertices of $\mathscr{E}_1$. It is interesting to concentrate first, solely on the behaviour of the pair H_p, V_p, by producing a sequence of sketches. For a very complete set of portraits, see Poston and Stewart (1978), p. 240.

Exercises 11.8.1

1. What happens to the portrait at D in Fig. 11.8.1(a) if p moves from there, through the left vertex F of $\mathscr{E}_1$ and into the exterior of $\mathscr{E}_1$?
2. Produce reliable computer graphics for such bifurcation points as A or F in Fig. 11.8.1(a).
3. Produce a complete gallery for the case $w = 0$ in (11.8.2).

Another catastrophe is the **parabolic umbilic**, used by Zeeman to discuss the process of gastrulation in embryology; it is the contact set in $\mathbb{R}^4_{abcd}$ of the family

$$F_{abcd}(x, y) = x^2y + y^4 + ax^2 + by^2 + cx + dy \tag{11.8.7}$$

and is quite complicated. For a partial description of the geometry, see Poston and Stewart (1978), p. 190, and the references cited there.

11.9 THOM'S THEOREM

We have seen that one useful feature of catastrophe theory is to give a standard form when functions have critical points. Thus, if a single function f: $\mathbb{R}^k \to \mathbb{R}$ (with $k = 3$, say) has a critical point, p, then we first ask if p is non-degenerate. If yes, then Morse's lemma (D.7.1) tells us that there is a coordinate transformation (= diffeomorphism) on a neighbourhood of p such that in the new coordinates, f takes the form $\pm x^2 \pm y^2 \pm z^2$. If no, then f may take a manageable form if it lies in a structurally stable family, and it is here that catastrophe theory comes in, as in the examples above.

The hard mathematics that gave rise to catastrophe theory lies in the famous theorem of Thom, which is often summarized by saying that *if we use no more than four parameters, then there are just seven corresponding catastrophes.* They are the standard forms we have met above—quadratic, cusp, swallowtail, butterfly, and the three umbilics. The subject matter concerns k-parameter families of smooth functions. In more detail, given any such structurally stable family with $k \leq 4$, then by suitable coordinate transformations, the entire family can always be brought to the form of one of the seven standard families—at least when the variables are all sufficiently small. (Thom's result has since been extended to families with more than four parameters, but the work is less complete.) Thom thought that the four parameters would be important as coordinates in space–time, and he directed attention accordingly. Moreover, he suggested that in applications we should look for structurally stable models, partly because those are more likely to fit the facts (including our ignorance) as well as being more tractable mathematically. The simplifying coordinates predicted by the theory are often more 'natural' than the obvious ones that may just lie at hand, just as the standard (x, y) coordinates were not invariant in Fig. 11.8.2.

Although we have already developed enough language to be able to make a precise statement of Thom's theorem, a good deal of additional explanation of the finer points would still be necessary. Since readers are now ready for the excellent discussion in Poston and Stewart (1978), Chapter 6, we do not repeat it here because of limitations of space.

This, then, concludes our journey from the population model of the eighteenth-century Rev. Thomas Malthus, via much significant classical and modern mathematics (illustrated by modern computer graphics), to the developments in catastrophe theory, chaos, and fractals of current times. We hope that readers who have made this journey with us will have acquired the techniques that will enable them to pursue their curiosity further.

A

Using technology to help gain mathematical insights

A.1 INTRODUCTION

A major problem with producing any book which encourages the use of current technology is the speed with which new developments occur—and so such a book is bound to be out of date even by the time it first appears in print. However, we are convinced that the technology holds such a key position both in helping the learner of mathematics to gain a better feel for what is going on than has hitherto been possible, and in helping to carry out the mathematical research itself, that we must attempt to take as full advantage of its benefits as possible.

Another major problem is the great diversity of hardware and software available now. We must, then, make some minimum assumptions about the technology available to the reader. First, we assume access to some convenient calculating device—such as a 'scientific calculator'—for doing rapid calculations. Second, we assume access to some sort of programmable device for carrying out numerical algorithms. This might range from a programmable calculator, through domestic and pocket microcomputers with BASIC, to sophisticated computers with spreadsheets and/or a range of programming languages. Lastly we assume access to some sort of graphical device which will allow us to draw the pictures which help us to gain insight into the mathematics. This might range from a graphical calculator, through microcomputers with a programming language which allows points to be plotted and lines to be drawn, to sophisticated computers with a range of graph plotting and other mathematical software, as well as, perhaps, programming languages supporting graphics. Of course it will be an advantage to have access to software which will perform the manipulations appropriate to the part of the book you are reading, such as complex numbers, matrices, line integrals, and so on. We say more about this in section A.4.

It is impossible to try to spell out ways in which varieties of different software (such as spreadsheets or graph plotters) may be used at appropriate points in the text. Our approach is to encourage readers with access to such software to familiarize themselves with its facilities and to make use of it as frequently as possible while reading this book. On the other hand, we have worked with many students of mathematics over recent years who have had access either to a programmable, graphical calculator (such as those from Casio, Texas Instruments, Hewlett-Packard, and Sharp) or to a domestic or educational microcomputer (such as BBC/Acorn, PC, Atari, Commodore, and Macintosh) with a programming language (usually a version of BASIC) with graphical facilities. Again we cannot spell out programs which will run on all of that variety of hardware.

Finally we have had to make a decision about how to convey programs in the book. Opinion seems to be divided between (a) using a 'pseudo-code', which is really a particular form of mathematical notation, and will not execute on any known machine—leaving readers to work out how to convert it to any language to which they have access, and (b) using a particular dialect, such as GW-Basic, which will run on an appropriate machine, but which, again, many readers will have to adapt for themselves. We have chosen to follow (b), but also to give, in this chapter, as much information as possible to allow readers with access to other systems to convert the programs. We have chosen TrueBASIC, from Kemeny and Kurtz (1990), as the 'base language' for two main reasons. Firstly, it is widely, and inexpensively, available for a range of current personal computers, including both IBM compatibles, and Apple Macs. Secondly, it contains a good range of mathematical functions, and structured programming facilities, and so allows mathematical algorithms to be adapted in as intelligible way as possible.

A.2 TECHNIQUES FOR ITERATION

Chapter 2 makes extensive use of sequences formed by repeated iteration of a simple rule from a given starting number. For example, the Verhulst model of growth ΔP in a population of size P in a given period of time is: $\Delta P = bP - cP^2$. Here the quantity P is a variable which may change during the iteration, whereas b, c are taken as constants for the iteration (though they may well take different values for another iteration). In fact the initial value of P, P0 say, also needs to be specified, and hence b, c, P0 are **parameters** for the iteration. We may also want to count the number of iterations performed, and so introduce another variable, n say. This gives a simple algorithm:

```
Algorithm: Verhulst
Initialize parameters b, c, P0
Set counter n to zero, and variable P to P0
repeatedly
  output current values of n and P
  compute increment ΔP
  add ΔP to P and store as new P
  increment the counter n by 1
until some test condition is met
end.
```

One of the simplest means of implementing this algorithm is on a modern calculator with buttons marked 'Ans' and 'Exe': 'Ans' allows the result of a previous calculation to be included in the current calculation, and 'Exe' performs the calculation. Thus entering a number on the display, for example 100 Exe is equivalent to setting P0 to 100. Now the calculation:

$$\text{Ans} + 0.1\text{Ans} - 0.0001\text{Ans}^2\ \text{Exe}$$

replaces each occurrence of Ans with the result of the previous calculation, that is, 100, to give 109 as the current value. The calculation is stored in memory, so just pressing Exe again performs the same calculation, but now with 109 replacing Ans, giving 118.7119 as the next value. Pressing Exe repeatedly gives a chain of values which tend to 1000; here the stopping condition is either 'until your finger is sore', or 'until two successive values are sufficiently close'.

We shall now show a variety of ways that this can be programmed in TrueBASIC to bring out some points about the programming language, before returning to a calculator program. It is a good idea to include some comments about what the program does, and the REM statement (short for 'REMark') allows this. Values for variables can be assigned with a LET statement, or, if they are constants (as with the parameters b, c, P0) fed in by READ and DATA statements. Output is performed by the PRINT statement, and the simplest form of repeated structure is the DO ... LOOP. Thus a possible implementation of the Verhulst algorithm is:

```
REM Program A.2.1, Verhulst table
READ b  , c     , P0
DATA 0.1, 0.0001, 100
LET n = 0
LET P = P0
DO
     PRINT n, P
     LET deltaP = b*P - c*P^2
     LET P = P + deltaP
     LET n = n + 1
LOOP
END
```

We have followed the convention that 'keywords' are entered in upper-case, and that we try to keep the notation for the variables and parameters as consistent as possible with the mathematics. (Note that TrueBASIC does not need line numbers, that it does need the LET command for assignments, and that every program must have an END.)

We now exhibit some variations on the looping structure. One of these is the DO WHILE ... LOOP form. Thus, suppose we only want to continue when iterating while deltaP is not negligible. Then we can replace the DO line with:

```
DO WHILE deltaP > 0.001   .
```

We might alternatively wish to carry on iterating until (or if) P exceeds some value, and this uses the DO ... LOOP UNTIL ... structure, in which the LOOP line is replaced with, for example,

```
LOOP UNTIL P > 500
```

The other main form, found in all dialects of BASIC, is the counted loop using the FOR ... NEXT ... structure. Suppose we want to apply the iteration 50 times. Then we can remove the lines:

`LET n = 0` and `LET n = n + 1`

and replace the DO line with

```
FOR n = 0 TO 50
```

and the LOOP line with

```
NEXT n
```

Returning to calculators, many have simple programming languages which allow such iterations to be conveniently formed. Some only have one simple looping structure of the form: Lbl .... Goto .. , while others include analogues of the DO ... LOOP (with the WHILE and UNTIL variants) and the FOR ... NEXT loop. A typical simple form is:

```
100 → P
0 → N
Lbl 1
Disp P
0.1P - 0.0001P² → D
P+D → P
N+1 → N
If N<50
Goto 1
```

in which the statement (or clause) following the If statement is only obeyed if the test is true.

Of course it is often preferable to have the results produced as graphs rather than tables or lists of numbers. TrueBASIC makes this incredibly easy: the PRINT statement is replaced by a PLOT statement, and two additional commands are needed to set up the display screen. However, it is always necessary to plan ahead when producing graphs, just as it is when you do one by hand.

In the current example we shall have the value of n on the horizontal axis and that of P on the vertical axis. If we take n to run from 0 to 50 we shall need to guess a suitable range of P, say from 0 to 500, since we can always adjust the scales and run the program again. We now introduce new parameters: nmin, nmax, Pmin, Pmax which are again READ in from a DATA statement. The command: SET MODE ''graphics'' automatically detects the kind of graphics available (e.g., CGA, EGA, VGA, etc.), and the command:

```
SET WINDOW nmin, nmax, Pmin, Pmax
```

fixes the coordinate system to be used.

```
REM Program A.2.2, Verhulst graph
READ b   , c     , P0
DATA 0.1, 0.0001, 100
READ nmin, nmax, Pmin, Pmax
DATA 0   , 50  , 0   , 1000
LET P = P0
SET MODE "graphics"
SET WINDOW nmin, nmax,  Pmin, Pmax
FOR n = nmin TO nmax
     PLOT n, P
     LET deltaP = b*P - c*P^2
     LET P = P + deltaP
NEXT n
END
```

The output from this program is a discrete set of points, which are fairly widely spaced across the screen. In this example it is not mathematically sound to 'join the dots' to give a continuous graph, since the process is a discrete one. But if you want to join each successive point to its predecessor with a line segment, then just terminate the `PLOT n,P` command with a trailing semi-colon: as `PLOT n,P;` which means that an initial point should have been plotted. In fact TrueBASIC is smart enough to realize that in the first case, no line is required, and so it will just `PLOT 0,P0` anyway, though it might be 'better form' to put that line in explicitly just before the `FOR n = nmin TO nmax` line. Similarly you might want to draw axes. Putting the lines:

```
PLOT nmin,0 ; nmax,0
PLOT 0,Pmin ; 0,Pmax
```

immediately after the `SET WINDOW` command will do the trick.

Returning to the programmable calculator, if it is a graphical calculator then it is likely that the programming language will allow you to plot results as with the BASIC program. You will need to set the plotting range, and this is mostly easily done outside the program. Then you just need to change the line which displays `P:Disp P` to a line which plots a point, for example, `PT-On(n,P)`.

Caution: some graphical calculators change the values of certain variables, such as X and Y when graphics mode is used—so try to avoid using the letters X and Y for variables in programs which do graphic plots.

TrueBASIC includes a full set of functions, prefixed with `MAT` for the manipulation of matrices—including inverses, determinants, and so on. Similarly, several programmable graphic calculators now include functions for the manipulation of matrices, vectors, and complex numbers, as well as numerical routines for equation solving, derivatives, and integrals, which can be accessed for programs.

A.3 CONVERTING PROGRAMS BETWEEN VERSIONS OF BASIC

In order to make this as clear as possible we will take a program to plot the graph of a function, and include within it many of the features, such as plotting, testing, looping, functions, and procedures, which vary between dialects. The following program will superimpose the graphs of two functions. The quadratic $f(x)$ is defined in a single line, while the function $g(x)$ illustrates how a 'piecewise defined' function can be incorporated into a multi-line function.

```
REM Program A.3.1a, Graph plotter, TrueBASIC
FUNCTION f(x) = a*x^2 + b*x + c
FUNCTION g(x)
      IF x < 0 THEN g = -x
      IF x >= 0 THEN g = x
END DEF

READ a, b ,c
DATA 1, -2, 1
READ xmin, xmax, ymin, ymax, xstep
DATA -3  , 3   , -2  , 5   , 0.1
SET MODE "graphics"
SET WINDOW xmin,xmax, ymin,ymax
CALL axes(xmin,xmax,ymin,ymax)
FOR x = xmin TO xmax STEP xstep
     LET y = f(x)
     PLOT x,y;
NEXT x
PLOT
FOR x = xmin TO xmax STEP xstep
     LET y = g(x)
     PLOT x,y;
NEXT x
STOP

SUB axes(x1,x2,y1,y2)
     PLOT x1,0 ; x2,0
     PLOT 0,y1 ; 0,y2
END SUB
END
```

The second line is a 'single line function definition'—and this means that any subsequent reference to $f(x)$ will be computed according to this rule. The next four lines are a 'multi-line function definition' of the function $g(x)$ which ensure that a value is returned for g for any value of x in its domain. (Actually this defines the `ABS` function, which is already built-in.) The next four lines read and set the values of the parameters to be used for the program.

After two lines to set the graphics screen display, a procedure is called to draw axes, which passes the parameters `xmin`, `xmax`, `ymin`, `ymax` to the procedure definition, which appears in the penultimate four lines in the `SUB ... END SUB` structure. The body of the program consists of a pair of sequential `FOR ... NEXT` loops, and finishes with a `STOP` statement, which separates the main program from its 'appendices'—in this case just the definition of the `axes` procedure. Finally the `END` statement marks the completion of all the program statements. Note how blank lines can be included to aid readability.

A.3.1 BBC BASIC

The main snag is that there is no equivalent to a `WINDOW` command, and all graphics assume a screen size of 1280 wide, 1024 high (800 for the PC version), with an origin (0, 0) at the bottom left-hand corner of the screen. Thus the origin must be moved, and scale factors `sx`, `sy` calculated for mapping the mathematics coordinate window onto the computer coordinate screen. The program uses line numbers, but note how blank lines (consisting of the colon '`:`' symbol) can be included for readibility. A single line function definition can be included in the main body of the program, which is terminated by the `END` statement, but a multi-line function, like a `PROCedure` definition, must come after the `END` statement.

Because each graphics command, like `MOVE` and `DRAW`, will need its arguments to be multiplied by the scaling constants `sx`, `sy`, it is convenient to build a little library of graphics procedures which can be merged into the program once it has been adapted from the TrueBASIC text.

```
  10 REM Program A.3.1b, Graph plotter, BBCBasic
 100 DEF FNf(x) = a*x^2 + b*x + c
 110 READ a, b ,c
 120 DATA 1, -2, 1
 130 READ xmin, xmax, ymin, ymax, xstep
 140 DATA -3  , 3   , -2  , 5   , 0.1
 500 MODE 1
 510 PROCwindow(xmin,xmax, ymin,ymax)
 520 PROCaxes(xmin,xmax,ymin,ymax)
 530 FOR x = xmin TO xmax STEP xstep
 540   LET y = FNf(x)
 550   IF x = xmin THEN PROCplot(x,y)
 560   PROCdraw(x,y)
 570 NEXT x
 580 FOR x = xmin TO xmax STEP xstep
 590   LET y = FNg(x)
 600   IF x = xmin THEN PROCplot(x,y)
 610   PROCdraw(x,y)
 620 NEXT x
 990 END
 999 :
1000 DEF FNg(x)
1010   IF x<0 THEN g = -x
1020   IF x>=0 THEN g = x
1090 = g
1099 :
1500 DEF PROCaxes(x1,x2,y1,y2)
1510   PROCplot(x1,0) : PROCdraw(x2,0)
1520   PROCplot(0,y1) : PROCdraw(0,y2)
1590 ENDPROC
1599 :
9000 DEF PROCwindow(xl,xr,yb,yt)
9010   LOCAL sw, sh, Ox, Oy
9020   LET sw = 1280 : LET sh = 800
9030   LET sx = sw/(xr - xl) : LET sy = sh/(yt - yb)
9040   LET Ox = -xl*sx : LET Oy = -yb*sy
```

Program A.3.1b continued

```
9050    VDU 29,Ox;Oy;
9090 ENDPROC
9099 :
9100 DEF PROCplot(x,y)
9110    PLOT 69, x*sx, y*sy
9190 ENDPROC
9199 :
9200 DEF PROCdraw(x,y)
9210    DRAW x*sx, y*sy
9290 ENDPROC
```

Note here that line numbers are compulsory. The word LET can be omitted. Several statements can be combined on a single line with the colon symbol ':'. Function definitions have to be prefixed with the letters FN, as in FNf(x). A multi-line function ends with an '=' statement which assigns the value to the function.

A.3.2 QBASIC

This is the version of BASIC provided with more recent versions of Microsoft's MS-DOS operating system. It closely resembles TrueBASIC and the equivalent program is:

```
REM Program A.3.1c, Graph plotter, QBASIC
  DECLARE SUB axes (x1!, x2!, y1!, y2!)
  DECLARE FUNCTION g (x!)
DEF FNf (x) = a * x ^ 2 + b * x + c
READ a, b, c
READ xmin, xmax, ymin, ymax, xstep
SCREEN 2
WINDOW (xmin, ymin)-(xmax, ymax)
CALL axes(xmin, xmax, ymin, ymax)
FOR x = xmin TO xmax STEP xstep
     LET y = FNf(x)
     IF x = xmin THEN PSET (x, y)
     LINE -(x, y)
NEXT x
FOR x = xmin TO xmax STEP xstep
     LET y = g(x)
     IF x = xmin THEN PSET (x, y)
     LINE -(x, y)
NEXT x
WHILE INKEY$ = "": WEND
STOP
DATA 1, -2, 1
DATA -3, 3, -2, 5, 0.1
END

SUB axes (x1, x2, y1, y2)
     LINE (x1, 0)-(x2, 0)
     LINE (0, y1)-(0, y2)
END SUB

FUNCTION g (x)
     IF x < 0 THEN g = -x
     IF x >= 0 THEN g = x
END FUNCTION
```

The QBASIC editor separates out the blocks of a program. Thus the above is held in three blocks: the main program, the subprogram `axes`, and the function `g`. In order to 'tell' the main program that the words `axes` and `g` are defined elsewhere, the main block has to include the `DECLARE SUB` and `DECLARE FUNCTION` statements, which give both the names, `axes` and `g`, and the *type* of the parameter. The exclamation mark '`!`' after the argument x of the function `g` is QBASIC's symbol for a *single precision floating point* number. The other types are *integer, long integer*, and *double precision floating point*. The single-line function definition uses the prefix `FN`, so `FNf` is used in the first `FOR ... NEXT` block. The graphics commands are fairly similar, with `PSET` for `PLOT`, `SCREEN` for `SET MODE`, etc. Note, too, that `DATA` statements have to appear *after* the main program block. The `WHILE ... END` line keeps the graphics display until a key is pressed.

A.3.3 GW-Basic (and BASICA)

These are the original programming languages provided with earlier versions of MS-DOS and PC-DOS. GW-Basic is the least structured of the four versions of BASIC considered here. The only 'block structure' is the subroutine and there is no neat way of passing parameters as arguments of procedures or functions when they are rewritten as subroutines. The equivalent program is:

```
  10 REM Program A.3.1d, Graph plotter, GWBasic
 100 DEF FNF(X) = A*X^2 + B*X + C
 110 READ A, B ,C
 120 DATA 1, -2, 1
 130 READ XMIN, XMAX, YMIN, YMAX, XSTEP
 140 DATA -3  , 3   , -2  , 5   , 0.1
 500 SCREEN 2
 510 WINDOW (XMIN,YMIN) - (XMAX,YMAX)
 520 GOSUB 1500 : REM CALL axes(xmin,xmax,ymin,ymax)
 530 FOR X = XMIN TO XMAX STEP XSTEP
 540    LET Y = FNF(X)
 550    IF X = XMIN THEN PSET (X,Y)
 560    LINE -(X,Y)
 570 NEXT X
 580 FOR X = XMIN TO XMAX STEP XSTEP
 590    GOSUB 1000 : REM FUNCTION g(x)
 595    LET Y = G
 600    IF X = XMIN THEN PSET (X,Y)
 610    LINE -(X,Y)
 620 NEXT X
 990 END
 999 :
1000 REM FUNCTION g(x)
1010      IF X < 0 THEN G = -X
1020      IF X >= 0 THEN G = X
1090 RETURN
1099 :
1500 REM SUB axes(xmin,xmax,ymin,ymax)
1510    LINE (XMIN,0) - (XMAX,0)
1520    LINE (0,YMIN) - (0,YMAX)
1590 RETURN
```

Unfortunately the GW-Basic editor automatically converts any variable name from lower-case to upper-case, which does not aid readability. To make some sort of structure apparent we have started the two blocks at lines 1000 and 1500 with `REM` statements to explain what they do. They take their arguments from variables which have values defined in the main program block. The `axes` subroutine just does some drawing, and does not need to `return` any values to the main program. However, the function definition block `g` calculates a value for `G` based on the value of the variable `X` from the main program, and this is imported for use in line 595.

Otherwise note that the single-line function, and the graphic commands are just like those in its successor, QBASIC.

We hope that this example, and these variations on it, are sufficient to allow readers to adapt the TrueBASIC programs of the text to their own systems.

A.4 OTHER SOFTWARE FOR MATHEMATICS

There are many packages now, ranging from free *public domain* software, through *shareware*, and *student editions*, to powerful (and expensive) mathematical tools. Some examples are now given.

A.4.1 Graphing

Effectively most graphic calculators are convenient embodiments of graph-plotting software. Computer versions, of course, can give access to greater resolution, speed, and colour. Whether on PC or hand-held there are packages which allow ordinary Cartesian plots, in the form $y = f(x)$, parametric plots in the form $x = f(t)$, $y = g(t)$, polar plots in the form $r = f(\theta)$, and polar parametric plots in the form $r = f(t)$, $\theta = g(t)$. Most of these will allow 'zooming' in or out to enable portions of the graph to be scanned—which makes the task of root-finding, or searching for local maxima and minima, very simple. Many allow the superimposing of other features such as tangents, normals, numerical derivatives, numerical integrals, and so on. Some also allow transformations of the plane by, for example, rotations, enlargements etc, and by a specified 2×2 matrix. Some also allow conic plots from the quadratic forms of Chapter 7, and implicit plots of the form $f(x, y) = c$.

As well as two-dimensional plotting programs there is software for the display and manipulation of three-dimensional images, which are useful for exploring surfaces of the form $z = f(x, y)$, etc.

A.4.2 Geometry

There are packages which allow the construction of geometric objects in a Euclidean form, free of a coordinate system. These allow the construction of points according to some rule, and the exploration of loci and envelopes by applying the rule to many cases. Thus they are particularly suited to exploring, for instance, some of the ideas in Chapter 5.

A.4.3 Modelling software and numerical toolkits

There are some powerful packages particularly designed for the exploration of discrete dynamic models, which allow the iterating rule, such as in Chapter 2, to be entered in an editor, and which take care of the screen layout and presentation through a menu system. Some sophisticated versions allow the user to specify the technique to be used for the step-by-step approximation to continuous change (such as a Runge–Kutta method). Similarly there are numerical toolkit packages which can also be used to explore approximations to the continuous models of Chapter 3.

A.4.4 Matrices

There are specific packages to perform matrix calculations, such as determinant, inverse, eigenvalues, and so on. Some of these are now built into graphic calculators. Thus it is possible to explore, say, the function $\det(A - xI)$ for a given matrix A, as a function of x and to display its graph, locate its roots, etc. Such packages are particularly useful to explore the ideas of Chapter 7.

A.4.5 Complex arithmetic and functions

Again there are a number of packages, and calculators, which will manipulate complex numbers, and evaluate complex functions. For example, on some calculators it is possible to perform complex iterations, such as those of Chapter 9, using the technique of section A.2 of this chapter.

A.4.6 All singing, all dancing packages

Two major types of mathematical software are currently converging into very powerful and versatile systems. One is the numerical system designed to help with all sorts of calculations, such as multiple integrals, partial derivatives, line integrals, Fourier transforms, spline fitting, and so on, that are the stock in trade of the mathematical computer-aided designer. Such systems are geared towards the production of printed reports and allow easy manipulation of text and graphics. The other type is the symbolic system designed to perform the manipulation of symbolic expressions in, for example, algebra and calculus, such as factorizing, expanding, finding sums, integrals, derivatives, etc. The merging of these facilities into unified packages represents a significant advance in the range of tools available to the mathematical learner and researcher.

Thus it is not, by any means, essential for a mathematician to be able to write computer programs to use a computer for mathematical exploration. However, there are bound to be some problems for which the reader may not have (or be able to afford) a suitable off-the-peg software package. Thus we still contend that it is *highly desirable* for a student, or practitioner, of mathematics to know enough about programming to be able to read, write, or adapt a program which embodies a mathematical algorithm. At least such a program is *transparent* in that its workings are open to scrutiny and can be checked for correctness, whereas nearly every software package contains a number of *bugs* which may or may not affect the results in a particular case. We try throughout to caution readers not to accept without question any computer output. Through some experience with elementary programming, readers can begin to appreciate the differences between the ways in which humans and computers can tackle mathematics, and see the source of some of the spurious results which may be obtained.

B

Linear mathematics in $\mathbb{R}^3$

Calculus was designed to study complicated functions, or at least those which, when we magnify their graphs, look like pieces of straight lines or planes. More correctly, they are 'locally' like linear functions. These are the simplest to handle, so this chapter lays out the essentials of linear mathematics, as a basis that we must understand before we can really grasp the modern approach to calculus.

We shall assume that readers have met some coordinate geometry in at least two dimensions. In this chapter, we shall gather some language and results that apply to three-dimensional space, and concentrate on what can be described by linear equations. Thus we shall need some vector algebra, and the manipulation of matrices of order 2 and 3. This algebraic and geometrical material is needed later for doing calculus, and we present only the minimum for that, our primary purpose. Therefore, we cannot give as much detail as a text dedicated to linear algebra; so we hope that readers will turn to such a text if they feel the need for further practice and deeper understanding.

B.1 MATRICES AND EQUATIONS IN THE PLANE

Readers will have begun coordinate geometry by dealing with points $P = (x, y)$. Each coordinate is a real number, the distance of P from one of the coordinate axes $\mathbf{0}x$, $\mathbf{0}y$; and the set of all such points (x, y) is called the 'coordinate plane', and denoted by $\mathbb{R}^2$. This is a 'model' of our idea of a flat sheet of paper, if we could extend it to infinity in each direction.

We assume familiarity with (at least) numerical matrix notation, in that readers know that a pair of linear simultaneous equations such as

$$-2x + 3y = 1, \qquad 5x + y = -2$$

can be written in the shorter form

$$\begin{pmatrix} -2 & 3 \\ 5 & 1 \end{pmatrix}\begin{pmatrix} x \\ y \end{pmatrix} = \begin{pmatrix} 1 \\ -2 \end{pmatrix}. \tag{B.1.1}$$

When we solve the equations, we perform certain manipulations on them. To see what is happening in a general case we now use literal coefficients (a move analogous to moving from 'sums' in arithmetic to formulae with an 'x' in algebra). Thus we consider a pair of linear equations

$$ax + by = u, \qquad cx + dy = v \tag{B.1.2}$$

and write them in matrix form

$$\mathbf{w} = A\mathbf{z}, \qquad A = \begin{pmatrix} a & b \\ c & d \end{pmatrix}, \qquad \mathbf{z} = \begin{pmatrix} x \\ y \end{pmatrix}, \qquad \mathbf{w} = \begin{pmatrix} u \\ v \end{pmatrix}. \tag{B.1.3}$$

Here, we are writing **z** and **w** as column vectors instead of the more usual row vectors (x, y), etc. For typographical reasons we henceforth write $(x, y)^{\mathrm{T}}$ instead of $\begin{pmatrix} x \\ y \end{pmatrix}$ when needed for matrix work, and otherwise use (x, y) for ordinary coordinates.

For example, consider the standard unit vectors

$$\mathbf{i} = (1, 0), \qquad \mathbf{j} = (0, 1).$$

Then with the usual operations on vectors (which we assume are familiar to readers) we have

$$x\mathbf{i} + y\mathbf{j} = (x, 0) + (0, y) = (x, y).$$

By direct substitution in (B.1.3) we see that $A\mathbf{i}$, $A\mathbf{j}$ are the columns of A, that is,

$$A\mathbf{i} = (a, c)^{\mathrm{T}} = \begin{pmatrix} a \\ c \end{pmatrix}, \qquad A\mathbf{j} = (b, d)^{\mathrm{T}} = \begin{pmatrix} b \\ d \end{pmatrix}. \tag{B.1.4}$$

To solve the equations (B.1.2) we can eliminate y by multiplying the first equation by d, the second by b, and subtracting, to get

$$\Delta x = du - bv, \qquad \text{where } \Delta = ad - bc.$$

Similarly $\Delta y = -cu + av$. The number Δ is called the **determinant** of the matrix A and we write

$$\Delta = \det(A) = ad - bc. \tag{B.1.5}$$

Therefore, if $\Delta \neq 0$ we can write the solution of (B.1.2) (or (B.1.3)) in matrix form as

$$\mathbf{z} = A^{-1}\mathbf{w}, \qquad \text{where } A^{-1} = \frac{1}{\det(A)}\begin{pmatrix} d & -b \\ -c & a \end{pmatrix} \tag{B.1.6}$$

and A is then called an **invertible** matrix.

Here we have a unique solution, as we expect because the equations in (B.1.1) represent straight lines L, M which 'usually' intersect in a single point. But if L is parallel to M then there is no intersection. It might happen that $L = M$ so one equation is a multiple of the other (in which case there is an infinity of solutions—every point on L). In either of these cases one row of A is a multiple of the other, so by (B.1.3) we have $\Delta = 0$. All this may seem unduly fussy, but the 'exceptional' case of a denominator being zero corresponds to quite reasonable geometry (see Fig. B.1.1).

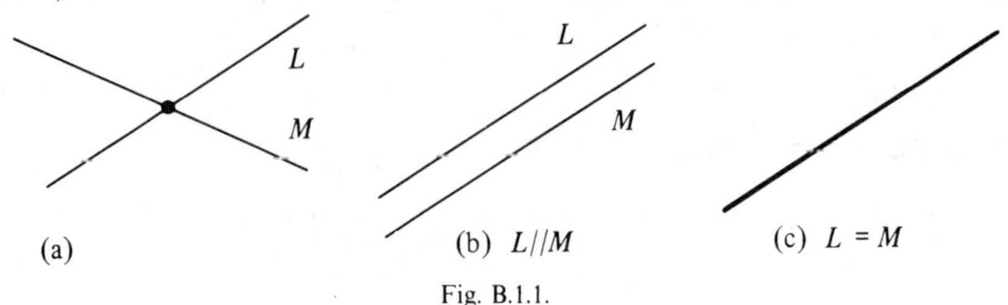

Fig. B.1.1.

The equations (B.1.2) are said to be **homogeneous** if $u = v = 0$. In that case there is always at least one solution, $(x, y) = \mathbf{0}$ (the origin), and this is known as the 'trivial' solution. If the two lines L, M coincide, then every point on that line is a solution and all but $\mathbf{0}$ are said to be 'non-trivial' solutions. Note that in this discussion we assume always that A is not the zero matrix (i.e., $A \neq 0$), but see exercise B.1.1(4) below.

Once we have done this algebra, we need not repeat it, provided we can later appeal to a summary; in the 'trade' (i.e., mathematics), such summaries are called variously theorems, lemmas or propositions (according to size), and they simply describe the facts of mathematics. Their proofs are the explanations of why these facts are so. In particular, we can summarize the above findings in the next theorem.

Theorem B.1.1

The equations (B.1.2) have the unique solution (B.1.4) if and only if $\det(A) \neq 0$. *In the homogeneous case, there is a non-trivial solution if and only if* $\det(A) = 0$.

Exercises B.1.1

1. Show that the matrix A of (B.1.1) is invertible. Use the formula (B.1.4) to write down A^{-1}. Notice how A^{-1} is obtained from A, and practice by writing down several 2×2 matrices of numbers, and calculating their inverses.
2. Use the formula (B.1.4) to calculate the solution of (B.1.2), and check your answer by substitution in the equations.
3. In the matrix of (B.1.1), show that there is just one number p that can be substituted for the '1' in the 'd' position, to give a zero determinant. Solve the resulting equations. Change u, v to zero and find all non-trivial solutions. Sketch the line upon which these lie.
4. To 'solve' the equations (B.1.2) means to find all points of the plane $\mathbb{R}^2$ that satisfy them. Hence show that if $A = 0$, then there are no solutions unless $(u, v) = 0$, in which case every point in $\mathbb{R}^2$ is a solution.
5. Find a 2×2 matrix A with all its entries non-zero for which $A^2 = 0$.

B.2 THE SPACE $\mathbb{R}^3$

After working with two variables x, y, we can try to work similarly with three variables x, y, z. Thus we can think of three axes $\mathbf{0}x$, $\mathbf{0}y$, $\mathbf{0}z$ in three-dimensional space

$\mathbb{R}^3$, each pair being mutually perpendicular (or **orthogonal**), and forming a coordinate plane. Then each point P in space has three coordinates (x, y, z), where x is the distance of P from the coordinate plane formed by $\mathbf{0}y$ and $\mathbf{0}z$, with y, z defined similarly (see Fig. B.2.1). Again this is a 'model' of the everyday space in which we live, but is not an accurate model of astronomical space. In particular, we can think of $\mathbb{R}^2$ as a subset with $z = 0$. Lines are the simplest objects in $\mathbb{R}_2$ after the points, but it turns out that in $\mathbb{R}^3$, a line has a more complicated algebraic description. After the points, the next most simple object in $\mathbb{R}^3$ is a plane, and this is described by a linear equation of the form

$$ax + by + cz = h, \tag{B.2.1}$$

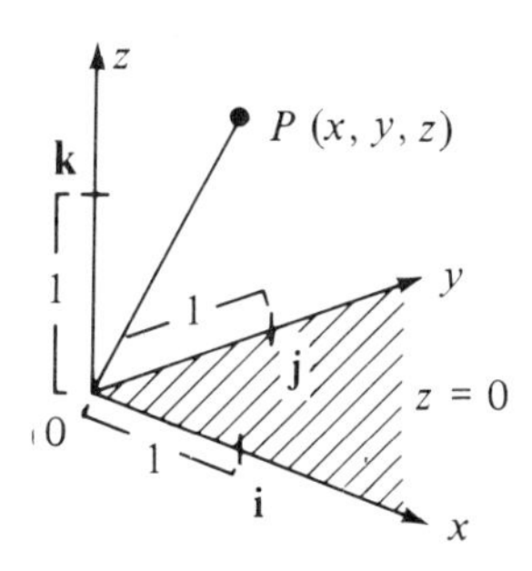

Fig. B.2.1.

where a, b, c, h are constants (not all zero). To see this, we shall work with a systematic notation, at the same time introducing some vector calculus which may be unfamiliar. It is always possible to retrieve results about the plane $\mathbb{R}^2$ by taking $z = 0$, as in Fig. B.2.1.

Because of these digressions, it will take us until (B.3.3) below, before we can establish (B.2.1).

We shall use systematically the notation $\mathbf{p} = (p_1, p_2, p_3)$ to give the coordinates of a point P, although we shall sometimes use $\mathbf{r} = (x, y, z)$. The triplets $\mathbf{p}, \mathbf{r}$ are often called 'vectors'. The reader is probably acquainted with the operations of vector addition, and scalar multiplication, which are given by the rules:

$$\begin{aligned} \mathbf{p} + \mathbf{q} = (p_1, p_2, p_3) + (q_1, q_2, q_3) &= (p_1 + q_1, p_2 + q_2, p_3 + q_3), \\ c\mathbf{p} &= (cp_1, cp_2, cp_3), \end{aligned} \tag{B.2.2}$$

when c is any real number (i.e., a 'scalar'). The first of these rules is called the 'parallelogram law of addition', and is illustrated in Fig. B.2.2. If each third coordinate is zero, then we have the usual rules for $\mathbb{R}^2$.

The **standard basis vectors** are the points $\mathbf{i} = (1, 0, 0)$, $\mathbf{j} = (0, 1, 0)$, $\mathbf{k} = (0, 0, 1)$ along the coordinate axes. Here $\mathbf{i}$ and $\mathbf{j}$ lie in the plane $z = 0$, and correspond to the unit vectors in (B.1.4). As a check on understanding, readers should verify the following claim.

Claim B.2.1
If $\mathbf{p} = (p_1, p_2, p_3)$ *then* $\mathbf{p} = p_1\mathbf{i} + p_2\mathbf{j} + p_3\mathbf{k}$.

The **norm** of the vector **p** is the length of the line segment **0**P and is denoted by $\|p\|$ (read 'norm p') or just by the letter p. By Pythagoras' theorem, this can be shown (see Fig. B.2.3) to be given by

$$p = \sqrt{(p_1^2 + p_2^2 + p_3^2)}. \tag{B.2.3}$$

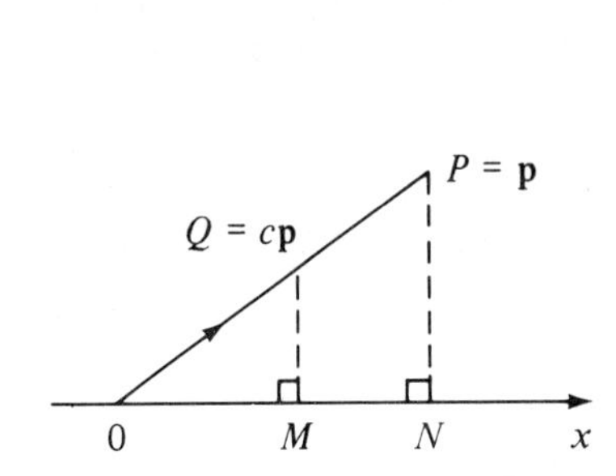

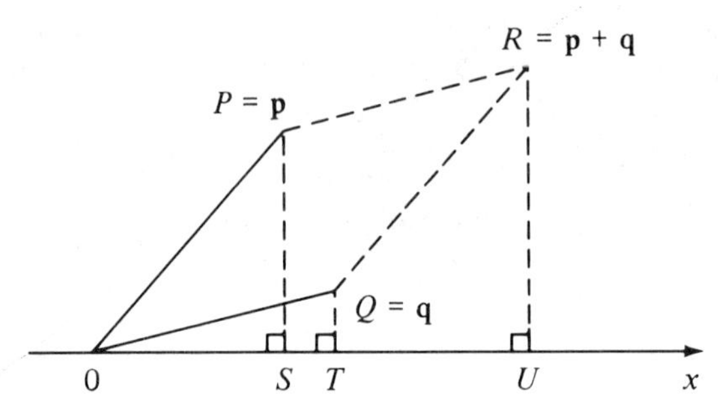

Fig. B.2.2.

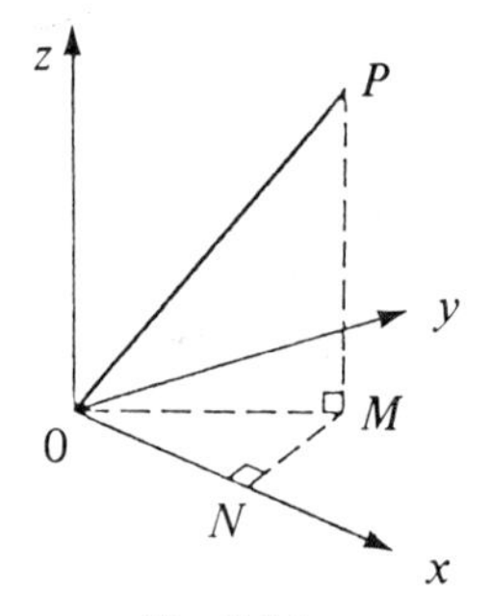

Fig. B.2.3.

Note that some books on mechanics use the term 'vector' to mean a directed line segment, sometimes calling those which start at **0** 'bound vectors'. The tip of such a bound vector is what we (and most mathematicians) call a 'vector'. The others are rather harder to handle, and we shall not be needing them.

Exercises B.2.1

1. Calculate $\|\mathbf{p}\|$, $\|\mathbf{q}\|$, $\mathbf{p} + \mathbf{q}$, $\mathbf{p} - \mathbf{q}$, $3\mathbf{p} - 5\mathbf{q}$, when $\mathbf{p} = (-1, 2, 3)$, and $\mathbf{q} = (7, 4, -5)$.
2. Show that each of the standard basis vectors $\mathbf{i}, \mathbf{j}, \mathbf{k}$ has norm 1. Let $\mathbf{a} = \mathbf{i} - 2\mathbf{j} + \mathbf{k}$, $\mathbf{b} = 3\mathbf{i} + \mathbf{j} - 4\mathbf{k}$, $\mathbf{c} = 2\mathbf{i} - \mathbf{j} + 3\mathbf{k}$. Find $\mathbf{a} - \mathbf{b}$, $\mathbf{a} + \mathbf{c}$, $\|\mathbf{a} - 2\mathbf{b} + \mathbf{c}\|$.
3. In $\mathbb{R}^2$, the circle of centre **a** and radius $s \geq 0$ is defined to be the set of all points $\mathbf{r} = (x, y)$ such that $\|\mathbf{a} - \mathbf{r}\| = s$. By squaring and simplifying, show that (x, y) must satisfy an equation of the form $x^2 + y^2 + 2gx + 2fy + c = 0$. Conversely, show that if (x, y) satisfies this equation and $d = f^2 + g^2 - c \geq 0$, then $\|\mathbf{a} - \mathbf{r}\| = \sqrt{d}$, where $\mathbf{a} = (-g, -f)$.
4. In $\mathbb{R}^3$ the **sphere** of centre **a** and radius $s \geq 0$ is defined to be the set of all points $\mathbf{r} = (x, y, z)$ such that $\|\mathbf{a} - \mathbf{r}\| = s$. By squaring and simplifying, show that (x, y, z) must satisfy an equation of the form $x^2 + y^2 + z^2 + 2gx + 2fy + 2hz + c = 0$. Conversely, show that if (x, y, z) satisfies this equation and $d = f^2 + g^2 + h^2 - c \geq 0$, then $\|\mathbf{a} - \mathbf{r}\| = \sqrt{d}$ where

$\mathbf{a} = (-g, -f, -h)$. Show that the plane $z = 0$ cuts this sphere if, and only if, $t = d - (a_3 - z)^2 \geq 0$; and the intersection is a circle, of radius $s - \sqrt{t}$.

5. Assuming equation (B.2.1), find the equation of the plane through the points $(-1, 2, 3)$, $(7, 4, 5)$, $(1, 3, 1)$. Compare your method with that of exercise B.6.1(9) below.

B.3 THE SCALAR PRODUCT

We shall need to show how a pair of vectors can be 'multiplied'. There are two different ways of multiplying, and one of these gives a third vector, whereas the other gives not a vector but a scalar. The latter is easiest: it is the **scalar product**, given by:

$$\mathbf{p} \cdot \mathbf{q} = p_1 q_1 + p_2 q_2 + p_3 q_3. \tag{B.3.1}$$

Thus the first equation in (B.1.2), and the equation (B.2.1), can be written in the brief forms

$$(a, b) \cdot (x, y) = u \qquad \text{and} \qquad \mathbf{v} \cdot \mathbf{r} = h \qquad (\text{with } \mathbf{v} = (a, b, c),\ \mathbf{r} = (x, y, z)).$$

Note that $\mathbf{p} \cdot \mathbf{p} = p^2$ by (B.2.3); we shall see below that

$$\mathbf{p} \cdot \mathbf{q} = pq \cos \theta, \tag{B.3.2}$$

where θ is the angle $\angle P\mathbf{0}Q$ between the line segments $\mathbf{0}P, \mathbf{0}Q$. Therefore these segments are **orthogonal** (i.e., $\theta = 90^\circ$) iff* $\mathbf{p} \cdot \mathbf{q} = 0$. This orthogonality condition is very important.

B.3.1 Planes

We can now derive the equation (B.2.1) of a plane Π. That equation is an algebraic description of a geometric object, so first we must give a geometric description of what a plane is. Several such descriptions are possible, and we choose the one that gives the quickest route to equation (B.2.1). Thus, by a 'plane with normal $\mathbf{0}P$' we shall mean the set of all points R in $\mathbb{R}^2$ for which the line segments $\mathbf{0}P$ and PR are perpendicular (see Fig. B.3.1). Readers may feel that a plane ought to have other properties, but these may be deducible (as in the exercises B.3.1) from the definition just given. Therefore we ask that the above definition be granted, to see what happens.

Let then Π be a plane in the sense of this definition, and let $\mathbf{0}P$ be the normal from $\mathbf{0}$ to Π. To obtain the equation (B.2.1) we must translate the geometric description into algebra. Let P have coordinates $\mathbf{p} = (p_1, p_2, p_3)$. Then $\mathbf{u} = (1/p)\mathbf{p}$ has norm 1 (so we call it a **unit** vector). Now, for any point $R = \mathbf{r} = (x, y, z)$ in Π, $\mathbf{0}P$ is perpendicular to PR, so we have (as in Fig. B.3.1)

$$\mathbf{0}P = \mathbf{0}R \cos \theta = \mathbf{u} \cdot \mathbf{r}, \tag{B.3.3}$$

so the coordinates of R satisfy the linear equation $u_1 x + u_2 y + u_3 z = p$. If we multiply through by h/p, we see that $\mathbf{r}$ satisfies the equation $\mathbf{v} \cdot \mathbf{r} = h$, where $\mathbf{v} = (h/p)\mathbf{u}$, and this is of the form (B.2.1) with $(a, b, c) = \mathbf{v}$. Conversely, if $\mathbf{r}$ satisfies (B.2.1), then

* iff is an abbreviation for 'if, and only if'.

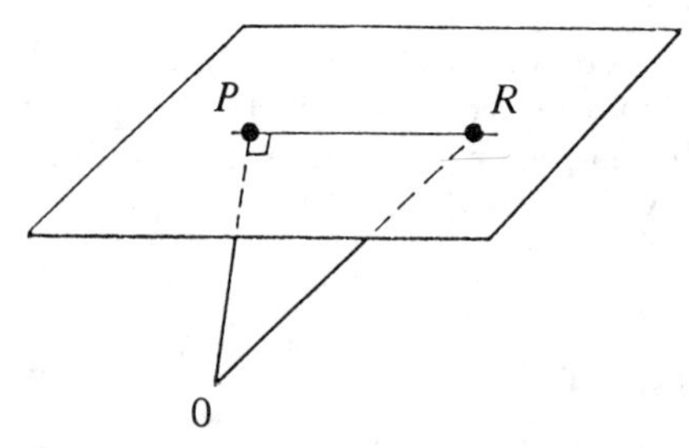

Fig. B.3.1.

$\mathbf{v}\cdot\mathbf{r} = h$, and $\mathbf{v} \neq \mathbf{0}$; so dividing through by v we have $\mathbf{u}\cdot\mathbf{r} = h/v$, with the same unit vector $\mathbf{u}$ as before. Therefore, $\mathbf{0}R\cos\theta = u/v = p = \mathbf{0}P$, so PR is perpendicular to $\mathbf{0}P$, whence R lies on the plane Π.

Exercises B.3.1

1. Find the scalar products of each of the pairs of vectors in exercise B.2.1(1). Now find the angles between the pairs. Find the unit vector, in the direction of each vector in the exercise.
2. Express the plane $x - 2y + 3z = 1$ in the form $\mathbf{v}\cdot\mathbf{r} = h$. Find the equation of the line in which it cuts $\mathbb{R}^2$. Find the lines in which it cuts the other two coordinate planes.
3. Show that any two of the standard basis vectors are orthogonal.
4. Verify that the scalar product is 'distributive over addition', that is, that $\mathbf{p}\cdot(\mathbf{q}+\mathbf{r}) = \mathbf{p}\cdot\mathbf{q} + \mathbf{p}\cdot\mathbf{r}$. (Try it first in $\mathbb{R}^2$.) Use this to economize the working out of $(2\mathbf{i} - 5\mathbf{j} + 3\mathbf{k})\cdot(-7\mathbf{i} - \mathbf{j} + 4\mathbf{k})$.
5. Use (B.3.2) to show that $|\mathbf{p}\cdot\mathbf{q}| \leq \|\mathbf{p}\|\,\|\mathbf{q}\|$ with equality only when $\mathbf{0}, \mathbf{p}, \mathbf{q}$ are collinear. Verify the inequality when $\mathbf{p} = (2, 9, -1), \mathbf{q} = (-4, 7, 6)$. Hence prove the **triangle inequality**:

   $$\|\mathbf{p} - \mathbf{q}\| \leq \|\mathbf{p} - \mathbf{r}\| + \|\mathbf{r} - \mathbf{q}\|.$$

   (This is obvious from a geometrical picture (draw one!), but it is a good exercise to find an algebraic proof: put $\mathbf{a} = \mathbf{p} - \mathbf{r}$, $\mathbf{b} = \mathbf{r} - \mathbf{q}$, and show that $\|\mathbf{a}+\mathbf{b}\|^2 = (\mathbf{a}+\mathbf{b})\cdot(\mathbf{a}+\mathbf{b}) \leq \mathbf{a}^2 + \mathbf{b}^2 + 2\,\|\mathbf{a}\|\,\|\mathbf{b}\|$, where $\mathbf{a}^2 = \mathbf{a}\cdot\mathbf{a}$.)
6. If each coordinate of $\mathbf{p}$ and $\mathbf{q}$ is a function of t, show that $\mathrm{d}/\mathrm{d}t\,(\mathbf{p}\cdot\mathbf{q}) = \mathbf{p}\cdot\mathbf{q}' + \mathbf{p}'\cdot\mathbf{q}$.
7. A particle, whose position in $\mathbb{R}^3$ at time t is $\mathbf{x}(t)$, is constrained to move on the unit sphere $S = \{\mathbf{x} \in \mathbb{R}^3 : \|\mathbf{x}\| = 1\}$. Show that its velocity vector $\mathbf{x}'(t)$ is orthogonal to $\mathbf{x}(t)$ for all time. (Differentiate $\mathbf{x}^2$ as a function of t.)
8. Use (B.2.3) in the cosine formula $c^2 = p^2 + q^2 - 2pq\cos C$, when (as in Fig. B.3.2), $\mathbf{c} = \mathbf{p} - \mathbf{q}$, to obtain (B.3.3) with $\theta = C$.
9. Let Π be a plane with equation $\mathbf{v}\cdot\mathbf{r} = h$, and let S be any point. Show that the distance d from S to Π is $(\mathbf{v}\cdot\mathbf{s} - h)/v$. (In the notation of (B.3.3), use the fact that $d = \mathbf{0}S\cos\theta - \mathbf{0}\mathrm{P}$.)
10. The planes $\mathbf{a}\cdot\mathbf{r} = l$, $\mathbf{b}\cdot\mathbf{r} = m$ are said to be **parallel** if they have no common point. Show that they are parallel if there is a number t such that $\mathbf{a} = t\mathbf{b}$ but $l \neq tm$; and that they are equal if $\mathbf{a} = t\mathbf{b}$ and $l = tm$. What set would the equation $\mathbf{a}\cdot\mathbf{r} = l$ represent if $\mathbf{a} = \mathbf{0}$ and (i) $l \neq 0$, (ii) $l = 0$?

B.4 MATRIX MULTIPLICATION IN $\mathbb{R}^2$

Readers will probably have met the row-by-column multiplication that is needed for multiplying two 2×2 matrices; but whether or not, we now explain it in terms of the scalar product. In this section we work in $\mathbb{R}^2$, and later we shall see how the results can be extended to $\mathbb{R}^3$.

The rule for multiplying matrix A by matrix B is:

$$AB = \begin{pmatrix} A_1 B_1^T & A_1 B_2^T \\ A_2 B_1^T & A_2 B_2^T \end{pmatrix}, \tag{B.4.1}$$

where A_i is the ith row of A and B_j^T is the jth column of B. At first sight, this rule may seem peculiar, but we shall see below how it arises quite naturally. Also it has the merit of ensuring that 'det is multiplicative', which is to confirm what readers will probably know in numerical cases—the general rule for the determinant of a product:

$$\det(AB) = \det(A)\det(B). \tag{B.4.2}$$

Curiously, there is no known elegant proof of this quite old result, and we leave the reader to do what we all must do at least once in our lives—namely slog it out. That is to say, evaluate the product AB, then work out its determinant by the definition (B.4.1), and check that the resulting expression agrees with that obtained by multiplying out the expression for $\det(A)$ with that for $\det(B)$.

With the matrix A in (B.1.3) we associate its **transpose**:

$$A^T = \begin{pmatrix} a & c \\ b & d \end{pmatrix}, \tag{B.4.3}$$

obtained by interchanging rows and columns. Hence the rth row of A^T is the rth column of A itself, and similarly for columns. We can apply this process also to (x, y) (when we regard it as a row vector), and transpose it to the column vector $\begin{pmatrix} x \\ y \end{pmatrix} = (x, y)^T$; similarly we can transpose the column vector $(s, t)^T$ to (s, t). It now follows from the product rule (B.4.2) that *transposition reverses products*:

$$(AB)^T = (B^T)(A^T). \tag{B.4.4}$$

Also we can write the equations (B.1.2) in the alternative form

$$(u, v) = (x, y)(A^T). \tag{B.4.5}$$

Moreover, it follows from (B.1.3) and (B.4.3) that

$$\det(A^T) = \det(A). \tag{B.4.6}$$

All these operations and equations are valid for more variables than two, as we shall see later. Readers should absorb them as absolutely standard mathematics.

Exercises B.4.1

1. Let A be the matrix in (B.1.1) and let $B = \begin{pmatrix} -8 & 2 \\ 7 & -3 \end{pmatrix}$. Calculate AB from the formula (B.4.1) and then transpose your resulting matrix. Also calculate $(A^T)(B^T)$ directly and thus

verify (B.4.4) in this case. By calculating determinants numerically, verify equation (B.4.2).
2. Perform the algebraic manipulation required to establish that det is multiplicative in (B.4.2).
3. By multiplying $(x, y)^{\mathrm{T}}$ by A and then multiplying the result by (x, y), show that the result is $ax^2 + (c + d)xy + dy^2$.
4. Express $3x^2 - 4xy + 5y^2$ in the form $(x, y)(A(x, y)^{\mathrm{T}})$.
5. Use (B.4.1) to multiply the matrix A in (B.1.3) by its inverse A^{-1} in (B.1.6) to show that $AA^{-1} = I = (A^{-1})A$, where I is the **identity matrix** $I = \begin{pmatrix} 1 & 0 \\ 0 & 1 \end{pmatrix}$. Show also that $A = IA = AI$, that $\det(I) = 1$, $\det(A^{-1}) = 1/\det(A)$, and $(A^{\mathrm{T}})^{-1} = (A^{-1})^{\mathrm{T}}$.
6. Let $P = (\mathbf{u} \quad \mathbf{v})$ be a matrix with columns $\mathbf{u}$, $\mathbf{v}$. It is sometimes convenient to think of $\det(P)$ as a product and write $\det(P) = \mathbf{u} \wedge \mathbf{v}$. Show that $\mathbf{u} \wedge \mathbf{v} = -\mathbf{v} \wedge \mathbf{u}$, and this product is linear, that is, $\mathbf{u} \wedge (\mathbf{v} + \mathbf{w}) = \mathbf{u} \wedge \mathbf{v} + \mathbf{u} \wedge \mathbf{w}$.

An example of the use of equation (B.1.2) occurs when we rotate the coordinate axes anticlockwise through an angle θ. For then the new coordinates (x', y') of (x, y) are given (see Fig. B.4.1) by the equations

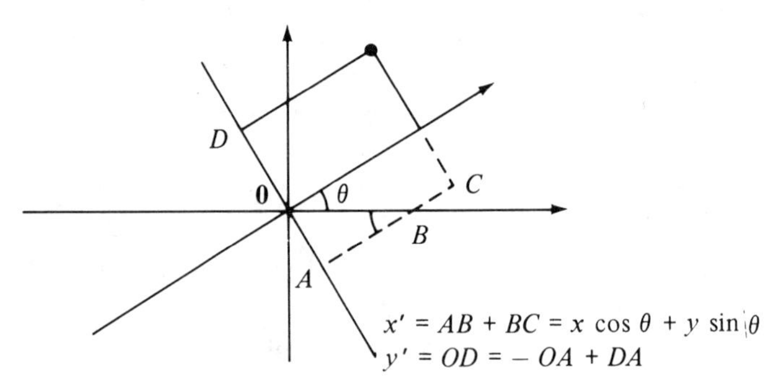

Fig. B.4.1.

$$x' = x \cos\theta + y \sin\theta, \qquad y' = -x \sin\theta + y \cos\theta,$$

or, in matrix form,

$$\mathbf{z}' = R_\theta \mathbf{z}, \qquad R_\theta = \begin{pmatrix} c & -s \\ s & c \end{pmatrix}, \qquad (c = \cos\theta, s = \sin\theta). \tag{B.4.7}$$

The matrix R_θ is called a **rotation matrix** (through the angle θ). If we follow R_θ by a second rotation R_ϕ, then the result is clearly the anticlockwise rotation $R_{\theta+\phi}$ through the angle $\theta + \phi$. But

$$R_{\theta+\phi} = (R_\phi)(R_\theta),$$

and if we multiply the product on the right and compare entries we have the famous addition formulae

$$\begin{aligned} \sin(\theta + \phi) &= \sin\theta \cos\phi + \cos\theta \sin\phi \\ \cos(\theta + \phi) &= \cos\theta \cos\phi - \sin\theta \sin\phi. \end{aligned} \tag{B.4.8}$$

Putting $\theta = \phi$ gives

$$\sin 2\theta = 2 \sin \theta \cos \theta; \qquad \cos 2\theta = \cos^2\theta - \sin^2\theta, \tag{B.4.9}$$

the **double angle formulae**. (Formulae (B.4.8) and (B.4.9) are worth committing to memory, but study their structure closely first to make the memorization easier!)

From (B.4.7) we see that the transpose of R_θ is $R_{-\theta}$, which is a rotation matrix $-\theta$. Hence $R_{-\theta}$ is the inverse of R_θ, so

$$R_\theta^{-1} = R_\theta^{\mathrm{T}}. \tag{B.4.10}$$

B.5 LINES IN SPACE

If $\mathbf{p}, \mathbf{q}$ are two vectors in 3-space $\mathbb{R}^3$, then the line L joining them consists of all vectors $\mathbf{r}$ of the form

$$\mathbf{r} = \mathbf{p} + t(\mathbf{q} - \mathbf{p}) \tag{B.5.1}$$

as we see from Fig. B.5.1. The equation says that to reach $\mathbf{r}$ from $\mathbf{0}$, we must first move to $\mathbf{p}$, and then (using the parallel rule) move parallel to $\mathbf{q} - \mathbf{p}$ for a distance given by the number t (called a **parameter**). Thus L is parallel to the direction of $\mathbf{q} - \mathbf{p}$ (i.e., to the line joining $\mathbf{0}$ to $\mathbf{q} - \mathbf{p}$) and this direction is called the **direction** of L). Note that (B.5.1) is a *vector equation*, and not a scalar equation like (B.1.1) or (B.1.2), but if we compare first coordinates on each side, we have $x = p_1 + t(q_1 - p_1)$, and two similar scalar equations for y and z.

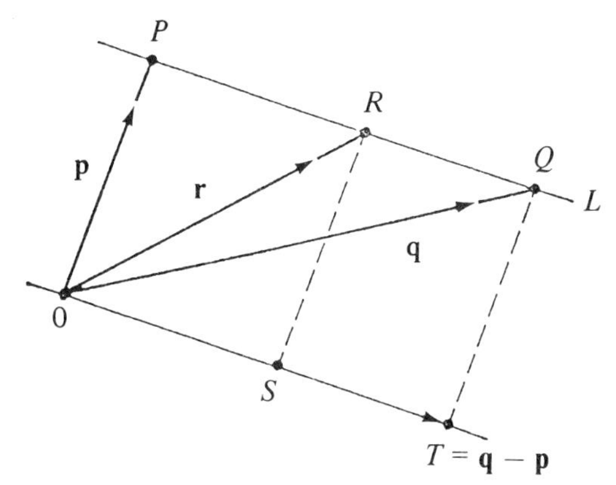

Fig. B.5.1.

Exercises B.5.1

1. Find the vector equations of the lines joining the pairs $(1, 2, 3)$, $(3, 2, 1)$, and $(-1, 2, -3)$, $(7, 5, -2)$. Show that these lines do not intersect.
2. Find the vector equation of the line through $(1, 2, 3)$ with direction $(3, 2, 1)$. Where does it intersect the plane $x - 2y + 7z = 5$?
3. Show that the line (B.5.1) intersects the plane $\mathbf{v} \cdot \mathbf{r} = h$ in just one point, namely when $t = (h - \mathbf{v} \cdot \mathbf{p})/d$, where $d = \mathbf{v} \cdot (\mathbf{q} - \mathbf{p})$, provided $d \neq 0$. (Thus a plane is not the whole of $\mathbb{R}^3$.) What can you say when $d = 0$? (Draw a sketch.)

4. Let **a**, **b** be two points in the plane $\Pi: \mathbf{v}\cdot\mathbf{r} = h$. Show that the whole (infinite) line through **a** and **b** lies in Π. Now let **c** be a third point in Π, and let **r** be a point in the triangular disc with vertices **a**, **b**, **c**. Show that **r** lies in Π. (Show that $\mathbf{r} = u\mathbf{a} + v\mathbf{b} + w\mathbf{c}$ for some positive u, v, w with $u + v + w = 1$, by noting that AR meets BC in some S with $\mathbf{s} = l\mathbf{b} + (1 - l)\mathbf{c}$, with $0 \neq l - 1$, and then $\mathbf{r} = m\mathbf{a} + (1 - m)\mathbf{s}$ with $0 \neq m - 1$.)
5. The plane Π contains the points $(0, 2, 3), (2, -1, -1)$ and $(4, 0, 3)$. Find a vector perpendicular to Π and hence, or otherwise, find the vector equation of the line L through the point $(-3, 1, 5)$ perpendicular to Π. Determine the coordinates of the point at which the line L intersects Π. [SU]

B.6 THE VECTOR PRODUCT

We mentioned above that vectors **p** and **q** can be multiplied to form a new vector. At first sight, the definition seems rather arbitrary, but it is simplest to give a description, and then to establish its inherent reasonableness. We therefore introduce the **vector product** $\mathbf{p} \wedge \mathbf{q}$ as a vector with norm

$$\|\mathbf{p} \wedge \mathbf{q}\| = pq \sin\theta \tag{B.6.1}$$

which is orthogonal to both **p** and **q**, and has coordinates

$$(p_2q_3 - p_3q_2, p_3q_1 - p_1q_3, p_1q_2 - p_2q_1). \tag{B.6.2}$$

Note how each coordinate changes to the next under the cyclic interchange 1-2-3-1, and that the first one does not involve the first subscript 1. By trying the following exercises, readers will (we hope) gain fluency and come to see the reason for choosing such apparently bizarre coordinates.

Exercises B.6.1

1. Calculate $\mathbf{p} \wedge \mathbf{q}$ when $\mathbf{p} = (-1, 2, 3)$ and $\mathbf{q} = (7, 4, -5)$.
2. With the standard basis vectors **i**, **j**, **k** show that $\mathbf{i} \wedge \mathbf{j} = \mathbf{k}$ with similar expressions obtained by cyclic interchange.
3. Verify by substitution in the scalar product formula that $\mathbf{p} \wedge \mathbf{q}$ is orthogonal to each of **p** and **q**.
4. Verify that the vector product is (like the scalar product in exercise B.3.1(4)) 'distributive over addition', that is, that $\mathbf{p} \wedge (\mathbf{q} + \mathbf{r}) = \mathbf{p} \wedge \mathbf{q} + \mathbf{p} \wedge \mathbf{r}$. Use this to economize the working out of $(2\mathbf{i} - 5\mathbf{j} + 3\mathbf{k}) \wedge (-7\mathbf{i} - \mathbf{j} + 4\mathbf{k})$. Also show that $\mathbf{p} \wedge \mathbf{q} = -\mathbf{q} \wedge \mathbf{p}$. Use these properties to show that $(\mathbf{p} + \mathbf{r}) \wedge (\mathbf{q} + \mathbf{r}) = \mathbf{p} \wedge \mathbf{q} + \mathbf{p} \wedge \mathbf{r} + \mathbf{r} \wedge \mathbf{q}$.
5. If each coordinate of **p** and **q** is a function of t, show that $\mathrm{d}/\mathrm{d}t\,(\mathbf{p} \wedge \mathbf{q}) = \mathbf{p} \wedge \mathbf{q}' + \mathbf{p}' \wedge \mathbf{q}$.
6. Using (B.3.2) with the formula $\sin^2\theta = 1 - \cos^2\theta$, obtain (B.6.1) from (B.6.2).
7. Show that $\mathbf{p} \wedge \mathbf{q} = 0$ iff **q** is a scalar multiple $c\mathbf{p}$ of **p**.
8. Suppose that, in (B.6.2), $p_1q_2 - p_2q_1 \neq 0$. Find a vector $\mathbf{r} = (x, y, z)$ which is orthogonal to each of **p** and **q** as follows: solve the equations $\mathbf{p}\cdot\mathbf{r} = 0 = \mathbf{q}\cdot\mathbf{r}$ for x, y in terms of z, to show that **r** must be a scalar multiple of $\mathbf{p} \wedge \mathbf{q}$ as given by (B.6.2).
9. Show that, in exercise B.5.1(4), the vector $\mathbf{q} = (\mathbf{a} - \mathbf{c}) \wedge (\mathbf{b} - \mathbf{c})$ is parallel to the normal to the plane Π through **a**, **b**, **c**; and that $\mathbf{q} = \mathbf{a} \wedge \mathbf{b} + \mathbf{b} \wedge \mathbf{c} + \mathbf{c} \wedge \mathbf{a}$. Hence find the plane through $(-1, 2, 3)$, $(7, 4, -5)$, and $(1, 3, 1)$. (The required plane is of the form $\mathbf{q}\cdot\mathbf{r} = h$, where $h = \mathbf{q}\cdot\mathbf{a}$.)

10. The vectors **a** and **b** are defined by $\mathbf{a} = 3\mathbf{i} - \mathbf{j}$, $\mathbf{b} = \mathbf{i} + 2\mathbf{j} - 3\mathbf{k}$. Find $\mathbf{c} = \mathbf{a} - \mathbf{b}$; $\mathbf{d} = \mathbf{a} \wedge \mathbf{b}$, and the angle between the vectors **c** and **d**. [SU]
11. If $\mathbf{a} = \mathbf{i} + \mathbf{j} + \mathbf{k}$, $\mathbf{b} = \mathbf{j} + \mathbf{k}$, $\mathbf{c} = \mathbf{i}$, find, with as little calculation as possible,

    (a) $\mathbf{a} \wedge \mathbf{b}$, (b) $(\mathbf{a} - \mathbf{b})\cdot(\mathbf{b} - 3\mathbf{a})$, (c) $(\mathbf{a} - \mathbf{c}) \wedge \mathbf{b}$,
    (d) $(\mathbf{b} + 2\mathbf{c}) \wedge \mathbf{a}$, (e) $\mathbf{a}\cdot(\mathbf{b} \wedge \mathbf{c})$, (f) $\mathbf{b}\cdot(\mathbf{a} \wedge \mathbf{c})$,
    (g) $\mathbf{c}\cdot(\mathbf{a} \wedge \mathbf{b})$, (h) $\mathbf{a} \wedge (\mathbf{b} \wedge \mathbf{a})$, (i) $\mathbf{a}\cdot(\mathbf{b} \wedge \mathbf{a})$. [SU]

12. (a) Let $\mathbf{a} = (1, -2, -7)$, $\mathbf{b} = (3, 5, -4)$. Find the angle $\angle\mathbf{a0b}$ and the area of the triangle **a0b**. If also $\mathbf{c} = (2, -1, 3)$ calculate $\mathbf{a}\cdot(\mathbf{b} \wedge \mathbf{c})$, and $\mathbf{a} \wedge (\mathbf{b} \wedge \mathbf{c}) - (\mathbf{a} \wedge \mathbf{b}) \wedge \mathbf{c}$ ($\neq \mathbf{0}$!).
    (b) Find the equataion of the plane Π through **a**, **b**, **c** and the distance between Π and $\mathbf{b} \wedge \mathbf{c}$.
    (c) If l is the line joining **a**, **b**, find its vector equation.
    (d) Find the vector equation of the line parallel to l and through $\mathbf{b} \wedge \mathbf{c}$. [SU]

13. Find unit vectors with the given property:

    (a) parallel to the line $x = 3t + 1$, $y = 12t - 3$, $z = -(t + 2)$;
    (b) orthogonal to the plane $x - 6y + z = 12$;
    (c) orthogonal to both $\mathbf{i} + 2\mathbf{j} - \mathbf{k}$ and to **k**. [SU]

B.7 DETERMINANTS

In (B.1.5), we saw that with every 2×2 matrix M of numbers we can associate the number $\det(M)$. A 3×3 matrix **N** will also be a square array, but we now need to use subscripts to denote its entries. Thus N is of the form

$$N = \begin{pmatrix} a_1 & a_2 & a_3 \\ b_1 & b_2 & b_3 \\ c_1 & c_2 & c_3 \end{pmatrix}$$

and its determinant is defined to be the alternating sum

$$\det(N) = a_1 \det\begin{pmatrix} b_2 & b_3 \\ c_3 & c_3 \end{pmatrix} - a_2 \det\begin{pmatrix} b_1 & b_3 \\ c_1 & c_3 \end{pmatrix} + a_3 \det\begin{pmatrix} b_1 & b_2 \\ c_1 & c_2 \end{pmatrix}. \tag{B.7.1}$$

Those who have met this definition before may find it strange, but it has evolved from experience and turns out to be useful, and also natural. Firstly, the reader should work out some determinants to become fluent.

Exercises B.7.1

1. Evaluate $\det(N)$ when N is each of the matrices

$$I = \begin{pmatrix} 1 & 0 & 0 \\ 0 & 1 & 0 \\ 0 & 0 & 1 \end{pmatrix}, \quad A = \begin{pmatrix} 1 & 1 & 1 \\ 1 & 1 & 1 \\ 1 & 1 & 1 \end{pmatrix}, \quad B = \begin{pmatrix} 1 & 2 & 3 \\ 3 & 2 & 1 \\ 1 & 5 & -7 \end{pmatrix}, \quad C = \begin{pmatrix} a & h & g \\ h & b & f \\ g & f & c \end{pmatrix}$$

$$\begin{pmatrix} 5 & 1 & 8 \\ 15 & 3 & 6 \\ 10 & 4 & 2 \end{pmatrix}, \quad \begin{pmatrix} 4 & 96 & 85 \\ 0 & 1 & -7 \\ 0 & 0 & 6 \end{pmatrix}, \quad \begin{pmatrix} 16 & 22 & 4 \\ 4 & -3 & 2 \\ 12 & 25 & 2 \end{pmatrix}, \quad \begin{pmatrix} 28 & 18 & 24 \\ 12 & 27 & 12 \\ 70 & 15 & 40 \end{pmatrix}, \quad \begin{pmatrix} 28 & 33 & 8 \\ 13 & 17 & 4 \\ 40 & 54 & 13 \end{pmatrix}.$$

You should use the definition (B.7.1) and calculate 'by hand' to get used to the pattern. Then check by computer. Also I is called the 3×3 **identity** matrix, and $\det(C) = bc + ca + ab - (f^2 + g^2 + h^2)$.)

2. 3×3 matrices are added, or multiplied by a number, by operating entry by entry. Thus in exercise 1,

$$3A = \begin{pmatrix} 3 & 3 & 3 \\ 3 & 3 & 3 \\ 3 & 3 & 3 \end{pmatrix}, \qquad A + C = \begin{pmatrix} a+a & 1+h & 1+g \\ 1+h & 1+b & 1+f \\ 1+g & 1+f & 1+c \end{pmatrix}.$$

Evaluate $2A - 7B$, and $\det(I + 3A - 5B)$ $(= X$, say). Is X equal to the sum: $\det I + 3 \det A - 5 \det B$?

3. Show that we may obtain (B.6.2) from a 'determinant':

$$\mathbf{p} \wedge \mathbf{q} = \det \begin{pmatrix} \mathbf{i} & \mathbf{j} & \mathbf{k} \\ p_1 & p_2 & p_3 \\ q_1 & q_2 & q_3 \end{pmatrix}$$

where the determinant is expanded as if $\mathbf{i}, \mathbf{j}, \mathbf{k}$ were numbers. (Some people find this an easier way of remembering the coordinates of $\mathbf{p} \wedge \mathbf{q}$.)

4. Why is it obvious that $(\mathbf{a} \cdot \mathbf{b}) \wedge \mathbf{c} \neq \mathbf{a} \cdot (\mathbf{b} \wedge \mathbf{c})$? (Hence $\mathbf{a} \cdot \mathbf{b} \wedge \mathbf{c}$ can only be interpreted in one way and it is usual to omit the parentheses.)

5. The three vectors $\mathbf{a}, \mathbf{b}, \mathbf{c}$ determine a parallelepiped P, as shown in Fig. B.7.1. The volume V of P is the product of the height h and the area A of the base. Show that $V = \mathbf{a} \cdot \mathbf{b} \wedge \mathbf{c}$. ($A = \|\mathbf{b} \wedge \mathbf{c}\|$, and the height is the projection of the remaining edge on the normal to the base.)

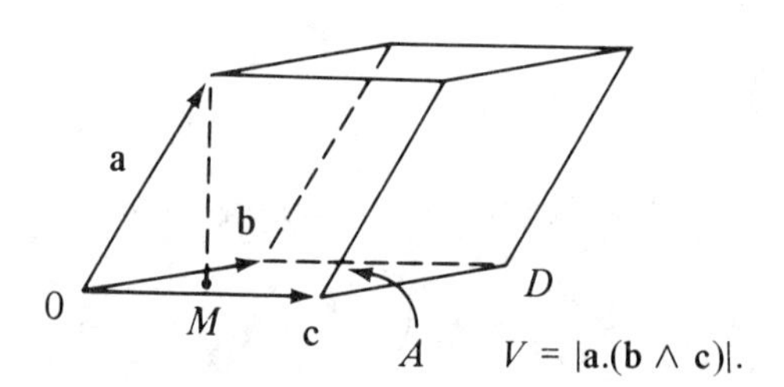

Fig. B.7.1.

6. Show that in (B.7.1), $\det(N) = \mathbf{a} \cdot \mathbf{b} \wedge \mathbf{c}$, where the rows of N are the vectors $\mathbf{a}, \mathbf{b}, \mathbf{c}$. Show that if two of these vectors are interchanged, then the determinant Δ of N changes sign, but if they are all interchanged cyclically, then Δ (unlike N) is unaltered. Hence show that if two rows of N are equal then $\det N = 0$, and if two rows of N are added, then the resulting matrix has determinant equal to $\det N$. Now show that if $A = \det \begin{pmatrix} 1 & p & p^2 \\ 1 & q & q^2 \\ 1 & r & r^2 \end{pmatrix}$ then $A = 0$ when $p = q$, and hence factorize A. Similarly, by subtracting appropriate rows to simplify matters, calculate

$$\det \begin{pmatrix} b^2+c^2 & a^2 & a^2 \\ b^2 & c^2+a^2 & b^2 \\ c^2 & c^2 & a^2+b^2 \end{pmatrix}, \quad \text{and} \quad \det \begin{pmatrix} b^2+c^2 & ab & ca \\ ab & c^2+a^2 & bc \\ ca & bc & a^2+b^2 \end{pmatrix}.$$

7. Show that $\det(N) = \det(N^{\mathrm{T}})$, where N^{T} is the transposed matrix of N, obtained by reflection in the diagonal, so its rows are the columnns of N.
8. If each vector in exercise 6 is a function of t, show that $\mathrm{d}\Delta/\mathrm{d}t$ is a sum of three determinants, each obtained from N by differentiating one row.
9. The **eigenvalues** of N in exercise 7 are by definition the roots of the cubic $f(\lambda) = \det(N - \lambda I)$, where I is the identity matrix, as in exercise 1. Show that

$$f(\lambda) = (\mathbf{a} - \lambda\mathbf{i})\cdot(\mathbf{b} - \lambda\mathbf{j}) \wedge (\mathbf{c} - \lambda\mathbf{k}) = -\lambda^3 + T\lambda^2 - P\lambda + D,$$

where $T = T(\mathbf{a}, \mathbf{b}, \mathbf{c})$ is the **trace** $a_1 + b_2 + c_3$ of N, and $P = T(\mathbf{b} \wedge \mathbf{c}, \mathbf{c} \wedge \mathbf{a}, \mathbf{a} \wedge \mathbf{b})$.
10. Use the pattern inherent in (B.7.1) to suggest how you expect the determinant of a 4×4 matrix to be defined. (Your suggestion ought to have properties similar to those given in exercise 6.) Evaluate

$$\det\begin{pmatrix} 1 & 2 & -1 & 2 \\ 3 & 0 & 1 & 5 \\ 1 & -2 & 0 & 3 \\ -2 & -4 & 1 & 6 \end{pmatrix}, \quad \det\begin{pmatrix} -2 & 8 & -3 & 7 \\ -4 & 1 & 4 & -2 \\ -5 & -8 & 4 & -2 \\ 8 & 16 & -6 & 15 \end{pmatrix}.$$

B.8 INTERSECTIONS

Recall that theorem B.1.1 summarizes what can happen when we solve two simultaneous linear equations in two unknowns. It corresponds to seeking the intersection of two lines in $\mathbb{R}^2$. Now consider the analogous problem for planes in $\mathbb{R}^3$. Let A, B be two planes with equations $\mathbf{a}\cdot\mathbf{r} = l$, $\mathbf{b}\cdot\mathbf{r} = m$. Their intersection I is now the set of points r that satisfy these equations simultaneously. We shall show that I is 'usually' an entire line. Now—dealing with the exceptions first—if $A = B$, then I is the whole plane A, while if A is parallel (but not equal) to B then I is empty. Apart from these two unusual cases, a sketch (or paper model) shows that I will be a line in $\mathbb{R}^3$, and we find its vector equation as follows. Since I lies in both A and B, it is orthogonal to each of $\mathbf{a}$ and $\mathbf{b}$ and hence has direction $\mathbf{a} \wedge \mathbf{b}$; also it cuts the plane $z = 0$ at $\mathbf{s} = (x, y, 0)$, where $a_1x + a_2y = l$, $b_1x + b_2y = m$. Hence the vector equation of I is $\mathbf{r} = \mathbf{s} + t\mathbf{a} \wedge \mathbf{b}$.

Exercises B.8.1

1. Find the lines of intersection of the pairs of planes $x + 2y + z = 1$, $3x - y - 4z = 2$; $3x + 2y - z = 7$, $x + 2y + z = 5$. Do these lines intersect?
2. In the discussion above, solve the equations to find $\mathbf{s}$. What happens if, instead, we find where I cuts the plane $x = 0$?
3. With A, B intersecting in the line I as above, let C be a third plane, with equation $\mathbf{c}\cdot\mathbf{r} = n$. Show that if $\mathbf{r}$ is in the intersection J of I and C, then its coordinates (x, y, z) satisfy the vector equation $x\mathbf{u} + y\mathbf{v} + z\mathbf{w} = \mathbf{s}$ where $\mathbf{u} = (a_1, b_1, c_1)$, $\mathbf{v} = (a_2, b_2, c_2)$, $\mathbf{w} = (a_3, b_3, c_3)$, $\mathbf{s} = (l, m, n)$. Now show that $x\Delta = \mathbf{s}\cdot\mathbf{v} \wedge \mathbf{w}$ where $\Delta = \mathbf{u}\cdot\mathbf{v} \wedge \mathbf{w}$, with similar equations for y and z. Thus, if $\Delta \neq 0$, then we can solve for x, y and z to see that J consists of just one point. Hence find J when A, B, C are, respectively, the first two planes in exercise 1, and C is the third.
4. Show that, in exercise 3, $\Delta = 0$ if either C contains the line I, or C is parallel to I, or two of the three planes coincide or are parallel, or all three are parallel. Draw sketches in each case to decide when J is empty or not.

B.9 LINEAR TRANSFORMATIONS

In exercises B.8.1, we had the problem of finding all solutions to a set of three equations. These we write, first in their scalar product form, and then in matrix form, as

$$\begin{aligned} \mathbf{a}\cdot\mathbf{r} &= l \\ \mathbf{b}\cdot\mathbf{r} &= m \\ \mathbf{c}\cdot\mathbf{r} &= n \end{aligned} \quad \text{or} \quad \begin{pmatrix} a_1 & a_2 & a_3 \\ b_1 & b_2 & b_3 \\ c_1 & c_2 & c_3 \end{pmatrix}\begin{pmatrix} x \\ y \\ z \end{pmatrix} = \begin{pmatrix} l \\ m \\ n \end{pmatrix} \tag{B.9.1}$$

or as

$$N\mathbf{r} = \mathbf{s} \tag{B.9.2}$$

where N is the 3×3 matrix, and the vectors $\mathbf{r}, \mathbf{s}$ are now thought of as columns rather than rows. If $\mathbf{s}$ happens to be zero, then the set of all solutions forms the intersection of three planes through $\mathbf{0}$, so $\mathbf{0}$ itself is a solution (called **trivial**). The three equations are then said to be **homogeneous**, as with two variables. Therefore a full summary of the conditions for a solution is more complicated than theorem B.1.1, and we shall give only the part that is most useful in practice.

Theorem B.9.1
If $\det(N) \neq 0$ *in (B.9.2), then there is a unique solution* $\mathbf{r}$ *for each* $\mathbf{s}$. *Hence, for the homogeneous equations to have a non-trivial solution, we need* $\det(N)$ *to be zero.*

(This result should be committed to memory, and it helps to note a common source of confusion: to have a 'nice' solution, that is, either unique or non-trivial, we need $\det(N)$ to behave in quite opposed ways. The logic of mathematics often conflicts with our psychological tastes.)

Exercises B.9.1

1. The matrix N in (B.9.2) has columns $\mathbf{u}, \mathbf{v}, \mathbf{w}$ as in exercise B.8.1(3), and $\det(N) = \Delta$. Use that exercise to show that if $\Delta \neq 0$, then the solution of (B.9.2) is $\mathbf{s} = Q\mathbf{r}$, where Q is the matrix with rows $(\mathbf{v} \wedge \mathbf{w})/\Delta$, $(\mathbf{w} \wedge \mathbf{u})/\Delta$, $(\mathbf{u} \wedge \mathbf{v})/\Delta$ (in that order). Q is the **inverse** of N; calculate the inverse of each matrix in exercise B.7.1(1) whenever an inverse exists.
2. Three (distinct, non-zero) vectors $\mathbf{a}, \mathbf{b}, \mathbf{c}$ in $\mathbb{R}^3$ are said to be **linearly independent** if the only solution of the equation $\lambda\mathbf{a} + \mu\mathbf{b} + \nu\mathbf{c} = \mathbf{0}$ is $\lambda = \mu = \nu = 0$. Otherwise $\mathbf{a}, \mathbf{b}, \mathbf{c}$ are **dependent.**

   (a) If $\mathbf{a} = (1, 3, 1)$, $\mathbf{b} = (0, 2, 2)$, $\mathbf{c} = (-1, -1, 1)$, show that these points are dependent (by finding s, t with $1\cdot\mathbf{a} + s\mathbf{b} + t\mathbf{c} = \mathbf{0}$). Hence show that the planes

   $$x - z = 1, \qquad 3x + 2y - z = 7, \qquad x + 2y + z = 5$$

   intersect in a line, and find its equation.

   (b) Prove that $\mathbf{i}, \mathbf{j}, \mathbf{k}$ are linearly independent.

   (c) If $\mathbf{a}, \mathbf{b}, \mathbf{c}$ are linearly independent, show that the matrix D with columns $\mathbf{a}, \mathbf{b}, \mathbf{c}$ has non-zero determinant, and every $\mathbf{d} \in \mathbb{R}^3$ can be expressed in the form $\mathbf{d} = \lambda\mathbf{a} + \mu\mathbf{b} + \nu\mathbf{c}$.

   (d) If $\det D \neq 0$ show that $\mathbf{a}, \mathbf{b}, \mathbf{c}$ are linearly independent, and there is a unique plane through $\mathbf{a}$, $\mathbf{b}$, and $\mathbf{c}$.

3. For what values of λ (if any) are the vectors

$$\mathbf{a} = (1, 3, \lambda), \qquad \mathbf{b} = (1, 2, 2), \qquad \mathbf{c} = (-1, -1, 1)$$

linearly independent? Show that, if **a**, **b**, **c** are linearly dependent, then the planes

$$\mathbf{r}\cdot\mathbf{a} = 1, \qquad \mathbf{r}\cdot\mathbf{b} = 2, \qquad \mathbf{r}\cdot\mathbf{c} = \mu$$

intersect in a line, for just one value of μ. [SU]

4. The triangle ABC has vertices $(1, 2)$, $(3, 2)$, $(5, -1)$ and its circumcircle has equation of the form

$$x^2 + y^2 + 2gx + 2fy + c = 0.$$

Write down three equations in g, f, c in the matrix form $A\mathbf{u} = \mathbf{v}$, where $\mathbf{u}$ has components g, f, c. Solve for g, f, c, and calculate the centre and radius of the circle. [SU]

5. In $\mathbb{R}^3$ let ABC, $A'B'C'$ be triangles such that the lines AA', BB', CC' meet in $\mathbf{0}$. Pairs of sides such as AB, $A'B'$ are said to **correspond**, and no pairs of corresponding sides are parallel. Show that each such pair intersects, and the three points thus obtained lie on one line. (Represent A, A' by vectors $\mathbf{a}$, $u\mathbf{a}$, with similar notation $b, \ldots, w\mathbf{c}$ for $B, \ldots, C'$. Find the vector equations of AB, $A'B'$ and show that the intersection is $N = (\mathbf{a}u(v-1) - \mathbf{b}v(u-1))/(v-u)$, with similar formulae for the other two points L, M obtained by permuting the notation cyclically. (Why is $v - u \neq 0$?) Now use theorem B.9.1 since **a**, **b**, **c** are linearly independent. This is the famous Theorem of Désargues (1591–1661); the converse of the theorem holds. The diagram is rather beautiful.)

6. An **eigenvector** of a matrix M is any non-zero vector $\mathbf{v}$ for which $M\mathbf{v} = \lambda\mathbf{v}$ for some number λ. Show that $\mathbf{v}$ satisfies the homogeneous equation $(M - \lambda I)\mathbf{r} = \mathbf{0}$, and hence that λ is a root of the equation $\det(M - \lambda I) = 0$ (which is known as the **characteristic equation** of M). Using exercise B.7.1(9), find the eigenvalues and corresponding eigenvectors of the following matrices:

$$\begin{pmatrix} 1 & 3 & 0 \\ 0 & 2 & 0 \\ 0 & 0 & 2 \end{pmatrix} \qquad \begin{pmatrix} 3 & -3 & 1 \\ 2 & -2 & 1 \\ 0 & 0 & 1 \end{pmatrix} \qquad \begin{pmatrix} 1 & 1 & -1 \\ -2 & 4 & -1 \\ -4 & 4 & 1 \end{pmatrix}.$$

A change of view about (B.9.2) is now necessary. When solving (B.9.2) previously, we took **s** as given and asked for **r**; but we could regard (B.9.2) as a rule for generating a new vector **s** from a given vector **r**. In this latter form, the rule is a function F which assigns to each **r** in $\mathbb{R}^3$ the vector $\mathbf{s} = N\mathbf{r}$, and then we write $\mathbf{s} = F(\mathbf{r})$, displaying matters as:

$$F\colon \mathbb{R}^3 \to \mathbb{R}^3, \qquad F(\mathbf{r}) = N\mathbf{r}. \tag{B.9.3}$$

Here $\mathbb{R}^3$ happens to be both the domain and co-domain of F, and F is often called a 'transformation' or 'mapping' of $\mathbb{R}^3$, because it 'maps' each point **r** to another, $F(\mathbf{r})$. Such mappings arise also when we transform from one set of coordinates to another, and then $F(\mathbf{r})$ tells us the new coordinates of **r** in terms of its old ones.

The problem: 'find **r**, with **s** given' is then the problem of finding the *image* of F. Exercise B.9.1(1) shows that if $\det(N) \neq 0$, then the image is all $\mathbb{R}^3$ (so F is 'onto'); and in that case also the solution for each **s** is unique (being given by the formulae above) so F is also an injection. (With these two properties, F is a 'bijection'.) Mappings like F are rather special, and very important, because they satisfy the following two properties.

Definition B.9.1 (Linearity properties)
Let $T: \mathbb{R}^3 \to \mathbb{R}^3$ be a mapping. Then T is said to be **linear**, if for any two vectors $\mathbf{p}, \mathbf{q}$, and scalar t,

$$T(\mathbf{p} + \mathbf{q}) = T(\mathbf{p}) + T(\mathbf{q}), \qquad T(t\mathbf{p}) = t \cdot T(\mathbf{p}). \tag{B.9.4}$$

(The first equality says that if we first add $\mathbf{p}$ and $\mathbf{q}$ and then map the result, it has the same effect as mapping $\mathbf{p}$ and $\mathbf{q}$ separately, and then adding. Similarly for the second equality.) Linearity is defined similarly if we change $\mathbb{R}^3$ to $\mathbb{R}^2$ or to $\mathbb{R}$ or have mappings $\mathbb{R}^2 \to \mathbb{R}^3$ (for example).

Most mappings are *not* linear, for example, the functions $\surd : \mathbb{R} \to \mathbb{R}$, $\cos : \mathbb{R} \to \mathbb{R}$, and most polynomials. Why not? Well, for example, it isn't true that for all x, y we have $\sqrt{(x + y)} = \sqrt{x} + \sqrt{y}$, for example, when $x = y = 1$. The particular mapping F in (B.9.3) *is* linear, but many beginners are puzzled to know how to start verifying such a statement, so we demonstrate in (possibly too much) detail—just once!

First, let us try a simpler case. We shall show that the mapping $f: \mathbb{R} \to \mathbb{R}$, given by $f(x) = 7x$, is linear. We have to show that each of the two equations in (B.9.4) holds when T is f. For the first equation in (B.9.4) we have

$$f(p + q) = 7(p + q) = 7p + 7q = f(p) + f(q)$$

as required. To verify the second equation, we must show that $f(tp) = tf(p)$; but by definition of f we know that $f(tp) = 7(tp) = t(7p) = tf(p)$, and we are done.

Turning next to the problem of verifying the linearity of F in (B.9.3), we go through the same procedure. We have to show that each of the two equations in (B.9.4) holds when T is F. The second is easiest; we must show that $F(t\mathbf{p}) = tF(\mathbf{p})$, and by definition of F in (B.9.3), we know that $F(t\mathbf{p}) = N \cdot t\mathbf{p}$, where the dot is inserted for clarity. Now if, in each scalar product in (B.9.1), we change $\mathbf{r}$ to $t\mathbf{r}$, then we can bring t to the front and get $N \cdot t\mathbf{p} = tN \cdot \mathbf{p}$, which equals $tF(\mathbf{p})$. Thus we have the required equation $F(t\mathbf{p}) = tF(\mathbf{p})$. Now for the first equation in (B.9.4); we have $F(\mathbf{p} + \mathbf{q}) = N(\mathbf{p} + \mathbf{q})$, and again from the scalar products in (B.9.1) we have $N \cdot (\mathbf{p} + \mathbf{q}) = N \cdot \mathbf{p} + N \cdot \mathbf{q}$, and this is $F(\mathbf{p}) + F(\mathbf{q})$. Thus $F(\mathbf{p} + \mathbf{q}) = F(\mathbf{p}) + F(\mathbf{q})$ as required; the entire verification is now complete.

Exercises B.9.2

1. According to Mr. Gradgrind in Dickens's novel *Hard Times*, children should be taught that a horse is a 'gramnivorous quadruped'. (He assumed that each word was understood.) What checks must a child then apply, in order to decide whether an approaching animal is a horse? Is this process similar to that of verifying the conditions of (B.9.1)? Now verify that the following mappings are linear: $(x, y) \mapsto (-x, -y)$; $(x, y) \mapsto (3x + 5y, x - 7y)$; $(x, y) \mapsto 0$.
2. By writing $\mathbf{r} = x\mathbf{i} + y\mathbf{j} + z\mathbf{k}$, show that if T is linear as in (B.9.4), then $T(\mathbf{r}) = xT(\mathbf{i}) + yT(\mathbf{j}) + zT(\mathbf{k})$. Hence we know all about T as soon as we know the three values $T(\mathbf{i}), T(\mathbf{j}), T(\mathbf{k})$. Show that when T is F as in (B.9.3), these values are the columns of the matrix N. (For a general linear T, we form the matrix of T in the same way, using $T(\mathbf{i}), T(\mathbf{j}), T(\mathbf{k})$ (in that order) as columns.)
3. If T is a linear mapping of $\mathbb{R}^2$, with matrix A, show that the axes are mapped to lines

through $\mathbf{0}$; and the unit square S, formed by the unit vectors $\mathbf{i}, \mathbf{j}$ with remaining vertex at $(1, 1)$, is mapped to a parallelogram P. Verify that P has area given by $\det(A)$. (It can be shown that T distorts all areas in the ratio $1 : \det(A)$.) Similarly for volumes in $\mathbb{R}^3$.

4. Suppose F is a mapping of $\mathbb{R}^2$ that sends each line $x = c$ to one branch of the hyperbola $x^2 - y^2 = c^2$. Can F possibly be linear? What if, instead, F maps each such line to the circle $x^2 + y^2 = c^2$? Get a feeling for linear mappings by observing the effects of various linear mappings on circles, etc. (Produce some computer graphics.)
5. The mapping $E: \mathbb{R}^3 \to \mathbb{R}^3$ which sends each $\mathbf{p}$ in $\mathbb{R}^3$ to itself (so $E(\mathbf{p}) = \mathbf{p}$) is called the **identity mapping**. Show that it is linear, and that its matrix is the identity matrix I in exercise B.5.7(1).
6. Let $\mathbf{a}$ be a fixed vector in $\mathbb{R}^3$, and let $G: \mathbb{R}^3 \to \mathbb{R}^3$ be the mapping that sends each $\mathbf{p}$ to $\mathbf{a} \wedge \mathbf{p}$. Verify that G is linear, and find the matrix of G. Solve (for $\mathbf{p}$) the equation $G(\mathbf{p}) = \mathbf{q}$. (Draw a sketch to help you, and try $\mathbf{p} = \mathbf{a} \wedge \mathbf{q}$.)
7. With T as in (B.9.4), let $S: \mathbb{R}^3 \to \mathbb{R}^3$ be a second linear mapping. Show that the composition $S \circ T: \mathbb{R}^3 \to \mathbb{R}^3$ is linear, where $S \circ T(\mathbf{p}) = S(T(\mathbf{p}))$. (Verify that $S \circ T$ satisfies the two equations (B.9.4).)
8. With E the identity mapping as in exercise 5, show that $E \circ T$ and $T \circ E$ are both equal to T. (For each $\mathbf{p}$, show that $E \circ T(\mathbf{p}) = T(\mathbf{p})$.)
9. Show that the only linear functions $f: \mathbb{R} \to \mathbb{R}$ are those of the form $f(x) = kx$, for some constant k. (Try $f(t) = f(t \cdot 1)$.)
10. If we think of the points of the plane $\mathbb{R}^2$ as complex numbers in the Argand diagram, let $w = u + jv$ be a *fixed* complex number. Let $h: \mathbb{R}^2 \to \mathbb{R}^2$ map each complex number $z = x + jy$ to the number wz. Find the coordinates of $h(z)$, and hence show that h is a linear mapping.

B.10 MATRIX MULTIPLICATION

Exercise B.9.2(7) above is important, because is suggests a way of combining matrices that corresponds to the the composition of mappings. Thus let M, N be two 3×3 matrices; these correspond to linear mappings S, T, respectively, as with (B.9.3). By the exercise, $S \circ T$ is also a linear mapping, so it has a matrix, $\mathrm{mat}(S \circ T)$. We then call $\mathrm{mat}(S \circ T)$ the **product** of M and N, and write

$$MN = \mathrm{mat}(S \circ T) = \mathrm{mat}(S)\,\mathrm{mat}(T). \tag{B.10.1}$$

If $L = MN$ what do the entries of L look like? By (B.9.2) we have to write $\mathbf{v} = S \circ T(\mathbf{r})$ as $L\mathbf{r}$, but as a set of equations in x, y, z. Now $T(\mathbf{r}) = \mathbf{s}$ with the coordinates l, m, n of $\mathbf{s}$ given by (B.9.2). Hence if the rows and the columns of M, N are, respectively, the vectors $\mathbf{a}', \mathbf{b}', \mathbf{c}'$ and $\mathbf{u}, \mathbf{v}, \mathbf{w}$, then we leave the reader to verify that

$$\mathbf{v} = M\mathbf{s} = \begin{pmatrix} \mathbf{a}' \cdot \mathbf{s} \\ \mathbf{b}' \cdot \mathbf{s} \\ \mathbf{c}' \cdot \mathbf{s} \end{pmatrix} = \begin{pmatrix} \mathbf{a}' \cdot \mathbf{u} & \mathbf{a}' \cdot \mathbf{v} & \mathbf{a}' \cdot \mathbf{w} \\ \mathbf{b}' \cdot \mathbf{u} & \mathbf{b}' \cdot \mathbf{v} & \mathbf{b}' \cdot \mathbf{w} \\ \mathbf{c}' \cdot \mathbf{u} & \mathbf{c}' \cdot \mathbf{v} & \mathbf{c}' \cdot \mathbf{w} \end{pmatrix} \begin{pmatrix} x \\ y \\ z \end{pmatrix}. \tag{B.10.2}$$

The matrix is L, as required. It may seem complicated, but note that the entry in the ith row and jth column is simply the scalar product of the ith row of M with the jth column of N. This agrees with the rule (B.4.1) for multiplying 2×2 matrices, and arises for $n \times n$ matrices in the same natural way, being simply the algebraic description of the composition of two linear transformations. It arises from the very natural equation (B.10.1). Compatibly with this row-and-column multiplication, a scalar product of vectors $\mathbf{p}$, $\mathbf{q}$ is written as a matrix product $\mathbf{p}^T\mathbf{q}$, in which $\mathbf{p}^T$ is a row and $\mathbf{q}$ a column.

Exercises B.10.1

1. Let I be the identity matrix, and N be as in (B.9.3). Show that $IN = N = NI$.
2. Practice matrix multiplication by multiplying out the pairs of matrices in exercise B.7.1(1). Hence show that in general, $MN \neq NM$ (so matrix multiplication is not *commutative*).
3. Show that when the matrix N has non-zero determinant in exercise B.9.1(1), then the matrix Q there satisfies $QN = I = NQ$. Thus N is **invertible** and $A^{-1} = Q$. Show that N can have only one inverse; that is, if also $RN = I = NR$ then $R = Q$.
4. Prove that matrix multiplication is *associative*, that is, that $N(MP) = (NM)P$, as with numbers. (Composition of functions is associative: now use (B.10.1).)
5. Let A and B be invertible matrices. Show that $(AB)^{-1} = B^{-1}A^{-1}$.
6. It can be shown that det AB = det A det B (where A, B are both $n \times n$ matrices). Illustrate this result for the following determinants:

$$A = \begin{bmatrix} 1 & 0 & 2 \\ -5 & 1 & 3 \\ 6 & -2 & 4 \end{bmatrix}, \quad B = \begin{bmatrix} 2 & 1 & 3 \\ 0 & 5 & -6 \\ 7 & -2 & 5 \end{bmatrix}.$$

   Use the result to show that in exercise 3, $\det(N^{-1}) = (\det N)^{-1}$.
7. Use (B.10.2) to show that $(ST)^T = T^TS^T$.
8. Let λ, μ be two unequal eigenvalues of a *symmetric* matrix M (for which $M = M^T$), with eigenvectors $\mathbf{v}$, $\mathbf{w}$. Show that $\mathbf{v}^T(M\mathbf{w}) = \mu\mathbf{v}^T\mathbf{w} = \mathbf{w}^TM\mathbf{v} = \lambda\mathbf{v}^T\mathbf{w}$ and hence that $\mathbf{v}$, $\mathbf{w}$ are orthogonal.
9. Let $\mathbf{v}$ be an eigenvector of N in (B.9.2), and F the mapping in (B.9.3). Show that F maps the entire line L through $\mathbf{0}$ and $\mathbf{v}$ to itself (so L is 'invariant' under N). Let P be the plane through $\mathbf{0}$, orthogonal to L. Show that P is invariant under the transpose N^T. (For each $\mathbf{w}$ in P consider the scalar product $\mathbf{v}^TN\mathbf{w}$.) Hence show that N has, besides L, an invariant plane.
10. Show that the mapping F in (B.9.3) is a bijection if, and only if, its matrix $\mathbf{N}$ is invertible (see exercise 3). If F is a bijection then it has an inverse mapping, F^{-1}; show that F^{-1} is linear. (The same result holds for mappings of $\mathbb{R}^2$.)
11. Give examples to show that a mapping T of $\mathbb{R}^2$ or $\mathbb{R}^3$ with a general invertible matrix B will not usually keep the coordinate axes at right angles, but it maps lines to lines, and parallelograms to parallelograms so it preserves these items. It therefore does not distort too badly, and in particular it preserves the *linear* structure. (For such reasons, T is often called a **linear isomorphism** (of the coordinate plane)—from the Greek 'iso' and 'morphos', referring to 'same' and 'structure', respectively.)

Exercise B.10.2 (Further miscellaneous exercises)

1. Let $\mathbf{a} = \mathbf{i} + 2\mathbf{j} + 5\mathbf{k}$, $\mathbf{b} = 3\mathbf{i} - \mathbf{j} + \mathbf{k}$, $\mathbf{c} = 2\mathbf{i} + 7\mathbf{j} - \mathbf{k}$. Write down $\mathbf{b} + \mathbf{c}$ and verify that

$$\mathbf{a} \wedge (\mathbf{b} + \mathbf{c}) = \mathbf{a} \wedge \mathbf{b} + \mathbf{a} \wedge \mathbf{c}.$$

   Calculate $\mathbf{b} \wedge \mathbf{c}$ and hence $(\mathbf{a} - \mathbf{b}) \wedge (\mathbf{c} - \mathbf{b})$, to find the equation of the plane Π through $\mathbf{a}$, $\mathbf{b}$, $\mathbf{c}$. Find the distance between Π and the point $\mathbf{a} \wedge \mathbf{b}$.

   Let l be the line joining the points $\mathbf{a} \wedge \mathbf{b}$ and $\mathbf{a} \wedge \mathbf{c}$. Find where it cuts Π.

2. Do the following planes intersect? If so, find where; if not, give a geometrical explanation as to why not

$$(\mathbf{r} - 4\mathbf{i} + \mathbf{j} + 3\mathbf{k})\cdot(\mathbf{i} - \mathbf{j} + 2\mathbf{k}) = 0,$$
$$(\mathbf{r} + \mathbf{i} - 2\mathbf{j} - \mathbf{k}) \wedge (3\mathbf{i} + \mathbf{j} - \mathbf{k}) = \mathbf{0}.$$

3. Prove the identities:

(a) $(\mathbf{a} \wedge \mathbf{b})\cdot(\mathbf{c} \wedge \mathbf{d}) = \begin{vmatrix} \mathbf{a}\cdot\mathbf{c} & \mathbf{a}\cdot\mathbf{d} \\ \mathbf{b}\cdot\mathbf{c} & \mathbf{b}\cdot\mathbf{d} \end{vmatrix}$;

(b) $(\mathbf{a} \wedge \mathbf{b}) \wedge (\mathbf{c} \wedge \mathbf{d}) = \{\mathbf{a}\cdot(\mathbf{c} \wedge \mathbf{d})\}\mathbf{b} - \{\mathbf{b}\cdot(\mathbf{c} \wedge \mathbf{d})\}\mathbf{a}$, and also
$(\mathbf{a} \wedge \mathbf{b}) \wedge (\mathbf{c} \wedge \mathbf{d}) = \{\mathbf{d}\cdot(\mathbf{a} \wedge \mathbf{b})\}\mathbf{c} - \{\mathbf{c}\cdot(\mathbf{a} \wedge \mathbf{b})\}\mathbf{d}$.

How is the direction of this vector related to the planes of **a** and **b** and **c** and **d**?

4. Evaluate the determinant of the matrix

$$A = \begin{pmatrix} 1 & 1 & 0 \\ 0 & 2 & 0 \\ 1 & 0 & \alpha \end{pmatrix}$$

where α is a constant. For what values of α does A^{-1} exist?

5. If $A = \begin{pmatrix} 2 & 3 & 0 \\ 1 & 2 & 1 \\ 1 & 0 & 1 \end{pmatrix}$, evaluate det A and find A^{-1}. Hence find the solution to the system of equations

$$\begin{aligned} 2x + 3y &= 5 \\ x + 2y + z &= 5 \\ x + z &= 3 \end{aligned}$$

6. If $B = \begin{pmatrix} 1 & 3 & 0 \\ 3 & -1 & 0 \\ 0 & 0 & 3 \end{pmatrix}$ and I denotes the unit matrix, express $\det(B - \lambda I)$ in the form

$$\det(B - \lambda I) = a\lambda^3 + b\lambda^2 + c\lambda + d$$

for suitable constants a, b, c and d.

(a) Find the values of λ such that $\det(B - \lambda I) = 0$.
(b) Evaluate B^2 and B^3 and show that

$$aB^3 + bB^2 + cB + dI = 0$$

where 0 is the 3×3 zero matrix.

7. Find the eigenvalues and the corresponding eigenvectors of the matrix

$$B = \begin{pmatrix} 4 & -5 \\ 1 & -2 \end{pmatrix}.$$

Calculate the product CD for the matrices

$$C = \begin{pmatrix} 1 & 2 \\ -1 & 1 \end{pmatrix} \quad \text{and} \quad D = \begin{pmatrix} -1 & 0 & 1 \\ 1 & 3 & 0 \end{pmatrix}.$$

8. For what value of the constant α does the system of equations

$$\begin{aligned} x + \ y &= 0 \\ y + \ z &= 0 \\ x + \alpha z &= 0 \end{aligned}$$

have non-trivial solutions?

9. The rectangular Cartesian coordinates of the points A and B are $(1, 2, 3)$ and $(5, 4, -1)$, respectively. Show that the plane which passes through the mid-point of AB and which is perpendicular to AB has the equation

$$2x + y - 2z = 7.$$

Find the perpendicular distance from the origin $\mathbf{0}$ to the plane.

10. The vertices of a tetrahedron $ABCD$ have position vectors $\mathbf{a}, \mathbf{b}, \mathbf{c}, \mathbf{d}$, respectively, relative to an origin $\mathbf{0}$. If edge AB is perpendicular to edge CD and edge AC is perpendicular to edge BD prove that

$$\mathbf{a}\cdot\mathbf{c} + \mathbf{b}\cdot\mathbf{d} = \mathbf{a}\cdot\mathbf{d} + \mathbf{b}\cdot\mathbf{c} = \mathbf{a}\cdot\mathbf{b} + \mathbf{c}\cdot\mathbf{d}.$$

Deduce that edge AD is perpendicular to edge BC.

11. A line L passes through the two points $(1, -2, -1)$ and $(2, 3, 1)$ and a plane P through the three points $(2, 1, -3)$, $(4, -1, 2)$, and $(3, 0, 1)$.

   (a) Find the vector equation of the line L.
   (b) Show that the plane P has the vector equation

   $$\mathbf{r}\cdot(1, 1, 0) = 3.$$

   (c) Find the coordinates of the point of intersection of the line L and the plane P.
   (d) Derive the vector equation of the plane that contains the line L and is perpendicular to the plane P.

12. Points A and B have position vectors $\mathbf{a}$ and $\mathbf{b}$, respectively, relative to an origin $\mathbf{0}$. A straight line through A has equation

$$\mathbf{r} = \mathbf{a} + \lambda\mathbf{u}$$

where $\mathbf{r}$ is the position vector of a point on the line, $\mathbf{u}$ is a given fixed vector, and λ is a parameter. Show that the equation of the perpendicular from B to the line may be expressed as:

$$\mathbf{r} = \mathbf{b} + \mu\mathbf{u} \wedge \{(\mathbf{a} - \mathbf{b}) \wedge \mathbf{u}\}.$$

If C is a third point, show that the foot D of the perpendicular from C to AB is given by

$$\mathbf{d} = [\mathbf{a}(\mathbf{b} - \mathbf{a})\cdot(\mathbf{b} - \mathbf{c}) - \mathbf{b}(\mathbf{b} - \mathbf{a})\cdot(\mathbf{a} - \mathbf{c})]/(\mathbf{b} - \mathbf{a})^2.$$

(Find t such that $\mathbf{d} = \mathbf{a} + t(\mathbf{b} - \mathbf{a})$ and $(\mathbf{c} - \mathbf{d})\cdot(\mathbf{a} - \mathbf{b}) = 0$.)

13. Let ABC be a triangle and $\mathbf{r}_l = l\mathbf{b} + (1 - l)\mathbf{c}$ a point on BC, with $\mathbf{r}_m$, $\mathbf{r}_n$ defined similarly on CA, AB. Show that a necessary and sufficient condition for $\mathbf{r}_l$, $\mathbf{r}_m$, $\mathbf{r}_n$ to be collinear is that $lmn + (1 - l)(1 - m)(1 - n) = 0$. (Take C at $\mathbf{0}$, and find a linear transformation that maps A, B onto the axes; verify the condition for the resulting figure.)

14. Write the equations (B.1.2) as $x\mathbf{p} + y\mathbf{q} = \mathbf{w}$, where $\mathbf{p} = (a, c)^{\mathrm{T}}$, $\mathbf{q} = (b, d)^{\mathrm{T}}$. Let $\mathbf{p}^{\perp} = c, -a)^{\mathrm{T}}$, and show that $\mathbf{p}\cdot\mathbf{p}^{\perp} = 0$ while $\Delta = \det\ A = \mathbf{p}\cdot\mathbf{q}^{\perp} = -\mathbf{p}^{\perp}\cdot\mathbf{q}$. If $\Delta \neq 0$ show that $x = \mathbf{w}\cdot\mathbf{q}^{\perp}/\Delta, \qquad y = -\mathbf{w}\cdot\mathbf{p}^{\perp}/\Delta$.

C

The exponential function and its relatives

In this chapter we give derivations of several functions such as the logarithm and exponential. We also give the Taylor–MacLaurin expansion, and a treatment of the inverse trigonometrical functions.

C.1 THE LOGARITHM AND EXPONENTIAL FUNCTIONS

The notion of area allows us to fill in a gap in the list of derivatives given by the extension of (4.2.1) to negative powers. If (as in the seventeenth century) we limit our notion of 'function' to the obvious ones like polynomials and the trigonometric functions, then there does not seem to be a function f such that $f'(x) = x^{-1}$. But by the fundamental theorem of calculus (4.5.4) we know that one function that does the trick is the area under the graph of $y = 1/x$, and we call it the 'natural' logarithm, defined by

$$\ln(p) = \int_1^p \frac{1}{x}\,\mathrm{d}x, \qquad (p > 0) \tag{C.1.1}$$

where we need to keep p positive to avoid the explosion at the origin. Since x^{-1} is differentiable, then its integral exists, by general theorems; and by the fundamental theorem (4.5.4), $\mathrm{d}/\mathrm{d}p(\ln(p)) = 1/p$, as we require. This derivative is always positive on the domain $\{p > 0\}$ so the graph of ln constantly rises, with positive gradient. It has no asymptote as a 'ceiling' for the following reasons.

(1) ln has the remarkable algebraic property

$$\ln(pq) = \ln(p) + \ln(q), \tag{C.1.2}$$

because

$$\int_1^{pq} \frac{1}{x}\,\mathrm{d}x = \int_1^{p} \frac{1}{x}\,\mathrm{d}x + \int_p^{pq} \frac{1}{x}\,\mathrm{d}x$$

and the last integral is $\int_1^q 1/t\,\mathrm{d}t$, where $pt = x$. Hence

$$\ln(p^2) = 2\ln(p),$$
$$\ln(p^3) = \ln(p \cdot p^2) = \ln(p) + \ln(p^2) = \ln(p) + 2\ln(p) = 3\ln(p),$$

and similarly, by induction on n,

$$\ln(p^n) = n\ln(p), \qquad n = 1, 2, \ldots,$$

a result which can be extended to negative n as well (see exercise C.1.1).

(2) Next, since $1/x \geq 1/2$ if x lies between 1 and 2, then $\ln(2) = \int_1^2 1/x\,\mathrm{d}x > \int_1^2 1/2\,\mathrm{d}x = 1/2$. Thus by (1), $\ln(2^n)$ exceeds $1\frac{1}{2}^n$. Hence the graph of ln crosses and rises above any horizontal line $y = c\ (c > 0)$ because $\ln(2^m) > c$ for all integers m with $\frac{1}{2}m > c$.

From (1) also, we have $\ln(1/p) = -\ln(p)$; for,

$$0 = \ln(1) = \ln\left(p \cdot \frac{1}{p}\right) = \ln(p) + \ln\left(\frac{1}{p}\right),$$

and this accounts for the shape of the graph of ln, when x lies between 0 and 1. In particular $\ln(x)$ approaches $-\infty$ as x approaches 0.

Although he did not define them in this way, Napier in the seventeenth century invented logarithms so that a seasick navigator (among others) could convert a multiplication (hard) into an addition of logarithms (easy)—using property (C.1.2)—and find the required answer by taking the antilogarithm. For him, there were no calculator buttons to press, and if the answer came out wrong, then the ship might be lost. The antilogarithm appears in calculus as the famous **exponential** function, which we now discuss.

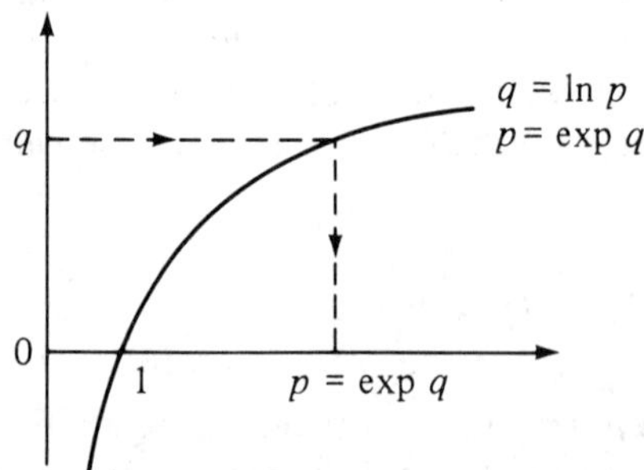

Fig. C.1.1.

Given the point $(0, q)$ on the y-axis (see Fig. C.1.1), the horizontal line $y = q$ hits the graph of ln exactly once (because by (2) above, the curve bursts through every ceiling) at some point (p, q). Then p is here a function of q, and we name it 'exp' for a reason to appear. Thus

$$p = \exp(q), \qquad q = \ln(p),$$

so each of the two functions exp and ln 'undoes' the work of the other, in the sense that the last two equations give:

$$p = \exp(\ln(p)), \qquad q = \ln(\exp(q)). \tag{C.1.3}$$

Therefore, if we use exp on (C.1.2) we see that it works like an antilogarithm, and gives

$$pq = \exp(\ln(pq)) = \exp(\ln(p) + \ln(q)).$$

From this we obtain the vital *algebraic* property of exp, namely by writing $a = \ln(p)$, $b = \ln(q)$ we have

$$\exp(a) \cdot \exp(b) = \exp(a + b). \tag{C.1.4}$$

By (C.1.3), $\exp(0) = \exp(\ln(1)) = 1$, so putting $b = -a$ in (C.1.4) shows that $\exp(a)$ is *never zero*, and $\exp(-a) = 1/\exp(a)$. Arguments like those for ln then show eventually that

$$(\exp(a))^r = \exp(ar),$$

for all rational fractions r. In particular, the number exp(1) is denoted by e, so the last equation says that the rth power of e is, in fact, $\exp(r)$. Hence the name of the function! Consequently we often write e^x for $\exp(x)$ when x is any real number. This causes no ambiguity when x is rational, and *defines* what we mean by 'raising e to the xth power' when x is not rational. Finally we can now define a^x, for any positive a, to mean

$$a^x = e^{x \cdot \ln(a)}. \tag{C.1.5}$$

The exponential function also has an immensely important 'calculus' property, in that it is its own derivative. To see this, we differentiate the second equation in (C.1.3) with respect to q by the chain rule to get

$$1 = \left(\frac{1}{\exp(q)}\right) \cdot \exp'(q)$$

whence

$$\exp'(q) = \exp(q). \tag{C.1.6}$$

Applying the chain rule again, we have for any constant k:

$$\exp'(kx) = k \cdot \exp(kx). \tag{C.1.7}$$

How big is the number e? By (C.1.3) we have $\ln(e) = 1$, and we prove that $2 < e < 4$ as follows. If we look at Fig. C.1.2 we see that ln(2) is less than the area of the trapezium shown there, so $\ln(2) < (1 + 1/2)/2 < 1$, whence $2 < e$. On the other hand, the sum of the three rectangles shown is $(1/2 + 1/3 + 1/4) = 13/12 < \ln 4$; so $e < 4$. In fact $e = 2.71\ldots$. It is quite hard to prove that e is not rational, but a way of calculating e as accurately as desired follows from the technique of the next section.

Exercises C.1.1

1. Prove in detail that if $r = u/v$ and u, v are non-zero integers, then $(\exp(u))^{1/v} = \exp(u/v)$. Prove that, for negative n, $\ln(p^n) = n \ln(p)$.
2. Show that if $a < b$ then $e^a < e^b$.

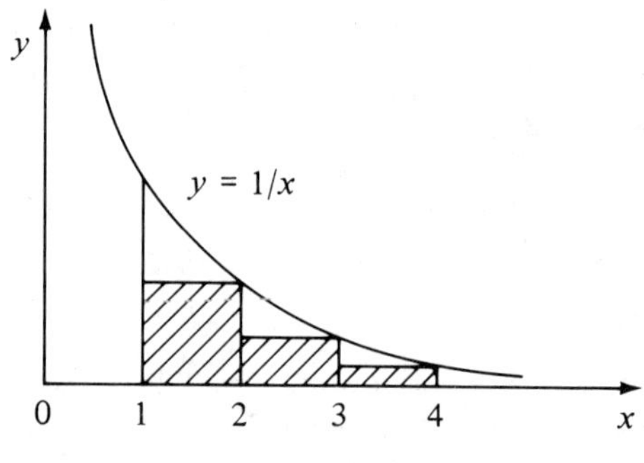

Fig. C.1.2.

3. When $a > 0$, we *define* a^x to be $\exp(x \ln a)$. From this definition show that $a^0 = 1$, $a^{x+y} = a^x \cdot a^y$, $a^{-x} = (1/a^x)$, $a^{1/2} = \sqrt{a}$, $a^3 = a \times a \times a$.
4. Use narrower rectangles than those in the text to get a better approximation to e. Show that $\ln(3) > 1$, so $3 > \mathrm{e}$.
5. Show that if $y' = 1/x$, then $y = \ln|x| + \text{const.}$ (on $\mathbb{R} - \{0\}$). (Check the two cases $x < 0, x > 0$.)
6. Sketch on the same diagram, graphs of the two functions

$$y = (1 - x^2) \qquad \text{and} \qquad y = (1 - x^2)\mathrm{e}^{-x},$$

identifying in each case the x coordinates of any local maxima and minima. [SU]
7. Sketch the graph of the function

$$y = \mathrm{e}^{-px^2} - \mathrm{e}^{-(p+1)x^2},$$

where p is a positive constant, and show in particular that the maximum value of y is given by $p^p/(p+1)^{(p+1)}$. [SU]
8. Find appropriate domains in $\mathbb{R}$ for the functions
(a) $f(x) = \ln(\cos x)$
(b) $g(x) = \ln(\cos x - \sin x)$.
Caculate the maxima and minima (if any) of these functions, and sketch their graphs.
9. Let $I_n = \int_0^x \mathrm{e}^{at} t^n \, \mathrm{d}t$, $n = 0, 1, 2, \ldots$, where $a \in \mathbb{R}$ and $a \neq 0$. Using integration by parts, show that

$$aI_n + nI_{n-1} = \mathrm{e}^{ax}x^n - 1$$

and hence evaluate I_n.
10. Sketch the graph of the function f defined by

$$f \colon x \mapsto \mathrm{e}^{-x^2} \sin(x^2).$$

Show that $f'(x)$ may be written $f'(x) = -2\sqrt{2}\, x \sin(x^2 - \pi/4)$ and hence determine the general formula for the critical points of f. By using the sketch graph of the function, or otherwise, determine which are maxima, minima, or points of inflection. [SU]
11. Given that $\int_{-\infty}^{\infty} \mathrm{e}^{-x^2} \, \mathrm{d}x = \sqrt{\pi}$ determine the values of:
(a) $\int_{-\infty}^{\infty} \mathrm{e}^{-x^2/2a^2} \, \mathrm{d}x$
(b) $\int_{-\infty}^{\infty} x\mathrm{e}^{-x^2/2a^2} \, \mathrm{d}x$
(c) $\int_{-\infty}^{\infty} x^2\mathrm{e}^{-x^2/2a^2} \, \mathrm{d}x$.

Find the points of inflection (which need not be critical points) of the function $f(x) = e^{-x^2/2a^2}$ and show they are a distance $2a$ apart. [SU]

12. Evaluate the following integrals:

$$J = \int_1^x e^{-2x}(x^2 - 1)\,dx \qquad \text{and} \qquad K = \int_1^2 \ln(y + \sqrt{y^2 - 1})\,dy.$$ [SU]

C.2 THE TAYLOR–MACLAURIN EXPANSION

We assume that readers have had some practice in the technique of integration by parts, by which we mean that if we wish to integrate a product of two functions f, g and we know an antiderivative F for f, then

$$\int_a^p f(x)\cdot g(x)\,dx = [F(x)\cdot g(x)]_a^p - \int_a^p F(x)\cdot g'(x)\,dx \tag{C.2.1}$$

and we may find the second integral easier to evaluate than the first. The formula (C.2.1) is a direct consequence of the product rule of differentiation. We shall apply it to obtain the 'Taylor–Maclaurin expansion' of a function f, which is a polynomial that approximates f, and is easier to handle. Such a polynomial is, so to speak, a rough model of f. Thus suppose $a > 0$. We integrate $f'(a - x)$ in two ways, first directly to get (using the chain rule to deal with $-x$):

$$I = \int_0^a f'(a - x)\,dx = [-f(a - x)]_0^a = f(a) - f(0),$$

and then by parts to get

$$I = \int_a^p 1\cdot f'(a - x)\,dx = [x\cdot f'(a - x)]_0^a - \int_0^a x\cdot(-f''(a - x))\,dx.$$

Hence

$$f(a) = f(0) + a\cdot f'(0) + \int_0^a x\cdot f''(a - x)\,dx \tag{C.2.2}$$

assuming f'' exists and is continuous.

Suppose then that we know bounds for f'', that is, (using the modulus $|x|$, see Fig. C.4.1 below) that $|f''|$ never exceeds some number M. Then the modulus of the last integral cannot exceed $M\cdot\int_0^a x\,dx = M\cdot a^2/2$. Therefore we have a useful approximation to f, namely

$$f(a) = f(0) + a\cdot f'(0) + \text{error}, \qquad |\text{error}| \le M\cdot a^2/2. \tag{C.2.3}$$

Examples C.2.1

(1) Recall from (4.2.3) the equation of the tangent to the graph of f at $P = (p, f(p))$. Then if x is close to p, let Q, R be the points on the graph and on the tangent, given respectively by $Q = (x, f(x))$, and $R = (x, y(x))$, where by (4.2.3),

$$y(x) = f(p) + f'(p)\cdot(x - p).$$

Thus Q and R are almost indistinguishable, in the sense that by (C.2.2), their ordinates differ only by $y(p) - f(p)$, of magnitude $M \cdot (x - p)^2/2$. This is a 'second-order' quantity, being smaller than $x - p$ if x is near enough to p; and we see that the tangent is a good approximation to the graph, near P.

(2) If $f(x) = \ln(1 + x)$, then $f'(x) = (1 + x)^{-1}$ and $f''(x) = -(1 + x)^{-2}$. Then $f'(0) = 1$, but f'' becomes arbitrarily large as x approaches -1 which we therefore cordon off by restricting x to be $\geq -1/2$ (say). Then the greatest value of f'' occurs when x is least, so M here is $1/(1/4) = 4$. Hence

$$\ln(1 + x) = x + \text{error}, \qquad |\text{error}| \leq 2x^2 \qquad \left(\text{if } x \geq -\frac{1}{2}\right). \tag{C.2.4}$$

Of course, this error is small only when x is small; if we want it to be $< x/10$, then we must have $x < 1/20$.

(3) If $f(x) = (1 + x)^{-1}$, then $f'(0) = -1$ and $f''(x) = 2 \cdot (1 + x)^{-3}$. The same restriction, $x \geq -1/2$, as before gives $M = 16$ and

$$(1 + x)^{-1} = 1 - x + \text{error}, \qquad |\text{error}| \leq 8x^2. \tag{C.2.5}$$

We can improve on (C.2.3) by using integration by parts on the integral in (C.2.2) to get

$$f(a) = f(0) + a \cdot f'(0) + \left[\frac{x^2}{2} \cdot f''(a - x)\right]_0^a + \int_0^a \frac{x^2}{2} \cdot f'''(a - x)\,dx$$

$$= f(0) + a \cdot f'(0) + \frac{a^2}{2} \cdot f''(0) + \int_0^a \frac{x^2}{2} \cdot f'''(a - x)\,dx.$$

Thus we can iterate this process. We use the notation $f^{(n)}$ for the nth derivative of f, and $n!$ (factorial n) for the denominators, to obtain by induction,

$$f(a) = P_n(a) + R_n(a),$$

where $P_n(a)$ is the polynomial

$$P_n(a) = f(0) + a \cdot f'(0) + \frac{a^2}{2!} \cdot f''(0) + \frac{a^3}{3!} \cdot f'''(0) + \ldots + \frac{a^n}{n!} \cdot f^{(n)}(0) \tag{C.2.6}$$

and $R_n(a)$ is the **remainder**

$$R_n(a) = \int_0^a \frac{x^n}{n!} \cdot f^{(n+1)}(x)\,dx. \tag{C.2.7}$$

Just as with (C.2.3), we can say that if there is some number N such that $|f^{(n+1)}(z)|$ never exceeds N, then

$$R_n(a) \leq N \cdot \int_0^a \frac{x^n}{n!}\,dx = N \cdot \frac{a^{n+1}}{(n + 1)!}, \tag{C.2.8}$$

which gives a simple estimate of the error in replacing $f(a)$ by the polynomial $P_n(a)$ in (C.2.6).

Note that if $f(a)$ is polynomial of degree n, then $f^{(n+1)} = 0$; so $f(a)$ must equal $P_n(a)$, and $R_n(a)$ is zero. In particular, if $f(a) = (a + y)^n$, then this is clearly a polynomial of degree n in a, so we have at once the **binomial theorem** for positive integral index:

$$(a + y)^n = y^n + n \cdot a \cdot y^{n-1} + \frac{n(n-1)}{2!} \cdot a^2 \cdot y^{n-2} + \cdots + a^n.$$

As another example, let $f(x) = \sin(x)$. Then all its derivatives exist and equal $\pm\sin(x)$ or $\pm\cos(x)$ which never exceed 1; so here N in (C.2.8) is 1 (which happens to be independent of n). Also at $x = 0$, we have $\sin(0) = 0$, $\sin'(0) = \cos(0) = 1$, $\cos'(0) = -\sin(0) = 0$, etc., so

$$\begin{aligned}\sin(a) &= 0 + a \cdot 1 + \frac{a^2}{2!} \cdot 0 + \frac{a^3}{3!} \cdot (-1) + \ldots \\ &= a - \frac{a^3}{3!} + \frac{a^5}{5!} - \cdots + (-1)^m \cdot \frac{a^{2m+1}}{(2m+1)!} + R_{2m+2}(a)\end{aligned}$$

and, by (C.2.8),

$$|R_{2m+2}(a)| \le \frac{a^{2m+3}}{(2m+3)!},$$

so the error when we stop at one term is no worse than the magnitude of the next term. (Note how rapidly these particular terms decrease to zero.)

Exercises C.2.1

1. Show that $x = -2$ is a solution of

   $$x^4 + 5x^3 + 9x^2 + 7x + 2 = 0$$

   and find the other solutions. (Use the binomial theorem.) [SU]
2. Show that if $|x| < 1$ then

   $$\ln\left(\frac{1+x}{1-x}\right) = 2\left(x + \frac{x^3}{3} + \frac{x^5}{5} + \ldots + \frac{x^{2n+1}}{2n+1} + \cdots\right).$$

   By taking $x = 1/(1 + 2y)$ with $y > 0$ show that

   $$\ln(1 + y) = \ln y + 2\left\{\frac{1}{1+2y} + \frac{1}{3}\left(\frac{1}{1+2y}\right)^3 + \cdots\right\}.$$

   This is a rapidly convergent series: use it to calculate ln 2, ln 3, and so on.
3. Write down the first three non-zero terms in the Taylor–Maclaurin expansion of $\cos x$, $\sin x \sin(x^2)$, $\tan x$, $\ln\{\sqrt{t^2 + 1} - t\}$. [SU]
4. Show that $e^a = 1 + a + a^2/2! + a^3/3! + \ldots + a^n/n! + R_n(a)$. Prove that if $a < 0$ then $R_n(a) \le p = |a|^{n+1}/(n+1)!$ if $a < 0$, but if $a > 0$ then $R_n(a) \le \exp(a) \cdot p$, hence calculate e. Since $e < 3$ (see exercise C.1.1(4)) then $R_n(1) < 3p$, and $R_n(a) < 3^m p$ where m is the least integer such that $a \le m$. Hence find how large to take n if we want $R_n(1) < 10^{-6}$, $R_n(20) < 10^{-6}$.

5. Show that

$$f(a+x) = f(a) + xf'(a) + \frac{x^2}{2!}f''(a) + \ldots + \frac{x^n}{n!}f^{(n)}(a) + \frac{x^{n+1}}{(n+1)!}S_n$$

   and calculate S_n.
6. Using exercise 5, show that if a is a critical point of f, then f has a local maximum or minimum at a according as $f''(a) < 0$ or $f''(a) > 0$. What can $f'''(a)$ tell you if $f''(a) = 0$? (Examine the functions $f(x) = x^4$, $f(x) = x^5$, etc.)
7. Show that the tangent to the graph G of $y = f(x)$ at $(a, f(a))$ is the graph T of the function $y = t(x)$, where $t(a+h) = f(a) + hf'(a)$. Hence show that if $f''(a) = 0$ then $f(a+h) - t(a+h) = h^3 g(h)$ where $g(h)$ is strictly positive or negative when h is small, provided the same holds for $f'''(a)$. Now show that G crosses T if $f''(a) = 0 \neq f'''(a)$. (a is then a **point of inflexion** of f.)
8. Prove the **second mean value theorem** for integrals: let f, g be two continuous functions on $[a, b]$ with $g > 0$. Then there is some $c \in [a, b]$ such that $J = f(c) \cdot K$, where

$$J = \int_a^b f(x)g(x)\,dx, \qquad K = \int_a^b g(x)\,dx.$$

   (For each $x \in [a, b]$, $m \leq f(x) \leq M$, where m, M are the least and greatest values of f. Since $g > 0$, we can multiply the inequalities by $g(x)$; and they are still preserved if now we integrate to obtain $mK < J < MK$. But mK, MK are values of the continuous function $Kf(x)$, which must take the intermediate value J.)

   Hence show that the remainder term $R_n(a)$, in the Taylor–Maclaurin expansion (C.2.7), equals $a^{n+1}f^{(n+1)}(c)/(n+1)!$ for some $c \in [0, a]$; and hence also if $a < 0$. Since c depends on a, we can therefore write:

$$f(a) = f(0) + Aa^n(1 + ag(a)), \qquad A = a_n/n!$$

   where a_n is the first non-zero coefficient after $f(0)$, and $g(a) = f^{(n+1)}(c)$ as before. By solving the above equation for $g(a)$, show that g is differentiable if $a \neq 0$; now calculate $g(a)/a$ to show that g is differentiable at $a = 0$.

C.3 THE FUNCTION cis

In calculus, it is useful to allow the x in e^x to be a complex number, and here we show how to arrange this. We assume that readers are familiar with the plane $\mathbb{R}^2$ as the Argand diagram, so that we can write the point $z = (x, y)$ (of $\mathbb{R}^2$) as the complex number $x + jy$, where j is the square root of -1 with coordinates (0, 1). Thus z is called 'real' if $y = 0$, and the real numbers fill out the x-axis. In $\mathbb{R}^2$, the unit circle S has equation

$$S: \quad x^2 + y^2 = 1,$$

and if z lies on S, then for some angle θ, $z = \cos\theta + j\sin\theta$, of modulus 1. Thus z is a function of θ, to which we (temporarily) give the name 'cis': thus, $z = \text{cis}(\theta)$. We show that cis behaves like exp in (C.1.4) as follows.

By the addition formulae for sin and cos (see Chapter B) we have, on multiplying out,

$$\text{cis}(\theta) \cdot \text{cis}(\phi) = A + jB = \text{cis}(\theta + \phi), \tag{C.3.1}$$

with

$$A = \cos(\theta)\cdot\cos(\phi) - \sin(\theta)\cdot\sin(\phi)$$

$$B = \sin(\theta)\cdot\cos(\phi) + \cos(\theta)\cdot\sin(\phi).$$

Now let k be a real constant. Then we have

$$\frac{\mathrm{d}}{\mathrm{d}\theta}(\mathrm{cis}(k\theta)) = -k\cdot\sin(k\theta) + k\cdot\cos(k\theta)\cdot\mathrm{j} = \mathrm{j}k\cdot\mathrm{cis}(k\theta),$$

since $\mathrm{j}^2 = -1$. Therefore, the function cis behaves almost as if $\mathrm{cis}(k\theta)$ were $\exp(k\theta)$, which as yet is undefined. This leads us to *define* e^z, for any complex $z = x + \mathrm{j}y$, by the formula

$$\mathrm{e}^z = \mathrm{e}^x\cdot\mathrm{cis}(y), \tag{C.3.2}$$

which agrees with the original definition when z is real, since $\mathrm{cis}(0) = 1$; and if $x = 0$ then $\mathrm{e}^{\mathrm{j}y} = \mathrm{cis}(y)$, as was suggested above. This new definition therefore 'extends' the domain of exp from the axes of $\mathbb{R}^2$ to the whole plane.

We claim that the nice behaviour, recorded in (C.1.4), (C.1.6), still holds. For if $w = u + \mathrm{j}v$, then

$$\mathrm{e}^z\cdot\mathrm{e}^w = (\mathrm{e}^x\cdot\mathrm{e}^u)\cdot(\mathrm{cis}(y)\cdot\mathrm{cis}(v)) = \mathrm{e}^{x+u}\cdot\mathrm{cis}(y+v)$$

by (C.1.3) for real numbers, and by (C.3.1). Also

$$\begin{aligned}\frac{\mathrm{d}}{\mathrm{d}t}(\mathrm{e}^{zt}) &= \frac{\mathrm{d}}{\mathrm{d}t}(\mathrm{e}^{xt}\cdot\mathrm{cis}(yt)),\\ &= (x\cdot\mathrm{e}^{xt})\cdot\mathrm{cis}(yt) + \mathrm{e}^{xt}\cdot\mathrm{j}y\cdot\mathrm{cis}(yt) \qquad \text{by the calculation prior to (C.3.2)}\\ &= z\cdot\mathrm{e}^{xt}\cdot\mathrm{cis}(yt) = z\cdot\mathrm{e}^{zt}.\end{aligned}$$

Thus, our claim is established, and this extended function (C.3.2) behaves like the original, real, exponential function.

Example C.3.1

To evaluate the integrals

$$C = \int_0^x \mathrm{e}^{at}\cos(bt)\,\mathrm{d}t, \qquad S = \int_0^x \mathrm{e}^{at}\sin(bt)\,\mathrm{d}t, \qquad (a \neq 0 \neq b)$$

form

$$C + \mathrm{j}S = I = \int_0^x \mathrm{e}^{at}\mathrm{cis}(bt)\,\mathrm{d}t = \int_0^x \mathrm{e}^{zt}\,\mathrm{d}t$$

where $z = a + \mathrm{j}b$. Then

$$I = \left[\frac{\mathrm{e}^{zt}}{z}\right]_0^x = \frac{\mathrm{e}^{zx} - 1}{z},$$

and

$$\frac{1}{z} = \frac{a - jb}{z \cdot (a - jb)} = \frac{a - jb}{m}, \qquad m = a^2 + b^2 \neq 0.$$

Therefore the real and imaginary parts of I are

$$C = \frac{e^{ax}(ac + bs) - a}{m}, \qquad S = \frac{e^{ax}(as - bc) + b}{m},$$

where $s = \sin(bx)$ and $c = \cos(bx)$.

For our purposes, the principal use of e^{zt} is in the solution of differential equations, as we see in the next section.

Exercises C.3.1

1. Evaluate $\int_0^x t^n \cos(bt)\,dt$, $\int_0^x t^n \sin(bt)\,dt$, $(n = 1, 2, 3, \ldots)$.
2. By comparing $\text{cis}(x)$ with $\text{cis}(-x)$, prove that

$$\cos(x) = (e^{jx} + e^{-jx})/2, \qquad \sin(x) = (e^{jx} - e^{-jx})/2j. \qquad \text{(C.3.3)}$$

By analogy, we define the **hyperbolic functions**

$$\cosh(x) = (e^x + e^{-x})/2, \qquad \sinh(x) = (e^x - e^{-x})/2. \qquad \text{(C.3.4)}$$

Prove that $\cosh^2(x) - \sinh^2(x) = 1$, and hence that the point $(a \cdot \cosh(x), b \cdot \sinh(x))$ lies on the hyperbola $x^2/a^2 - y^2/b^2 = 1$.

3. Prove the analogues of the trigonometrical addition formulae, e.g.,

$$\cosh(a + b) = \cosh(a) \cdot \cosh(b) + \sinh(a) \cdot \sinh(b)$$
$$\cosh(2x) = \cosh^2(x) + \sinh^2(x).$$

Note the ways in which the plus and minus signs differ between the trigonometrical and the hyperbolic functions.

4. By differentiating the formulae (C.3.4) show that

$$\frac{d}{dx}(\sinh(x)) = \cosh(x), \qquad \frac{d}{dx}(\cosh(x)) = \sinh(x).$$

Now define $\tanh(x) = \sinh(x)/\cosh(x)$, $\text{sech}(x) = 1/\cosh(x)$ and prove that $\text{sech}^2(x) + \tanh^2(x) = 1$; and

$$\frac{d}{dx}(\tanh(x)) = \text{sech}^2(x), \qquad \frac{d}{dx}(\text{sech}(x)) = -\text{sech}(x) \cdot \tanh(x).$$

5. Show that $\cosh(0) = 1$, $\sinh(0) = 0$, and also that $\cosh(x) = \cosh(-x) \geq 1$, but $\sinh(-x) = -\sinh(x)$. If $\sinh(x) = a$, show that e^x satisfies a quadratic equation with roots $a \pm \sqrt{(1 + a^2)}$. Why then can we say that $x = \ln(a + \sqrt{(1 + a^2)})$?
   Hence show that $\sinh(x)$ can be arbitrarily large.
6. Draw the graphs of the hyperbolic functions, and of the functions $\tanh(x)$, $\text{sech}(x)$. Compare your sketches with the output of a calculator.
7. For each integral in exercise 1 and in example C.3.1, change cos, sin to cosh, sinh, respectively, and evaluate the results. Use the substitutions $t = \tanh\theta$, $t = \sinh\theta$ to evaluate.

$$\int_0^x (1 - t^2)^{-1}\,dt, \qquad \int_0^x (1 + t^2)^{-1/2}\,dt.$$

8. For each of the functions

$$h(x) = \ln(\cosh x), \qquad k(x) = \ln(\cosh x - \sinh x).$$

give the maximum domain and determine which, if any, are periodic. Calculate the derivative of each function and give a rough sketch of their graphs. [SU]

9. Differentiate $e^{x^3}\cosh(x^2)$.

10. Derive the Taylor series expansion

$$\cosh^3 x = 1 + \frac{3x^2}{2} + \frac{7}{8}x^4 + R(x)$$

where $R(x)$ is a remainder you need not find. [SU]

11. (a) From the definitions of $\cosh y$ and $\sinh y$ show that
   (i) $\cosh y + \sinh y = e^y$,
   (ii) $\cosh^2 y - \sinh^2 y = 1$,
   (iii) $\sinh^{-1} x = \ln(x + \sqrt{x^2 + 1})$,
   (iv) $d/dx(\sinh^{-1} x) = 1/\sqrt{x^2 + 1}$.

   (b) Prove that

$$\frac{d}{dx}(\sinh^{-1} x) = \frac{1}{\sqrt{x^2 + 1}}$$

   and hence show that $z = x \sinh^{-1}(x^2)$ satisfies the differential equation

$$x^2(x^4 + 1)\frac{d^2 z}{dx^2} - x(x^4 + 3)\frac{dz}{dx} + (x^4 + 3)z = 0.$$ [SU]

12. Expand the following expressions as power series in t up to and including terms in t^4:

   (a) $\displaystyle\int_0^t \frac{x^2}{(e^x - 1)}\, dx$;

   (b) $\displaystyle\int_0^t \frac{x^2}{(e^x + 1)}\, dx$.

   You may assume that $|t|$ is sufficiently small for the series expansions to be convergent and use may be made of the standard expansions of e^x, $\ln(1 + x)$ and $(1 + x)^a$. [SU]

13. Let z be a complex number. Define $\sin z$ and $\cos z$ in terms of complex exponentials of z and show that

$$\sin(jz) = j \sinh z, \qquad \cos(jz) = \cosh z.$$

   Show that, if x and y are real, then

$$\cos(x + jy) = \cos x \cosh y - j \sin x \sinh y.$$

   Hence deduce that, if θ is real and $z = x + jy$, then

$$\cos z = e^{j\theta}$$

   implies that

$$\sinh^2 y - \sin^2 x = 0.$$

   Show that, for fixed x, this last equation has either one or two solutions for y, and sketch the region in the z-plane where you would expect to find solutions of

$$\cos z = e^{j\theta}.$$ [SU]

14. Show that for $t > 0$

$$\int_t^\infty \left(x^2 + \frac{1}{x^2}\right) e^{-x^2/2}\, dx = \left(t + \frac{1}{t}\right) e^{-t^2/2}$$

and

$$\int_t^\infty (4x^2 - 1) e^{-x^2} \sin x \, dx = (2t \sin t + \cos t) e^{-t^2}$$

by applying the method of integration by parts to integrals of the form

$$\int x^p e^{-x^2/2}\, dx \qquad \text{and} \qquad \int x^p e^{-x^2} \sin x \, dx$$

for judiciously chosen values of p. (You may find it convenient to write $x^p = x^{p-1} \cdot x$.)

[SU]

15. The *Laplace transform* of $f(t)$ is the function $F(s)$ where

$$F(s) = \int_0^\infty e^{-st} f(t)\, dt,$$

provided the integral exists. Find $F(s)$ when $f(t)$ is $1, t^p$, and show that $F(s) = 1/(s - p)$ when $f(t) = e^{pt}$. Now find F when $f(t)$ is $\sin t$ or $\cos t$.

Also show that the Laplace transform of $f'(t)$ is $f_0 + sF(s)$ provided $e^{-st} f(t)$ converges to 0 as t converges to ∞. Hence show that if f satisfies the differential equation $f' + kf = g$, then F satisfies the algebraic equation $(s + k)F = G(=\text{Laplace transform of } g)$. (Thus, taking Laplace transforms makes algebraic models of linear differential equations.)

16. If

$$\frac{d^2 y}{dt^2} + y = t$$

and if $y = 1$ and $dy/dt = -2$ at $t = 0$, show that the Laplace transform of y is

$$\bar{y}(s) = \frac{1}{s^2(s^2 + 1)} + \frac{s - 2}{s^2 + 1}.$$

Hence show that

$$y(t) = t + \cos t - 3 \sin t.$$

[SU]

17. Find the Laplace transform of the differential equation

$$\frac{dy}{dt} + 2y = f(t),$$

where $y = 0$ when $t = 0$, and

$$f(t) = \begin{cases} 1, & 0 < t < 1, \\ 0, & t > 1. \end{cases}$$

(Thus y can now be found from a table of Laplace transforms in many books on differential equations.)

Deal similarly with the equation

$$\frac{d^2y}{dx^2}+\frac{dy}{dx}-12y=\begin{cases}1 & \text{if } 0 \le x < 2\\ 0 & \text{otherwise,}\end{cases}$$

$y(0)=0$ and $y'(0)=0$. [SU]

18. Use Laplace transforms to solve the simultaneous differential equations

$$\dot{x}-x+\dot{y}=e^{-t}$$

$$2\dot{x}+\dot{y}+y=2,$$

where $x(0)=1$, $y(0)=0$ and the dot denotes differentiation with respect to t. [SU]

19. Two simultaneous differential equations are:

$$\frac{dx}{dt}+2x+3y=t, \qquad \frac{dy}{dt}+3x+2y=1;$$

with $x(0)=y(0)=0$. Show that $X(s)$, $Y(s)$, the Laplace transforms of $x(t)$, $y(t)$ are as follows:

$$X(s)=-\frac{2}{s^2(s+5)}; \qquad Y(s)=\frac{s+3}{s^2(s+5)}.$$

Invert these transforms to find the solutions $x(t)$, $y(t)$. [SU]

C.4 CONTINUOUS FUNCTIONS

Experience leads us to feel that the graph of a differentiable function can be drawn without lifting the pen from the paper. Now other quite reasonable functions have this property, but their graphs may have points with a sharp kink rather than a tangent. The functions in this wider class are said to be 'continuous'. Thus, f is **continuous** at p if we make the assumption that

$$\lim(f(p+h)-f(p))=0 \qquad \text{when } h \text{ tends to } 0. \tag{C.4.1}$$

If f is continuous at each point in its domain then we say simply that f is continuous. Thus *every differentiable function is continuous*. Analysis books prove the following **boundedness theorem**: if f is continuous at each point of the closed interval I with end points a, b inclusive as above, then f is *bounded*—even more, f has a *least* value m, and a *greatest* value M. This means that there are points p, q in I such that for all x in I,

$$m=f(p)\le f(x)\le f(q)=M. \tag{C.4.2}$$

Moreover, the **intermediate value theorem** holds, which is to say that if k is any number between the values m, M, then there is some point c in I such that $k=f(c)$ (see exercise 4.4.1(1)). This is, of course, the justification of the familiar method of locating zeros of functions.

Not all continuous functions are differentiable, as is shown by the abs function, given (see Fig. C4.1a) by

$$\text{abs}(x)=x\ \text{sgn}(x) \tag{C.4.3}$$

where sgn(x) is 0 if $x = 0$, 1 if $x > 0$, and -1 if $x < 0$. (Often we write $|x|$, the modulus of x, for abs(x).) At $p = 0$ we have abs$(0 + h) -$ abs$(0) = h$ and this tends to 0 as h tends to 0, so abs is continuous at 0. But in this case $u(h) = (\text{abs}(0 + h) - \text{abs}(0))/h = \pm 1$, according as h is positive or negative; so $u(h)$ does not tend to a limit as h tends to 0. So there is no tangent to the graph at the origin, and abs is not differentiable there.

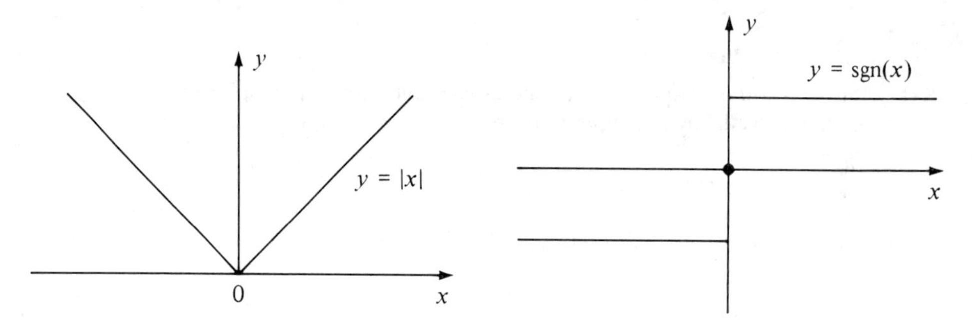

Fig. C.4.1.

Used together with the mean value theorem (MVT for short) the abs function is valuable for making estimates of the values of a function which we may not be able to calculate precisely. For example, suppose f is such that f'' is bounded on the interval I. This means that there is some $B > 0$ such that for all x in I we have $-B \leq f''(x) \leq B$, abbreviated to: $|f''| \leq B$. We assert that when $a < x < b$ and x changes by an amount h, then f changes by an amount equal to $hf'(X)$ plus an error that we can estimate. More precisely we have the following proposition.

Proposition C.4.1

If $x + h$ lies in I, and f is such that $|f''| \leq B$ then $f(x + h) = f(x) + hf'(x) + E$, where $|E| \leq Bh^2$.

Proof

By the MVT, there is some c between x and $x + h$ such that $f(x + h) = f(x) + hf'(c)$. Thus c is of the form $x + sh$, for some s between 0 and 1. Now apply the MVT to f', to get $f'(c) = f'(x) + shf''(d)$ for some d between x and c. Hence $f(x + h) = f(x) + hf'(x) + E$, where $E = sh^2f''(d)$. But $|s| \leq 1$ and, since f'' is bounded by B, $|f''(d)| \leq B$. Hence $|E| \leq Bh^2$ as required.

We can obtain a better estimate of E once the Taylor expansion is available, but the above estimate is good enough for many purposes.

Exercises C.4.1

1. Prove from the formula, rather than from the sketch, that the sgn function in (C.4.3) is not differentiable at $x = 0$. Why is it differentiable, with derivative zero, everywhere else? Show that $\int_a^b \text{sgn}(x)\,dx = |b| - |a|$. What is $\int_a^b |x|\,dx$?

2. Use the MVT to show that if $f'(x) > 0$ and $h > 0$, then $f(x - h) < f(x) < f(x + h)$ provided $x - h$ and $x + h$ lie in I.
3. The equation of the tangent to the graph of f at $(p, f(p))$ gives the function $y(x) = f(p) + (x - p)f'(p)$. Show that $|y(x) - f(x)| = K(x - p)^2$ when x lies in an interval I centred at p, and K is a constant depending on I and f''. (Older books express this by saying that the tangent approximates the graph 'to the second order'.)
4. By applying the mean value theorem to $\ln(1 + x)$ prove that

$$\frac{x}{1+x} < \ln(1 + x) < x$$

when $-1 < x < 0$ and when $x > 0$. Illustrate these inequalities with a diagram. [SU]

5. Determine the domains for which these functions are continuous:

(a) $f(x) = x + |x|$, (b) $f(x) = 1/(x^2 - 1)$,

(c) $f(x) = \dfrac{x^5 + x^2 - 1}{4 + \sin x - 2\cos x}$, (d) $f(x) = \dfrac{x^3 + 4x^2 + x - 6}{(x - 1)(x + 4)}$. [SU]

6. Let $f(x) = (1 - \cos x)/x^2$ when $x \neq 0$. Show that f can be defined at $x = 0$ so as to be continuous there. Prove that f is then differentiable at $x = 0$. Sketch the graph of f. Show that f' is a decreasing function in $[0, \pi/2]$ and hence find the largest and smallest values of f in the interval $[0, \pi/2]$. [SU]
7. Let $A(1, 0)$, $B(0, 1)$, $C(-1, 0)$, $D(0, -1)$ be the vertices of a square and consider a point $P(x(t), y(t))$ which moves from A at time $t = 0$ and follows the square in an anticlockwise path with constant velocity. Show that $|x(t)| + |y(t)| = 1$ for all t, and sketch graphs of both $x(t)$ and $y(t)$ against t. The functions $x(t)$ and $y(t)$ are thus the 'square analogues' of the circular functions $\cos(t)$ and $\sin(t)$ (as cosh and sinh are their 'hyperbolic analogues'). Thus denote $x(t) = \text{cosq}(t)$ and $y(t) = \text{sinq}(t)$ and investigate the similarities and differences between the square and the circular functions. For example, continuity, differentiability, periodicity, oddness, evenness, derivatives, integrals, Taylor expansions, compound formulae, etc.)

   Locally the graph of each of cosq and sinq seems to contain portions which resemble the modulus (or abs) function $|t|$. Can you find a way of defining the cosq and sinq functions in terms of the abs function?

C.5 MONOTONICITY

There is a principle which in the past was useful for making progress, although it is false! This is the expectation that if y is a function of x then x is a function of y. Because of the usefulness, it is an important practical problem to make precise the grain of truth in the principle; and this section is a sustained exercise in using the relevant techniques. These lead us to such functions as the inverse tangent and cosine.

A basic fact about f' is that it can tell us how f is growing. More precisely, if $f'(p) > 0$ [or <0] then f is **strictly increasing** (or **decreasing**) at p. This is useful for drawing the graph of f, which rises or falls according as $f' > 0$ (or <0). Briefly, we say that f is **monotonic** on an interval I if its graph constantly rises (or constantly falls) on I (i.e., its behaviour is monotonous). It is then that the above principle turns out to be correct. This will be clarified through the following examples.

Example C.5.1

If $y = x^3 = c(x)$, say, then $c'(c) = 3x^2$, which is strictly positive everywhere except when $x = 0$. Therefore if $u < v$ then $u^3 < v^3$ so the graph climbs from $-\infty$ to $+\infty$ as in Fig. C.5.1(a). This has the important consequence that the cubing function c is invertible in the following sense. To 'read' the graph of c at p, we move our eye from $(p, 0)$ on the x-axis, vertically upwards to the point $P = (p, c(p))$ on the graph, and then horizontally from P until we hit the y-axis at $(0, q)$. Then we conclude that $q = c(p)$. In the present case, we can reverse the process. Starting with $(0, y_0)$ we move horizontally to hit the graph at the unique point Q, then vertically to $(x_0, 0)$. This correspondence from y_0 to x_0 is here the cube root function, valid for every y, and we write $x = y^{1/3}$. It is the 'inverse' of the cubing function, and its graph is obtained by reflecting $\mathbb{R}^2$ in the line $y = x$ to give Fig. C.5.1(b). This graph now has a vertical tangent at $y = 0$; but $\mathrm{d}x/\mathrm{d}y = 1/3\,y^{-2/3}$, provided $y \neq 0$ (in order for the division to be possible), so here x is *not a differentiable function* of y at $y = 0$.

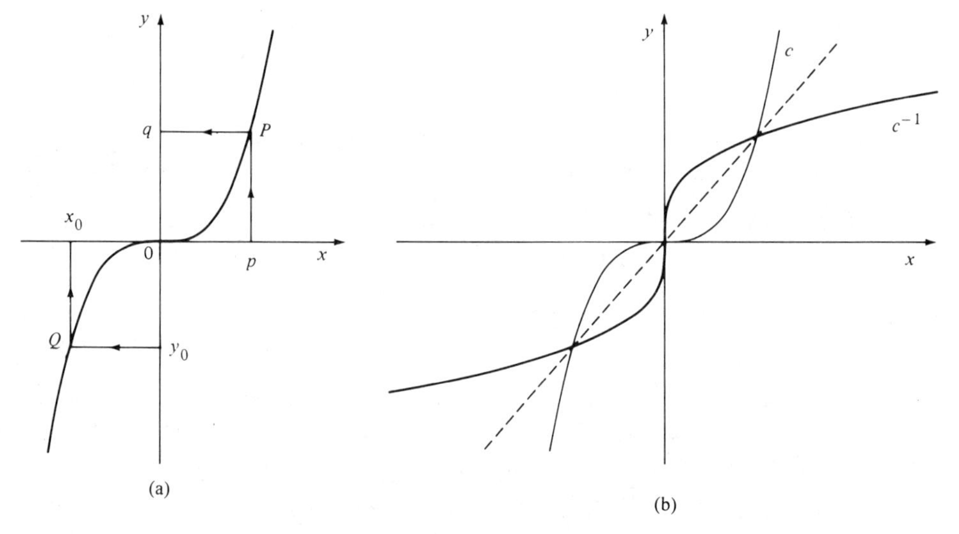

Fig. C.5.1.

Example C.5.2

Consider, instead, the squaring function, $y = x^2 = s(x)$, say. Here, $s'(x) = 2x > 0$ if $x > 0$, and $s'(x) < 0$ if $x < 0$. Therefore the graph falls from $-\infty$ to 0, and then rises to $+\infty$, as in Fig. C.5.2(a). As before, we can read the graph to go from $(p, 0)$ to $P = (p, s(p))$ on the graph, and then horizontally to $(0, q)$. But when we try to reverse the process, we cannot start with just any q, because the horizontal line through $(0, 1)$ does not hit the graph of s unless q is already a value of s. (We then say that q must be **in the image of** s.) And now the line hits the graph at two points like P and Q in Fig. C.5.2(a), unlike the situation with the cubing function c above. Thus we say that c is **one–one** (another term is 'injective'), so s is *not* one–one. With s then we must make a choice; it is customary to take the right-hand point P, then to go down to the $x-$axis to $(p, 0)$. Then the correspondence $q \mapsto p$ (i.e., start at q, end at p) is the function we call 'the positive square root', sqr (which corresponds to the computer notation SQR), and then $p = \text{sqr}(q)$. Had we decided on the left-hand choice Q, we wo... got the negative square root. (Observe what a graphical calculator or gr...

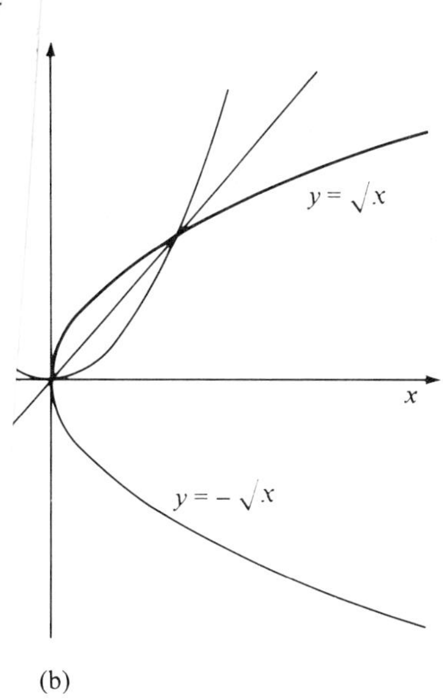

(b)

...ify the **domain** and **co-domain** of
...when we refer to $f(x)$, we should
...ome rule, and a set Y in which the
...we also write $f: X \to Y$ to indicate
...As we have seen, we then need to
...hen know that all the values lie in
...nction $f: Z \to V$, and the notation
...ften called 'f' when such precision

...ction s is the whole set $\mathbb{R}$ of real
...$\mathbb{R}$. The value $s(x)$ lies at first sight
...main $\mathbb{R}$. But s is not **onto** $\mathbb{R}$, in the
...that the image of s is the set $\mathbb{R}_+$ of
...that for every q in the co-domain,
...now, the right-hand choice we made
...e set $\mathbb{R}_+$, to force s to be one–one.
...onto, and one–one, so we call it a
...nverting process was actually applied,
...$\mathbb{R}_+$. Also $\mathrm{d}x/\mathrm{d}y = 1/2\, y^{-1/2}$ if $y \neq 0$,
..., even though the graph (obtained by
...vertical tangent there as shown in

Example C.5.3

Suppose $y = \tan x$. Here the domain is the set D that remains when the odd multiples of $\pi/2$ are removed from $\mathbb{R}$; and since $d/dx(\tan x) = \sec^2 x \geq 0$ the values increase from $-\infty$ to $+\infty$ (exclusive). Thus we have a function tan: $D \to \mathbb{R}$. Certainly, its graph is cut by every horizontal line, so the function is onto. But each such line cuts more than once, so the function is not one–one. To make it so, we must restrict it to a suitable smaller domain I, and it is customary to take I to be the interval $\{-\pi/2 < x < \pi/2\}$, because then the function tan: $I \to \mathbb{R}$ is a bijection— as shown by its graph that lies in a vertical strip of $\mathbb{R}^2$ (see Fig. C.5.3(a)). We then apply the inversion process to this function, to obtain its inverse denoted by $\tan^{-1}$: $\mathbb{R} \to I$ (but commonly called just $\tan^{-1}$ or arctan, and ATN in computer notation). The graph of this is again obtained by reflection in the line $x = y$, and is shown in Fig. C.5.3(b). It has no vertical tangent, since the graph of tan has no horizontal tangent, and so $\tan^{-1}$ is differentiable.

Caution: do not confuse $\tan^{-1}x$ with $(\tan x)^{-1}$; the latter is the reciprocal of the number $\tan(x)$, usually called cot(x).

These three examples illustrate the general process: given a function f: $J \to K$ where J, K are portions of $\mathbb{R}$, we use the derivative f' to find the image K' of f, and then we restrict f to a convenient subset J' of J to force it to be one–one. The resulting bijection f: $J' \to K'$ is then inverted to form f^{-1}: $K' \to J'$, and its graph G is obtained from the graph F of f by reflection in the line $y = x$. We now prove a result that justifies our previous assertions about the derivatives of $y^{1/3}$ and sqr(y).

Theorem C.5.1

Suppose $p \in J'$, *and* $f(p) = q \in K'$. *If* $f'(p) \neq 0$, *then*

$$(f^{-1})'(q) = 1/(f'(p)). \tag{C.5.1}$$

Proof

For each x in J', we have $f^{-1}(f(x)) = x$. Differentiating this by the chain rule gives: $(f^{-1})'(f(x)) \cdot f'(x) = 1$, and when we put $x = p$ we get $(f^{-1})'(q) \cdot f(p) = 1$ so (C.5.1) follows.

Our proof is 'algorithmic', but confirms a geometrical observation that we can see in Fig. C5.4. There, the tangent T to the graph of f at $P = (p, q)$ makes the angle θ with $\mathbf{0}x$, and $\tan\theta = f'(p) \neq 0$ (given). Under the reflection in the line $y = x$, T becomes the tangent T' to the graph of f^{-1} at $P' = (q, p)$; also the angle ϕ between T' and $\mathbf{0}x$ is the angle between T and $\mathbf{0}y$, which is $\pi/2 - \theta$. But

$$\tan\phi = (f^{-1})'(q) = \cot\theta = 1/\tan\theta = 1/(f'(p))$$

as required.

Applying theorem C.5.1 to the function $\tan^{-1}$: $\mathbb{R} \to I$, we find that if $p = \tan^{-1}(q)$, then $q = \tan p$, so the derivative of $\tan^{-1}(q)$ is given by

$$(\tan^{-1}(q))' = \frac{1}{\tan'(p)} = \frac{1}{\sec^2(p)} = \frac{1}{1 + \tan^2(p)} = \frac{1}{1 + q^2}.$$

For some purposes (such as measuring the angle θ we have just mentioned) we

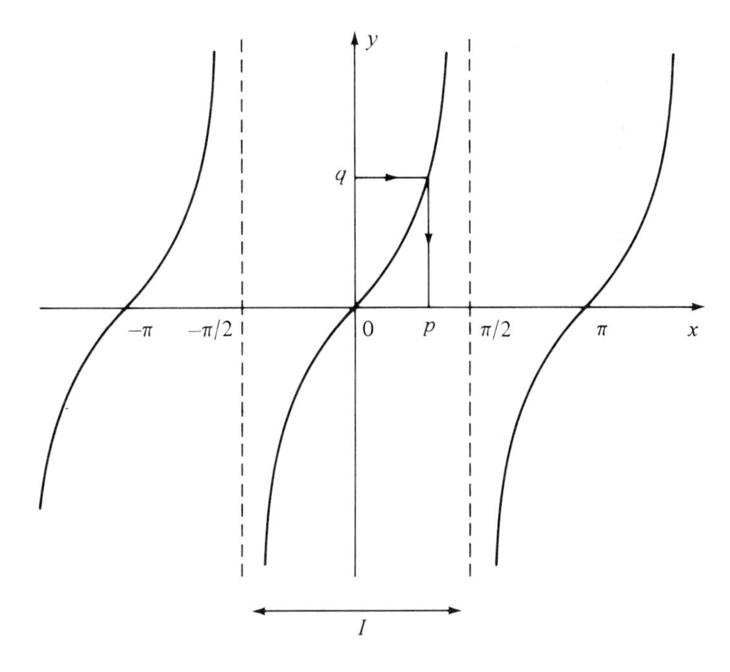

(a)

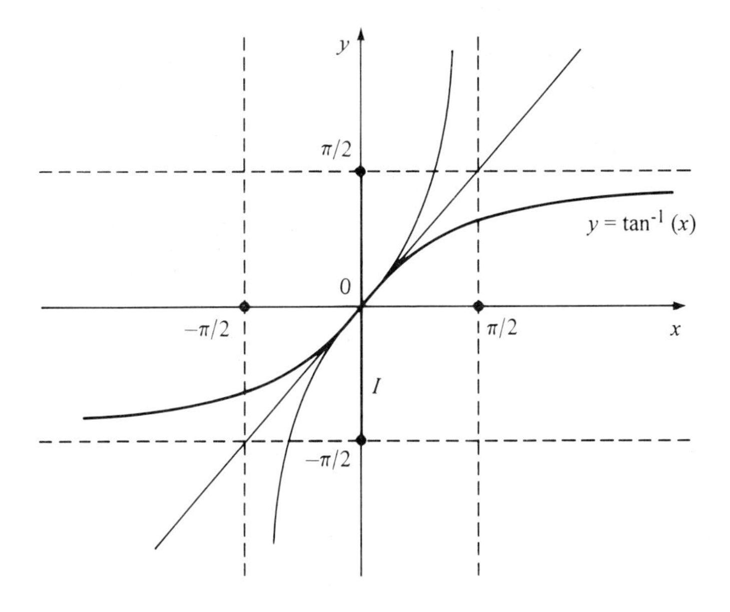

(b)

Fig. C.5.3.

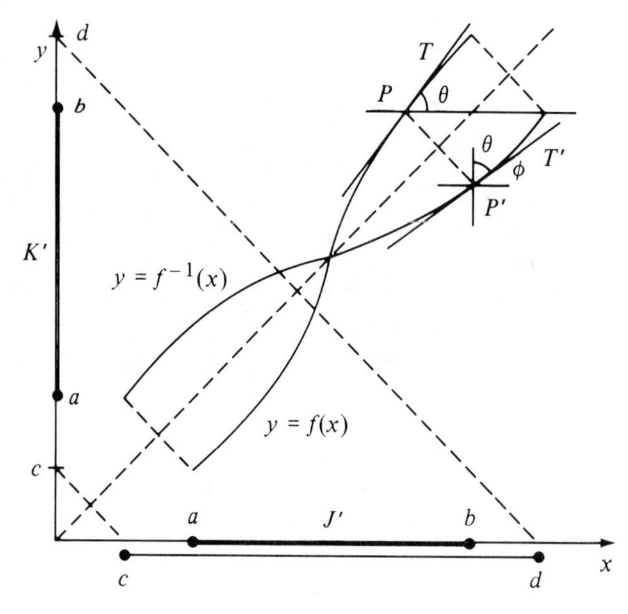

Fig. C.5.4.

need angles that lie in the interval P between 0 and π inclusive; but P is *not* the co-domain of $\tan^{-1}$. Therefore we need to use the inverse cosine, which is the function

$$\cos^{-1}\colon N \to P, \qquad N = [-1, 1] \tag{C.5.2}$$

constructed as follows. The usual cosine function has domain $\mathbb{R}$ and image N, and when we draw its graph (see Fig. C.5.5), we find that it is monotonic decreasing on the interval P, because $\cos' = -\sin \leq 0$ on P. Hence if we restrict its domain to P, we obtain a bijection $\cos\colon P \to N$, with inverse $\cos^{-1}\colon N \to P$ as required.

To find an inverse of the sine function we work similarly (see Fig. C.5.6). Briefly, we note that the image of sine is the same interval N in (C.5.2) as for cos. To summarize, there are inverses

$$\sin^{-1}\colon N \to J \qquad \text{with } J = \{-\pi/2 \leq x \leq \pi/2\} \tag{C.5.3}$$

$$\cos^{-1}\colon N \to P \qquad \text{with } P = \{0 \leq x \leq \pi\}, \tag{C.5.4}$$

such that if $y = \sin(x)$ and $x \in J$ then $x = \sin^{-1}(y)$; and similarly for $\cos^{-1}$.

The method for sin is most like that for tan: from the graph we see that we can restrict the domain to give a function $\sin\colon J \to N$, which is increasing and a bijection; so it has the inverse given in (C.5.3). (We refer informally to 'the inverse sine'; other notations such as arcsin or ASN are used. Intervals other than J can also be used, but it is the largest possible choice that contains the origin. For this reason it is the commonest choice, and older books call it the 'principal branch of the inverse sine'.)

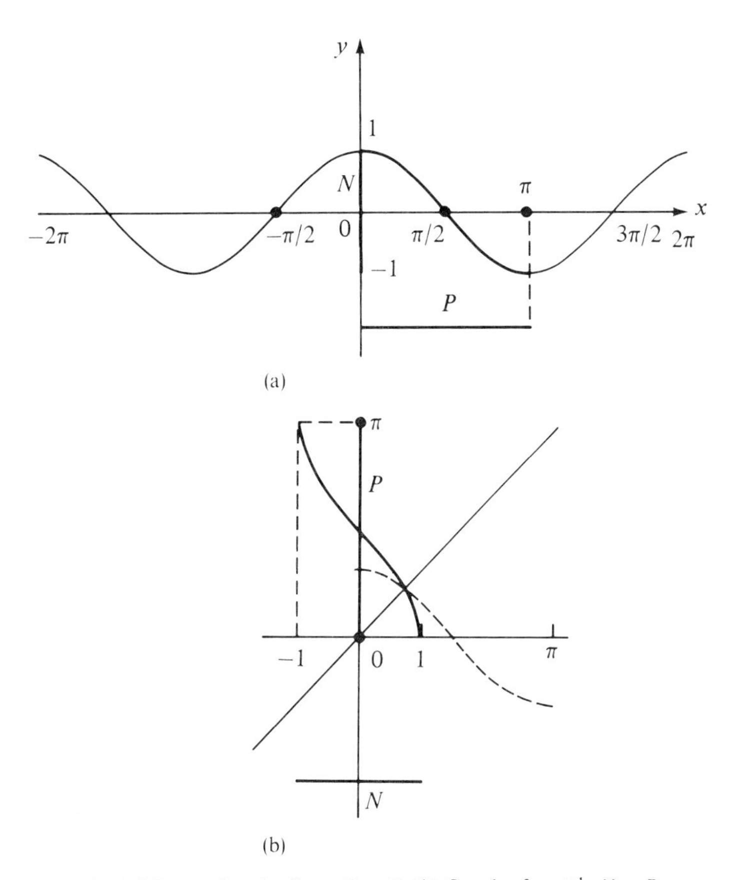

Fig. C.5.5. (a) Graph of cos: $\mathbb{R} \to N$. (b) Graph of $\cos^{-1}$: $N \to P$.

Exercises C.5.1

1. Show that for each non-end-point of N, the derivative of $\cos^{-1}(q)) = -1/\sqrt{(1-q^2)}$, and use a sketch to show that the graph of $\cos^{-1}$ has vertical tangents at these end points.
2. Show that sin: $J \to N$ is a bijection, and that the derivative of $\sin^{-1}$: $N \to [-\pi/2, \pi/2]$ is $1/\sqrt{(1-q^2)}$ for all q in N except at ± 1, where again the tangents are vertical.
2. Find 'good' inverses for the functions sec, cosec, and cot.
4. In Fig. C.5.4, $\tan\theta = f'(p)\,(=g$, say); so when θ is obtuse, $g < 0$ so $\phi = \tan^{-1}(g) < 0$. Hence $\phi \neq \theta$. Show that in this case, $\theta = \pi + \phi$. However, show that for all θ in P (except when the tangent is vertical, and so g is not defined), $\theta = \cos^{-1}(\operatorname{sgn}(g)/\sqrt{(1+g^2)})$.
5. Show that for each x in $\mathbb{R}$, $\tan^{-1}(-x) = -\tan^{-1}(x)$; and if $x \neq 0$, then $\tan^{-1}(1/x) = \pi/2 - \tan^{-1}(x)$. Hence show that if y tends to $+\infty$ or $-\infty$, then $\tan^{-1}(y)$ tends to $\pi/2$ or $-\pi/2$, respectively.
6. Specify the maximum domain and image of the real-valued function defined by

$$f: x \mapsto \left(\frac{x+1}{x-1}\right)^{1/2}$$

and find the inverse function. [SU]

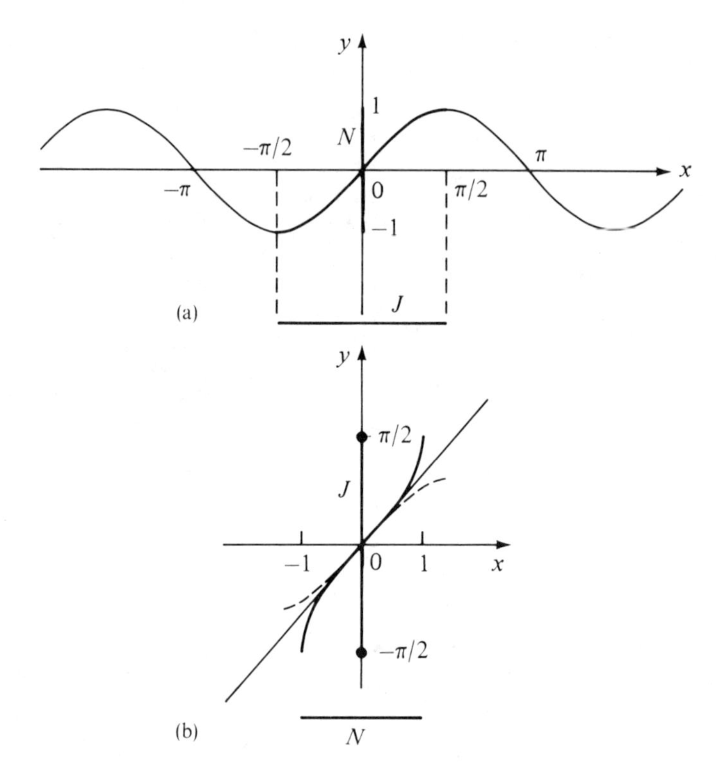

Fig. C.5.6. (a) Graph of sin: $\mathbb{R} \to N$. (b) Graph of $\sin^{-1}$: $N \to J$.

7. The functions f and g are defined by
   (a) $y = f(x) = x^2 - 4x + 5$ for $2 \leq x \leq 4$, and
   (b) $y = g(x) = e^{1/x}$, for $x > 0$.
   Determine the corresponding inverse functions f^{-1} and g^{-1}, in each case giving their domain and image.
   Sketch graphs of f and f^{-1} on one diagram, and g and g^{-1} on another.
   Evaluate the derivatives of f^{-1} and g^{-1} both by making use of the derivatives of f and g, and directly. [SU]
8. Evaluate

$$\lim_{x \to \infty} \coth x, \quad \lim_{x \to -\infty} \coth x, \quad \lim_{x \to 0+} \coth x \quad \text{and} \quad \lim_{x \to 0-} \coth x.$$

   Also calculate $d/dx(\coth x)$ and use all this information to sketch the graph of $\coth x$.
   Specify a domain on which $\coth x$ is monotonic and hence specify a domain and range for $\coth^{-1} x$. Sketch the graph of $\coth^{-1} x$, and calculate $d/dx(\coth^{-1} x)$. [SU]

These ideas can be used to show that any smooth function $y = h(x)$ can, by a change of variables, be written in the form $z = ku^n$, at least 'locally". To be precise, suppose we have h: $I \to \mathbb{R}$ given on an interval I centred at b. First write $x = b + a$, and $h(b + a) = f(a)$. By exercise C.2.2(8) we can write f in the form $y = f(a) = f(0) + Aa^n(1 + ag(a))$, where $A \neq 0$ and $|ag(a)| < M|a|$, for some constants A, M. Hence, changing y to $z = y - f(0)$ we have $z = Aa^n(1 + ag(a))$, where a lies in an interval J centred at 0. To study z near $a = 0$, it is not necessary to 'neglect terms of order higher than n' (and then wonder what bugs such neglect might let in), as we now show.

Suppose first that n is odd. Then we can define u:

$$u = a \cdot [A(1 + ag(a))]^{1/n} \tag{C.5.5}$$

whence $z = u^n$ as required. But to ensure that the function $u(a)$ is invertible, we need to have it monotonic. However, if we evaluate $\mathrm{d}u/\mathrm{d}x$ at $a = 0$, we find $u'(0) = A$ so by the Inertia Principle, u' remains non-zero on some interval K in J, centred at 0. Hence the mapping $u: K \to R$ is monotonic with image $u(K)$, another interval containing 0. Therefore, the graph of z on J is mapped bijectively and smoothly onto the graph of the function u on K: each qualitative feature of the one is present in the other. If n is even, then we proceed as before if $A > 0$, with a small enough to ensure that $1 + ag(a) > 0$. *If* $A < 0$, we must use $-A$ in the definition of u.

Exercises C.5.2

Near $x = 0$, express in the form ku^n the functions $x - \sin x$, $\cosh x - \cos x$, $\sin^3 x - \tan^3 x$.

At various points in the foregoing, we have taken certain points to be 'obvious', such as the differentiability of f^{-1} in Theorem C.5.1. Our arguments throughout have been based on 'obvious' statements like the mean value theorem, the Inertia Principle, existence of area under a graph, and so on. After a long historical process, mathematicians needed to be more certain, and these 'obvious' results were modelled in a more precise language that allowed acceptable proofs of them to be constructed. The resulting body of theorems and techniques is known as Mathematical Analysis, and is not needed for the purposes of this book. Nevertheless, readers will often be doing bits of Analysis without perhaps realizing it. For a first organized exposition, see Bryant (1990).

D

Functions of several variables

D.1 PARTIAL DIFFERENTIATION

Functons of a single variable are not sufficient for describing or modelling many of the processes we observe in the outside world. Often, we must study a variable V which depends on several, rather than only one, other variables; and V can change—usually gradually but sometimes drastically—with changes in these other variables. Therefore, we must inevitably study functions of several variables. For example, each point in a room has three rectangular coordinates x, y, z relative to some choice of axes; and then the temperature at that point is a function $T(x, y, z, t)$, where t is the time. Here, we consider all the variables as being on the same footing, and not (as with the functions met in Chapter 6) with one as the main variable, and the rest as 'control' variables.

If T is the room temperature, then we may wish to study its variation as just one of its variables changes. For example, we might be interested in changes of the height z, and keep x, y, t fixed for the moment. In that case, T then varies like an ordinary function of a single variable, and has a derivative in the ordinary way. To show that this derivative is calculated relative to z, we use the curved ∂ symbol to denote it by $\partial T/\partial z$ (pronounced 'partial dee tee by dee zed'), and calculate it according to the original rule (4.2.2):

$$\frac{\partial T}{\partial z} = \lim_{h \to 0} \left(\frac{T(x, y, z + h, t) - T(x, y, z, t)}{h} \right). \tag{D.1.1}$$

Similarly for the other variables, so that, for example,

$$\frac{\partial T}{\partial t} = \lim_{h \to 0} \left(\frac{T(x, y, z, t + h) - T(x, y, z, t)}{h} \right).$$

The usual rules apply as before, so that, for example, if

$$T(x, y, z, t) = x^2 \cdot z^3 + \sin(t + y) \cdot x^{-1} \tag{D.1.2}$$

then $\partial T/\partial y = \cos(t + y) \cdot x^{-1}$, just as if x, z, t were constant numbers; and

$$\frac{\partial T}{\partial z} = 3x^2 \cdot z^2, \qquad \frac{\partial T}{\partial t} = \cos(t + y) \cdot x^{-1}, \qquad \frac{\partial T}{\partial x} = 2xz^3 - \sin(t + y) \cdot x^{-2}.$$

To indicate when we are evaluating these derivatives at a specific point P, we write, for example, $\partial T/\partial x\,|_P$. Thus with (D.1.2) we have at the point $P = (1, 2, 3, 4)$,

$$\left.\frac{\partial T}{\partial t}\right|_{(1,2,3,4)} = \cos(6).$$

It is convenient sometimes to think of all the first derivatives of T together as a vector, and then we have the **gradient** of T, writing:

$$\operatorname{grad} T = \left(\frac{\partial T}{\partial x}, \frac{\partial T}{\partial y}, \frac{\partial T}{\partial z}, \frac{\partial T}{\partial t}\right); \tag{D.1.3}$$

and if we evaluate each of these derivatives at P, then we get a (clumsy) quadruple of numbers that we write for brevity as:

$$(\operatorname{grad} T)|_P = \left(\left.\frac{\partial T}{\partial x}\right|_P, \left.\frac{\partial T}{\partial y}\right|_P, \left.\frac{\partial T}{\partial z}\right|_P, \left.\frac{\partial T}{\partial t}\right|_P\right)$$

or

$$(\operatorname{grad} T)|_P = \left.\left(\frac{\partial T}{\partial x}, \frac{\partial T}{\partial y}, \frac{\partial T}{\partial z}, \frac{\partial T}{\partial t}\right)\right|_P.$$

Similarly for 2, 3 or n variables.

Exercises D.1.1

1. With T as in (D.1.2), work out $(\operatorname{grad} T)|_{(1,2,3,4)}$.
2. Let $x(t)$ be the solution of the differential equation $x'' + ax' + b = 0$ with $x_0 = 1$, $x_0' = 2$. Calculate $\partial x/\partial a$, $\partial x/\partial b$.
3. The definition of $T(x, y, z, t)$ in (D.1.2) makes sense provided $x \neq 0$, since we cannot divide by zero. Thus the **domain** of T is $\mathbb{R}^2 - \{\mathbf{0}\}$. Similarly the domain of $f(x, y) = (x^2 + y^2)/xy$ is $\mathbb{R}^2$ minus the axes. Find suitable domains when $f(x, y)$ is:

   $$xy/(x^2 + y^2); \qquad (x^2 + 3y^2)/(1 + x^2 - y^2); \qquad \ln((x^2 + y^2)/xy),$$
   $$\tan^{-1}(x/y); \qquad ax^2 + 2hxy + by^2; \qquad x^2 + y^2 + z^2 - 3xyz.$$

   Calculate $\operatorname{grad} f$ in each case.
4. For each of the functions in exercise 3, calculate $\mathbf{z} \cdot \operatorname{grad} f|_{\mathbf{z}}$, where $\mathbf{z} = (x, y)$, and compare the answer with $f(\mathbf{z})$. Can you see a general rule here? (See exercise D.4.1.) [SU]
5. The **direction** of a vector $\mathbf{v}$ is the unit vector $\hat{\mathbf{v}} = \mathbf{v}/\|\mathbf{v}\|$. With T as in D.1.2, its **directional derivative** at P **along** $\mathbf{v}$ is defined to be the scalar product $\hat{\mathbf{v}} \cdot (\operatorname{grad} T)|_P$. Evaluate this when $P = (1, 2, 3, 4)$ and $\mathbf{v} = (-7, -5, -3, -1)$. Evaluate the directional derivative of each two-variable function f in exercise 3, when $P = (1, 2)$ and $\mathbf{v} = (3, 4)$.
6. Find the rate of change of $x^2 + xy + y^2 + z^2$ at $(1, 1, 1)$ in the direction of the line $x = y = z$.

7. For the function $F(\mathbf{r}) = xyz + x^3 + y^2 + z$, and point P with coordinates $(1, 0, 3)$, find:

   (a) grad F at P;
   (b) the unit normal at P to the surface $F(\mathbf{r}) = 4$;
   (c) the equation of the plane tangential to $F(\mathbf{r}) = 4$ at P;
   (d) the rate of change of F at the point P in the direction of $\hat{\mathbf{l}}$, where $\mathbf{l} = \mathbf{j} + \mathbf{k}$. [SU]

Just as in the case of a single variable, we can have higher derivatives. For these we have the natural analogues of $\mathrm{d}^2 f/\mathrm{d}x^2$:

$$\frac{\partial^2 T}{\partial x^2} = \frac{\partial}{\partial x}\left(\frac{\partial T}{\partial x}\right), \tag{D.1.4}$$

that is, use the operator $\partial/\partial x$ on the function $\partial T/\partial x$, and similarly for the other variables. But we also have mixed derivatives, such as

$$\frac{\partial^2 T}{\partial x \partial y} = \frac{\partial}{\partial x}\left(\frac{\partial T}{\partial y}\right), \quad \frac{\partial^2 T}{\partial y \partial x} = \frac{\partial}{\partial y}\left(\frac{\partial T}{\partial x}\right), \tag{D.1.5}$$

and fortunately these two can be proved to be equal, provided each is continuous (as happens in most practical cases). For example, if T is given by (D.1.2) then:

$$\frac{\partial^2 T}{\partial x^2} = 2z^3 + \sin(t + y)\cdot x^{-3}, \qquad \frac{\partial^2 T}{\partial y \partial x} = -\cos(t + y)\cdot x^{-2}.$$

The formulae at (D.1.5) are beginning to look complicated, and look even worse if we insert the point $P = (x, y, z, t)$ at which they are to be evaluated; so we reduce the clutter by using the abbreviated form $T_1(P)$ for $\partial T/\partial x\,(P)$. Thus the '1' here refers to the 'slot' occupied by the variable x, as the y-slot is numbered '2', and so on. The first of the mixed derivatives in (D.1.5) is then $((T_2)_1)(P)$, which we simplify to $T_{21}(P)$; and then the equality of the derivatives in (D.1.5) can be written as

$$T_{12}(P) = T_{21}(P), \tag{D.1.6}$$

which we always assume to hold, because (unless we specify to the contrary) we shall always assume that the functions we deal with have continuous derivatives of order as high as needed. As a brief reminder, we shall speak of such functions as **smooth**.

Exercises D.1.2

1. For each function f in exercise D.1.1(3), verify that $\partial^2 f/\partial x \partial y = \partial^2 f/\partial y \partial x$.
2. The famous equation of Laplace is (in two variables) $\partial^2 V/\partial x^2 + \partial^2 V/\partial y^2 = 0$. Show that this is satisfied when V is:

   (a) $\ln r$, $(r = (x^2 + y^2)^{1/2})$;
   (b) $e^x (x \cos y - y \sin y) + 4xy$;
   (c) the real or imaginary part of z^n for any integer n and complex z.

3. In plane polar coordinates, Laplace's equation is $\partial^2 V/\partial r^2 + 1/r\, \partial V/\partial r + 1/r^2\, \partial^2 V/\partial \theta^2 = 0$. Show that this is satisfied by $V = (Ar^n + Br^{-n})\cos(n\theta + \varepsilon)$, where A, B, n and ε are arbitrary constants.

4. If $f(x)$ is smooth, then for any constant c, $\partial/\partial t\,(f(x-ct)) = -cf'(x-ct)$. Hence verify that if $g(x)$ is also smooth, then $V = f(x-ct) + g(x+ct)$ satisfies the one-dimensional wave equation:

$$\frac{\partial^2 V}{\partial x^2} - \frac{1}{c^2}\frac{\partial^2 V}{\partial t^2} = 0.$$

   Sketch graphs of $f(x-ct)$, $f(x+ct)$ when $f(x)$ is (i) x^3, (ii) $\sin x$.
5. In three variables, the *Laplacian* of $V(x, y, z)$ is the function $\partial^2 V/\partial x^2 + \partial^2 V/\partial y^2 + \partial^2 V/\partial z^2$, usually denoted by $\nabla^2 V$. Show that $\nabla^2 V = 0$ when $V = r^{-1}$ and $r = (x^2 + y^2 + z^2)^{1/2}$. Calculate $\nabla^2 V$ when $V = \sin ax \cdot \cos by \cdot \sin cz$, where a, b, c are constants. (∇ is 'del'.)
6. Given the function $F(\mathbf{r}) = x(y^2 + z^2) + z(x^2 + y^2) + y(x^2 + z^2)$ find:

   (a) grad F at the point P with coordinates $(1, 2, -1)$;
   (b) a unit normal $\hat{\mathbf{n}}$ to the surface $F(\mathbf{r}) = 14$ at P;
   (c) $\cos\theta$, where θ is the angle between $\hat{\mathbf{n}}$ and the z-axis;
   (d) the directional derivative of F in the direction $\mathbf{l} = (1, 1, 1)$;
   (e) $\nabla^2 F$ at P. [SU]

D.2 CONTINUITY

For expository purposes, we usually work with a function $F(x, y)$ of just two variables. Readers should then try rewriting the theory for three variables as an active way of gaining understanding and fluency, but we shall often spell out appropriate results for clarity. The theory for n variables should then be clear.

Several basic results are direct analogues of results we have already seen for one variable, and we shall consider these first. With two variables, the pair (x, y) corresponds to a point $\mathbf{z}$ in the plane $\mathbb{R}^2$, and $\mathbf{z}$ can be any point in the domain D of F, which is some specified set (usually but not necessarily the largest) in $\mathbb{R}^2$ for which the rule for calculating F is valid. For example, if $F(x, y) = 1/(xy)$, then the largest D is $\mathbb{R}^2$ minus the axes; for, we cannot divide by zero, and $xy = 0$ exactly when (x, y) lies on one or other axis.

If $r > 0$, then an r-**neighbourhood** of $\mathbf{z}$ means a square disc, centre $\mathbf{z}$, with sides of length $2r$, parallel to the axes. Then we suppose that the domain D of the function $F(x, y)$ is an **open** set, which means that for every $\mathbf{z}$ in D, some r-neighbourhood of $\mathbf{z}$ lies entirely in D (see Fig. D.2.1). With three variables (x, y, z), D would be a subset of 3-space $\mathbb{R}^3$, and an r-neighbourhood is a solid cubical box (with edges parallel to the axes).

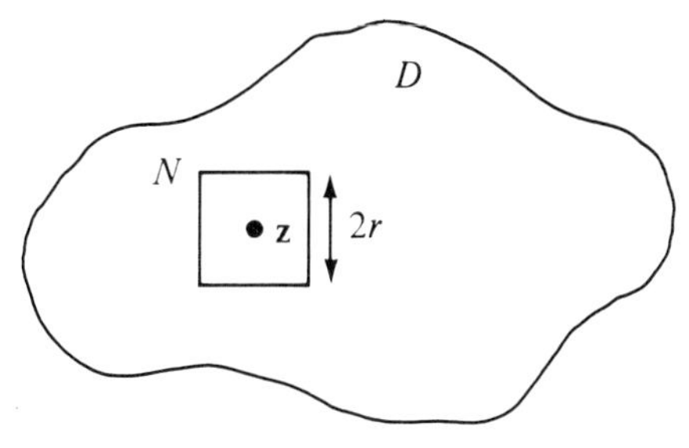

Fig. D.2.1.

We call F 'continuous' if the criterion (C.4.1) holds, because that still makes sense. It can be shown that the standard differentiable functions, their sums, products, and compositions, are all continuous. Just as with one variable, (C.4.2), F is also *bounded*, this time on every rectangle of the form

$$\{(x, y) \mid a \leq x \leq b, \quad c \leq y \leq d\}.$$

Books on Analysis prove these facts, as well as the following, which is valid for $n \geq 1$ variables.

Theorem D.2.1 (Inertia principle)
If $F: D \to \mathbb{R}$ is continuous at $\mathbf{z}$, and $F(\mathbf{z}) > 0$, then there is some neighbourhood N of $\mathbf{z}$ on which F remains positive, that is, for every $\mathbf{w} \in N$, we have $F(\mathbf{w}) > 0$.

(By changing F to $-F$ if necessary, we may reverse the inequalities.) The principle gets its name by analogy with mechanics. If a heavy body has positive velocity at time t, then its velocity stays positive for a whole interval of time: the body cannot stop dead, because of its mechanical inertia.

Exercises D.2.1

1. For each function f in exercise D.1.1(3), find the set of points where it is (i) positive, (ii) zero. Show that f does not stay constantly zero on any neighbourhood.
2. Let $f_a(x)$ be a polynomial in x, which depends smoothly on a control variable a, and let u be a simple root of f_a. Show that when b is near a, f_b has no multiple root near u. (This will be made more precise later; for the moment try to see what it means to you.)
3. Let $\Delta(a, b, c)$ be the determinant of the matrix $A = \begin{pmatrix} a & b \\ b & c \end{pmatrix}$ and suppose that $\Delta(a_0, b_0, c_0) = p > 0$. Show that Δ remains positive on the r-neighbourhood of (a_0, b_0, c_0) provided $r < p/(|a_0| + 2|c_0| + |b_0|)$. (If all we need is *some* $r > 0$, then the inertia principle allows us to avoid such tricky calculations of an explicit r.) Sketch the surface on which Δ is zero. Now do a similar calculation for the general case when the second row of A is replaced by (c, d).
4. Can $[\sin(x + y)]/(x + y)$ be made continuous by suitably defining it when $x + y = 0$? Can $xy(x^2 + y^2)$ be made continuous by suitably defining it when $(x, y) = (0, 0)$? [SU]
5. Let $f: \mathbb{R}^2 \to \mathbb{R}$ be given by $f(x, y) = xy(x^2 + y^2)^{-1/2}$ if $(x, y) \neq (0, 0)$ and $f(0, 0) = 0$.
   (a) Show that f is continuous at $(0, 0)$.
   (b) Calculate the partial derivatives $\partial f/\partial x$ and $\partial f/\partial y$ at the origin and at $(x_0, y_0) \neq (0, 0)$.
   (c) Show that the partial derivatives are not continuous at $(0, 0)$.

D.3 THE FORMULA FOR SMALL INCREMENTS

With a function f of one variable, we proved in proposition C.4.1 the **formula of small increments**:

$$f(x + h) = f(x) + hf'(x) + O(h^2), \tag{D.3.1}$$

(where $O(h^2)$ indicates that the error is of magnitude $h^2 \times$ constant). The formula allows an easy estimate of the change in f when x is changed by h; the estimate becomes increasingly accurate as h gets smaller. We now prove an analogue of the formula for a smooth function $F(x, y)$, and the analogue for more variables is obtained in exactly the same way.

Proposition D.3.1

Let $F: \mathrm{D} \to \mathbb{R}$ be smooth on the open domain D in $\mathbb{R}^2$, and let N be an r-neighbourhood of $\mathbf{z} = (x, y)$ in D. If $|h|, |k| < r$, then

$$F(x + h, y + k) = F(z) + hF_1(z) + kF_2(z) + O(r^2). \tag{D.3.2}$$

Before giving the proof we supply some exercises.

Exercises D.3.1

1. In (D.3.2) we denote the portion $hF_1(\mathbf{z}) + kF_2(\mathbf{z})$ by $d_z f$. Calculate $d_z f$ when $\mathbf{z} = (6, 7)$, $(h, k) = (0.2, -0.3)$ and f is:

   (i) $x^2 + y^2$, (ii) $5x^3 + 3x^2y$; (iii) $e^x \sin(2x/y)$.

   The difference $F(x + h, y + k) - F(z)$ is often denoted by ΔF. For each of (i)–(iii) calculate $\Delta F - d_z f$.
2. In physical experiments, one can locate a point such as z only by approximations $(x + h, y + k)$ to its real coordinates (x, y), but knowing that $|h|$ and $|k|$ do not exceed certain limits, say α, β. By the triangle inequality,

   $$|d_z f| \leq \alpha \cdot |F_1(\mathbf{z})| + \beta \cdot |F_2(\mathbf{z})|.$$

Estimate $|d_z f|$ in exercise 1 if $(\alpha, \beta) = (0.25, 0.35)$.

3. In exercise 2, we could not, in practice, calculate $F_1(\mathbf{z}), F_2(\mathbf{z})$ because we know only $\mathbf{w} = (x + h, y + k)$. Use (D.3.2) on F_1, F_2 to show that

   $$\Delta F = hF_1(\mathbf{w}) + kF_2(\mathbf{w}) + O(r^2).$$

   What difference would this make to your calculations in exercise 1? Use this last equation to estimate the error in calculating the period P of a pendulum from the formula $P = 2\pi\sqrt{l/g}$ when there is an error of ± 3 cm in measuring a 3 m length l; and g, taken as 981 cm s^{-2}, may be in error by ± 0.5 cm s^{-2}.
4. Suppose we estimate $[(6.02)^2 + (7.79)^2]^{1/2}$ as $f(x, y) + d_2 f$, where $f(x, y) = (x^2 + y^2)^{1/2}$, $(x, y) = (6, 8)$ and $h = 0.02 = -k$. Show that $d_2 f = 0.004$. Now, the $O(r^2)$ term in D.3.2 means a quantity of modulus $\leq Ar^2$ for some constant $A > 0$. Show that if A is only as large as 5, then the $O(r^2)$ term is here as large as $d_z f$. (This shows the *practical* need for a good estimate of the $O(r^2)$ term, which we give later; but (D.3.2) is often sufficient for theoretical purposes.)

Proof of proposition D.3.1

By treating F as a function of the first variable, we have by (D.3.1):

$$\begin{aligned} F(x + h, y + k) &= F(x, y + k) + hF_1(x, y + k) + O(h^2) \\ &= (F(x, y) + kF_2(x, y) + O(k^2)) \\ &\quad + h(F_1(x, y) + kF_{12}(x, y) + O(k^2)) + O(h^2) \end{aligned}$$

by applying (D.3.1) to $F(x, y + k)$ and to $F_1(x, y + k)$. Thus

$$F(x + h, y + k) = F(x, y) + hF_1(x, y) + kF_2(x, y) + E,$$

where

$$E = O(k^2)(1 + h) + hkF_{12}(x, y) + O(h^2).$$

Since $h^2 < r^2$ then $O(h^2)$ is $O(r^2)$; and since F is assumed smooth, F_{12} is bounded on N, so $hkF_{12}(x, y + k)$ is $O(r^2)$. The three terms in E then collapse into one, $O(r^2)$, and we have the required assertion.

Exercises D.3.2

1. By using the term Br^2 in proposition C.4.1, find a suitable constant M to show that, for E above: $|E| \leq Mr^2$.
2. In the last proof, check that we have assumed only the continuity of first derivatives and the boundedness of the second ones.
3. Modify the last proof so as to obtain the three-variable analogue of formula (D.3.2): if $\mathbf{X} = (x, y, z)$, then $F(x + h, y + k, z + l) = F(\mathbf{X}) + hF_1(\mathbf{X}) + kF_2(\mathbf{X}) + lF_3(\mathbf{X}) + O(r^2)$, when $r^2 = x^2 + y^2 + z^2$. Formulate an analogue for n variables.

D.4 THE CHAIN RULE

With $F: D \to \mathbb{R}$ as before, then using the grad notation of (D.1.3), we note that the middle term in (D.3.2) is the scalar product of the two vectors $(\text{grad } F)|_{\mathbf{z}}$ and (h, k). Thus we write (D.3.2) as:

$$F(\mathbf{z} + (h, k)) = F(\mathbf{z}) + (\text{grad } F)|_{\mathbf{z}} \cdot (h, k) + O(r^2). \tag{D.4.1}$$

We can now formulate a chain rule, analogous to the one for a single variable. The simplest version is the following, which tells us how to differentiate $F(x, y)$ when each variable x, y is a differentiable function of a single variable t (like time). Let

$$\mathbf{z} = (x(t_0), y(t_0)), \qquad \text{and} \qquad \mathbf{w} = \mathbf{z} + (x, y). \tag{D.4.2}$$

Then $F(\mathbf{w})$ is a function f of t, and $F(\mathbf{w}(t_0)) = f(\mathbf{z} + \mathbf{0}) = f(\mathbf{z})$. We shall prove an important rule, which older books write in the simple form:

$$\frac{\mathrm{d}f}{\mathrm{d}t} = \frac{\partial F}{\partial x}\frac{\mathrm{d}x}{\mathrm{d}t} + \frac{\partial F}{\partial y}\frac{\mathrm{d}y}{\mathrm{d}t}.$$

But this simplicity is misleading, as we soon find if we ask a computer to evaluate the right-hand side at time t_0; the first 'x' on the right is a *variable*, the second is a *function.* In fact the equation means (when written in full with $\mathbf{z}$ as in (D.4.2)):

$$\left.\frac{\mathrm{d}f}{\mathrm{d}t}\right|_{t_0} = F_1(\mathbf{z})u'(t_0) + F_2(\mathbf{z})v'(t_0), \tag{D.4.3}$$

which abbreviates to

$$\left.\frac{\mathrm{d}f}{\mathrm{d}t}\right|_{t_0} = (\text{grad } F)|_{\mathbf{z}} \cdot (u', v')|_{t_0}. \tag{D.4.4}$$

We postpone the proof until the end of this section. The function f in (D.4.3) is the composition of the two functions F and $\mathbf{w}$, and we denote it by $F \circ \mathbf{w}$. Therefore we can regard (D.4.3) as another chain rule about *functions*:

$$(F \circ \mathbf{w})' = u' \cdot F_1 \circ \mathbf{w} + v' \cdot F_2 \circ \mathbf{w} \tag{D.4.5}$$

which, with the variable t_0 inserted, gives (D.4.3) itself.

Exercises D.4.1

1. The function $\mathbf{w}$ above is called a **smooth path**, and as t changes, the point $\mathbf{w}(t)$ moves along a curve called the **track** of $\mathbf{w}$. (Thus the track is a *set*, and the path is a pair of *functions*.) Find two different paths $\mathbf{w}_1, \mathbf{w}_2$, which have the same track.
2. Suppose $F(x, y) = 3x^2y + \sin(x + 2y)$. Find $f'(t)$ when $f = F \circ \mathbf{w}$, and $\mathbf{w}$ is the path with $\mathbf{w}(t)$ given by

   (i) (t, t^2); (ii) (t^2, t); (iii) $(a \cos t, b \sin t)$.

3. Let D be a **connected** domain in $\mathbb{R}^2$ (i.e. any two points in D are the ends of a smooth path in D). Suppose $f(x, y)$ has gradient zero everywhere on D. Show that $f =$ constant on D. (For any two points X, Y in D, take a smooth path $\mathbf{w}$ from X to Y in D, that is, for each t between 0 and 1, $\mathbf{w}(t)$ is a point $(u(t), v(t))$ in D, and $\mathbf{w}(0) = X$, $\mathbf{w}(1) = Y$. Now use (D.4.3) to show that $f(X) = f(Y)$.)
4. Prove the three-variable analogue of (D.4.5).
5. (*Euler's theorem for homogeneous functions.*) Call a function $f: \mathbb{R}^n \to \mathbb{R}$ *homogeneous* if for all $t \in \mathbb{R}$, and $\mathbf{x} \in \mathbb{R}^n$ we have $f(t\mathbf{x}) = t^m f(\mathbf{x})$ for some constant integer m. Apply the chain rule and then put $t = 1$ to show: $f(\mathbf{x}) = m\mathbf{x} \cdot \operatorname{grad} f$.
6. If $f(x, y, z)$ is a homogeneous function of degree n, prove that

$$x^2 \frac{\partial^2 f}{\partial x^2} + y^2 \frac{\partial^2 f}{\partial y^2} + z^2 \frac{\partial^2 f}{\partial z^2} + 2xy \frac{\partial^2 f}{\partial x \partial y} + 2yz \frac{\partial^2 f}{\partial y \partial z} + 2zx \frac{\partial^2 f}{\partial z \partial x} = n(n-1)f.$$

7. If $y = \phi(x)$ is substituted in $F(x, y)$ to form $f(x) = F(x, \phi(x))$ show that $f'(x) = (F_1 + F_2 \cdot \phi')|_x$. (Take $\mathbf{w}(x)$ in D.4.5 to be $(x, \phi(x))$.)

Given the smooth function F on D as before, suppose that the point $\mathbf{z} = (x, y)$ is a smooth function of two variables u, v, so $x = x(u, v)$, $y = y(u, v)$. Then we can regard F as a function G of (u, v), and we need a chain rule analogous to (D.4.3) for calculating the derivatives of G in terms of those of F. This rule is given in classical notation in the simple-looking form:

$$\frac{\partial F}{\partial u} = \frac{\partial F}{\partial x}\frac{\partial x}{\partial u} + \frac{\partial F}{\partial y}\frac{\partial y}{\partial u}, \qquad \frac{\partial F}{\partial v} = \frac{\partial F}{\partial x}\frac{\partial x}{\partial v} + \frac{\partial F}{\partial y}\frac{\partial y}{\partial v}, \tag{D.4.6}$$

but again this is not sufficient as a rule for computation. To free it from ambiguity we write it as follows. First $\mathbf{z} = (x, y)$ is a function μ of the point $\mathbf{q} = (u, v)$; so F 'as a function of u, v' is the composition $G = F \circ \mu$, and $G(\mathbf{q}) = F(\mu(\mathbf{q}))$. Also $\mu(\mathbf{q}) = (\alpha(\mathbf{q}), \beta(\mathbf{q}))$, say, whence $x = \alpha(\mathbf{q}) = \alpha(u, v)$, and $y = \beta(u, v)$. We suppose that α, β are each differentiable on some domain X of $\mathbb{R}^2_{uv}$. Thus $\partial F/\partial u$, $\partial F/\partial v$ above mean G_1, G_2, and the rule (D.4.6) means that the first derivatives of $G = F \circ \mu$ are given by:

$$G_1(\mathbf{q}) = (\operatorname{grad} F)|_{\mathbf{z}} \cdot (\alpha_1, \beta_1)|_{\mathbf{q}}, \qquad G_2(\mathbf{q}) = (\operatorname{grad} F)|_{\mathbf{z}} \cdot (\alpha_2, \beta_2)|_{\mathbf{q}}, \tag{D.4.7}$$

each bit of which is now clearly specified since $\mathbf{z} = \mu(\mathbf{q})$. (Equations (D.4.7) can now 'tell' a computer what G is: humans can 'regard' F as a function of $\mathbf{q}$ but a computer needs a specification.) Before we give a proof of (D.4.7), some familiarization exercises may well be helpful.

Exercises D.4.2

1. Let $F(x, y) = 3x^2y + \sin(x + 2y)$. In polar coordinates $x = r\cos\theta$, $y = r\sin\theta$, so take $\mathbf{q} = (r, \theta)$ in (D.4.7) and show $\partial_1 G(= \partial F/\partial r) = (6xy + \cos(x + 2y)\cdot\cos\theta + (3x^2 + 2\cos(x + 2y))\cdot\sin\theta$. Calculate $\partial F/\partial\theta$ $(= \partial_2 G)$ similarly. By substituting $r\cos\theta$, $r\sin\theta$ for x, y, we now have $\partial_1 G$ as a function of (r, θ). Evaluate $\partial_{11} G$ $(= \partial^2 G/\partial r^2)$, $\partial_{12} G = \partial^2 G/\partial r\partial\theta$, $\partial_{22} G = \partial^2 G/\partial\theta^2$.
2. Use (D.4.7) in the form $G_1 = \alpha_1\cdot(F_1\circ\mu) + \beta_1\cdot(F_2\circ\mu)$ to show: $G_{11}(\mathbf{z}) = A(\mathbf{z}) + (\alpha_1\cdot(F_1\circ\mu)_1 + \beta_1\cdot(F_2\circ\mu)_1)|_{\mathbf{z}}$ where $A(\mathbf{z}) = (\alpha_{11}\cdot F_1 + \beta_{11}\cdot F_2)|_{\mathbf{z}}$. Hence show that $G_{11}(\mathbf{z}) = A(\mathbf{z}) + (\alpha_1^2\cdot F_{11} + 2\alpha_1\beta_1 F_{12} + \beta_1^2 F_{22})|_{\mathbf{z}}$ and similarly that
$$G_{12}(\mathbf{z}) = B(\mathbf{z}) + (\alpha_1\alpha_2 F_{11} + (\alpha_1\beta_2 + \alpha_2\beta_2)F_{12} + \beta_1\beta_2 F_{22})|_{\mathbf{z}},$$
$$G_{22}(\mathbf{z}) = C(\mathbf{z}) + (\alpha_2^2 F_{11} + 2\alpha_2\beta_2 F_{12} + \beta_2^2 F_{22})|_{\mathbf{z}}$$
where $B(\mathbf{z}) = (\alpha_{12}\cdot F_1 + \beta_{12}\cdot F_2)|_{\mathbf{z}}$, $C(\mathbf{z}) = (\alpha_{22}\cdot F_1 + \beta_{22}\cdot F_2)|_{\mathbf{z}}$.
3. Obtain the formulae of exercise 2 by using (D.4.6). (You will need to express $\partial/\partial u\,(\partial F/\partial x)$ by substituting $\partial F/\partial x$ for F in (D.4.6).)
4. Show, using the chain rule, that if $u = x^2 - y^2$ and $v = 2xy$ then:
$$\frac{\partial^2 f}{\partial x^2} + \frac{\partial^2 f}{\partial y^2} = 4(x^2 + y^2)\left(\frac{\partial^2 f}{\partial u^2} + \frac{\partial^2 f}{\partial v^2}\right).$$
Hence or otherwise show that
$$f = \cosh(x^2 - y^2)$$
is a solution of the equation:
$$\frac{\partial^2 f}{\partial x^2} + \frac{\partial^2 f}{\partial y^2} = 4(x^2 + y^2)f.$$
[SU]
5. A function $f(x, y)$ is to be regarded as a function of u, v, where $x = u(1 + v^2)^{1/2}$, $y = uv$. Show that
$$\frac{\partial f}{\partial u} = (1 + v^2)^{1/2}\frac{\partial f}{\partial x} + v\frac{\partial f}{\partial y},$$
$$\frac{\partial f}{\partial v} = uv(1 + v^2)^{-1/2}\frac{\partial f}{\partial x} + u\frac{\partial f}{\partial y},$$
and deduce
$$\frac{\partial^2 f}{\partial x^2} - \frac{\partial^2 f}{\partial y^2} = \frac{\partial^2 f}{\partial u^2} - \frac{(1 + v^2)}{u^2}\frac{\partial^2 f}{\partial v^2} + \frac{1}{u}\frac{\partial f}{\partial u} - \frac{v}{u^2}\frac{\partial f}{\partial v}.$$
[SU]
6. If $z = \theta(x, y)$ is substituted in $F(x, y, z)$ to form $g(x, y) = F(x, y, \theta(x, y))$ show that $g_i(x, y) = (F_i + F_3\cdot\theta_1)|_{(x,y)}$, $i = 1, 2$. ($g = F\circ\mu$, where $\mu(x, y, z) = (x, y, \theta(x, y))$.)
7. Suppose $f(t)$ and $h(x, y)$ are smooth functions and $z = f(h(x, y))$. Show that $\partial z/\partial x = f'(h_1(x, y))$. (Keep y fixed and regard $h(x, y)$ as a function $h_y(x)$, so $h_y'(x) = \partial h/\partial x$.)

Similarly show that $\partial z/\partial y = f'(h_2(x, y))$. Take $h(x, y) = y/x^2$ and show that

$$x\frac{\partial z}{\partial x} + 2y\frac{\partial z}{\partial y} = 0.$$

8. Show that if $f(x, y) = F(u, v)$, where u, v are new variables defined by

$$u = \ln(xy) \qquad v = \ln\left(\frac{x}{y}\right)$$

then

$$x\frac{\partial f}{\partial x} + y\frac{\partial f}{\partial y} = 2\frac{\partial F}{\partial u}.$$

Hence show that

$$4\frac{\partial^2 F}{\partial u^2} = x^2\frac{\partial^2 f}{\partial x^2} + 2xy\frac{\partial^2 f}{\partial x \partial y} + y^2\frac{\partial^2 f}{\partial y^2} + x\frac{\partial f}{\partial x} + y\frac{\partial f}{\partial y}$$

and thus that $f(x, y) = (x/y)\ln xy$ is a solution of

$$x^2\frac{\partial^2 f}{\partial x^2} + 2xy\frac{\partial^2 f}{\partial x \partial y} + y^2\frac{\partial^2 f}{\partial y^2} + x\frac{\partial f}{\partial x} + y\frac{\partial f}{\partial y} = 0.$$ [SU]

9. The transformation between Cartesian coordinates (x, y) and plane polar coordinates (r, θ) is defined by $x = r\cos\theta$ and $y = r\sin\theta$. Use the chain rule to show that for any function f:

$$\frac{\partial f}{\partial r} = \cos\theta\frac{\partial f}{\partial x} + \sin\theta\frac{\partial f}{\partial y},$$

and obtain a similar expression for $\partial f/\partial\theta$. Let u and v be two functions related by the equations

$$\frac{\partial u}{\partial x} = \frac{\partial v}{\partial y} \qquad \text{and} \qquad \frac{\partial u}{\partial y} = -\frac{\partial v}{\partial x}.$$

Show that

$$\frac{\partial u}{\partial r} = \frac{1}{r}\frac{\partial v}{\partial \theta} \qquad \text{and} \qquad \frac{\partial v}{\partial r} = -\frac{1}{r}\frac{\partial u}{\partial \theta}.$$

Use these equations to show that:

$$\frac{\partial^2 u}{\partial r^2} + \frac{1}{r}\frac{\partial u}{\partial r} + \frac{1}{r^2}\frac{\partial^2 u}{\partial \theta^2} = 0.$$ [SU]

Proof of (D.4.6)

If we fix v and allow only u to vary, then the point $\mu(u, v)$ will trace out a smooth path of points $\mathbf{z}_v(u)$ in D. Hence $G(u, v) = F \circ \mathbf{z}_v(u)$; so as a function of u, G is the composition $F \circ \mathbf{z}_v$, and by (D.4.4) (with u now playing the role of t):

$$\frac{\mathrm{d}G}{\mathrm{d}u} = (\text{grad } F)\circ \mathbf{z}_v \cdot \left(\frac{\mathrm{d}\alpha}{\mathrm{d}u}, \frac{\mathrm{d}\beta}{\mathrm{d}u}\right). \tag{D.4.8}$$

But now we ought to write $\partial\alpha/\partial u$, etc., to signify that although α depends on v (as well as u), we are keeping v fixed. Hence, using the 'slot' notation (in which $\partial\alpha/\partial u$ is α_1, etc.), we can express (D.4.8) without ambiguity as

$$G_1 = F_1 \circ \mathbf{z}_v \cdot \alpha_1 + F_2 \circ \mathbf{z}_v \cdot \beta_1. \tag{D.4.9}$$

If we evaluate each side at $\mathbf{q}$, then we obtain the first equation in (D.4.7).The second equation is obtained similarly by allowing v to vary instead of u.

Since $\mathbf{z} = \mu(\mathbf{q})$ and we always write grad F as a row vector, the two equations in (D.4.7) can be written in matrix form as:

$$\text{grad } G = (\text{grad } F) \circ \mu \cdot J(\alpha, \beta), \qquad J(\alpha, \beta) = \begin{pmatrix} \alpha_1 & \alpha_2 \\ \beta_1 & \beta_2 \end{pmatrix} \tag{D.4.10}$$

where μ and $J(\alpha, \beta)$ are evaluated at $\mathbf{q}$. This matrix is called the **Jacobian matrix** of (α, β) at $\mathbf{q}$. Observe that its rows are the gradient vectors of the functions α, β. The matrix is important because it tells us how derivatives transform when variables (such as u, v) are transformed by the mapping μ to variables (x, y).

Exercises D.4.3

1. Suppose $\mu(x, y) = (x^3 y, y^2 x + 2x^2)$. Thus α, β here are the functions $\alpha(x, y) = x^3 y$, $\beta(x, y) = y^2 x + 2x^2$. Evaluate $J(\alpha, \beta)$. Similarly when α, β map (r, θ) to (x, y) by the rule: $\alpha(r, \theta) = r \cos \theta$, $\beta(r, \theta) = r \sin \theta$, show that

$$J = \begin{pmatrix} \cos \theta & -r \sin \theta \\ \sin \theta & r \cos \theta \end{pmatrix} \text{ with determinant } r.$$

2. If $f: \mathbb{R}^2 \to \mathbb{R}$ is smooth, show that $J(\text{grad} f) = \begin{pmatrix} f_{11} & f_{12} \\ f_{12} & f_{22} \end{pmatrix}$
3. In general, if we have n smooth functions $f^i(\mathbf{x})$, where $\mathbf{x} \in \mathbb{R}^m$, then we often write $(x_1, \ldots, x_m) \mapsto (f^1(\mathbf{x}), \ldots, f^n(\mathbf{x}))$. The n functions have a Jacobian matrix J defined by taking its ith row to be grad f^i. (In particular, if $n = 1$ then J is just grad f^1.) Calculate the Jacobian matrices of the following functions:

   (a) $f: \mathbb{R}^2 \to \mathbb{R}^2$, $(x, y) \mapsto (\cos(x + y) + y, xy^2)$
   (b) $f: \mathbb{R} \to \mathbb{R}^3$, $x \mapsto (ze^x, -ye^z, xz)$
   (c) $f: \mathbb{R}^3 \to \mathbb{R}$, $(x, y, z) \mapsto x^2 + y^2 + z^2$
   (d) $f: \mathbb{R}^2 \to \mathbb{R}^3$, $(x, y) \mapsto (x^2 + e^{x+y}, ye^x, z^2 + xy)$
   (e) $f: \mathbb{R}^3 \to \mathbb{R}^2$, $(x, y, z) \mapsto (x + e^y + z, xy)$.

4. If F is a function of three variables α, β, γ, each of which is a function of u, v, obtain an analogue of (D.4.10) with a rectangular matrix $J(\alpha, \beta, \gamma)$ that has three rows of which the first is grad $\alpha = (\alpha_1, \alpha_2)$. If, instead, we have three variables u, v, w, obtain a corresponding Jacobian matrix which is square.
5. In (D.4.10), suppose that u, v are each functions of a single time variable, t, so that G can be regarded as a function g of t. Show that $g' = ((\text{grad } F) \circ \mu \cdot J(u', v)) \cdot (u', v')$, and evaluate each side at t. If instead, u, v are functions of two variables s, t, regard G as a function H of (s, t). Show that grad $H = (\text{grad } F) \circ \mathbf{w} \cdot J(\alpha, \beta) \cdot J(u, v)$, for a certain $\mathbf{w}$, and explain how to evaluate grad H at (s, t).

D.4.1 Mappings

With a slight change of notation, suppose that on the open domain D, two differentiable functions $f, g : D \to \mathbb{R}$ are defined. Then for each point $\mathbf{z}$ in D we obtain a point $\mathbf{w}$ in $\mathbb{R}^2$, with coordinates $(f(\mathbf{z}), g(\mathbf{z}))$; and as $\mathbf{z}$ runs through D, $\mathbf{w}$ moves in $\mathbb{R}^2$ and fills out a set D' (say). The process of calculating $\mathbf{w}$ from the given functions f and g is called a **mapping** F of D into $\mathbb{R}^2$, with image D'. We write $F: D \to \mathbb{R}^2$ or $F: D \to D'$ and F is said to be a **differentiable mapping**, because f and g are differentiable. (The two functions f, g are called the **components of** F.) Various examples are given in exercise D.4.3(3).

For each $\mathbf{z}$ in D', $F(\mathbf{z})$ has coordinates: $(u, v) = (f(\mathbf{z}), g(\mathbf{z}))$ and $\mathbf{z} = \mathbf{z}_0 + (h, k)$, say. By the basic relation (D.4.1), we have

$$u = f(\mathbf{z}_0) + (\operatorname{grad} f)\cdot(h, k) + O(r^2), \qquad r^2 = h^2 + k^2,$$

where $\operatorname{grad} f$ is evaluated at $\mathbf{z}_0$; and similarly for v. Thus, transposing (h, k),

$$\mathbf{w} = F(\mathbf{z}) = (u, v) = F(\mathbf{z}_0) + K\cdot(h, k)^{\mathrm{T}} + O(r^2), \tag{D.4.11}$$

where, in the notation of (D.4.10), K is the Jacobian matrix $J(f, g)$ of F, also denoted by $JF(\mathbf{z}_0)$. Thus

$$JF(\mathbf{z}_0) = \begin{pmatrix} (\operatorname{grad} f)|_{\mathbf{z}_0} \\ (\operatorname{grad} g)|_{\mathbf{z}_0} \end{pmatrix} = \begin{pmatrix} f_1(\mathbf{z}_0) & f_2(\mathbf{z}_0) \\ g_1(\mathbf{z}_0) & g_2(\mathbf{z}_0) \end{pmatrix}. \tag{D.4.12}$$

and (D.4.11) says that if r^2 is small, then at $\mathbf{z}_0$, F is approximately equal to the linear mapping given by $JF(\mathbf{z}_0)$.

Exercises D.4.4

1. Let M be a differentiable mapping on domain E in $\mathbb{R}^2$, such that M maps E into D, the domain of F in (D.4.11). Then we can form the composition $G = F \circ M$. Use (D.4.1) to show that G is differentiable, and use D.4.10 to obtain the 'ultimate' form of the chain rule as a matrix product:

   $$J(F \circ M)|_{\mathbf{p}} = JF(\mathbf{q})\cdot JM(\mathbf{p}), \qquad \mathbf{q} = M(\mathbf{p}). \tag{D.4.13}$$

   (This gives the reason for the word 'chain': from the chain of mappings $E \xrightarrow{M} D \xrightarrow{F} \mathbb{R}^2$, we derive the chain of *linear* mappings $\mathbb{R}^2 \xrightarrow{P} \mathbb{R}^2 \xrightarrow{Q} \mathbb{R}^2$, where P, Q have matrices $JM(\mathbf{z}_0)$, $JF(\mathbf{z}_0)$, respectively. In other words, 'the J of the composition $F \circ M$ is the composition of the J's'. Similarly for longer chains of mappings.
2. Verify the chain rule for the mappings $f, g, h : \mathbb{R}^2 \to \mathbb{R}^2$ where $h = g \circ f$ and

   $$f(s, t) = (s \cos t, s \sin t), \qquad g(u, v) = (u^2 + v^2, u^2 - v^2).$$ [SU]
3. The mapping f: $\mathbb{R}^2 \to \mathbb{R}^2$ is defined by $f(x, y) = (u(x, y), v(x, y))$ where $u(x, y) = x - y$ and $v(x, y) = x^2 + x + y$. Let F be the composition of f with itself, $F = f \circ f$. Find JF both by the chain rule and by direct calculations.

We conclude this section with a proof of (D.4.3).

Proof of (D.4.3)
We must calculate $f'(t_0)$ as the limit, as h tends to zero, of

$$\frac{f(t_0+h)-f(t_0)}{h}. \tag{D.4.14}$$

Applying (D.3.1) to x and y separately, we have

$$x(t_0+h) = x(t_0)+a, \qquad y(t_0+h) = y(t_0)+b$$

where

$$a = hu'(t_0)+O(h^2), \qquad b = hv'(t_0)+O(h^2). \tag{D.4.15}$$

Then in (D.4.14), the numerator, N, is

$$\begin{aligned} F(\mathbf{w}(t_0+h)) - F(\mathbf{w}(t_0)) &= F(x(t_0+h), y(t_0+h)) - F(\mathbf{z}) \\ &= F(\mathbf{z}+(a,b)) - F(a,b) \\ &= (\text{grad } F)|_{\mathbf{z}}\cdot(a,b) + O(a^2+b^2) \qquad \text{by (D.4.1)} \\ &= h\{F_1(\mathbf{z})u'(t_0) + F_2(\mathbf{z})v'(t_0) + O(h^2)\} \end{aligned}$$

by (D.4.15), because if we add the O terms, we still have an O term of the same magnitude. Now $O(h^2)$ tends to 0 as h tends to 0, so N/h tends to $F_1(\mathbf{z})u'(t_0) + F_2(\mathbf{z})v'(t_0)$, which is just the right-hand side of (D.4.3) as required.

D.5 THE TAYLOR–MACLAURIN EXPANSION IN SEVERAL VARIABLES

There is a version of the Taylor–Maclaurin expansion that allows us to obtain simple approximations to a smooth function of several variables. Recall that we call f 'smooth' if all derivatives exist for the task in hand (so before carrying out the task in a specific case, the reader must verify that the f in question really is smooth enough).

Although the basic idea is the same as before, the description is naturally more complicated to write out, so we work now with a function $F(x, y)$ of just two variables (x, y) in a domain D; readers need fluency with this theory, before they can be at ease with that in section D.7 for three or more variables.

Our immediate aim is to find an expansion for $F(\mathbf{z}+\mathbf{p})$ in terms of $\mathbf{p}$, when $\mathbf{z}$ is fixed in D and $\mathbf{p}=(k,l)$ is small enough for $\mathbf{z}+\mathbf{p}$ still to lie in D. We shall obtain an expansion

$$F(\mathbf{z}+\mathbf{p}) = F(\mathbf{z}) + L(\mathbf{p}) + Q(\mathbf{p}) + O(\|\mathbf{p}\|^3), \tag{D.5.1}$$

where L is a *linear* function of $\mathbf{p}$ and Q is a *quadratic* function. More precisely,

$$L(\mathbf{p}) = (kF_1(\mathbf{z}) + lF_2(\mathbf{z})) = (\text{grad } F)|_{\mathbf{z}}\cdot\mathbf{p}, \tag{D.5.2}$$

$$Q(\mathbf{p}) = \frac{1}{2}(k^2F_{11}(\mathbf{z}) + 2klF_{12}(\mathbf{z}) + l^2F_{22}(\mathbf{z})). \tag{D.5.3}$$

Although L and Q have a complicated appearance, it is important to realize that L and Q are used as functions of k and l, and the coefficients are regarded as constants, depending only on $\mathbf{z}$. Then, if k and l are small and L is not zero, L will dominate Q, and then F behaves near $\mathbf{p}$ approximately like the linear function L.

If $L = 0$, then F behaves approximately like the quadratic function Q, and it is rarely necessary to go beyond these second-order terms, for reasons explained later. Before reading the proof of (D.5.1), readers will probably prefer to get a 'feel' for it by working the next exercise.

Exercise D.5.1

Find $L(\mathbf{p})$ and $Q(\mathbf{p})$ when $F(x, y)$ is

(i) $e^x \cos y$, (ii) $e^x \sin y$, (iii) $e^x \ln(x + y)$, (*iv*) $\sqrt{(1 - x)/(1 + y^2)}$.

Proof of (D.5.1)

The clever move is to define a path in D from $\mathbf{z}$ to $\mathbf{z} + \mathbf{p}$. Thus take $\mathbf{w}(t) = (u, v)$ in D (when $0 \leq t \leq 1$), given by

$$u = x + kt, \qquad v = y + lt, \qquad \text{so} \qquad \mathbf{w}(0) = \mathbf{z}, \qquad \mathbf{w}(1) = \mathbf{z} + \mathbf{p}, \tag{D.5.4}$$

and consider the Taylor–Maclaurin expansion of the one-variable function $f = F \circ \mathbf{w}$. Thus we now keep both $\mathbf{z}$ and $\mathbf{p}$ fixed and look at $f(t)$, where for each t,

$$f(t) = (F \circ \mathbf{w})(t) = F(\mathbf{z} + t\mathbf{p}), \tag{D.5.5}$$

so

$$f(t) = f(0) + tf'(0) + \frac{t^2}{2} f''(0) + R, \tag{D.5.6}$$

where R is the remainder term and we must calculate the derivatives of f. First, we have $f(0) = (F \circ \mathbf{w})|_0 = F(\mathbf{w}(0)) = F(\mathbf{z})$. Next, we apply the chain rule (D.4.4), to get

$$\begin{aligned} f'(t) = \frac{\mathrm{d}f}{\mathrm{d}t} &= (\operatorname{grad} F)|_{\mathbf{p}(t)} \cdot (u', v')|_t \\ &= kF_1(x + kt, y + lt) + lF_2(x + kt, y + lt), \end{aligned} \tag{D.5.7}$$

so

$$f'(0) = (F \circ \mathbf{w})'|_0 = kF_1(\mathbf{z}) + lF_2(\mathbf{z}).$$

To work out $f''(0)$, we differentiate (D.5.7) with respect to t, and by using (D.5.7) itself on the compositions $F_1 \circ \mathbf{w}$, $F_2 \circ \mathbf{w}$, we get

$$(F \circ \mathbf{w})'' = k(kF_{11} \circ \mathbf{w} + lF_{12} \circ \mathbf{w}) + l(kF_{21} \circ \mathbf{w} + lF_{22} \circ \mathbf{w}).$$

Now evaluate at $t = 0$, using (D.5.5), and obtain (since $F_{12} = F_{21}$)

$$f''(0) = k^2 F_{11}(\mathbf{z}) + 2klF_{12}(\mathbf{z}) + l^2 F_{22}(\mathbf{z}). \tag{D.5.8}$$

Having calculated these derivatives of f at $t = 0$, we substitute them into (D.5.6), and then (D.5.1)–(D.5.3) follow.

If $f'''(t)$ is worked out by the same technique as above, we see that $|R| \leq M\|\mathbf{p}\|^3$, where M is a bound on the size of all the third derivatives of F. Clearly, with more work, we could continue the expansion of F as far as we like, although the labour increases rapidly.

The form (D.5.1) is the most useful in practice, at least for suggesting what might be true, but we can often avoid the remainder term as we shall see in the next section. As indicated previously, (D.5.1) allows us to treat F 'locally' (i.e., in a neighbourhood of $\mathbf{z}$) as if it is a polynomial. Indeed, if $(\text{grad } F)|_{\mathbf{z}} \neq 0$ then, as we shall see, F behaves as if it *is* the linear function $F(\mathbf{z}) + ah + bk$, with $(a, b) = (\text{grad } F)|_{\mathbf{z}}$. In that case $\mathbf{z}$ is called an **ordinary point** of F, whereas $\mathbf{z}$ is a **critical point** if grad F is zero at $\mathbf{z}$. At such a critical point, it will turn out that we can normally expect F to be locally no worse than a quadratic function. In this way of doing calculus, bits of local information about F are obtained fairly easily (because we know how to handle linear and quadratic functions), and then we piece these bits together to obtain 'global' information. This should forewarn readers of our general strategy.

One way of getting rid of the remainder term is this: we can always express F as an 'almost linear' function in the following way. To simplify notation, assume $\mathbf{z} = \mathbf{0} = F(\mathbf{z})$, and in the domain D let N be an r-neighbourhood of $\mathbf{0}$. Then we claim: *we can find functions A, B with domain N such that*

$$F(x, y) = xA(x, y) + yB(x, y), \tag{D.5.9}$$

without a remainder, *and* $A(0) = F_1(0)$, $B(0) = F_2(0)$. For, since N is a square disc, then for any point $\mathbf{z} = (x, y)$ in N the line segment $\mathbf{0}z$ lies in N, and consists of all points of the form $t\mathbf{z}$, with $0 < t < 1$. Hence, by the chain rule,

$$F(x, y) = \int_0^1 \frac{\mathrm{d}}{\mathrm{d}t} F(tx, ty)\,\mathrm{d}t = \int_0^1 xF_1(tx, ty)\,\mathrm{d}t + \int_0^1 yF_2(tx, ty)\,\mathrm{d}t$$
$$= xA(x, y) + yB(x, y)$$

where $A(x, y) = \int_0^1 F_1(tx, ty)\,\mathrm{d}t$ and similarly for B. Then

$$A(0) = \int_0^1 F_1(0)\,\mathrm{d}t = F_1(0)\,[t]_0^1 = F_1(0),$$

and similarly $B(0) = F_2(0)$, so our claim is established.

The form (D.5.9) will be refined in the next section, after some practice exercises.

Exercises D.5.2

1. If $F(x, y) = ux + vy + ax^2 + 2hxy + by^2$, then this can be rearranged as $xA(x, y) + yB(x, y)$ without using the above procedure. Nevertheless, check that $A(x, y) = \int_0^1 (u + 2axt + 2hyt)\,\mathrm{d}t = u + ax + hy$ and similarly for y.
2. If $F(x, y) = \mathrm{e}^x \cos y$, show that

$$A(x, y) = \int_0^1 \mathrm{e}^{tx} \cos ty\,\mathrm{d}t = \frac{\mathrm{e}^x(x + y \sin y) - x}{x^2 + y^2} \qquad \text{if } x^2 + y^2 \neq 0,$$

   and find a similar formula for $B(x, y)$. (Use example C.3.1.)
3. Find A, B when $F(x, y)$ is (i) $\mathrm{e}^x \sin y$, (ii) $\sin x \cdot \cos y$.

D.6 MAXIMA AND MINIMA

In this section our object is to locate local maxima and minima of a smooth function $F(x, y)$, defined on the open domain D. These are suggested when we think of the graph of F in $\mathbb{R}^3$, because it looks like a mountainous surface, with mountain peaks and valley bottoms that correspond respectively to local maxima and minima. But mountains also display 'saddle points' (see Fig. D.6.1), which are highest points in one direction and lowest in another (thus allowing travellers to rise out of valleys with least effort). In algorithm D.6.1 below we shall summarize the theory into a practical procedure, and readers may prefer to understand its use before grasping the justificatory theory.

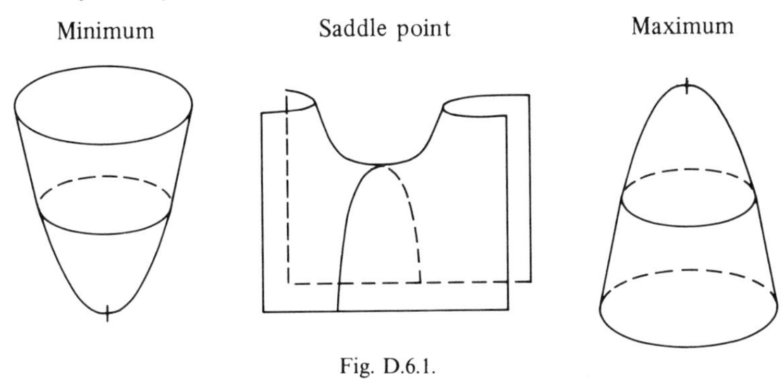

Fig. D.6.1.

As in the one-variable case, we say that $\mathbf{z}$ is a **local maximum** of F provided that, for all $\mathbf{p}$ in some neighbourhood N' of $\mathbf{z}$ (possibly smaller than N) we have

$$F(\mathbf{p}) \leq F(\mathbf{z}). \tag{D.6.1}$$

If the inequality is reversed, then $\mathbf{z}$ is a local minimum of F. For brevity we call a local maximum a 'max', a local minimum a 'min' and a neighbourhood a 'nbd'. As with one variable, the value of F at a max need not be the greatest among all the values taken by F on its entire domain D—hence the qualification 'local': the test (D.6.1) refers only to N'.

Our first observation is: *if* $\mathbf{z}$ *is a max or min of* F, *then* $(\text{grad } F)|_{\mathbf{z}} = \mathbf{0}$. For, suppose (D.6.1) holds. Then if $\mathbf{p}$ is confined to the x-axis (so $\mathbf{p} = (x, 0)$), we may regard F as a function of the one variable x, with a max at $\mathbf{z}$. Hence by the single-variable theory, we must have $F_1(\mathbf{z}) = 0$. Similarly, $F_2(\mathbf{z}) = 0$, so $(\text{grad } F)|_{\mathbf{z}} = \mathbf{0}$. Similarly also for a min. This establishes the above observation.

Therefore, in the terminology following (D.5.8), we must look among the critical points of F, to find the max and min. Now, if $\mathbf{z}$ is a critical point, then the linear function L in (D.5.2) is zero, and the quadratic Q is dominant. We examine its behaviour as follows.

If $\mathbf{z}$ is a max of F, then (D.6.1) must hold when $\mathbf{z}$ is replaced by $t\mathbf{z}$, so $t = 0$ is a max of the function $f(t) = F(\mathbf{z} + t\mathbf{p})$ in (D.5.5). This will happen if (from the ordinary theory for a single variable) we have $f''(0) < 0$; and by (D.5.8), $f''(0) = Q(k, l)$. Therefore if $Q(k, l) < 0$, then $F(\mathbf{z})$ is larger than $F(\mathbf{z} + t\mathbf{p})$ for all t near 0; unfortunately this may not include $t = 1$, corresponding to our 'test' point $\mathbf{p}$. It turns out, nevertheless, that this leaky argument has given us the right criterion, and we patch up the leak as follows. Impatient readers may now prefer to use their knowledge of quadratic forms and try exercises D.6.1 below, and then return to read the following 'accurate' theory.

As before, we simplify notation by assuming that $\mathbf{z}$ is the origin $\mathbf{0}$, with $F(\mathbf{0}) = 0$. Suppose then that grad $F = \mathbf{0}$ at $\mathbf{0}$. We shall show that with N as in (D.5.9): *we can express F on the nbd N in the form*

$$F(x, y) = x^2 R(x, y) + 2xyT(x, y) + y^2 V(x, y) \tag{D.6.2}$$

where

$$R(0) = \frac{1}{2} F_{11}(0), \qquad V(0) = \frac{1}{2} F_{22}(0), \qquad T(0) = \frac{1}{2} F_{12}(0). \tag{D.6.3}$$

Proof: we know that (D.5.9) holds on N, but (since grad $F = \mathbf{0}$ at $\mathbf{0}$) with the associated functions A, B both zero at $\mathbf{0}$. Hence we can apply (D.5.9) to each of them to get

$$F(x, y) = xA(x, y) + yB(x, y) = x(xR(x, y) + yS(x, y)) + y(xU(x, y) + yV(x, y))$$

where the functions R, S, U, V are obtained from A, B in the same way as the latter are obtained from F. In particular

$$R(0) = A_1(0), \qquad S(0) = A_2(0), \qquad U(0) = B_1(0), \qquad V(0) = B_2(0).$$

Thus, (D.6.2) follows with $T = \frac{1}{2}(S + U)$. But then

$$F_1(x, y) = 2xR + x^2 R_1 + 2yT + xyT_1 + y^2 V_1,$$

so $F_{11}(x, y)$ is of the form $2R + xX + yY$, say, whence $F_{11}(0) = 2R(0)$, as asserted in (D.6.3). Similarly for the other equations in (D.6.3).

The quadratic (D.6.2) can be used to decide the nature of a critical point of the function F as follows. We shall establish the following convenient test, which is expressed in terms of the function

$$\Delta(x, y) = F_{11}(x, y)F_{22}(x, y) - (F_{12}(x, y))^2, \tag{D.6.4}$$

where Δ is the determinant of the matrix $\mathscr{H}$ called the **Hessian of** F. We call F **non-degenerate** if, at each of its critical points P, the value $\Delta(P)$ of the Hessian is non-zero. It can be shown that 'most' functions are non-degenerate, in that F can always be perturbed slightly so as to become, as we assume from now on, non-degenerate.

D.6.1 (The max–min test)

If $\Delta(0) > 0$, then $\mathbf{0}$ is a max or min of F according as $F_{11}(0)$ is negative or positive.

Before giving the proof, we look at the following example.

Example D.6.1

Let $F(x, y) = x^3 + 2y^2 - 3x$. Then $F_1(x, y) = 3x^2 - 3$, $F_2(x, y) = 4y$. Hence the critical points are $P = (-1, 0)$, $Q = (1, 0)$. Also $\Delta(x, y) = 6x \cdot 4 - 0^2 = 24x$, so $\Delta(P) < 0 < \Delta(Q)$. By the above test, Q is a local minimum. We shall investigate P later.

Exercises D.6.1

1. Find all critical points of the functions

   (i) $x^3 + x^2y + xy^2 - x$; (ii) $x^3 + 2xy^2 - y^3 - x$; (iii) $4x^2e^y - 2x^4 - e^{4y}$,

   and calculate Δ for each point.
2. Show that all the critical points of $g(x, y) = \cos xy$ are degenerate, but that $h(x, y) = xg(x, y)$ has no critical points at all.
3. Show that $f(x, y) = \sin xy$ has a degenerate critical point at $\mathbf{0}$, and non-degenerate critical points on the hyperbolae $xy = (2k + 1)\pi/2$ ($k \in \mathbb{Z}$).

   On each line $y = mx$ in $\mathbb{R}^2$, examine the function $g_m(x) = \sin mx^2$, and hence build up a picture of the graph of f.
4. Does the function $Q\colon \mathbb{R}^3 \to \mathbb{R}$ given by

   $$Q(x) \equiv x_1x_2 + 2x_1x_3 + 2x_2x_3 \qquad (\mathbf{x} = (x_1, x_2, x_3))$$

   have a minimum at the origin? Could the addition of higher-order terms affect the answer? Justify your answers. Sketch or describe the contours of Q near the origin in $\mathbb{R}^3$. [SU]
5. Show that with three variables, the quadratic form analogous to (D.6.2) can be written in matrix form as $\mathbf{r}^{\mathrm{T}}M\mathbf{r}$, where $\mathbf{r}$ is the column vector $(x, y, z)^{\mathrm{T}}$ and M is a matrix of functions, which (at the origin) is the symmetric (Hessian) matrix

   $$\mathscr{H} = \begin{pmatrix} F_{11}(\mathbf{0}) & F_{12}(\mathbf{0}) & F_{13}(\mathbf{0}) \\ F_{21}(\mathbf{0}) & F_{22}(\mathbf{0}) & F_{23}(\mathbf{0}) \\ F_{31}(\mathbf{0}) & F_{32}(\mathbf{0}) & F_{33}(\mathbf{0}) \end{pmatrix}$$

Proof of theorem D.6.1

We are given that $\Delta(\mathbf{0}) > 0$, so by the Inertia Principle (theorem D.2.1), there is a nbhd M of $\mathbf{0}$ within the nbd N such that Δ remains positive on M. Therefore, in (D.6.2) the function R is not zero at any $\mathbf{z}$ in M (otherwise $\Delta(\mathbf{z}) = 0 - (F_{12}(\mathbf{z}))^2 \leq 0$). Therefore $R(\mathbf{0}) \neq 0$ and by the inertia principle again we may assume M so small that $\operatorname{sgn}(R(\mathbf{z}))$ always equals $\operatorname{sgn}(R(\mathbf{0}))$. By (D.6.3), $\operatorname{sgn}(R(\mathbf{0})) = \operatorname{sgn}(F_{11}(\mathbf{0})) \neq 0$).

Hence we may divide (D.6.2) by R, and the complete the square to get

$$\begin{aligned} F(x, y) &= R\left(\left(x + y\frac{T}{R}\right)^2 + y^2\left(\frac{V}{R} - \frac{T^2}{R^2}\right)\right) \\ &= R\left(u^2 + y^2\frac{W(x, y)}{R^2}\right) \end{aligned} \tag{D.6.5}$$

where $u = x + y\,T/2$ and $W(x, y) = VR - T^2$.

Now by (D.6.3), $W(0, 0) = \Delta(0, 0) > 0$, so by the inertia principle, W remains positive on some (possibly smaller) nbhd L of $\mathbf{0}$ in M. Thus, for every point $\mathbf{z} = (x, y)$ in L, we see that $F(\mathbf{z})$ is $R(\mathbf{z})$ multiplied by a sum of squares (since W is positive). Therefore, if $R(\mathbf{z}) < 0$, then $F(\mathbf{z}) < 0$ and (D.6.1) is satisfied: $\mathbf{z}$ is a max of F. Similarly, if $R(\mathbf{z}) > 0$ then $\mathbf{z}$ is a min. But in the previous paragraph, we saw that

$$\operatorname{sgn}(R(\mathbf{z})) = \operatorname{sgn}(R(\mathbf{0})) = \operatorname{sgn}(F_{11}(\mathbf{0})).$$

Therefore if we know that $F_{11}(\mathbf{0}) < 0$, then we know that $\mathbf{0}$ is a max of F; if $F_{11}(\mathbf{0}) > 0$ we have a min and the proof is complete.

Remark D.6.1

The proof shows that if $\mathbf{z}$ is not $\mathbf{0}$ itself, then $F(\mathbf{z})$ is non-zero. Therefore $\mathbf{0}$ is the only point in the nbhd M at its own level; such a point is said to be an **isolated critical point** of F.

The question remains: what happens if $\mathbf{0}$ is a critical point of F, but the Hessian $\Delta < 0$ at $\mathbf{0}$? We meet this with the function:

$$h(x, y) = x^2 - y^2, \tag{D.6.6}$$

for then grad $h = (2x, -2y)$ so the only critical point is $\mathbf{0}$, while in (D.6.4), $\Delta(x, y) = 2(-2) - 0^2 = -4 < 0$. With this in mind, it is natural to simplify (D.6.5) to the form

$$F(x, y) = \pm(w^2 + sv^2), \tag{D.6.7}$$

where (recalling the meaning of u there) if $r = \sqrt{|R|}$ then

$$s = \operatorname{sgn}(\Delta), \qquad w = \left(x + y\frac{H}{2}\right)r, \qquad v = \frac{y}{r}\sqrt{(s\Delta)}. \tag{D.6.8}$$

Since F was assumed to be non-degenerate, there is no need to investigate the case $\Delta = 0$.

We might guess from (D.6.7) that the contours of F near the critical point $\mathbf{0}$ will resemble ellipses or hyperbolae as they appear on a geographical map. To justify this guess, we shall need the notion of a diffeomorphism, to be explained in the next chapter.

Meanwhile, note that when $s = 1$ or -1 in (D.6.7), we obtain the simplest non-linear functions of two variables, with a non-degenerate critical point at $\mathbf{0}$. These are $\pm\text{el}(x, y)$, $\pm\text{hp}(x, y)$, where

$$\text{el}(x, y) = x^2 + y^2, \qquad \text{hp}(x, y) = x^2 - y^2. \tag{D.6.9}$$

Here, it is easily checked that el has a minimum at $\mathbf{0}$, and its contours are all circles with $\mathbf{0}$ as centre, and the same holds for those of $-$el, which has $\mathbf{0}$ as a maximum. Also the contours of hp are rectangular hyperbolae, so the portraits of el and hp are as shown in Figs. D.6.2(a),(b).

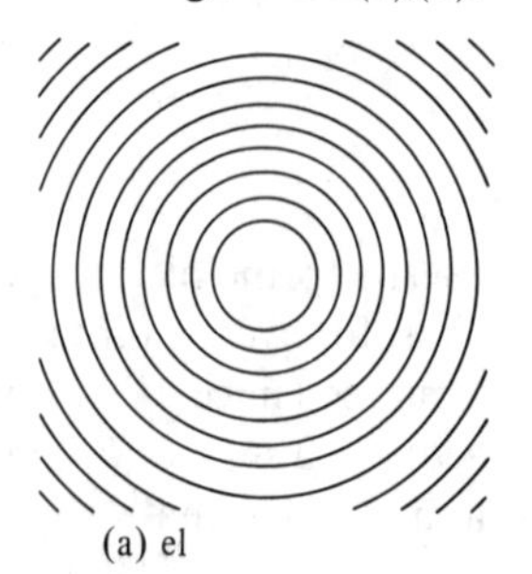

(a) el

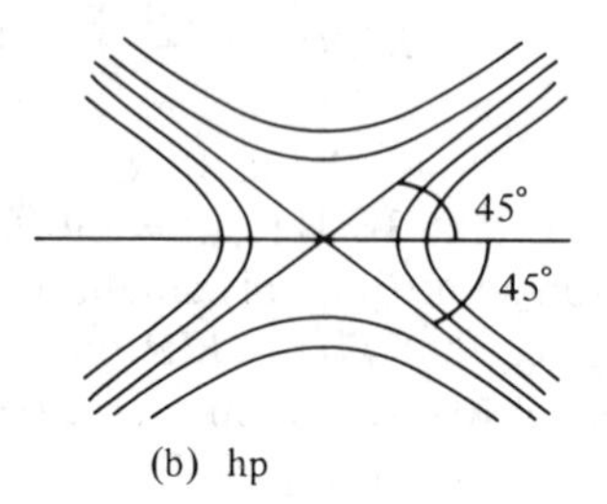

(b) hp

Fig. D.6.2.

After (D.6.8), we said essentially that near $\mathbf{0}$, the contours of the function $F(x, y)$ there would 'resemble' those of el or hp. 'Near $\mathbf{0}$' meant 'on the neighbourhood L' defined after (D.6.5). It is usual to say that F 'displays' a **centre** or a **saddle**, according as the contours near $\mathbf{0}$ resemble those of el, or of hp. (Clearly then, $-$el displays a centre near $\mathbf{0}$.) Using this language, we can then summarize the results of this section in the form: *the portrait of F near* $\mathbf{0}$ (i.e., on L) *displays a centre or a saddle according as the Hessian* $\Delta(\mathbf{0})$ *is positive or negative.* Combining this with the max–min test (theorem D.6.1), we get the following theorem.

Theorem D.6.2

Suppose P is a critical point of F. If $\Delta(P) > 0$ then the portrait of F displays a centre at P, and P is a max or min according as $F_{11}(P)$ is negative or positive. If $\Delta(P) < 0$, then the portrait of F displays a saddle at P.

Observe that the theorem says nothing about the case when $\Delta(P) = 0$, because, as explained after (D.6.4), we are assuming that F is non-degenerate. Thus the theorem is dealing with the most usual cases.

We summarize our work into a procedure for analysing a given non-degenerate function F.

Algorithm D.6.1

Step 1
Find the critical points P of F, by solving the simultaneous equations $F_1(x, y) = 0 = F_2(x, y)$.

Step 2
For each critical point P, work out the values $F_{11}(P)$, $F_{12}(P)$, $F_{22}(P)$ of the second partial derivatives of F.

Step 3
Use these values to calculate the Hessian function $\Delta(P)$, given by (D.6.4).

Step 4
Classify P as a saddle if $\Delta(P) < 0$. If $\Lambda(P) > 0$, classify P as a max or min according as $F_{11}(P)$ is negative or positive. (If $\Delta(P) = 0$, then F is degenerate.)

Exercises D.6.2

1. If $f(x, y) = x \sin y + y^3$ when $(x, y) \in \mathbb{R}^2$, find all the critical points of f and show that they are all saddle points.
2. For each function in exercise D.6.1(1) determine the nature of the critical points, stating whether each is a local maximum, a local minimum, or a saddle point. [SU]
3. Find all the critical points of the function $f: \mathbb{R}^2 \to \mathbb{R}$ given by

$$f(x_1, x_2) \equiv (x_1^2 + x_2^2)^2 - 14(x_1^2 + x_2^2).$$

Sketch the contours of f, and sketch or describe the graph of f. (Hint: exploit symmetry.) Show that when $a = 6$ the function

$$g_a(x_1, x_2) = f(x_1, x_2) + 4ax_1$$

has precisely one minimum, one maximum, and one saddle point. Sketch the contours of g. (They need not be numerically accurate, but should be consistent with the main geometrical features of g.)

Show that as a increases, the maximum and saddle point move towards each other, and coalesce as a passes through a certain positive value a_*. Find a_*. (It may be helpful to sketch the graph of $g_a(x_1, 0)$.) [SU]

4. Show that the critical points of the function

$$f(x, y) \equiv (x - 2y)(x^2 - y^2 - 1)$$

are $\pm(1/\sqrt{3})(2, 1)$. Decide if they are non-degenerate and describe their nature (max/min/saddle, or otherwise). Show that each contour $f^{-1}(c)$ is a smooth curve (1-manifold) when $c \neq 0$, stating any result you use. Sketch the contour $f^{-1}(0)$, indicating those points at which it cannot be expressed locally as the graph of a function $y = g(x)$. Sketch also the contour $f^{-1}(1)$, paying attention to the behaviour of $f(x, y)$ along the x-axis and along the lines $y = \pm x$.

If the function f is changed to

$$g(x, y) \equiv (x - 2y)(x^2 + y^2 - 1)$$

does it remain true that every contour except the 0-contour is a smooth curve? Justify your answer. [SU]

5. *Calculate* R, T, V in (D.6.2) when $F(x, y) = \sin x \cdot \sin y$. (Note: $F(x, y) = (x - x^3/6 + \cdots) \times (y - y^3/6 + \cdots) = xy + \cdots$ which tells us at once that F has a saddle point at $\mathbf{0}$, but we cannot easily find R, T, V from that expansion.)

D.7 THREE VARIABLES

The previous considerations were based on the transformation of F near $\mathbf{0}$ to the form (D.6.7). If there are three or n variables, we recall from Chapter 7, section 6, Lagrange's reduction process and modify it slightly. By exercise D.6.1(5), there is a nbhd N of $\mathbf{0}$ in $\mathbb{R}^3$, such that if $\mathbf{r} = (x, y, z)^{\mathrm{T}}$ lies in N, then (using now f to denote the function) $f(\mathbf{r})$ is of the form

$$\mathbf{r}^{\mathrm{T}} M \mathbf{r} = Ax^2 + By^2 + Cz^2 + 2Fyz + 2Gxy + 2Hyz$$

where the entries in the matrix M are $A, B, \ldots, H$ which are smooth functions of $\mathbf{r}$ on N, and at $\mathbf{r} = \mathbf{0}$, $M(\mathbf{0})$ is the Hessian matrix $\mathscr{H}$. We now carry out the Lagrange process, working as if these functions were constant. Thus if, for example $A(\mathbf{0}) \neq 0$, then there is in N a nbhd L of $\mathbf{0}$, such that A is never zero on L; so the function g is defined on L, where:

$$g = f(\mathbf{r}) - (Ax + Gy + Hz)^2/A.$$

It is easily checked that g is of the form $Py^2 + Qyz + Rz^2$, where P, Q, R are functions which may still depend on x as well as y, z. If P and R are zero at $\mathbf{0}$, we write—as in the algebraic case—$y = u - v$, $z = u + v$, to obtain a quadratic form containing squares. Therefore, we may now assume that (say) P is never zero on a nbhd K of $\mathbf{0}$ in L. Then

$$g = P\left(y + \frac{Q}{2P}\right)^2 + \left(R - \frac{Q^2}{P}\right)z^2.$$

Hence

$$f = al^2 + bm^2 + cn^2, \tag{D.7.1}$$

where a, b, c are, respectively, the signs of A, P, and $S = (R - Q^2/P)z^2$, and

$$l = (Ax + Gy + Hz)/\sqrt{|A|}, \qquad m = \sqrt{|P|}\,(y + Q/2P), \qquad n = \sqrt{|S|}\,z.$$

Clearly, this diagonal form is what we would obtain by using the same steps on the ordinary quadratic form $\mathbf{r}^{\mathrm{T}}M(\mathbf{0})\mathbf{r}$; and since $\mathscr{H} = M(\mathbf{0})$, the numbers of non-zero, and of negative, terms respectively equal rank$(\mathscr{H})$, ind$(\mathscr{H})$, respectively. These numbers can therefore be calculated directly from $\mathscr{H}$ in the first place. We call a critical point $\mathbf{0}$ of f **non-degenerate** if $\mathscr{H}$ is of maximum rank (3 in the case of (D.7.1)); and, as with two variables, f itself is non-degenerate if all its critical points are non-degenerate. Once more, 'most' functions are non-degenerate, so we confine the discussion to such functions.

By the above change of coordinates, then, we have expressed f in the form (D.7.1); so if ind$(\mathscr{H}) = 0$, then f is a sum of squares, so $\mathbf{0}$ is a min. If ind$(\mathscr{H}) = 3$, then $-f$ is a sum of squares and $\mathbf{0}$ is then a max. For the other values, 1, 2 of ind$(\mathscr{H})$, f has a three-dimensional version of a saddle point; and near $\mathbf{0}$, f looks like the appropriate algebraic function:

$$-l^2 + m^2 + n^2 \qquad \text{or} \qquad -l^2 - m^2 - n^2.$$

These are illustrated in Fig. 7.7.2. Therefore, both here and with the guess following (D.6.8), it is time to deal with the question of what 'looks like' means, which we answer in the next chapter.

Meanwhile, we conclude the present section by remarking that the above analysis of a critical point extends directly to n variables, and then leads to an extensive theory known as Morse theory, constructed by Marston Morse and others after 1925. The fact that f can be reduced to a sum of positive or negative squares is called Morse's lemma.

Exercises D.7.1

1. Find the critical points of the following functions. In each case decide whether the point is non-degenerate, and if it is, find its index.

   (a) $T(x, y, z, t)$ in D.1.2 with t kept constant.
   (b) $Q(x, y, z) = x^2 + 5y^2 + 3z^2 - 8xz + 8yz$.
   (c) $xy + yz + zx$.
   (d) xyz.
   (e) $P(x, y, z) = 3x^2 + y^2 + z^2 - 2yz + 2xz - 2xy$.

In (e) show that the Hessian has eigenvalues 0, 1, and 4, and find an orthogonal transformation to show that P has a line of critical points, and that all levels but $P = 0$ are elliptical cylinders.

2. Let $\mathbf{a}, \mathbf{b}, \mathbf{c}$ be three points in $\mathbb{R}^3$, at which electrical charges u, v, w, respectively, are placed. The electrical potential at $\mathbf{r}$ is $V(\mathbf{r}) = u \cdot \|\mathbf{r} - \mathbf{a}\|^{-1} + v\|\mathbf{r} - \mathbf{b}\|^{-1} + w\|\mathbf{r} - \mathbf{c}\|^{-1}$. Investigate its critical points and level surfaces. (This problem has not yet been completely solved, let alone the problem of n points and charges. V satisfies Laplace's equation (see exercise D.1.2(5).)

D.8 WHEN IS F INDEPENDENT OF x?

In this final section, we consider when a function, which appears at first sight to depend on x and y, really depends only on x. The simplest function F on D which depends on neither of x and y, is the constant c. Thus $F(x, y) = c$ for all (x, y) in D, and for brevity we write $F \equiv c$ ('F is identically equal to c') on D.

We shall now consider a two-variable analogue of (D.6.1), in which we shall see that the shape of D can be important. In $\mathbb{R}$ the main domains are intervals, with essentially the same, simple, shape. A domain D in $\mathbb{R}^2$ can be connected but have holes, and even greater complexity is possible in $\mathbb{R}^3$. Given the smooth function $F(x, y)$ with domain D, consider the following question.

Question D.8.1
Suppose that $\partial f/\partial x = 0$ on D. Must F be a function of y only?

The answer seems likely to be 'Yes', but we must be careful! For, if we fix y at b, say, let us write $f(x) = F(x, b)$. Then $f'(x) = \partial F/\partial x$ evaluated at (x, b) and this is zero. Hence, by (D.6.1), f is a constant provided that the line L_b: $y = b$ cuts D in a connected interval J_b. For simplicity suppose that there is some a such that for each b with J_b non-empty, (a, b) lies in J_b (see Fig. D.8.1). Thus, changing b back to y we have $F(x, y) = f(x) = f(a) = F(a, y)$, so the required function y is $F(a, y)$. It is not necessary to be so restrictive about D but we cannot allow D to be like E in Fig. D.8.1; see exercise D.8.1(2) below.

It may be as well here to state, without proof, three theoretical results that answer questions that may have occurred to readers. Proofs can be found in Courant (1952), Vol 2, Chapter 2. They refer to a function defined in a domain D of $\mathbb{R}^2$.

Theorem D.8.1
If the first derivatives F_1, F_2 exist at each point of D and are bounded, then F is continuous.

Theorem D.8.2
If F_1 and F_2 are continuous, then F is differentiable.

Theorem D.8.3
If F_{12} and F_{21} are continuous, then they are equal.

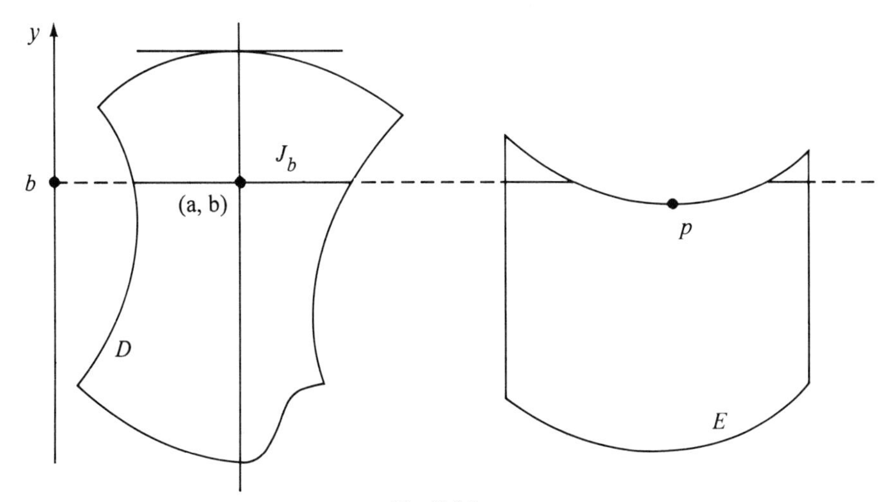

Fig. D.8.1.

Most functions satisfy the above hypotheses, which should be checked when a function F is given. Awkward functions do exist, however. For example, let $r^2 = x^2 + y^2$ and $g(x, y) = xy/r^2$ when $(x, y) \neq \mathbf{0}$, but $g(0, 0) = 0$. Then g_1 and g_2 always exist yet g is not continuous at $\mathbf{0}$. If $q(x, y) = rq(x, y)$ then q_1, q_2 exist and are bounded but not continuous at $\mathbf{0}$. Smooth functions, as explained earlier, always satisfy the hypotheses.

Exercises D.8.1

1. Show that the answer to Question D.8.1 is 'yes' if D is convex; and also if D can be divided into convex regions by lines $x = a_1, \ldots, x = a_n$. Show that E in Fig. D.8.1 cannot be subdivided in this way.
2. Suppose that in Fig. D.8.1, the tangent t at p in E is $0x$, and divides E into E_0 (below t) and the two upper regions E_1, E_2. Define $F(x, y)$ on E by the rules: $F(x, y) = 0$ if $y < 0$; $F(x, y) = \sin^2 y$ or $\sin^3 y$ according as $(x, y) \in E_1$ or E_2. Show that all the second derivatives of F exist and are continuous, so that the answer to Question D.8.1 is 'no' for this example.
3. Sketch the graphs of g and q in the text above. (Polar coordinates are helpful, here.)
4. Calculate the derivatives of g and q at $\mathbf{0}$. If $r \neq 0$, the usual rules apply: show that then $\partial_1 g = y(y^2 - x^2)/r^4$, which is $1/y$ when $x = 0$. But for $\mathbf{0}$ itself, we must return to the definition, and find $\lim_{h \to 0} (g(h, 0) - g(\mathbf{0}))/h$ since h here is expressly not zero. Show that $\partial_1 g|_{\mathbf{0}} = 0$, and similarly for $\partial_2 g$. Now prove that g is not continuous at $\mathbf{0}$, both directly and by using Theorem D.8.1. Using a similar technique for q, show that $\partial_1 q = y^3/r^3$ if $r \neq 0$ and $\partial_1 q|_{\mathbf{0}} = 0$; so q is continuous at $\mathbf{0}$. Show that if $k \neq 0$, then $(\partial_1 g(0, k) - \partial_1 g(0, 0))/k = 1/k$, so $\partial_{12} g|_{\mathbf{0}}$ does not exist.

E

Smooth mappings, diffeomorphisms, and integration

In this chapter we give a slight introduction to the problem of integrating functions on paths or regions of $\mathbb{R}^2$. First, however, we introduce more formal language to allow greater precision when we use the notion of a 'mapping', and especially the concept of a 'diffeomorphism'.

E.1 SMOOTH MAPPINGS

Several times, we have mentioned the word 'mapping', and have seen various examples without having an organized discussion. We now tidy up matters, and then we can give precision to the notion that we left in a tentative way in the previous chapter, when we talked of sketches 'resembling' each other. This will bring us to the technical notion of a 'diffeomorphism' which is a special kind of mapping that can be described succinctly as a 'differentiable bijection with a differentiable inverse'. For functions of a single variable we have already met the idea in section C.5, but we now explain it for functions of two variables, where the variety is much richer.

We recall (D.4.5), but now drop the use of bold type for denoting points of $\mathbb{R}^2$ and mappings. Thus suppose we have two smooth functions f, g defined on the open domain D in $\mathbb{R}^2$. Then for each point z in D we obtain a point w in $\mathbb{R}^2$, with coordinates $(f(z), g(z))$; and as z runs through D, w moves in $\mathbb{R}^2$ and fills out a set D' (say). The process of calculating w, from the given functions f and g, is called a **mapping** F of D into $\mathbb{R}^2$, with **image** D'; and F is said to be smooth because f and g are smooth. Conversely, given $w = (u, v)$ we often need to solve for z the equation $F(z) = w$, and thus the equations:

$$f(z) = u, \qquad g(z) = v. \tag{E.1.1}$$

This recalls the problem of solving equation (8.1.2). Now in (E.1.1), z is at an intersection of the two curves $\{f = u\}$ of f and $\{g = v\}$ of g, so we would expect solutions of (E.1.1) to be isolated points in D. Let us consider the nature of F near a solution z of (E.1.1). Firstly, we show that z has a nbd (neighbourhood) N on which F is approximately a linear function.

In the open set D, each point z_0 has some r-nbd N such that if $\|(h, k)\| = r$ then $z = z_0 + (h, k)$ lies in N. Also, $F(z) = (u, v) = (f(z), g(z))$ and by (D.4.11), we have

$$w = F(z) = F(z_0 + (h, k)) = (u, v) = F(z_0) + M(h, k)^{\mathrm{T}} + O(r^2), \tag{E.1.2}$$

where M is the Jacobian matrix $JF(z_0)$ of F. Here

$$JF(z_0) = \begin{pmatrix} (\operatorname{grad} f)|_{z_0} \\ (\operatorname{grad} g)|_{z_0} \end{pmatrix} = \begin{pmatrix} f_1(z_0) & f_2(z_0) \\ g_1(z_0) & g_2(z_0) \end{pmatrix} \tag{E.1.3}$$

Let $(h, k) = p$. Thus (E.1.2) says that if r^2 is small, then near z_0, $F(z_0 + p) - F(z_0)$ is approximately equal to the linear mapping L: $\mathbb{R}^2 \to \mathbb{R}^2$ given by $L(p) = Mp$; and the smaller we take r, the better the approximation. We emphasize that z_0 here is fixed, and we are considering how $F(z)$ varies when $z = z_0 + p$ and p stays small, so z stays in the neighbourhood N.

E.2 BIJECTIONS

We saw above that on N, F is approximately equal to the linear mapping L with matrix $M = JF(z_0)$. Now L is a bijection if M is invertible, so it is reasonable to guess that if $JF(z_0)$ is an invertible matrix, then (N being sufficiently small) F itself will map N in a bijective way. This guess turns out to be justified by the 'inverse mapping theorem' stated below, after some further preparatory discussion.

First let us consider how F might map D as a whole. How drastically does F distort D? For example, does F crush points together by mapping distinct points to a single point? Consider the mapping H of $\mathbb{R}^2$ into itself, which sends (x, y) to (x^2, y). If $x \neq 0$, then the points (x, y) and $(-x, y)$ are distinct, yet H maps them both to the same point (x^2, y). The effect of H on $\mathbb{R}^2$ is like folding a page about its centre line; here the 'page' is $\mathbb{R}^2$, and the line is the y-axis. In Fig. E.2.1(a) we illustrate this effect, and in Fig. E.2.1(b) that of the mapping which sends (x, y) to (x^2, y^2).

Exercises E.2.1

1. Write down the matrix $JF(p)$ for the mapping F: $\mathbb{R}^n \to \mathbb{R}^m$ when $p \in \mathbb{R}^n$ and

   (a) F: $\mathbb{R}^3 \to \mathbb{R}^2$: $F(x_1, x_2, x_3) = (x_2^2 + 2x_2, 2\sin^2 x_1 x_2 x_3)$
   (b) F: $\mathbb{R}^3 \to \mathbb{R}^2$: $F(x_1, x_2, x_3) = (x_1 + 2x_2 + 3x_3, 4x_1 + 5x_2 + 6x_3)$
   (c) F: $\mathbb{R}^2 \to \mathbb{R}$: $F(x_1, x_2) = 3x_1^2 + 4x_1 x_2 + 5x_2^2$

   Verify the chain rule (D.4.13): $J(F_3 \circ F_2)(p) = JF_3(q) \cdot JF_2(p)$ where $q = F_2(p)$ and F_2, F_3 are the above examples (b), (c), respectively.
2. A point $p \in \mathbb{R}^2$ is a **singular point** of the mapping F: $\mathbb{R}^2 \to \mathbb{R}^2$ if $JF(p)$ is not invertible; and then the point $q = F(p)$ is a **singular value** of F. If $\mathbf{r} \in \mathbb{R}^2$ we write $F^{-1}(\mathbf{r})$ for the set of

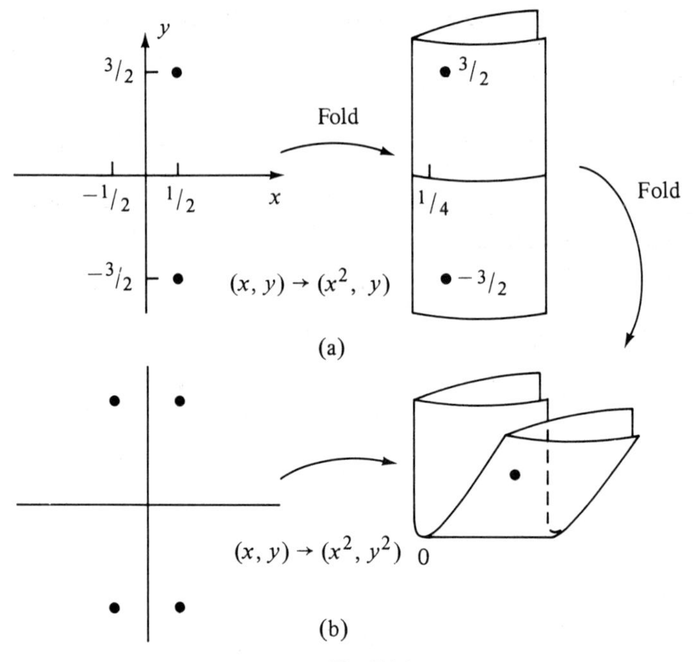

Fig. E.2.1.

points s such that $F(s) = r$. Sketch the set of singular points of each of the following mappings. Sketch also the set of singular values. (Call these Σ, Δ, respectively.)

(a) $F(x, y) = (x, x^2 + y^2)$
(b) $F(x, y) = (x, x^3 - x + y^2)$
(c) $F(x, y) = (x^2 - y^2, 2xy)$
(d) $F(x, y) = (x, y^4 + xy^2)$

In each case try to indicate the geometrical structure of the map F, and label the various regions of $\mathbb{R}^2 - \Delta$ according to the number of points of $F^{-1}(\mathbf{q})$, for $\mathbf{q}$ in each region. (For example, (a), if $F(x, y) = (u, v)$, then $u = x$, $v = x^2 + y^2$, so $y^2 = v - u^2$; hence there are two possibilities for y if $v > u^2$, and none if $v < u^2$.)

Next, suppose that the mapping $F: D \to \mathbb{R}^2$ does not crush together points that are distinct in D. Then F maps every pair of points in D to a *pair* of points of D', and we call F an **injection**. Now, if z moves on a curve K in D, then $w = F(z)$ moves on a curve J in D', and J is the image $F(K)$ of K; and if K is closed like a loop, then so is $F(K)$. But K could be a line, yet $F(K)$ has a loop in it, with a cross-over point X as in a figure-eight curve. That would mean that two different points z, p of K have been mapped to X; so then F could not be an injection. If F is an injection and K is a circle, then $F(K)$ is the *injective image* (under the mapping F) of a circle, and we call it a *simple closed curve* (more often a *Jordan* curve, after the French mathematician who first studied such curves). Thus, injective mappings 'preserve' Jordan curves (and much else, as well as pairs of points).

We denoted the image of D by D', so F maps D *onto* D'. If also F is injective, then as in the case of one variable, we say that $F: D \to D'$ is *bijective*, or that F is a *bijection*. There is then an inverse mapping $G: D' \to D$ which 'undoes' the effect of F; that is to say

$$G(F(z)) = z \qquad \text{and} \qquad F(G(w)) = w \tag{E.2.1}$$

for every z in D and w in D'. (In fact, to calculate G at w in D', we know that there is some z in D such that F maps z to w; and there cannot be more than one since F is an injection. Then $G(w)$ is *that* z.) By symmetry, F is also the inverse of G, and we write $G = F^{-1}$, $F = G^{-1}$. It need not happen that G is smooth, just as on $\mathbb{R}$ the cubing function $y = x^3$ is a smooth bijection, with the cube root as inverse, but that inverse is not differentiable at 0.

Suppose, however, that we are in a nice situation, with D' an open region of $\mathbb{R}^2$, and G has been verified to be a smooth mapping. Then we call F a **diffeomorphism** of D onto D' (abbreviating 'differentiable homeomorphism' from the Greek morphos = structure, and homeo = same). Thus, also, G is a diffeomorphism of D' onto D. We have therefore explained the succinct description mentioned at the start of this chapter, but we now need to work with diffeomorphisms, to see their usefulness. It should be clear how to work with (E.2.1) and to define diffeomorphisms with three or more variables. Clearly, every invertible linear mapping $\mathbb{R}^2 \to \mathbb{R}^2$ is a diffeomorphism. A non-linear example is: $(x, y) \mapsto (\sinh x, \tanh y)$, with differentiable inverse $(u, v) \mapsto (\sinh^{-1} u, \tanh^{-1} v)$.

E.3 LOCAL DIFFEOMORPHISMS

Let us take Jacobians in the first equation of (E.2.1). Then by exercise D.4.4(1), we have:

$$JG(w) \cdot JF(z) = I, \qquad (w = F(z)),$$

where I is the inverse matrix; so $JF(w)$ is an invertible matrix, with inverse $JG(w)$. This gives an easily checked sufficient condition for deciding whether a given mapping is a bijection: if at some point the Jacobian is not invertible, then the mapping cannot be bijective.

Exercises E.3.1

1. Let $F: \mathbb{R}^2 \to \mathbb{R}^2$ be the mapping given by $F(x, y) = (u, v)$ where u, v are the real and imaginary parts of z^3, when $z = x + jy$. Show that JF is invertible everywhere except at $\mathbf{0}$; and the image of F is the whole of $\mathbb{R}^2$, but F is not injective. (Hint: compare $(1, 0)$ with $e^{2\pi j/3}$.) What happens if we change z^3 to $z^3 + 5z$?
2. Let $F: \mathbb{R}^2 \to \mathbb{R}^2$ be the mapping given by

$$F(x, y) = (e^x \cos y, e^x \sin y).$$

Show that $JF(x, y)$ is always an invertible matrix, with determinant e^x. To check whether F is an injection, suppose $F(x, y) = F(u, v)$. Prove that $x = u$, $\cos y = \cos v$, $\sin y = \sin v$, and hence that $v = y + 2k\pi$, $(k = 0, \pm 1, \pm 2, \cdots)$. Thus F is not an injection, but JF is invertible everywhere. Show that F maps each line $x = \text{constant}$ to a circle; and that it maps the rectangular strip R: $\{x \geq 0 \leq y < 2\pi\}$ bijectively onto the outside of the circle $x^2 + y^2 = 1$.

3. The change of variables: $x = r\cos\theta$, $y = r\sin\theta$ can be regarded as a mapping $P\colon \mathbb{R}^2 \to \mathbb{R}^2$, from the point with **Cartesian** coordinates (r, θ) to the point (x, y). Show that JP has determinant 1 everywhere, and its image is all $\mathbb{R}^2$, but it is not a bijection, even on the strip R in exercise 2. However, show that if $R^+ = \{(r, \theta) \in R \mid r \geq 0\}$, then P maps R^+ diffeomorphically onto $\mathbb{R}^2$ minus the semi-axis $x > 0$. (This is a justification for the convention of taking $r > 0$ in polar coordinates.)

Most smooth mappings F are not injective, but it may happen that F does not crush together points of some smaller open set U in D. Thus, if we regard F as mapping only U, then the resulting mapping is an injection. For this resulting mapping we use the notation $F|U$: it is the mapping obtained from F by forgetting about any points not in U. We then call $F|U$ the **restriction** of F to U, and we can shorten the previous statement to 'F may not be an injection on D, but $F|U$ is an injection'. Clearly, $F|N$ is (like F itself) a smooth mapping. All this may seem pedantic, but such language has to be constructed to describe quite common situations; we cannot expect ordinary literary language always to be adequate—otherwise it would never have been necessary to have invented algebra, let alone calculus!

In particular, if F on D is not an injection, what happens locally, near a point $z_0 = (x_0, y_0)$? Since D is open, there is some neighbourhood N of z_0 that lies in D, and as suggested earlier, we might expect $F|N$ to be injective when confined to a well-chosen N, provided $JF(z_0)$ is invertible. This turns out usually to be the case, and even better, we can then expect $F|N$ to map N diffeomorphically onto an open region U of $\mathbb{R}^2$; for brevity we then say that F is a **local diffeomorphism** at z. It is important for readers to get a good intuitive understanding of what diffeomorphisms can and cannot do. On N here, F is 'close' to the linear mapping with matrix $JF(z_0)$, so it distorts rather like a lens, with straight lines looking nearly straight in the centre (at z_0) but growing more curved and distorted as we move further away. (F need not be close to the same linear mapping everywhere, but only to one in each locality.)

Before going further, the reader may like to investigate local diffeomorphisms using computer graphics. The following program maps (x, y) to its image (x', y') where $x' =$ `f1`(x, y), $y' =$ `f2`(x, y). The screen is divided vertically in half to give two windows L, R. In the left window L a square grid (like graph paper) is drawn in the rectangular region with `xmin` $\leq$ `x` $\leq$ `xmax`, `ymin` $\leq$ `y` $\leq$ `ymax`. As it is being drawn, so its image is constructed in R. Then a family of curves are defined by the parametric equations `x` $=$ `fx(t)`, `y` $=$ `fy(t)` for `tmin` $\leq$ `t` $\leq$ `tmax`. In the example the family are circles whose radii depend upon a parameter `fmin` $\leq$ `f` $\leq$ `fmax`. As each curve of the family is superimposed on the square grid in L, so its image is drawn in R. Show that, for the given mapping, lines of the form $bx + ay + c = 0$ are invariant, but that the Jacobian determinant is zero on the line $bx - ay + 1 = 0$. Try other functions. Adapt the program to explore functions of a complex variable.

```
REM Program E.3.1, Diffeomorphism
FUNCTION f1(x,y) = x - a*x*y
FUNCTION f2(x,y) = y + b*x*y
FUNCTION fx(t)   = r*COS(t)
FUNCTION fy(t)   = r*SIN(t)
READ a   , b   , hmin, hmax, vmin, vmax, fmin, fmax, fstep
DATA 0.05, 0.05, -7.5, 7.5 , -10 , 10  , 0.1 , 0.9 , 0.1
READ xmin, xmax, xstep, ymin , ymax, ystep, tmin, tmax, tstep
DATA -5  , 5   , 1    , -5   , 5   , 1    , 0   , 6.4 , 0.2
DECLARE DEF arc
SET MODE "graphics"
OPEN #1 : SCREEN 0,0.5 , 0,1
SET COLOR 1
SET WINDOW hmin,hmax, vmin,vmax
OPEN #2 : SCREEN 0.5,1 , 0,1
SET COLOR 1
SET WINDOW hmin,hmax, vmin,vmax
FOR y = ymax TO ymin STEP -ystep
    LET xs = xmin
    LET ys = y
    LET us = f1(xs,ys)
    LET vs = f2(xs,ys)
    FOR x = xmin TO xmax STEP xstep/4
      CALL arc(xs,ys,x,y,us,vs)
    NEXT x
NEXT y
FOR x = xmin TO xmax STEP xstep
    LET xs = x
    LET ys = ymin
    LET us = f1(xs,ys)
    LET vs = f2(xs,ys)
    FOR y = ymin TO ymax STEP ystep/4
      CALL arc(xs,ys,x,y,us,vs)
    NEXT y
NEXT x
WINDOW #1
SET COLOR 2
WINDOW #2
SET COLOR 2
FOR f = fmin TO fmax STEP fstep
    LET r = f*xmax
    CALL curve
NEXT f
STOP


SUB arc(xs,ys,x,y,us,vs)
    WINDOW #1
    PLOT xs,ys ; x,y
    LET u = f1(x,y)
    LET v = f2(x,y)
    WINDOW #2
    PLOT us,vs ; u,v
    LET xs = x
```

Program E.3.1. continued

```
        LET ys = y
        LET us = u
        LET vs = v
    END SUB

    SUB curve
        LET xs = fx(tmin)
        LET ys = fy(tmin)
        LET us = f1(xs,ys)
        LET vs = f2(xs,ys)
        FOR t = tmin TO tmax STEP tstep
          LET x = fx(t)
          LET y = fy(t)
          CALL arc(xs,ys,x,y,us,vs)
        NEXT t
    END SUB

    END
```

Readers should now find the following result meaningful; it is one of the most important theoretical results of the subject, and is known as the *inverse mapping* (or function) theorem, but for brevity we refer to it as the IMT. It concerns a differentiable mapping $F: D \to \mathbb{R}^2$, near a point z in D, and says that if the Jacobian matrix of F at z is invertible, then F is a local diffeomorphism at z_0.

Theorem E.3.1 (The inverse mapping theorem)
Let $F: D \to \mathbb{R}^2$ be a smooth mapping, and in D let z be such that the matrix $JF(z)$ is invertible. Then there is in D a neighbourhood N of z, an open set U in $\mathbb{R}^2$, and a smooth mapping $G: U \to N$, such that $F(z)$ lies in U while $F|N$ is a diffeomorphism onto U, with inverse G.

For a proof, see Webb (1991). There is an analogue with n variables.

Exercises E.3.2

1. Derive the implicit function theorem from theorem E.3.1 as follows. We want to find the solutions $z = (x, y)$ of the equation $f(z) = c$, in a neighbourhood N of a particular solution $z_0 = (x_0, y_0)$, when $f_2(z)$ is never zero on N. Let F be the mapping given by $F(z) = (u, v)$, where $u(z) = x$, $v(z) = f(z)$. Show that $F(z_0) = (x_0, c)$ ($= P$, say), and that $JF(z_0)$ is invertible. Hence find a square disc U containing P and a mapping G of U into N, and show that the required set of solutions is $G(l)$, where l is the set of points $\{v = c\}$ in U. Hence show that this set is the graph of the function θ, where $\theta(x) = G(x, f(x, c))$. Specify domains for all the functions involved.
2. In theorem E.3.1 let $w = F(z)$ when $z \in N$. Show that $JG(w) = (JF(z))^{-1}$.
3. Let $F: \mathbb{R}^3 \to \mathbb{R}^3$ be the mapping given by $F(\alpha, \beta, \gamma) = (a, b, c)$ where $x^3 + ax^2 + bx + c$ is the cubic polynomial with roots α, β, γ. By expressing (a, b, c) as symmetric functions of (α, β, γ) calculate the Jacobian matrix J of F, and show that its determinant is

$$g = -(\alpha - \beta)(\beta - \gamma)(\gamma - \alpha).$$

Show that F is everywhere a local diffeomorphism except on the planes $\alpha = \beta$, $\beta = \gamma$, $\gamma = \alpha$.

For each point $P \in \mathbb{R}^3$ that does not lie on any of these planes, regard α at P as a function of (a, b, c), and prove that

$$\frac{\partial \alpha}{\partial a} \cdot \frac{\partial \alpha}{\partial b} \cdot \frac{\partial \alpha}{\partial c} = \left(\frac{c}{g}\right)^2. \qquad \text{[SU]}$$

4. Show that the mapping

$$\phi\colon \mathbb{R}^2_{k,\varepsilon} \to \mathbb{R}^2_{p,q}, \qquad \phi(k, \varepsilon) = (k - \varepsilon, 1 - k\varepsilon)$$

is a local diffeomorphism everywhere except on the line $L\colon k = -\varepsilon$. Given that $\phi(k, \varepsilon) = (p, q)$, express k in terms of p, q, to show that $\phi(L)$ is the parabola $q = 1 + p^2/4$ and that $\phi(\mathbb{R}^2)$ is the region $Y\colon q \leq 1 + p^2/4$. [SU]

E.4 SADDLES AND CENTRES REVISITED

For an application of the IMT, recall what we did during the proof of the max–min test (theorem D.6.1) in the previous chapter. We showed that if F has a non-degenerate critical point at $\mathbf{0}$, then we can transform the x, y coordinates to new coordinates v, w such that for each point P on the particular neighbourhood L of $\mathbf{0}$, F takes the simple form (D.6.7), that is,

$$F(P) = \pm(sv^2 + w^2) \qquad \text{(E.4.1)}$$

where $s = \pm 1$. Thus in the new coordinates, F really does display a centre or saddle—the contours $\{F = c\}$ are ellipses or hyperbolae—but what exactly does that tell us about life in the old coordinates? To explain, it is probably easiest to think of two copies $\mathbb{R}^2_{vw}$, $\mathbb{R}^2_{xy}$ of $\mathbb{R}^2$, and then the transformation of coordinates given by (D.6.8) yields a mapping $G\colon L \to \mathbb{R}^2_{vw}$, where $G(x, y) = (v, w)$ and v, w are as in (D.6.8). Here, P lies in L which lies in $\mathbb{R}^2_{xy}$, and G maps origin to origin.

We now compute the Jacobian matrix of G at $\mathbf{0}$; so we must calculate grad v and grad w at $\mathbf{0}$. But when we differentiate the formulae in (D.6.8), we find that all but two terms are multiplied by x or y, so all but these are zero when we evaluate at $\mathbf{0}$. Therefore the Jacobian at $\mathbf{0}$ is

$$JG(\mathbf{0}) = \begin{pmatrix} r(\mathbf{0}) & 0 \\ 0 & \sqrt{(s\Delta(\mathbf{0}))} \end{pmatrix} \qquad \text{(E.4.2)}$$

with determinant $r(\mathbf{0})\sqrt{(s\Delta(\mathbf{0}))} \neq 0$, since neither factor is zero. Therefore $JG(\mathbf{0})$ is an invertible matrix, whence by the IMT, there is a nbhd N of $\mathbf{0}$ such that the restriction $G|N$ is a diffeomorphism onto an open region U of $\mathbf{0}$ in $\mathbb{R}^2_{vw}$. Thus $G|N$ has a smooth inverse mapping $M\colon U \to N$.

To see why this knowledge helps us, suppose first that in (E.4.1), the signs are such that the right-hand side is $v^2 + w^2$. Using the el function as in (D.6.9), this is el(v, w). Then (E.4.1) comes from (D.6.6), which says that $F(x, y) = \mathrm{el} \circ G(x, y)$. Thus, if we watch (x, y) moving along the contour $C\colon \{F = c\}$ in $\mathbb{R}^2_{xy}$, we shall see $G(x, y)$ moving along the contour $K\colon \{\mathrm{el} = c\}$ in $\mathbb{R}^2_{vw}$, since

$$c = F(x, y) = \mathrm{el} \circ G(x, y) = \mathrm{el}(v, w). \qquad \text{(E.4.3)}$$

We know the structure of the contours K, and we use the mapping M to show that the contours C are Jordan curves of the type mentioned prior to (E.2.1). For, if K is sufficiently near $\mathbf{0}$, it will lie in U, and hence its image $M(K)$ is a Jordan curve in N. Conversely, if C is near $\mathbf{0}$ in N, then $G(C)$ lies in U, and is a contour K of el; now $C = M(G(C))$ so C is the Jordan curve $M(K)$. In this way, M transfers all the significant features of the portrait of el on U to that of F on N.

Similarly, when the function el is replaced by the function hp, so that in (E.4.1), $v^2 - w^2 = \text{hp}(v, w)$. Each hyperbolic arc in U is mapped to a smooth arc in N. Since the asymptotes in U cross at $\mathbf{0}$, then M maps them to smooth (but not necessarily straight) arcs in N that cross at $\mathbf{0}$ and form the separatrices displayed at a saddle.

It should be evident from this that by using diffeomorphisms, we can make much sharper statements than (for example) our earlier attempt to explain the word 'resembles' in relatively literary language. However, some readers may have been repelled by some of the above technical description, feeling that they still do not understand what a diffeomorphism really is. Some mathematicians would insist that all that needs to be said is contained in the formal definition surrounding (E.2.1) and we should let it speak for itself; understanding will come if the concept is used. But that definition is a very abstract (and clever) distillation arising from earlier mathematical experiences, which students have not had. We therefore now try to supply some background for readers who prefer to be informed of it.

E.5 MODELLING RESEMBLANCES BY DIFFEOMORPHISMS

Since, historically, a background of experience led to the formal definition of a diffeomorphism, its description has to be in the looser language that was available before the definition was formulated. Consider first the 'watching' process, mentioned prior to (E.4.3). Our eye movements are in fact controlled by the mapping G, because our eye associates each (x, y) on the contour C with $G(x, y)$ on K. Differentiability is required, in order that K shall be smooth if C is, and conversely. Now, we would hardly admit resemblance unless, conversely, each (v, w) on K corresponded to some point (x, y) on C, and to no more than one in all $\mathbb{R}^2_{xy}$. This is why (unconsciously perhaps) we require G to be bijective. This fundamental notion of a bijection appears not only in calculus, but throughout mathematics from the time we learn to count. It appears implicitly whenever we wish to decide whether two objects are 'the same', relative to some standard of comparison.

In the physical world, we usually decide that two objects X, Y (sets of sweets, animals, cars, letters of the alphabet, etc.) resemble each other by making a visual inspection, and using some form of pattern recognition. Our eyes pick out salient points of X, and then look in a similar place to find such points on Y. Thus we are looking for a particular kind of pattern or structure, and our scanning process first looks for a one–one correspondence F between the points of X and Y.

We thus require F to have the two properties we have met with more specialized mappings before: first F must be an *injection* (or one–one) if, whenever $P \neq P'$ in D, then $F(P) \neq F(P')$ in E. (To drive a car into a garage, without crushing any metal, is an injection.) Second, F must be a *surjection* (or onto), that is, every point Q of Y is the image $F(P)$ of some point of X. (If our car were splattered all over the end wall of the garage, then we would have a mapping onto the wall.) If F is both an injection and a surjection, then it is a *bijection*. It is then a standard theorem of mathematics that F has an inverse, denoted by F^{-1}, which maps Y to X and is also a bijection. Moreover, just as with logarithms and antilogarithms (i.e., exponentials), each of F and F^{-1} undoes the work of the other, in that if $Q = F(P)$, then $P = F^{-1}(Q)$.

Our scanning process of X and Y next requires certain types of structure (such as nose between eyes if X and Y are faces) to be preserved. That is to say, if certain sets of points of X are in a significant relationship, then we expect the corresponding points of Y to be also in that relationship; if they are not, then we reject resemblance of X and Y. It often does not matter whether the placing is exactly similar: if X has a conventional nose, then we would expect to find a similar sort of protuberance on Y; and this might be slightly off-centre, but it must lie between the two eyes, rather than to the left of both. (A Picasso painting can shock, precisely by affronting this ingrained expectation.) In his classic, *Growth and form*, D'Arcy Thompson notes the resemblances between the skulls of different animals and shows how they are diffeomorphs of each other (see Figs. E.5.1 and E.5.2). This was a very original idea at the time, and has interesting biological implications.

If our objects X, Y happen to be the portraits of two functions, as in Chapter 8, then the salient features (the relevant structure) that strike us are those such as whether the contours in each are smooth or pointed, whether they are closed loops or go off to infinity, and so on. Our eyes also note the centres and saddles, and the

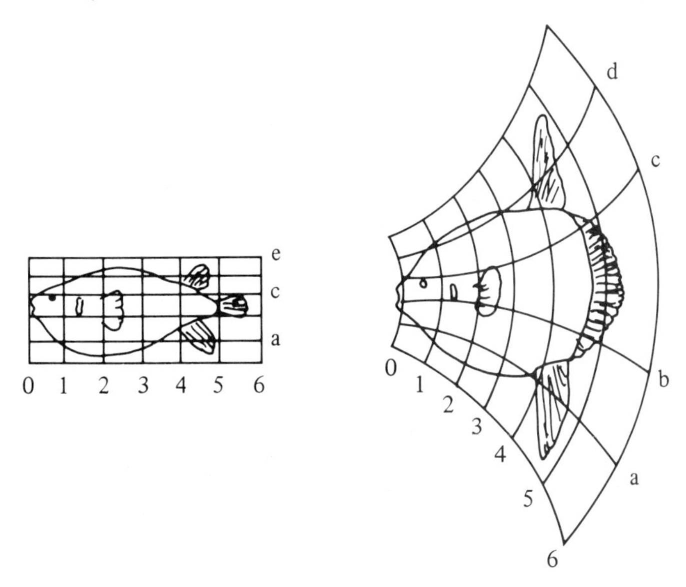

Fig. E.5.1. On the left is a typical species of the genus Diodon. On the right, says Professor D'Arcy Thompson, 'I have deformed its vertical coordinates into a system of concentric circles and its horizontal coordinates into a system of curves which approximately and provisionally are made to resemble a system of hyperbolas. The old outline transformed in its integrity to the new network appears as a manifest representation of the closely allied but very different-looking sunfish Orthagoriscus.'

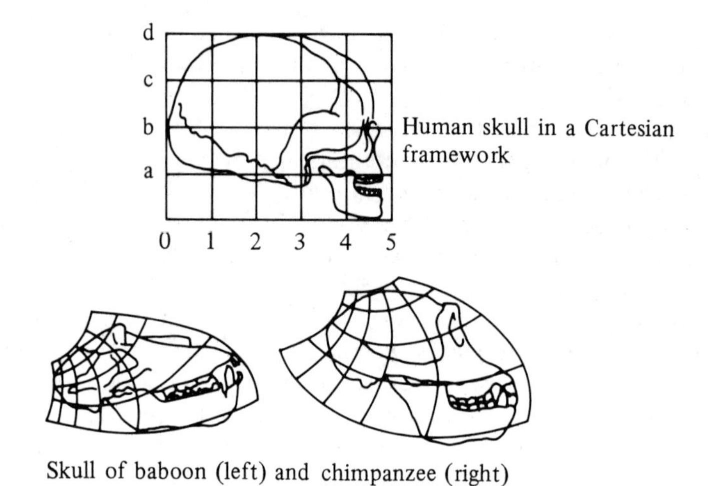

Skull of baboon (left) and chimpanzee (right)

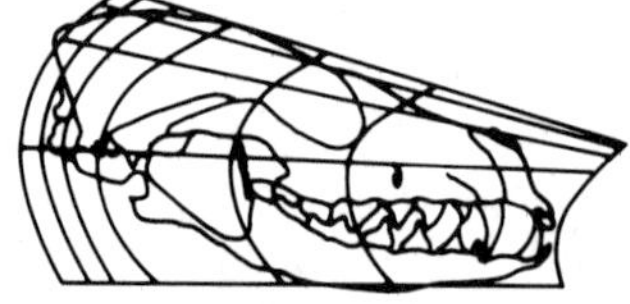
Skull of dog

Fig. E.5.2.

ways in which the separatrices 'separate' the other contours. To have words that capture these features with precision, mathematicians first invented the technical notion of a 'homeomorphism' (homeo = same), and later imposed the requirement of smoothness as in section E.2 above, which preserves differential (diffeo) structure—essentially the structure of tangents. Thus evolved the technical concept of diffeomorphism.

A very special case of a diffeomorphism is a **linear isomorphism** $M\colon \mathbb{R}^2 \to \mathbb{R}^2$ given by an invertible matrix. In that case, M preserves not only the differential structure of $\mathbb{R}^2$, but its algebraic (vector space) structure as well; but M will not necessarily preserve the Argand diagram structure of $\mathbb{R}^2$.

So much for the conceptual understanding of diffeomorphisms. For the activity of working with them, after all this description we supply some exercises, which, we hope, will suffice to let readers feel sufficiently at home with the idea of a diffeomorphism, for them to move on to later activities. When they feel the need for their idea to be made more precise, they should read again the technical discussion in section E.2.

Exercises E.5.1

1. Use the language of injections, etc., to show that any diffeomorphism F maps non-intersecting curves (in its domain D in $\mathbb{R}^2$) to non-intersecting ones (in its image), and hence that if H is a hyperbola, with centre $\mathbf{0}$, then F will map the two branches of H to two non-

intersecting curves, but the asymptotes are mapped to a pair of curves that intersect just once (at $F(\mathbf{0})$). Show that if F is linear, then these curves are straight lines.

2. Let F be a smooth mapping such that $D' = F(D)$, and let h be a non-degenerate function on D'. Now, if $z = (x, y)$, let $F(z) = w = (u, v)$. Show that F makes the portrait of h correspond to a portrait of the function $k(x, y) = h(F(x, y))$ contour by contour in the following sense. Suppose that w lies on the c-contour A (i.e., $\{h = c\}$) of h. Show as in (E.4.3) that then z lies on the c-contour C of k, and F maps C to A. If F is also a diffeomorphism show that each contour of h is the image by F of exactly one contour of k.
3. In exercise 2, suppose w is a critical point of h. Use the chain rule to show that z is a critical point of k. Show that if F is a local diffeomorphism at w, then w is a critical point of k if z is one for h, and each is non-degenerate if the other is. Show further that in this case, F maps saddles to saddles, but it may map a max to a min.
4. In exercise 2, let γ be a regular curve in D (see section 5.2). Show that if F is a diffeomorphism then $\delta = F \circ \gamma$ is a regular curve in D'. Now suppose γ is a trajectory of the differential equation $z' = R(z)$, so for each t we have $\gamma'(t) = R(\gamma(t))$. Show that δ is a trajectory of the differential equation $w' = S(w)$ where $S = K^{\mathrm{T}} \circ R \circ F^{-1}$ and $K = JF(z) = JF(F^{-1}(w))$.
5. Show that if $s \neq t$ then the normals at the points $(at^2, 2at)$, $(as^2, 2as)$ on the parabola $y^2 = 4ax$, meet at the point $(x, y) = f(s, t)$ where

$$f(s, t) = (2a + a(s^2 + st + t^2), -ast(s + t)).$$

Explain why the mapping $f\colon \mathbb{R}^2 \to \mathbb{R}^2$ sends the line $\{s = t\}$ to the evolute of the parabola. Show that f is a local diffeomorphism except on the lines $s = -\tfrac{1}{2}t$, $s = -2t$, and $s = -t$, and that f maps each line $\{s + t = \text{costant})$ to a normal of the parabola. [SU]

6. Let $f\colon \mathbb{R} \to \mathbb{R}$ be the function given by $f(x) = x^2 \sin x$. Show that there is a neighbourhood V of 0 in $\mathbb{R}$ and a function $\phi\colon V \to \mathbb{R}$ such that ϕ is a diffeomorphism onto its image, and for all $x \in V$, $f(x) = \phi(x)^3$. Show also that on U,

$$\phi(x) = x\left(1 - \frac{x^2}{18} + O(x^4)\right). \qquad \text{[SU]}$$

7. If $g(x, y) = (x^3 + 2y^2, -2x + y^5)$ evaluate Jg, and show that g is a local diffeomorphism everywhere except on the line $y = 0$ and the curve $y = (-4/15x^2)^{1/3}$.

E.6 HOW CAN WE INTEGRATE $F(X, Y)$?

Given a function F on a domain D in $\mathbb{R}^2$, there are various ways in which we might try to associate an 'integral' with F, by analogy with the case of one variable. We list three possibilities.

(6.1) The graph of F is a surface S in $\mathbb{R}^3$, and the height above the point $(x, y, 0)$ is $F(x, y)$. We could take the volume under S as the integral.

(6.2) Let $\gamma\colon I \to D$ be a path. Then $F \circ \gamma\colon I \to \mathbb{R}$ is an ordinary one-variable function, which we can then integrate in the ordinary way. (There is a snag, though.)

(6.3) Since grad F gives the pair (F_1, F_2) of functions, we might regard F as an antiderivative. Therefore we might regard the process of integration as starting with a pair (P, Q) of functions, and trying to find an antiderivative G such that grad $G = (P, Q)$.

Each possibility is worth investigating, as we shall show. We have not space to give many detailed examples, and readers are advised to look for these (with practice exercises) in the many books on vector calculus (see Courant (1952), Webb (1991).

Let us start with (6.1). Suppose first that D has a fairly simple shape, as in Fig. E.6.1. There, we replace the top edge of a rectangle B by the graph of a function $y = p(x)$. Then the slice of the graph of F, through the line $x = c$, has area $g(c) = \int_b^{p(c)} F(c, y)\, dy$. The volume V under the graph is composed of these slices, as x runs from a to a'. Then to 'integrate the function $F(x, y)$ over D' means to calculate V by evaluating the expression

$$\int_a^{a'} \left[\int_b^{p(c)} F(c, y)\, dy \right] dc \tag{E.6.1}$$

which for brevity is written $\int_D f(x, y)\, dxdy$. Note that what (E.6.1) requires is that:

(a) we keep x constant and integrate F (considered as a function of y) between $y = b$ and $y = p(x)$ to obtain a function $g(x)$; then
(b) we integrate with x running between a and a'.

The result is a number, and when D is the true rectangle B, it can be shown that we get the same result if we change the order and work with y first. We would work analogously if we had replaced the edge w of D in Fig. E.6.1 by the graph of a function $x = q(y)$. If a region G is cut up into subregions of these types, then we define the integral of F over G to be the sum of the separate integrals. Any two ways of cutting up G will give the same result. This is not the place to study the theory of integration, but interested readers can consult Webb (1991) for details.

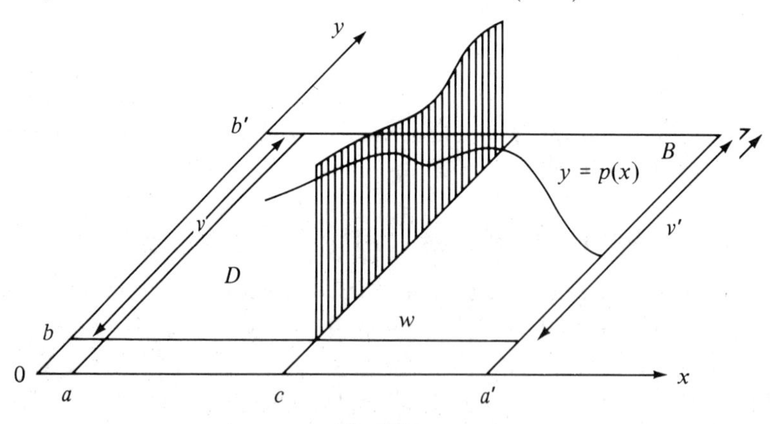

Fig. E.6.1.

Exercises E.6.1

1. Verify that

$$\int_0^1 \int_0^2 (x + 6y^2)\, dydx = \int_0^1 [(xy + 2y^3)]_{y=0}^{y=2}\, dx$$
$$= \int_0^2 [x^2/2 + 6xy^2]_{x=0}^{x=1}\, dy = \int_0^2 \int_0^1 (x + 6y^2)\, dxdy.$$

2. Evaluate:

(a) $\int_0^1 \int_0^x y^2 \, \mathrm{d}y\mathrm{d}x$,

(b) $\int_1^2 \int_y^{y^2} (x + 2y) \, \mathrm{d}x\mathrm{d}y$,

(c) $\int_{-1}^1 \int_0^{x^2} (x + 2y) \, \mathrm{d}y\mathrm{d}x$.

3. Sketch the region in the xy-plane corresponding to the double integral

$$\int_{1/2}^1 \int_{\sin^{-1} x}^{\pi/2} \cot y \, \mathrm{d}y\mathrm{d}x.$$

Reverse the order of integration and hence evaluate it. [SU]

4. Sketch the region bounded on the left by the circle whose equation is $(x - 1)^2 + y^2 = 1$, on the right by the line $y = 1 - x$, and below by the x-axis. Let R denote this region.

   By finding the point in the half-plane $y \geq 0$ where the circle $(x - 1)^2 + y^2 = 1$ intersects the line $y = 1 - x$, or otherwise, find the value of

$$\iint_R y \, \mathrm{d}x\mathrm{d}y.$$ [SU]

5. With B as in Fig. E.6.1, suppose that F happens to be of the form $\partial P/\partial x$. Show that

$$\int_B \frac{\partial P}{\partial x} \mathrm{d}x\mathrm{d}y = \int_b^{b'} P(a', y) \, \mathrm{d}y - \int_b^{b'} P(a, y) \, \mathrm{d}y. \qquad \text{(E.6.2)}$$

The first integral in (E.6.2) can be written for brevity as $\int_{v'} P$, meaning that we integrate P along the segment v' in Fig. E.6.1. Similarly for the opposite edge v, so we write

$$\int_B \frac{\partial P}{\partial x} = \int_{v'} P - \int_v P.$$

Check that if similarly F is of the form $\partial Q/\partial y$, then we obtain

$$\int_B \frac{\partial Q}{\partial y} = \int_{h'} Q - \int_h Q.$$

6. Suppose we have a domain E formed like a mosaic from several rectangles, as in Fig. E.6.2.

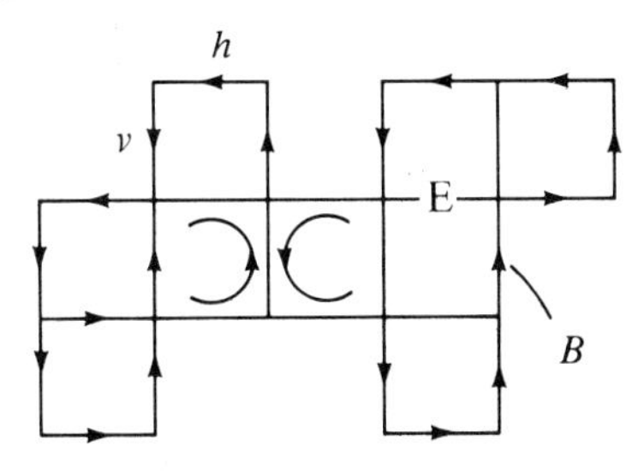

Fig. E.6.2.

Then by what was said above,

$$\int_E f\,\mathrm{d}x\mathrm{d}y = \sum \int_D f\,\mathrm{d}x\mathrm{d}y,$$

the sum being taken over each rectangle D of the mosaic. Take an anticlockwise arrow round the boundary B of E, to give a directional sense to each segment in B. Now use exercise 5 to show that integrals along interval edges cancel, so that

$$\int_E \left(\frac{\partial P}{\partial x} + \frac{\partial Q}{\partial y}\right) \mathrm{d}x\mathrm{d}y = \sum \int_v P + \sum \int_h Q, \tag{E.6.3}$$

the sums being taken over all vertical segments v and horizontal segments h in B, each with the assigned sense. (This is a primitive form of the *divergence* theorem which will be refined in exercise E.7.1 below.)

E.7 LINE INTEGRALS

Next, consider the possibility (6.2). If $\gamma = (u, v)$ is a path as there, then when the notation is unpacked, we get an ordinary integral of a function of the one variable, t, that depends on F and γ. We temporarily denote the result by $N(F, \gamma)$; thus

$$N(F, \gamma) = \int_a^b F(\gamma(t))\,\mathrm{d}t, \tag{E.7.1}$$

where a, b are the end points of the domain I of γ. However, a *useful* notion of integral along a curve like a circle or ellipse should depend only on the track, rather on which parametrization γ we use, provided we keep the same direction of I. For example, suppose $t = \theta(\tau)$, where τ still runs through I and from a to b rather than vice versa. Then $\delta(\tau) = \gamma(\theta(\tau))$ runs along the same track as γ itself, yet the integral in (E.7.1) changes, not to $N(F, \delta)$, but to one we write for brevity as

$$\int_a^b (F \circ \gamma) \circ \theta \cdot \theta'\,\mathrm{d}\tau,$$

with the factor $\theta'(\tau)$, and a possibly changed value. A way to remedy this defect is the following. Suppose that in the integrand of (E.7.1) we had used, not dt, but ds, where $s(t)$ is the arc-length of γ; the integral gives a new number $M(F, \gamma)$, say. Then $\mathrm{d}s = \|\gamma'(t)\|\,\mathrm{d}t$, so with θ, δ as before, we have $\|\delta'(\tau)\| \cdot \mathrm{d}\tau = \|\gamma'(t)\| \cdot \theta'(t)\,\mathrm{d}t$ (since $\theta' > 0$). Thus the change of variable from t to τ gives

$$\begin{aligned} M(F, \gamma) = \int_a^b F(\gamma(t))\,\mathrm{d}s &= \int_a^b F(\gamma(t)) \cdot \|\gamma'(t)\|\,\mathrm{d}t \\ &= \int_a^b F(\delta(\tau)) \cdot \|\delta'(\tau)\|\,\mathrm{d}\tau = M(F, \delta). \end{aligned} \tag{E.7.2}$$

We obtain a similar invariance if we define the integral of a mapping $\mathbf{G}: D \to \mathbb{R}^2$ along γ by

$$\int_a^b \mathbf{G} \cdot \gamma'(t) \, \mathrm{d}t$$

which is usually written $\int_\gamma \mathbf{G} \cdot \mathbf{ds}$. Thus if G, γ have components P, Q, and u, v, respectively, we have

$$\int_\gamma \mathbf{G} \cdot \mathrm{d}s = \int_a^b (P \circ \gamma \cdot u' + Q \circ \gamma \cdot v') \, \mathrm{d}t, \tag{E.7.3}$$

which is independent of the parametrization in the above sense. An integral of the form (E.7.3) is called a **line integral**, and often written

$$\int_a^b (P \circ w \, \mathrm{d}u + Q \circ w \, \mathrm{d}v).$$

Example E.7.1

To evaluate the line integral of $(xy, x - y)$ on the path $w(t) = (u(t), v(t)) = (t^2, 2t)$ when t runs on the interval I from 0 to 1, we substitute into (E.7.3):

$$F = x(t)y(t) = 2t^3, \qquad G = x(t) - y(t) = t^2 - 2t, \qquad u' = 2t, \qquad v' = 2,$$

to obtain the integral

$$\int_0^1 (4t^4 + 2(t^2 - 2t)) \, \mathrm{d}t = \left[t^5 + \frac{2}{3} t^3 - 2t^2 \right]_0^1 = \frac{-1}{3}.$$

Exercises E.7.1

1. Evaluate the line integral $\int_c (x + y) \, \mathrm{d}s$, where C consists of the polygonal path from $(1, 0)$ to $(2, 3)$ to $(4, 3)$.
2. Evaluate $\int_c \mathbf{F} \cdot \mathrm{d}\mathbf{s}$, where $\mathbf{F}(x, y) = (x + y)\mathbf{i} + xy\mathbf{j}$ and consists of line segments from $(0, 0)$ to $(1, 3)$ and $(1, 3)$ to $(2, -1)$.
3. Regard the ellipse $E: x^2/a^2 + y^2/b^2 = 1$ as the track of the path w, where $w(\theta) = (a \cos \theta, b \sin \theta)$. For each $P \neq 0$, let $F(P) = 1/\mathbf{0}P^2$. Show that, in the notation of (E.1),

$$N(F, w) = \int_0^{2\pi} (a^2 \cos^2 \theta + b^2 \sin^2 \theta)^{-1} \, \mathrm{d}\theta.$$

   Now write $t = \tan \theta$ to get

$$N = \frac{4}{b^2} \int_0^\infty \left(\frac{a^2}{b^2} + t^2 \right)^{-1} \mathrm{d}t = \frac{2\pi}{|ab|}.$$

   Show that if $a = b$, then $M(F, w)$ in (E.7.2) equals $a \cdot N$.
4. Show that in (3), N can be written $\int_0^{2\pi} (z^{\mathrm{T}} D z)^{-1} \, \mathrm{d}\theta$, where z is the column vector $(\cos \theta, \sin \theta)^{\mathrm{T}}$ and $D = L^2$ with L the diagonal matrix $\mathrm{diag}(|a|, |b|)$.

5. Let A be a real 2×2 matrix, and write $I(A) = \int_0^{2\pi} \mathrm{d}\theta/\|Az\|^2$. Show that with L as in (4), $I = I(L) = 2\pi/\det(L)$. Now let R be a rotation matrix. Show that we can replace z in the integrand of $I(A)$ by Rz, without affecting the value of the integral, so $I(AR) = I(A)$. Hence by using the fact that $A^{\mathrm{T}}A$ is symmetric and choosing R to diagonalize it, show that $I(A) = 2\pi/|\det(A)|$.
6. Explain why (E.6.3) allows us to write

$$\int_E \left(\frac{\partial P}{\partial x} + \frac{\partial Q}{\partial y}\right) \mathrm{d}x\mathrm{d}y = \int_B (Px' - Qy')\,\mathrm{d}t \tag{E.7.4}$$

where $x' = \mathrm{d}x/\mathrm{d}t$, $y' = \mathrm{d}y/\mathrm{d}t$. Here $\int_B$ is a sum of line integrals along the segments of B. If now the mosaic E consists of small squares so as to approximate a region G that has a curved boundary consisting of a single smooth curve C, then we would expect that the left and right sides of (E.7.2) will approximate the corresponding integrals $\int_G$, $\int_C$ over G, C, instead of over E, B. This proves to be the case (but only after a long proof) and gives us the famous **divergence theorem**:

$$\int_G \left(\frac{\partial P}{\partial x} + \frac{\partial Q}{\partial y}\right) \mathrm{d}x\mathrm{d}y = \int_C (Px' - Qy')\,\mathrm{d}t \tag{E.7.5}$$

(The name arises because in vector analysis, we call $\partial P/\partial x + \partial Q/\partial y$ the **divergence** of (P, Q), written $\mathrm{div}(P, Q)$.)

E.8 THE WINDING NUMBER

An important use of the line integral is to define the 'winding number' of a curve, as follows. Let γ be a smooth curve in $\mathbb{R}^2$ which does not pass through $\mathbf{0}$. Thus, if the domain of γ is the interval $K = [a, b]$, then $\gamma(t)$ is never zero, so the polar angle $\mathrm{Pol}(\gamma(t))$ is always defined (see (5.8.3)). One measure of the complexity of γ is to measure how Pol changes as t runs through K, and this change C is given in simple cases by

$$C = \mathrm{Pol}(\gamma(b)) - \mathrm{Pol}(\gamma(a)). \tag{E.8.1}$$

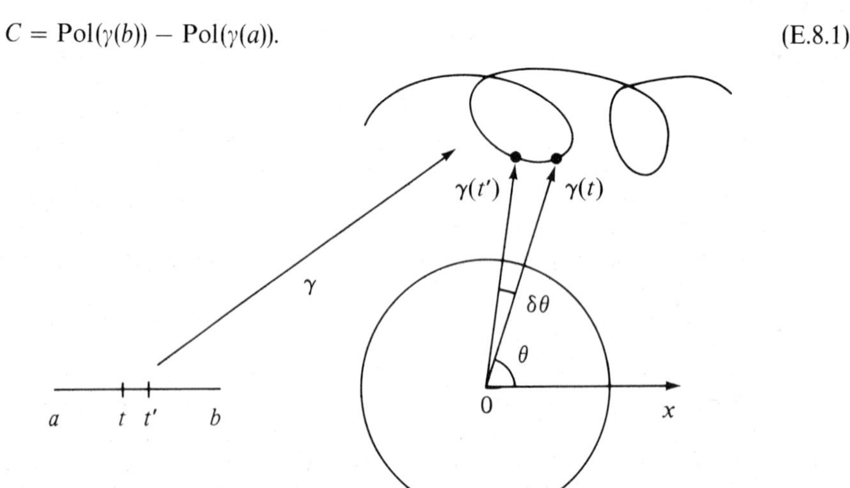

Fig. E.8.1.

This is meaningful provided γ does not cross the positive x-axis $\mathbb{R}_+$, since Pol is discontinuous there (see 5.8.3). But anyway

$$C = \int_a^b \frac{\mathrm{d}}{\mathrm{d}t} \mathrm{Pol}(\gamma(t))\,\mathrm{d}t, \tag{E.8.2}$$

and using the chain rule (D.4.4) with (5.8.2) we can work out the integrand $P(t)$ in terms of the coordinates γ_1, γ_2 of γ as

$$\begin{aligned} P(t) &= \mathrm{grad}(\mathrm{Pol}(\gamma(t))\cdot\gamma'(t)) \\ &= \frac{-\gamma_2(t)\gamma_1'(t) + \gamma_1(t)\gamma_2'(t)}{\gamma_1^2 + \gamma_2^2}. \end{aligned} \tag{E.8.3}$$

Hence, $P(t)$ is a continuous function of t, so it makes sense whether or not γ crosses $\mathbb{R}_+$ (but then the value of the integral of $P(t)$ is *not* given by (E.8.1), because of the discontinuity in Pol). In this more general case, if γ starts on $\mathbb{R}_+$ at X, and crosses $\mathbb{R}_+$ a finite number of times, then C (as given by (E.8.1)) jumps by 2π at each crossing; thus if γ is closed it returns to X and the resulting C is of the form $2n\pi$, for some integer n. We then call C the **winding number** of the closed curve γ and denote it by $W(\gamma)$ where, to summarize,

$$W(c) = \int_a^b P(t)\,\mathrm{d}t \tag{E.8.4}$$

with $P(t)$ given by (E.8.3). Note that by the first equation in (E.8.3), the integral (E.8.4) is the line integral of grad Pol along γ. In fact, $W(\gamma)$ counts the total number of times that γ winds around $\mathbf{0}$ after allowing for cancellations when steps are retraced. Hence we have proved the following proposition.

Proposition E.8.1
The winding number $W(\gamma)$ of a closed curve γ is an integral multiple of 2π.

Exercises E.8.1

1. If $K = [0, 2\pi]$, and $\gamma(t) = (a\cos(t), b\sin(t))$, use (E.8.3) to prove that the winding number of γ is 2π.
2. With the same K, let $q_n(t)$ be the complex number e^{jnt}, where n is an integer ≥ 0. Prove that the winding number is 2πn. If, instead, $\gamma(t)$ is any complex number (but γ is a smooth function) show that its complex conjugate has winding number equal to $-W(\gamma)$. What does this tell you about q_n above, when $n < 0$?
3. Write a computer program that will count the winding number of a curve, given parametrically.
4. Let p be a fixed point in $\mathbb{R}^2$, and γ a smooth closed curve that does not pass through p. Then we obtain a smooth curve d by defining $d(t) = \gamma(t) - p$. Show that d does not pass through 0. Hence we define the **winding number** $W(\gamma, p)$ of γ **with respect** to p to be $W(d)$. Show that $W(\gamma, 0) = W(\gamma)$, and calculate $W(\gamma, p)$ when γ is as in exercise 1 above and $p = (a/2, b/3)$ or $p = (2a, 3b)$. Calculate $W(q_n, p)$ when q_n is as in exercise 2 above and $|p| < 1$, or $|p| > 1$. We call γ **simple** if it never crosses itself (i.e., if $\gamma(s) \neq \gamma(t)$ and $s \neq t$, then $s = 0$

and $t = 2\pi$) (see Fig. E.8.2). Thus the track of c has an interior and an exterior. Show that if p is exterior, then $W(c, p) = 0$.

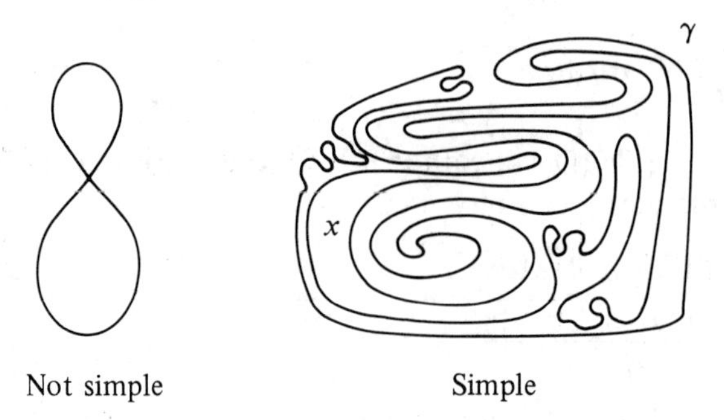

Fig. E.8.2. Closed curves. How would you decide, in the simple case, whether the point x is interior or exterior?

5. Show that the winding number of the closed curve $\gamma: K \to \mathbb{R}^2$ is a 'homotopy invariant' in the following sense. Suppose that for each s in the unit interval $U = \{s \mid 0 \le s \le 1\}$, $\gamma_s: K \to \mathbb{R}^2$ is a closed curve such that $\gamma_s(t)$ is a continuous function of the two variables s, t. Then we say that the curves γ_0, γ_1 are **homotopic** in $\mathbb{R}^2 - \{0\}$ if no curve γ_s passes through $\mathbf{0}$ (see Fig. E.8.3). By (E.8.4), $W(\gamma_s)$ is a continuous function of s. Show that $W(\gamma_0) = W(\gamma_1)$.

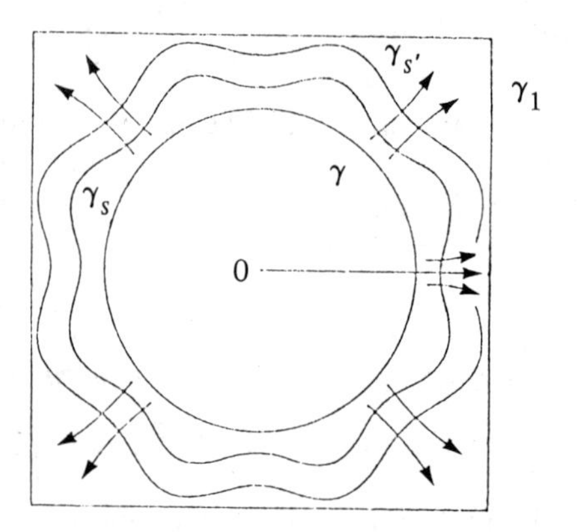

Fig. E.8.3. A homotopy of a circle (γ) to a square (γ_1). Two intermediate stages ($\gamma_s, \gamma_{s'}$) are shown, and the radial arrows indicate the tracks of points during the whole deformation.

6. If γ_1 in exercise 5 is a constant (so the track of γ_1 is a single point), show that $W(\gamma_0) = 0$.
7. Show that if γ_0 is the ellipse in exercise 1, then it is homotopic in $\mathbb{R}^2 - \{\mathbf{0}\}$ to any circle of centre $\mathbf{0}$, and also to any square or triangle with $\mathbf{0}$ in their interiors. (To find a homotopy between curves γ, δ, it is helpful to consider

$$\gamma_s(t) = s\delta(t) + (1 - s)\gamma(t).$$

Illustrate this equation in a sketch.) Show that each of these curves is homotopic to a constant in its interior. (This is also an important technique for forming transition curves in computer aided design (CAD).) Use exercise 5 to show that the winding number of a positively oriented *simple* closed curve is 1. (For a proof, see Hartman (1961), p. 147.)

E.9 VECTOR FIELDS

We now consider the suggestion (E.6.3) concerning antiderivatives. As we have seen several times before, a pair (P, Q) of smooth functions on a domain D defines a mapping $v: D \to \mathbb{R}^2$, which means that to each X in D, $v(X)$ is a point with coordinates $(P(X), Q(X))$. But also we can think of $v(X)$ in terms of an arrow with end $\mathbf{0}$ and tip $v(X)$ in $\mathbb{R}^2$. If now we move the end of this arrow to X without changing the direction, then we have an arrow $w(X)$ at X, with tip at $X + v(X)$. The set of all these arrows $w(X)$ is called a 'field' of vectors. By reversing this process, if we know the field we can reconstruct the mapping v.

For example, if we observe iron filings in a magnetic field, each one is a magnet with a southern end and northern tip; so we can think of each one as a (short) arrow attached to the point at its southern end. Similarly, if we imagine a wind blowing across D, then at each point X the wind has a velocity, $w(X)$, which can be represented by an arrow at X; again we have a vector field on D. For the linear system $\dot{z} = Az$, each point z of $\mathbb{R}^2$ is assigned the vector Az, and we have to find the trajectories as curves for which the velocity at $z(t)$ is exactly the vector Az. Each vector field can be represented by such an 'iron filings' diagram, and we have already seen examples with the linear systems of Chapter 3. In such diagrams, we can observe certain points at which the filings part their ways, like a parting of hair on the head. These points are 'singularities' of the field, and we shall see later how they can help to describe the field in a qualitative way.

One often regards a mapping as a vector field and vice versa, although in advanced work it is essential to make a conceptual distinction. We shall here use the terms more or less interchangeably, but the term 'vector field' is often a signal that it will be helpful to think of a point Y of $\mathbb{R}^2$ in terms of the arrow from $\mathbf{0}$ to Y.

Returning to the suggestion (E.6.3), the type of antiderivative we are looking for would give a useful description of a vector field, so we make a definition.

Definition E.9.1

If (P, Q) has the special property that there exists a function F on D for which $(P, Q) = \operatorname{grad} F$, then we say that (P, Q) is a **gradient field**, with **potential** F.

For example, if at each point X of $\mathbb{R}^2$, a planet at $\mathbf{0}$ exerts an attracting force per unit mass of $1/r^2$ units (where r is the distance $\mathbf{0}X$) then we obtain a gradient field with potential $1/r$; for then the force v is $-(x, y)/r^2$ (negative because it is attractive), and this is grad $(1/r)$.

If there is a potential, it greatly helps the analysis of a physical problem, and at once the following question arises. *Given two differentiable functions $P(x, y)$, $Q(x, y)$ on the domain D, when is there a function $F(x, y)$ on D such that $P = \partial F/\partial x$ and $Q = \partial F/\partial y$?* (Thus F would surely be a sort of 'antiderivative' of the pair (P, Q).)

By (D.1.6) a necessary condition is that $\partial P/\partial y = \partial Q/\partial x$, for if $P = \partial F/\partial x$ then

$$\frac{\partial P}{\partial y} = \frac{\partial}{\partial y}\frac{\partial F}{\partial x} = \frac{\partial}{\partial x}\frac{\partial F}{\partial y} = \frac{\partial Q}{\partial x}.$$

We now prove that the condition is also sufficient—at least if D is convex as shown in Fig. E.9.1. Fix a point (a, b) in D. Then, the convexity of D implies that the entire segment from (a, y) to (x, y) lies in D, so we may define $g(x, y) = \int_a^x P(u, y)\, du$. Then by the Fundamental theorem of calculus (4.5.4), we have $\partial g/\partial x = P$. Hence

$$\frac{\partial Q}{\partial x} = \frac{\partial P}{\partial y}\,(\text{given}) = \frac{\partial}{\partial y}\frac{\partial g}{\partial x} = \frac{\partial}{\partial x}\frac{\partial g}{\partial y}$$

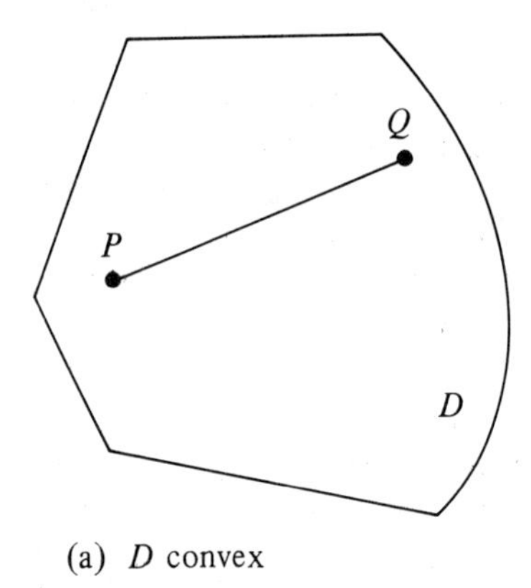

(a) D convex

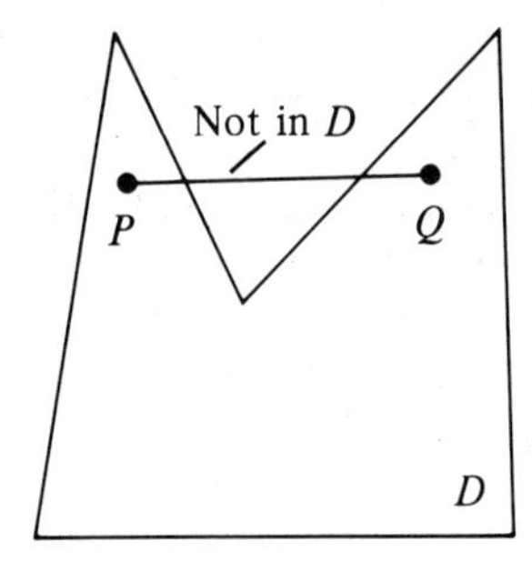

(b) D not convex

Fig. E.9.1.

by (D.1.6). Therefore $\partial/\partial x\,(Q - \partial g/\partial y) = 0$, so $Q - \partial g/\partial y$ is a function $h(y)$ of y alone, since D satisfies the conditions following (D.8.1). Now define the function F on D by: $F(x, y) = g(x, y) + \int_b^y h(v)\, dv$. Then $\partial F/\partial x = \partial g/\partial x = P$ (seen above) and $\partial F/\partial y = \partial g/\partial y + h(y) = Q$. Thus F is the function we require. We can summarize these observations in the following theorem.

Theorem E.9.1

Suppose P, Q are differentiable functions on the convex domain D. A necessary and sufficient condition for there to exist a function F on D with grad $F = (P, Q)$ *is that $\partial P/\partial y = \partial Q/\partial x$.*

The theorem holds for more general domains, but it is important that they should not have a hole in the middle (see exercise E.9.1(3) below).

Exercises E.9.1

1. Suppose $(P, Q) = (x^2 - y^2, -2xy)$ and $D = \mathbb{R}^2$. Referring to the above proof of theorem E.9.1, show that

$$g(x, y) = \frac{1}{3}(x^3 - a^3) - y^2(x - a), \qquad h(y) = -2ay,$$

and hence find $F(x, y)$ such that grad $F = (P, Q)$. In this example, without knowing the theory, we can see that $P = \partial F/\partial x$ so $F = x^3/3 - xy^2 + p(y)$ for some function p of y alone.

Similarly, use Q to obtain $F = -xy^2 + q(x)$, and compare the two results to find F. Use either method to find F when (P, Q) is

(i) $(6x^2 - 4y^2, 9y^2 - 8xy)$, (ii) $(2x + y^2 + ye^{xy}, 2xy + xe^{xy})$.

2. Find for what values of a, b the condition of theorem E.9.1 is satisfied, and find the corresponding function F when (P, Q) is:

   (i) $(x^2 y + ax \sin y, bx^3 - x^2 \cos y)$;
   (ii) $(ax \sinh y + \cos y, x^2 \cosh y + bx \sin y)$;
   (iii) $(2xy \exp(x^2 y) + by \cos xy + 2x, ax^2 \exp(x^2 y) + 2x \cos xy + 2)$. [SU]

3. With theorem E.9.1, show that we can allow D to be $\mathbb{R}^2$, or the quadrant $\{x, y \geq 0\}$. Why does the proof break down if D has a hole in it (e.g., when D is the region between two concentric circles)?
4. Recall the 'divergence', mentioned in exercise E.7.1(6). Show that if $\operatorname{div}(P, Q) = 0$, then $(Q, -P)$ is a gradient field.
5. Let $z, w: I \to D$ be two non-intersecting paths with the same end points a, b. Show that the line integrals of grad F along z and w are equal. (Use the divergence theorem, exercise E.7.1(6); D is convex.)

   Conversely, show that if the line integrals of (P, Q) along any two paths (with the same end points) are equal, then $\operatorname{div}(-Q, P) = 0$, so (P, Q) is a gradient field. (If $\operatorname{div}(-Q, P) > 0$ at some point z, consider two semicircles, centre z, and use the divergence theorem.)

The following questions use the language of physicists: 'equipotential curves' are our 'level curves', and the 'field lines' are curves that cut each level curve orthogonally (see exercise 9.4.1(2)).

6. For the vector field $\mathbf{F} = \dfrac{x}{\sqrt{x^2 + y^2}}\mathbf{i} + \dfrac{y}{\sqrt{x^2 + y^2}}\mathbf{j}$ on $\mathbb{R}^2$ $((x, y) \neq (0, 0))$, find:

   (a) $\operatorname{div}(\mathbf{F})$;
   (b) ϕ such that $\mathbf{F} = -\operatorname{grad} \phi$;
   (c) the equations of the field lines of $\mathbf{F}$.

   On a diagram in the xy-plane sketch the equipotential curves of ϕ and the field lines of $\mathbf{F}$. Indicate the direction of $\mathbf{F}$ at $(1, 1)$ and $(1, -1)$. [SU]
7. The vector field $\mathbf{E}(\mathbf{r})$ on $\mathbb{R}^2$ is defined by

   $$\mathbf{E} = \frac{x}{2}\mathbf{i} + y\mathbf{j}.$$

   Find Φ such that $\mathbf{E} = -\operatorname{grad} \Phi$. Calculate the equipotential curves $\Phi(\mathbf{r}) = \text{constant}$. Calculate the field lines of $\mathbf{E}(\mathbf{r})$. Sketch the field lines of $\mathbf{E}(\mathbf{r})$ and the equipotential curves of $\Phi(\mathbf{r})$ on the same diagram. [SU]
8. The vector field $\mathbf{E}(\mathbf{r})$ on $\mathbb{R}^2$ is defined by $\mathbf{E} = [x(x^2 - a^2)\mathbf{i} + y^3\mathbf{j}]$. Find Φ such that $\mathbf{E} = -\operatorname{grad} \Phi$. Show that the points $\mathbf{r} = (a, 0), (-a, 0)$ correspond to mountain peaks, and $\mathbf{r} = (0, 0)$ corresponds to a saddle point of the graph of $\Phi(x, y)$. What is $\mathbf{E}(\mathbf{r})$ at these points? Sketch the field lines of $\mathbf{E}(\mathbf{r})$ and the equipotential curves $\Phi(\mathbf{r})$. [SU]

Theorem E.9.1 tells us that most vector fields are not gradient fields, and hence cannot be described by means of the single 'potential' F. To obtain further information about the field (P, Q) we next consider its singularities, mentioned above. The reason why we observe such a singularity X is that we see no vector $w(X)$ there, and that is because $w(X)$ is zero. We say that a **zero** of the field is a point X at which P and Q are simultaneously zero. Thus the vector $v(X)$ is zero, and has no direction—a reason for calling a zero a 'singularity'.

Now suppose that γ is any smooth closed curve in the domain D, when no zero lies on γ. One source of information about the field $v = (P, Q)$ is to observe how v changes as we move round γ. In particular, we use the winding number (E.8.4) to measure the change in the polar angle, $\mathrm{Pol}(v)$. To do this, suppose the domain of γ is the interval $K = [a, b]$. Then as a mapping, v maps γ to another closed curve in $\mathbb{R}^2$ given by the composition $v \circ \gamma$, where $(v \circ \gamma)(t) = v(\gamma(t)) \neq 0$. Hence $v \circ \gamma$ has winding number $W(v \circ \gamma)$, defined by (E.8.4). We divide this by 2π for convenience, and then define the *index of γ with respect to v* by

$$I(\gamma, v) = \frac{1}{2\pi} W(v \circ \gamma). \tag{E.9.1}$$

This index I is a 'homotopy invariant' in two senses, similar to that of exercise E.8.1(5), which we now explain. (See Fig. E.9.2.)

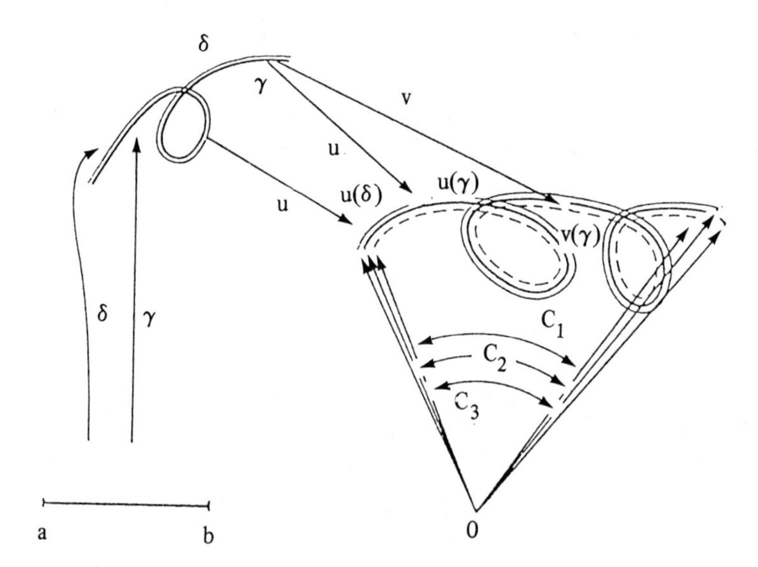

Fig. E.9.2. If γ is 'close' to δ or $u = v$, then C_1, C_2, C_3 (see E.8.1) are 'close'—a visual formulation of (E.9.2) and (E.9.3).

First suppose that u is a second vector field on G. Then for each s in the unit interval $U = \{s \mid 0 \leq s \leq 1\}$, we can form a new field $w(s) = sv + (1 - s)u$ with domain G. Then $w(0) = u$ and $w(1) = v$, and if $w(s)$ is never zero on the curve γ, we write $u \sim v\ (n \cdot z)$. In that case, $I(\gamma, w(s))$ is defined for each s in U, and we claim:

If the fields u, v satisfy $u \sim v\ (n.z)$ in G then

$$I(\gamma, u) = I(\gamma, v). \tag{E.9.2}$$

For, by (E.8.3), the integrand of $I(\gamma, w(s))$ in (E.8.4) is clearly a continuous function of s, and hence so is the index (because the integration is with respect to t, not s). But $I(\gamma, w(s))$ is an integer, so it must be constant. Thus (E.9.2) follows from:

$$I(\gamma, u) = I(\gamma, w(0)) = I(\gamma, w(s)) = I(\gamma, w(1)) = I(\gamma, v).$$

Second, suppose that δ is a closed curve which is homotopic to γ, by a homotopy $\gamma_s\colon K \to G$ as in exercise E.8.1(5). If no curve γ_s passes through a zero of the field v, we write $c \sim d\ (n.z)$ in G. Then the index $I(\gamma_s, v)$ is defined and is a continuous function of s. Hence an argument like that for $I(\gamma, w(s))$ above proves:

If the curves γ, δ satisfy $\gamma \sim \delta\ (n \cdot z)$ in G then

$$I(\gamma, v) = I(\delta, v). \tag{E.9.3}$$

Exercises E.9.2

1. Let $p_n\colon \mathbb{R}^2 \to \mathbb{R}^2$ be the vector field such that $p_n(z) = z^n$ when z is regarded as a complex number. With $q_n(t)$ as in exercise E.8.1(2), show that $(p_n \circ q_1)((t) = q_n(t)$. Hence show that $I(q_1, p_n) = n$. (Recall that q_1 parametrizes the unit circle.)
2. Show that if p, q are points in $\mathbb{R}^2$ such that $\|p\| > \|p - q\|$, then for no s in U can we have $sq + (1 - s)p = 0$. Now suppose that u, v are vector fields such that for every point z on the curve γ, $\|u(p)\| > \|u(p) - v(p)\|$. Prove that $I(\gamma, u) = I(\gamma, v)$. (Apply (E.9.2); this is Rouché's theorem.)
3. It can be shown (not easily) that if γ is any simple closed curve then γ is homotopic to a constant in its interior. (This is not difficult, however, for 'familiar' curves such as circles, etc.) Hence show that if the field v has no zeros interior to γ, then $I(\gamma, v) = 0$. Conversely, then, if $I(\gamma, v) \neq 0$, then γ has a zero in its interior.
4. Prove the **fundamental theorem of algebra**: if $f(z) = z^n + a_1 z^{n-1} + \cdots + a_n$ is a polynomial of degree $n > 0$, then it has a (possibly complex) root. (Regard f as a vector field on $\mathbb{R}^2$, and let $g(z) = f(z) - z^n$. Show that on a circle of radius r, $\|g(z)\| \leq Ar^{n-1}$, where A is the largest of the numbers $\|a_j\|$. Hence show that if $r > A$, then $\|z^n\| > \|g(z)\|$; and $I(q_1, f^n) = I(q_1, p_n)$ as in exercise 1. Now use exercise 3.

Next let us assume that the field v on D is such that all its zeros are **isolated**, which is to say that each one (say X) is the centre of a circle S that lies in G and encloses no other zero. Hence we can assign to X a number, its index, given by:

$$I(X, v) = I(S, v), \tag{E.9.4}$$

which is independent of the radius of S, by (E.9.3). Poincaré proved the following more precise (and very attractive) version of exercise E.9.2(3):

Theorem E.9.2

If γ is a simple closed curve in the domain of the field v, and γ contains only isolated zeros of v, then these are finite in number and the sum of all their indices equals $I(\gamma, v)$. $I(\gamma, v)$.

For brevity we give, in exercise E.9.3(6) below, an indication of the proof, so that the reader can complete the details.

Exercises E.9.3

1. If in (E.9.4), X is not a zero of v, and S contains no zeros of v, use exercise E.9.2(3) to show that $I(S, v) = 0$. Thus, it is consistent to define the index $I(Y, v)$ to be zero, in this case.
2. Let X be a zero of v as in (E.9.4), and let J be the Jacobian matrix of (P, Q), evaluated at X. Use exercise E.9.2(2) to show that if S has sufficiently small radius r, then $I(X, v) = I(X, w)$, where w is the field given by $w(Y) = J \cdot Y$.
3. In exercise 2, if J is invertible, show that $I(X, w)$ equals the sign of $\det(J)$ as follows. By (E.9.1) we must evaluate $W(w \circ S)$ where $S(t) = rz$ and $z = (\cos t, \sin t)^{\mathrm{T}}$, with t in $[0, 2\pi]$. Check that in (E.8.3), $(\gamma_1(t), \gamma_2(t))$ is now given by $rJ \cdot z$, and show that $P(t)$ simplifies to $(\det(J))/\|z^{\mathrm{T}} \cdot J^{\mathrm{T}} \cdot Jz\|$. Now use exercise E.7.1(5).
4. Let $F\colon D \to \mathbb{R}$ be a smooth non-degenerate function, as in (D.3.4). Then grad F is a vector field on D, and its singularities are the critical points of F, which we know to be saddles and local maxima and minima. Use exercise 2 to verify that each saddle has index -1, and the others have index 1. Similarly, show that with the linear system $\dot{z} = Az$ (when A is an invertible matrix), the only singularity is at $\mathbf{0}$, where the index is -1 for a saddle, and 1 otherwise.
5. Recall the vector field v given by the portrait in Fig. 9.1.2. Show that if γ is one of the ovals there, then $I(\gamma, v) = 1$; but if γ is the frame of the diagram then $I(\gamma, v) = 0$. Check some of the other portraits in Chapter 9.
6. Complete the proof of theorem E.9.2 as follows. The finiteness of the set of zeros follows from a general theorem of topology, because the zeros are isolated. Assuming this, surround each zero by a small circle S and join it to γ by a canal of small width w, none of the canals to intersect each other. Then we have a simple closed curve δ, that goes along the canal banks, with each zero on the left, and finally clockwise round part of γ. Thus $I(\delta, v) = 0$. But the contributions to the integral, of the opposite banks of each canal, approximately cancel; the portions along the small circles are approximately the indices of the zeros, and the rest is approximately $I(c, v)$. Now let h tend to zero to finish the proof.
7. Let A be an island in the sea, and let $h(P)$ be the height above sea level. On a flat contour map of A, grad h gives a gradient field. If h is non-degenerate, prove the 'mountaineer's equation', which says 'pits + peaks − passes = 1', where 'pits' means the number of local minima, etc. A 'pass' is a saddle point. This generalizes to other surfaces: see Griffiths (1981a).
8. Devise a model that will account for the contours of saddles and centres that can be seen in the grain of some polished wood.

F

Norms, sequences, and contracting mappings

In Chapter 2 and elsewhere, our descriptions of the behaviour of a sequence of vectors, starting from $\mathbf{z}$ under the repeated application of a matrix A as an orbit $\langle A^n\mathbf{z}\rangle$, used terms like 'converges' or 'winds outwards to infinity'. These imply judgements about the distance of $A^n\mathbf{z}$ from $\mathbf{0}$. In order to estimate such distances numerically, it is necessary to introduce more precise language, which is also needed whenever we need to make good estimates of errors in non-linear calculations. Readers meeting it for the first time always find difficulty—this is probably the most difficult chapter in the book. It might therefore be best for readers to skim the material first, as much of the easier work does not need it. But in certain contexts, this material cannot be escaped, and readers should then return: understanding comes as we see the point of using it. The chapter concludes with the Contracting Mapping theorem, so necessary material on metric spaces is then introduced for those readers who have not already taken a course of elementary topology.

F.1 NORMS

Recall from Chapter B the Euclidean length $\|\mathbf{z}\| = \sqrt{(\mathbf{z}\cdot\mathbf{z})}$ of the vector $\mathbf{z}$; we also call this the **norm** of $\mathbf{z}$. Then the distance between $\mathbf{z}$ and $\mathbf{w}$ is the norm of $\mathbf{z} - \mathbf{w}$, so

$$\text{dist}(\mathbf{z}, \mathbf{w}) = \|\mathbf{z} - \mathbf{w}\|. \qquad \text{(F.1.1)}$$

Since $(\mathbf{z} - \mathbf{w})^2 = (\mathbf{w} - \mathbf{z})^2$, then $\text{dist}(\mathbf{z}, \mathbf{w}) = \text{dist}(\mathbf{w}, \mathbf{z})$. Also, in triangle $\mathbf{0zw}$, the length of the side $\mathbf{zw}$ cannot exceed the sum of the lengths of the other two sides; so we have the **triangle inequality**:

$$\|\mathbf{z} - \mathbf{w}\| \le \|\mathbf{z}\| + \|\mathbf{w}\|, \qquad \text{(F.1.2)}$$

which we shall find very useful when making rough estimates. If we write $\mathbf{z} = \mathbf{a} - \mathbf{b}$ and $\mathbf{w} = \mathbf{b} - \mathbf{c}$, the triangle inequality takes the form:

$$\|\mathbf{a} - \mathbf{c}\| \le \|\mathbf{a} - \mathbf{b}\| + \|\mathbf{b} - \mathbf{c}\|. \qquad \text{(F.1.3)}$$

The idea of distance is needed, when we want to say that two things are 'near' to each other, or one thing is in another's 'neighbourhood'. Formally, we shall say that for any $r > 0$, the 'd-neighbourhood' of the point $\mathbf{z}$ is the set of all points $\mathbf{w}$ such that the distance $\|\mathbf{z} - \mathbf{w}\| < d$. This set is the circular disc, of centre $\mathbf{z}$ and radius d, and the circular boundary is not included; we abbreviate 'neighbourhood' to 'nbd', and denote the d-nbd of $\mathbf{z}$ by $N_d(\mathbf{z})$. Thus, when we said earlier that an orbit $\langle \mathbf{z}_n \rangle$ goes off to infinity, we can make this precise to anybody who challenges us, by saying that if she gives us a d-nbd $N_d(\mathbf{0})$ of $\mathbf{0}$ (however large), we can prove that the points of the orbit eventually stay outside $N_d(\mathbf{0})$. By 'eventually', we mean: we can state a number C, such that if $n > C$ then $\mathbf{z}_n$ is outside $N_d(\mathbf{0})$. We write: $\mathbf{z}_n \to \infty$ as $n \to \infty$.

This is a fairly elaborate test, but each rule is now clear. Let us apply it to the case, considered in section 2.6, wherein $\mathbf{w}(n) = r^n R_{n\theta}\mathbf{w}(0)$ and $R_{n\theta} = R_\theta^n$ is a rotation. Here, $\|\mathbf{w}(n)\| = r^n$, and this lies outside a given $N_d(\mathbf{0})$ as soon as $r^n > d$; so, taking logarithms, we need to have $n > C$ where $C = \ln(d)/\ln(r)$. The challenge is met.

The test also makes sense, not only for orbits, but for any sequence $u_1, u_2, \ldots, u_n, \ldots$, generated by some rule, for example,

$$u_n = (n^2, n + 1) \qquad \text{or} \qquad u_n = ((-1)^n, \sin(n)). \tag{F.1.4}$$

It is quite usual for students to find this test and its language hard to understand, so if you do not grasp it immediately, come back to it later. The next exercises will give you practice, and help you to see the point.

Exercises F.1.1

1. Let A be one of the standard matrices discussed in theorem 7.5.1, and suppose that each eigenvalue has modulus >1. Show that the above test is satisfied by $\langle \mathbf{z}_n \rangle$ when $\mathbf{z}_n = A^n\mathbf{z}_0(\neq\mathbf{0})$. (If $A = \operatorname{diag}(\lambda, \mu)$, then $\|\mathbf{z}_n\| = \|A^n\mathbf{z}\| = \sqrt{(\lambda^{2n}x^2 + \mu^{2n}y^2)}$. At least one of x, y is non-zero, say x, so $\|\mathbf{z}_n\| > |\lambda^n|\,|x|$. Now $\mathbf{z}_n$ lies outside a challenger's $N_d(\mathbf{0})$ if $\|\mathbf{z}_n\| > d$, so this will happen if $|\lambda^n|\,|x| > d$. We can arrange this since $|\lambda| > 1$, for then we need n to satisfy $n\ln(|\lambda|) > \ln(d/|x|)$. Thus, for the challenger's C required by the above test, we can use $\ln(d/|x|)/\ln(|\lambda|)$. Other types of A are dealt with similarly, but each has a separate twist that you should deal with.)
2. Show that the first sequence $\langle u_n \rangle$ in (F.1.4) satisfies the test, but the second one does not. (It 'oscillates boundedly'.)
3. Take $\mathbf{0}$ as the centre of your computer screen, and choose $d > 0$. Draw the circle of centre $\mathbf{0}$ and radius d to represent the boundary of $N_d(\mathbf{0})$. Now test the first sequence in (F.1.4) to find C such that u_n leaves $N_d(\mathbf{0})$ when $n > C$. Repeat with $d/2$ instead of d, and repeat again. Try the same exercise with the sequence $u_n = (n/\ln(n), n^{1/n})$.

F.2 CONVERGENCE

Next, what about a test to decide whether an orbit $\langle \mathbf{z}_n \rangle$ 'converges' to $\mathbf{0}$? Mathematicians have devised the following one, which turns the previous one inside out, so to speak. This time, given a d-nbd $N_d(\mathbf{0})$ (however small) we need to prove that the points of the orbit eventually stay inside $N_d(\mathbf{0})$. Again 'eventually' means: we can find a C, such that if $n > C$, then $\mathbf{z}$ lies in $N_d(\mathbf{0})$. Clearly this test makes sense if we apply it to sequences as well as orbits.

Exercises F.2.1

1. Do the analogue of exercises F.1.1(1), but when each eigenvalue of A has modulus <1. (If $A = \text{diag}(\lambda, \mu)$, then as before, $\|\mathbf{z}_n\| = \|A^n\mathbf{z}\| = \sqrt{(\lambda^{2n}x^2 + \mu^{2n}y^2)} \leq 2|\lambda^n|s$ where $s = \|\mathbf{z}\|$ and we assume $|\lambda| \geq |\mu|$. Now $\mathbf{z}_n$ lies inside a challenger's $N_r(\mathbf{0})$ if $\|\mathbf{z}_n\| < r$, so this will happen if $|\lambda^n|\,|s| < r$. We can arrange this since $|\lambda| < 1$, for then $\ln(|\lambda|) < 0$ and we need n to satisfy $n\ln(|\lambda|) < \ln(r/|s|)$. Thus, for the C required to counter the challenger, we can use the next integer after $\ln(r/|s|)/\ln(|\lambda|)$.)
2. Regard the number r as the point $(r, 0)$ on the x-axis of $\mathbb{R}^2$. Use the tests to establish the behaviour of the sequence $\langle r^n \rangle$ as follows.
   (a) If $|r| < 1$ then $\langle r^n \rangle$ converges to zero, and if also $-1 \leq r < 0$, then r^n oscillates alternately on either side of 0; it converges to 0 if $r \neq -1$.
   (b) If $r = 1$, then $r^n = 1$ and converges to 1 (contrary to the widely held belief that 'a convergent sequence approaches, but never reaches, its limit').
   (c) If $|r| > 1$ then $r^n \to \infty$ as $n \to \infty$, and oscillates on either side of 0.
   (If $r > 0$, then r^n stays on the positive side of 0, and we often write $r^n \to +\infty$ as $n \to \infty$.) This behaviour must be remembered, as it is a standard against which to assess other sequences.
3. Many students think it is obvious that the sequence $\langle 1/n \rangle$ in $\mathbb{R}$ converges to zero. Nevertheless, show that it passes the test. (If it didn't, we'd have to change the test, which was invented as a model of our psychological processes when we 'see' certain obvious limits. Verifying the test *is a check on the model.* Two disputants thus have a language with which to discuss those processes.)
4. Show that the first four of the following sequences pass the test for convergence to 0, while the last one does not: $\langle (0.9)^n \rangle$, $\langle (2 + 3(-1)^n)/n \rangle$, $\langle \sqrt{((1 + 2/n)} \rangle$, $\langle (\cos n)/n^2 \rangle$, $\langle \cos(n\pi/2) \rangle$.

Orbits of linear mappings can converge only to $\mathbf{0}$, whereas the non-linear mappings we shall consider later show more variety. For this purpose, we say that a sequence $\langle \mathbf{u}_n \rangle$ in $\mathbb{R}^2$ **converges** to $\mathbf{v}$, iff the sequence $\langle \mathbf{u}_n - \mathbf{v} \rangle$ converges to $\mathbf{0}$. We then write

$$\lim \mathbf{u}_n = \mathbf{v}. \tag{F.2.1}$$

Instead of testing with nbds of $\mathbf{0}$, we can achieve the same effect by requiring: given any d-nbd $N_d(\mathbf{v})$ of $\mathbf{v}$, then $\langle \mathbf{u}_n \rangle$ is eventually in $N_d(\mathbf{v})$. But now we need to show that $\langle \mathbf{u}_n \rangle$ cannot simultaneously converge to some other limit $\mathbf{w}$. We prove this as follows.

If $\mathbf{v} \neq \mathbf{w}$, then $\|\mathbf{v} - \mathbf{w}\| = s > 0$. Let $d = s/3$. Since $\langle \mathbf{u}_n \rangle$ converges to $\mathbf{v}$, then given $N_d(\mathbf{v})$, $\langle \mathbf{u}_n \rangle$ is eventually in $N_d(\mathbf{v})$. Similarly it is eventually in $N_d(\mathbf{w})$. But this is impossible, since $N_d(\mathbf{v})$ has no points in common with $N_d(\mathbf{w})$. This contradiction arises from assuming that $\mathbf{v} \neq \mathbf{w}$, so in fact $\mathbf{v} = \mathbf{w}$, and *the limit is unique.*

Exercises F.2.2

1. Using a calculator if necessary, guess limits for the following sequences, and then act as your own challenger to show that each sequence passes the test:

   $n \cdot \sin(1/n)$, $\quad 2^{1/n}$, $\quad (3n^2 + 1)/(n^2 + 5)$, $\quad (2^n + 5^n)^{1/n}$, $\quad (1 + 1/n)^7$,

   $(1 + 1/n)^n$, $\quad (\sqrt{(n + 1)} - \sqrt{n})$.

   (A systematic approach, to avoid using the test as much as possible, uses the 'algebra of limits': see any book on Mathematical Analysis.)

2. Use the triangle inequality (F.1.3) to prove that, in the last proof, $N_d(\mathbf{v})$ and $N_d(\mathbf{w})$ have no point $\mathbf{z}$ in common. (Use $\mathbf{z}$ to show that $\|\mathbf{v} - \mathbf{w}\|$ would be too small.)
3. Rewrite the last proof, to eliminate the word 'eventually' (by using the challenger's numbers 'C', one for each of $\mathbf{v}$ and $\mathbf{w}$: you must then make a certain decision!).
4. For any 2×2 real matrix A and point $\mathbf{z} = (x, y)^{\mathrm{T}}$, we often need to make careful estimates of the norm v of $A\mathbf{z}$, in terms of the eigenvalues λ, μ of A, where $|\lambda| \le |\mu|$. The following steps will establish that there are functions $p(\mu, n)$, $q(\lambda, n)$ such that, for all $\mathbf{z}$

$$p(\mu, n)\,\|\mathbf{z}\| \le \|A^n\mathbf{z}\| \le q(\lambda, n)\,\|\mathbf{z}\| \tag{F.2.2}$$

where, as the integer n increases, if $|\lambda| < 1$ then $q(\lambda, n) \to 0$, and if $|\mu| > 1$ then $p(\mu, n) \to \infty$.

(a) Show that $v^2 = \mathbf{z}^{\mathrm{T}}S\mathbf{z}$, where S is the symmetric matrix $A^{\mathrm{T}}A$. Now S has real eigenvalues, so let m^2, M^2 be the least and greatest of their moduli. Show that $m\,\|\mathbf{z}\| \le v \le M\,\|\mathbf{z}\|$.

(b) If A is in standard form (Theorem 7.5.1), and of type (i), show that $m = |\lambda|$ and $M = |\mu|$, and with type (iii), $m = M = |\lambda|$. Type (ii) matrices are more awkward, and we deal with them in step (f).

(c) If also, A is invertible, show that $\|\mathbf{z}\|/M \le \|A^{-1}\mathbf{z}\| \le \|\mathbf{z}\|/m$.

(d) Suppose that $B = PAP^{-1}$, and let the corresponding numbers m, M for P be m_p, M_p. Show that when $K = M_p/m_p$, then

$$m\,\|\mathbf{z}\|\,/K \le \|B\mathbf{z}\| \le M\,\|\mathbf{z}\|\,K.$$

(Note: $\|B\mathbf{z}\| = \|P(AP^{-1}\mathbf{z})\| \le M_p\,\|AP^{-1}\mathbf{z}\| \le M_p \cdot M \cdot (1/M_p)\,\|\mathbf{z}\|$, etc.)

(e) Show that, for each n,

$$m^n\,\|\mathbf{z}\|\,/K \le \|B^n\mathbf{z}\| \le M^n\,\|\mathbf{z}\|\,K.$$

This shows that $\|B^n\mathbf{z}\|$ constantly decreases or increases according as $M < 1$ or $m > 1$, and (F.2.2) follows with simple forms for p and q.

(f) To deal with the more difficult case when A is of type (ii), we first observe that $v^2 = \|\mathbf{z}\|^2 f(\theta)$ where $f(\theta) = \mathbf{u}^{\mathrm{T}}S\mathbf{u}$ and $\mathbf{u} = \mathbf{z}/\|\mathbf{z}\| = (\cos\theta, \sin\theta)^{\mathrm{T}}$. Use exercise 7.5.1(2) to obtain:

$$m\,\|\mathbf{z}\|\,\sqrt{((a' + b' - d')/2)} \le \|A\mathbf{z}\| \le M\,\|\mathbf{z}\|\,\sqrt{((a' + b' + d')/2)}$$

with $a' = a^2 + b^2$, $b' = c^2 + d^2$, $d'^2 = (a' - b')^2 + 4(ab + cd)^2$. Now let $B = P^{-1}AP$, with A standard of type (ii), and single real eigenvalue of modulus r. Recall that $B^n = P^{-1}A^nP$ and A^n was described in exercise 7.5.1(2). Show that if

$$s = \sqrt{(1 + 4r^2)} \qquad \text{and} \qquad t = 2r^2 + n^2$$

then $r^{n-1}\sqrt{((t - s)/2)} \le \|B^n\mathbf{z}\| \le r^{n-1}\sqrt{((t + s)/2)}$, and hence find the functions p, q required in (F.2.2).

(g) Suppose $\langle \mathbf{u}_n \rangle$ converges to $\mathbf{v}$. Show that if $r > 0$ is given, then we can find a C, such that, for all $n, m > C$,

$$\|\mathbf{u}_n - \mathbf{u}_m\| < r. \tag{F.2.3}$$

(Apply the test at $\mathbf{v}$ with $r = d/2$, and use the triangle inequality.)

Every sequence with property (F.2.3) is called a **Cauchy sequence** (after the French mathematician A.L. Cauchy (1789–1857), and the exercise shows that every convergent sequence is a Cauchy sequence. Note that (F.2.3) makes sense, whether or not we know there is a limit, and Cauchy used it to assert the existence of a limit. That assertion is not true if (like the classical Greeks) we know only rational numbers, and it led to a clarification of views about what real numbers 'ought' to be. Eventually Cantor and Dedekind constructed our system $\mathbb{R}$ of real numbers (cleverly using the rationals as raw material) which has the property (shown by Cauchy to be vital) known as 'completeness'. This can be stated as follows.

Theorem F.2.1
Every Cauchy sequence in $\mathbb{R}$ has a limit in $\mathbb{R}$.

Thus, for example, every Cauchy sequence of rationals does converge, but not necessarily to a rational number: with Cantor and Dedekind, you have to widen your concept of number to make the result true—just as you have had to widen it from integers to rationals to complex numbers in order to make various 'simple' algebraic results true.

Exercises F.2.3

1. The real number $\sqrt{2}$ is known to be not rational; let x_n denote its first n decimal places. Show that $\langle x_n \rangle$ is a Cauchy sequence of rationals, which does not converge in the rationals. Show that $\lim x_n = \sqrt{2}$ in $\mathbb{R}$. (Does it pass the test?)
2. Use the completeness of $\mathbb{R}$ to show that $\mathbb{R}^2$ and $\mathbb{R}^n$ are complete. (Let $\mathbf{z}_n = (x_n, y_n)$; show that $\langle \mathbf{z}_n \rangle$ is a Cauchy sequence iff the sequences $\langle x_n \rangle$, $\langle y_n \rangle$ are Cauchy sequences.)
3. Show that the unit interval $U = \{0 \le x \le 1\}$ in $\mathbb{R}$, is complete, (i.e., every Cauchy sequence $\langle x_n \rangle$ with $x_n \in U$, converges to a limit that lies in U). Now show that a rectangular disc E in $\mathbb{R}^2$ is complete. (Take $E = \{(x, y) | a \le x \le b, c \le y \le d\}$, and then use a rotation for other discs.)

Cauchy sequences arise especially in connexion with the simplest kind of non-linear mapping of $\mathbb{R}^2$: the 'contracting' type, which we now investigate.

F.3 THE CONTRACTING MAPPING THEOREM IN $\mathbb{R}^2$

Problems often arise in which there is a mapping F: $\mathbb{R}^2 \to \mathbb{R}^2$, which need not be linear but which has the property that there is a constant $k < 1$ such that for all points $\mathbf{z}, \mathbf{w}$ in $\mathbb{R}^2$, we have

$$\|F(\mathbf{z}) - F(\mathbf{w})\| \le k \, \|\mathbf{z} - \mathbf{w}\|. \tag{F.3.1}$$

This says that the distance between the F-images of $\mathbf{z}$ and $\mathbf{w}$ is less (by the factor k) than their original distance apart. We call F a k-**contraction**, or more loosely, a contracting mapping. Then the dynamics of F is very simple, in that we can prove the following result.

Proposition F.3.1
There is some point $\mathbf{v}$ (depending on F) such that every F-orbit not only converges, but converges to $\mathbf{v}$.

We summarize this as: the 'basin of attraction' of $\mathbf{v}$ is the whole of $\mathbb{R}^2$. Hence, to find $\mathbf{v}$, we can start with any convenient $\mathbf{z}$, and find its orbit $\langle \mathbf{z}_n \rangle$; moreover, this converges to $\mathbf{v}$ at a calculable rate. Also, the process is self-correcting, in that if we erroneously calculate a particular $\mathbf{z}_n$ as $\mathbf{w}$, then later iterations give us the orbit of $\mathbf{w}$ which still converges to $\mathbf{v}$.

Proof

We first estimate how close the terms of $\langle \mathbf{z}_n \rangle$ are to each other. By the contracting property (F.3.1) we have, for each n,

$$\|\mathbf{z}_n - \mathbf{z}_{n+1}\| = \|F(\mathbf{z}_{n-1}) - F(\mathbf{z}_n)\| \leq k \, \|\mathbf{z}_{n-1} - \mathbf{z}_n\|,$$

and by repetition,

$$\|\mathbf{z}_n - \mathbf{z}_{n+1}\| \leq k^2 \, \|\mathbf{z}_{n-2} - \mathbf{z}_{n-1}\| \leq \ldots \leq k^n \, \|\mathbf{z}_1 - \mathbf{z}_0\|.$$

Hence for each m,

$$\|\mathbf{z}_n - \mathbf{z}_{n+m}\| = \|(\mathbf{z}_n - \mathbf{z}_{n+1}) + (\mathbf{z}_{n+1} - \mathbf{z}_{n+2}) + \cdots + (\mathbf{z}_{n+m-1} - \mathbf{z}_{n+m})\|$$

and by the triangle inequality, this gives:

$$\|\mathbf{z}_n - \mathbf{z}_{n+m}\| \leq s(k^n + k^{n+1} + \cdots + k^{n+m-1}), \qquad s = \|\mathbf{z}_1 - \mathbf{z}_0\|,$$

$$\leq s\,k^n(1 + k + k^2 + \cdots + k^n + \cdots) = s\,k^n \frac{1}{1-k}, \tag{F.3.2}$$

on summing the geometric series. Now, given $r > 0$, there is a number C such that if $n > C$, then $s\,k^n\,1/(1-k) < r$, since $s/(1-k)$ is constant and $\langle k^n \rangle$ converges to 0. Hence (F.2.3) holds (with m there replaced by $n + m$ here). Therefore $\langle \mathbf{z}_n \rangle$ is a Cauchy sequence. Hence by the completeness of $\mathbb{R}^2$, $\langle \mathbf{z}_n \rangle$ converges to some limit, $\mathbf{v}$ (say).

Next, we prove that $\mathbf{v}$ is a *fixed point*: $\mathbf{v} = F(\mathbf{v})$. For, since $\lim \mathbf{z}_n = \mathbf{v}$, then for any $r > 0$ we can find C such that $\|\mathbf{z}_n - \mathbf{v}\| < r$ if $n > C$. In particular then, $\|\mathbf{z}_{n+1} - \mathbf{v}\| < r$. Also

$$\|\mathbf{z}_{n+1} - F(\mathbf{v})\| < k\,\|\mathbf{z}_n - \mathbf{v}\| < k\,r \qquad \text{by (F.3.1)},$$

so by the triangle inequality, $\|\mathbf{v} - F(\mathbf{v})\| < r(1 + k)$. Since this holds for every $r > 0$ (however small), we must have $\|\mathbf{v} - F(\mathbf{v})\| = 0$, so $\mathbf{v} = F(\mathbf{v})$ as required.

To complete the proof, we need to show that if we had started with $\mathbf{w}$ (say) then the limit $\mathbf{v}'$ of $\langle \mathbf{w}_n \rangle$ is actually $\mathbf{v}$. But by the contracting property (F.3.1) again, we have, since F fixes $\mathbf{v}$ and $\mathbf{v}'$:

$$\|\mathbf{v} - \mathbf{v}'\| = \|F(\mathbf{v}) - F(\mathbf{v}')\| < k\,\|\mathbf{v} - \mathbf{v}'\|, \tag{F.3.3}$$

so $\mathbf{v} = \mathbf{v}'$ since $k < 1$. Proposition F.3.1 is thus established.

Exercises F.3.1

1. The proof estimates $\|\mathbf{z}_n - \mathbf{z}_{n+m}\|$; from (F.3.1) show that $\|\mathbf{z}_n - F(\mathbf{v})\| \leq s\,k^n/(1-k)$. Find n so that $\mathbf{z}_n$ is within 10^{-6} of $\mathbf{v}$.
2. Let T: $\mathbb{R}^2 \to \mathbb{R}^2$ be the linear transformation given by $\mathbf{z} \to A\mathbf{z}$, where A is a real 2×2 matrix, and let $A\mathbf{z} = (u, v)^{\mathrm{T}}$. Use the triangle inequality to show that $u^2 \leq p^2\,\|\mathbf{z}\|^2$, where p is the maximum modulus of all the entries in A. Hence show that $\|A\mathbf{z}\| \leq \sqrt{2}\,p\,\|\mathbf{z}\|$, and therefore T is contracting if $p < 1/\sqrt{2}$.

3. Show that the mapping $x \mapsto \cos x$ maps the unit interval $U = [0, 1]$ into itself, and is a k-contraction, with $k = \sin(1)$.
4. Let $F(x, y) = (\frac{1}{5}(x^2 + y^2) + \frac{1}{2}, \frac{1}{3}xy + \frac{1}{4})$. Show that F maps the unit square into itself, but is not contracting on certain lines $y = mx$.
5. Let g: $F \to \mathbb{R}^2$ be a continuous mapping with domain $F \subseteq \mathbb{R}^2$. Explain what is meant by calling g a contraction. If g is differentiable explain how the mean value theorem (4.4.1) may be used (in appropriate circumstances) to verify that g is a contraction.

   Suppose that V is the unit square $\{0 \leqslant x \leqslant 1, 0 \leqslant y \leqslant 1\}$ in $\mathbb{R}^2$, and f: $V \to \mathbb{R}^2$ is the mapping given by $f(x, y) = (u, v)$ where

   $$u = Axy(1 - x^2y), \qquad v = 1 - B(x^2y - x^4y^3)$$

   and A, B are positive constants < 1.

   Show that f maps V into itself, and use the mean value theorem to prove that f is a contraction if $A, B < 1/4$. Hence show that for each such pair (A, B), f has a unique fixed point in V. [SU]
6. Show that the Newton mapping N_f in example 10.4.2 is a contracting mapping on a certain interval J. How would you find J in practice?

F.4 METRIC SPACES

We now do something peculiarly mathematical. Scrutinizing the last proof, we note that we never seriously used the fact that we were working in $\mathbb{R}^2$. It provided us with the points **z**, **w**, **v**, etc., but it would have worked just as well if we change $\mathbb{R}^2$ to $\mathbb{R}$ or $\mathbb{R}^3$, because the real story was about the properties of $\|\mathbf{z} - \mathbf{w}\|$, regarded as the *distance* between **z** and **w**. Indeed, the version for $\mathbb{R}$ is the basis of many computer algorithms for solving equations. But, without any more trouble, we can extract a much more general version which applies in many more situations as follows. Thus we need a set X, to provide us with such 'points' as **z**, **w** (but these need no longer refer to geometrical points, so we no longer use vector notation to represent them), and we need to be able to specify a numerical 'distance' dist(z, w), possibly quite different in form from the familiar Euclidean distance, which still satisfies the triangle inequality (F.1.3) in the form:

$$\text{dist}(z, w) \leq \text{dist}(z, v) + \text{dist}(v, w). \tag{F.4.1}$$

We also require it to be symmetric as in (F.1.3):

$$\text{dist}(z, w) = \text{dist}(w, z) \tag{F.4.2}$$

and to have the property used in proving $v = v'$ see (F.3.3)):

$$\text{dist}(z, w) = 0 \qquad \text{iff} \qquad z = w. \tag{F.4.3}$$

Such a distance function is called a **metric**; and X with this metric is called a **metric space**. Important examples occur when X is a circle with arc-length for distance, or when X is the surface of our own Earth, with distance measured along great circles.

We now have the language to define a d-nbd of a point p, as the set of all z in X for which dist$(p, z) < d$. Then we can translate word for word the earlier definition of a convergent sequence and the proof that limits are unique. The same proof as for (F.2.3), literally translated, shows that every convergent sequence in X is a Cauchy sequence. We then say that X is **complete** if every Cauchy sequence in X has a limit in X. Also we can say when a mapping $G: X \to X$ is a k-**contraction**: we simply require that (F.3.1) should hold in the form:

$$\text{dist}(G(z), G(w)) \le k \, \text{dist}(z, w). \tag{F.4.4}$$

Exercises F.4.1

1. Show that on $\mathbb{R}^2$, the following are metrics:
   (a) $d_1(\mathbf{z}, \mathbf{w}) = \max(|x - u|, |y - v|)$, $\quad (\mathbf{z} = (x, y), \mathbf{w} = (u, v))$;
   (b) $d_2(\mathbf{z}, \mathbf{w}) = (|x - u| + |y - v|)$;
   (c) $d_3(\mathbf{z}, \mathbf{w}) = 1$ unless $\mathbf{z} = \mathbf{w}$, in which case $d_3(\mathbf{z}, \mathbf{w}) = 0$.
   (You must check that each candidate satisfies conditions (F.4.1–F4.3).) What does $N_r(\mathbf{0})$ look like in each of these metrics?
2. Show that with either d_1 or d_2 as metric, the mapping T in exercise F.3.1(2) is contracting provided $p < 1$. (This means we can use a 'larger' matrix A than before.)
3. Let $\mathbf{z} = (x, y)$, and let dist $(\mathbf{0}, \mathbf{z})$ be r, s, t when 'dist' is, respectively, the Euclidean metric, d_1, or d_2. Thus $r = \|\mathbf{z}\|$. Show that $s \le r \le s\sqrt{2}$, and $r \le t \le r\sqrt{2}$. Hence show that if a sequence converges to $\mathbf{z}$ using any one of these metrics, then it converges to $\mathbf{z}$ in the others. (These metrics happen to be 'equivalent'.) Show that the only convergent sequences in the metric s_3 are those which are eventually constant.
4. Carry out a literal translation of the proof of (F.2.3) into the language of metric spaces.
5. Let X be a metric space with a metric $d(x, y)$. A mapping $f: X \to X$ is called **continuous** if, whenever a sequence $\langle x_n \rangle$ converges to x in X, the sequence $\langle f(x_n) \rangle$ converges to $f(x)$. Show that G in (F.4.4) is continuous (even if $k \ge 1$).
6. Consider the following 'big' metric space. Let X be the family of all finite non-empty sets $A = \{a_1, a_2, \ldots, a_n\}$ of points in $\mathbb{R}^2$. If also $B = \{b_1, b_2, \ldots, b_m\}$ is in X then define $w(A, B)$ to be the smallest number r such that every b_i is within a distance r of A. Let

   $$d(A, B) = \max(w(A, B), w(B, A)). \tag{F.4.5}$$

   Show that d is a metric on X, but first get the 'feel' of d by working out some examples:
   (a) If A consists of the points 0, 1, 2 on the x-axis, and $B = \{0, 1\}$ show that $w(A, B) = 0$, but $w(B, A) = 1$.
   (b) If A_n consists of the subdivision of the unit interval into the $2^n + 1$ points $k/2^n$, $(k = 0, 1, \ldots, 2^n)$, show that $d(A_n, A_{n+1}) = 1/2^{n+1}$.
   (c) Now verify that d satisfies the conditions (F.4.1–F.4.3). (Symmetry is obvious from (F.4.5). For the others, note that if $d(A, B) = s$, then by definition, given a_i it is 'within s of B' so:

   $$\text{there is some } b_j \text{ with } \|a_i - b_j\| \le s, \tag{F.4.6}$$

   and similarly with a's and b's interchanged. To prove d satisfies (F.4.1), let $d(B, C) = t$. Then there is some c_k with $\|b_j - c_k\| \le t$, whence by the triangle inequality for the norm, $\|a_i - c_k\| \le s + t$. Thus every a_i is within $s + t$ of C, so $w(C, A) \le s + t$. Interchange A and C in this argument to get $w(A, C) \le s + t$, and use (F.4.5) to obtain (F.4.1). For (F.4.3), we must prove that if $d(A, B) = 0$, then every $b_i \in A$, while every $a_i \in B$. But by (F.4.6), $w(A, B) = 0$, so given a_i there is some b_j with $\|a_i - b_j\| = 0$. Hence $a_i = b_j$ and $a_i \in B$. Similarly, every b_i in B lies in A, so $A = B$ as required.)

If we now translate the proof of Proposition F.3.1 into this language, we have at once (and with no further effort) a famous theorem, abstracted from earlier concrete results, by the Polish mathematician S. Banach.

Theorem F.4.1 (The contracting mapping theorem)
Let $G: X \to X$ be a k-contraction of the metric space X. Then there is a unique point p in X such that every orbit converges to p: the basin of attraction of p is all of X.

As a first application to show the extra power, consider the problem we meet in calculus, to find a function $f(x)$ such that for all x in $\mathbb{R}$,

$$f(x) = f'(x) \qquad \text{and} \qquad f(0) = 1. \tag{F.4.7}$$

We know of course that $f(x)$ must be e^x by traditional proofs, but let us see how to derive this from theorem F.4.1. First we work on a fixed interval $I = \{-a \leq x \leq a\}$ in $\mathbb{R}$. Next we remark that f has to be fixed under differentiation, but it is more convenient to work with integration. Then, for each continuous function $g: I \to I$, we construct a new one, h, given by $h(x) = 1 + \int_0^x g(t)\,dt$. Write $h = G(g)$. Since $h(0) = 1$, then by integrating (F.4.7) we see that we require a function f on I such that $f = G(f)$; so the required solution of (F.4.7) is a fixed point of G! This leads us to wonder whether theorem F.4.1 can be applied. If it could, then this fixed point is the unique solution (on I) to our problem, and we can calculate it by iterating from any starting point. Thus, starting from the constant function 1, its orbit is the sequence of functions $1, e_1, e_2, \ldots$, where $e_1 = G(1)$ and

$$\begin{aligned} e_1(x) &= 1 + \int_0^x 1\,dt = 1 + x, \\ e_2(x) &= 1 + \int_0^x (1 + t)\,dt = 1 + x + \frac{x^2}{2}, \end{aligned} \tag{F.4.8}$$

and so on, and we see the exponential series evolving.

It remains to set things up to meet the requirements of theorem F.4.1. For that we need a metric space for G to act on, but because of the above stipulation on g, G does act on the set X of all continuous functions on I, so now we find a metric for X. If $g_1, g_2 \in X$, define $\text{dist}(g_1, g_2)$ to be the maximum value of $\|g_1(x), g_2(x)\|$ on I (see Fig. F.4.1). This is a measure of the (ordinary) distance apart of the graphs of g_1, g_2. It is a standard fact (see Munkres (1975) p. 267) that X, with this metric, is complete. Thus it remains to show that $G: X \to X$ is contracting. But if $r = \text{dist}(g_1, g_2)$, then

$$\begin{aligned} \text{dist}(G(g_1), G(g_2)) &= \max \left| \int_0^a (g_1(t) - g_2(t))\,dt \right| \\ &\leq \int_0^a |g_1(t) - g_2(t)|\,dt \leq \int_0^a r\,dt \leq ra, \end{aligned} \tag{F.4.9}$$

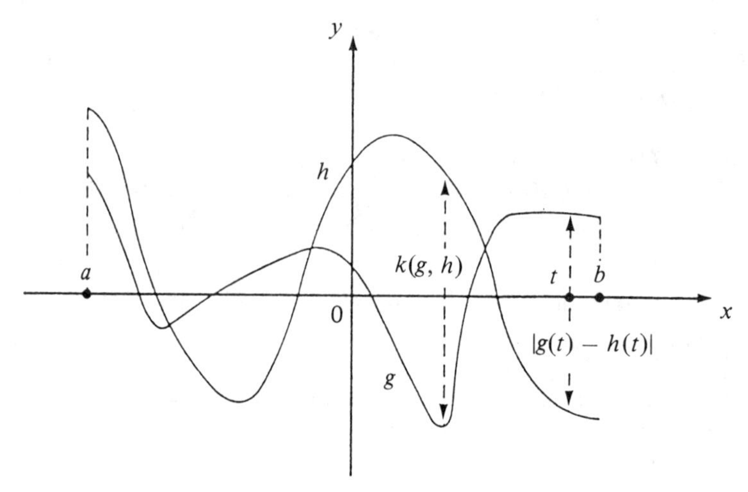

Fig. F.4.1.

so G is an a-contraction if $a < 1$. Therefore, theorem F.4.1 now applies, and we are in business: the sequence $\langle e_n \rangle$ defined above converges to the solution of (F.4.7). True, we have not obtained a solution on the whole of $\mathbb{R}$, but the present one can now be extended, by various strategies that need not concern us here.

Exercises F.4.2

1. Compute e_3, e_4, e_5 in (F.4.8).
2. Apply the above procedure to find the sine and cosine functions as solutions of the vector equation $(x', y') = (y, -x)$. (Let Y be the set of pairs (f, g) of functions in the previous space X, with $\text{dist}((f_1, f_2), (g_1, g_2)) = \max(\text{dist}(f_1, g_1), \text{dist}(f_2, g_2))$ and replace G by the mapping $(f, g) \mapsto (0, 1) + (\int_0^x f\,dt, \int_0^x g\,dt)$. First get the 'feel' by calculating $G^n(0, 1)$ for $n = 1, 2,3$.)
3. Let $U = \{0 \le x \le 1\}$. Calculate $\text{dist}(g_1, g_2)$ when $g_1(x), g_2(x)$ are:
   (a) $\sin(x), \cos(x)$;
   (b) x^2, x^3;
   (c) $0, x^n$;
   (d) $0, nx/(1 + n^2x^2)$.
4. In the notation of theorem F.4.1, show that for any orbit $\langle z_n \rangle$ $\text{dist}(z_n, p) \le sk^n/(1-k)$, where $s = \text{dist}(z_1, z_0)$. (See (F.3.2).)
5. Show that if $G: X \to X$ is a k-contraction as in theorem F.4.1, then its fixed point p depends continuously on G in the following sense. Let G': $X \to X$ be a second k-contraction, with fixed point p', and suppose that for all $x \in X$, $\text{dist}(G(x), G'(x)) \le r$. Prove that $\text{dist}(p, p') \le r/(1-k)$. (Use: $\text{dist}(p, p') \le \text{dist}(G(p), G(p')) + \text{dist}(G(p'), G'(p'))$.)

   Apply this to exercise F.3.1(6) to show that the equation $\phi(x, y) = 0$ has a continuous solution $y = p(x)$, near a solution (a, b). (Consider the Newton mapping for f_y when $f_y(x) = \phi(x, y)$, with ϕ a smooth function. This gives the continuity part of the implicit function theorem.)
6. Show that if $F(s, t)$ is a smooth function of (s, t), then the differential equation $dy/dx = F(x, y)$ has a unique solution near $x = 0$ if y is specified when $x = 0$. (A 'solution' is a function $y(x)$ such that $y(0)$ is given, and for all x in some interval I, $dy/dx = F(x, y(x))$. Take X as

for (F.4.9) and define $G_0(y) = h$ by: $h(x) = y(0) + \int_0^x F(t, y(t))\,dt$. Since F is smooth, we may assume there is a constant M such that $|F(x, s_1) - F(x, s_2)| \leq M\,|s_1 - s_2|$ when $(x, s_1), (x, s_2)$ lie in a rectangle centre $\mathbf{0}$. Hence, if a is chosen $< 1/m$, show that G_0 is k-contracting, where $K = Ma$.) This is *Picard's theorem*.

7. If in exercise 6 we change y_0 to y_1 to obtain a mapping G_1, show that for all $y, |G_0(y) - G_1(y)| = |y_0 - y_1|$. Hence use exercise 5 to show that the solution in exercise 6 depends continuously on the initial condition y_0.

F.5 APPLICATION TO THE CONSTRUCTION OF FRACTALS

For constructing fractal sets, we used a technique in section 10.10 which depends on a theorem that derives from attempting to apply the contracting mapping theorem to the space X of finite subsets of $\mathbb{R}^2$. Unfortunately X is not complete, and has to be enlarged to a complete space Y. In Y there is a unique fixed point P of a certain contracting mapping M, and P is a subset of $\mathbb{R}^2$ with fractal properties. To enable readers to grasp the details, we describe them through the steps of the following exercise.

Exercises F.5.1

1. Let F: $\mathbb{R}^2 \to \mathbb{R}^2$ be a k-contraction, and let X be the metric space of finite subsets as in exercise F.4.1(6). Define a mapping S: $X \to X$, by the rule that if $A = \{a_1, a_2, \ldots, a_n\}$ then $S(A) = \{F(a_1), F(a_2), \ldots, F(a_n)\}$. Show that S is a k-contraction. Unfortunately, theorem F.4.1 does not apply to X, because X is not complete. Prove this by showing that the sequence $\langle A_n \rangle$ in (b) of exercise F.4.1(6) cannot converge to a finite set. We show in (3) below how to put X into a bigger space Y, in which $\langle A_n \rangle$ does converge.
2. Additionally let G: $\mathbb{R}^2 \to \mathbb{R}^2$ be an l-contraction, with associated mapping T: $X \to X$. Using S, T we form a new mapping M: $X \to X$ by the rule that $M(A)$ is to be the union of the sets $S(A)$ and $T(A)$, that is, the set consisting of all the points $S(a_i)$ and $T(a_j)$, as a_i, a_j run through A. For example, on the x-axis, let $S(z) = z/3$, $T(z) = i + (z - i)/3$, where $i = (1, 0)$. Then writing $(u, 0)$ on the x-axis as u (so $\mathbf{0}$ becomes 0), we have

$$M(0) = \{S(0), T(0)\} = \{0, 1/3\},$$

$$M^2(0) = \{S(0), S(1), T(0), T(1)\} = \{M(0), M(1/3)\},$$

and so on. List the sets $M^3(0)$, $M^4(0)$. Show that M is an m-contraction, where $m = \max(k, l)$. Thus the M-orbit of 0 is a Cauchy sequence in X, which again does not converge, but which is closely related to Cantor's famous 'ternary' set (see Chapter 9).
3. So as to ensure convergence, we enlarge X to a metric space Y, as follows. We first need to mention the 'closed' subsets of $\mathbb{R}^2$. A set Z in $\mathbb{R}^2$ is **closed** if it contains all its limit points, that is, if, whenever $\langle x_n \rangle$ is a convergent sequence in Z, $\lim x_n$ also lies in Z. For example, every finite set is closed, as also is every finite polygon and every circular disc D with its boundary. If the boundary is not included, then D is not closed, because there are many sequences in D that converge to one of the missing boundary points. Also, a set is **bounded** if it lies in some nbd $N_r(0)$. For example, the x-axis is not bounded, whereas every finite polygon is bounded. Prove these various assertions. (To gain greater familiarity with these ideas, see Newman (1961) Chapter 2. Also try the next exercise.)
4. Let g: $\mathbb{R}^2 \to \mathbb{R}$ be continuous. Show that the set $\{(x, y) \in \mathbb{R}^2\colon g(x, y) > 0\}$ is open and that $\{(x, y) \in \mathbb{R}^2\colon g(x, y) = 0\}$ is closed.

Give an example of a continuous function ϕ: $\mathbb{R}^2 \to \mathbb{R}$ which is zero precisely on the set $\{(x, 0): x \geq 0\}$. Does there exist a continuous function which is zero precisely on the set $\{(x, 0): x > 0\}$? [SU]

5. Now take Y to be the set of all closed, bounded non-empty subsets of $\mathbb{R}^2$, so Y contains the previous space X. To define a metric on Y, we use (F.4.6) to define $w(A, B)$, and then define dist(A, B) as in (F.4.5). Essentially the same arguments as before show that this is a metric on Y, but it is somewhat harder to prove that Y is now complete, and we omit the proof. Hence the sequences $\langle A_n \rangle$, $\langle M^n(0) \rangle$, defined above, converge in Y. Prove that lim A_n is the unit interval $\{0 \leq x \leq 1\}$. The analogue of exercise 2 still holds: the mapping G gives a mapping S: $Y \to Y$ where $F(A)$ is now the image set of A (which is known to be closed and bounded, hence in Y). With T similarly associated with G, we define $M(A)$ to be the union of $S(A)$ and $T(A)$; and a similar proof to the above shows that M is an m-contraction of Y.

The point of doing all this is that we can now apply theorem 4.2 to obtain at once the theorem of Hutchinson (1981): *there is a unique P in Y such that $M(P) = P$.*

Thus by definition of M, P is 'invariant' under S and T, that is, S and T each map P into itself. Therefore we have, in a sense described below, the following proposition.

Proposition F.5.1

If S and T are also homeomorphisms, then P is self-similar.

Proof

Since S maps P into itself, P contains its own image P_1 under S, and this image is reduced in size by the contraction ratio k. Hence P_1 contains *its* image under S, and so on: we obtain within P successive copies P_n of itself, and P_n is $\leq k^n$ times the size of P. Similarly for the images under T if T is also a homeomorphism, and then for images under the homeomorphisms $S \circ T$, $T \circ S$, $S^2 \circ T^3 \circ S$, etc. Thus P is self-similar, in that it contains all these arbitrarily small copies of itself.

The fact that Y is the basin of attraction of P implies that P can be found as the limit of the M-orbit of any convenient starting-set A in Y (so for example, we can take A to be a single point of $\mathbb{R}^2$). As shown in Chapter 9, this gives a convenient way of constructing certain fractal sets.

The contracting mapping theorem is especially useful on infinite-dimensional spaces of functions, and is used for finding 'good' coordinate systems. It therefore tends to appear in rather advanced contexts beyond the scope of this book.

We conclude with some exercises that establish some familiar principles we have left unproved. Let f: $X \to \mathbb{R}$ be continuous, with X metric.

Exercises F.5.2

1. Prove the Inertia Principle: if $f(x) = a > 0$ then there is a nbd U of x in X such that for all $y \in U$, $f(y) > 0$. (If not, then for all integers $n > 0$, there exists y_n in the $(1/n)$-nbd of y_n with $f(y_n) \leq 0$. Obtain a contradiction since $f(y_n) = f(x)$.)
2. (The Boundedness Theorem). A subset A of X is **compact** if every infinite sequence $\langle a_n \rangle$ in A has a sub-sequence $\langle a_{n(j)} \rangle$ that converges to a limit in A. Show that A is closed and bounded, and $f(A)$ is compact in $\mathbb{R}$ (hence bounded). Since $\mathbb{R}$ is complete, it can be shown that every interval $\{a \leq x \leq b\}$ in $\mathbb{R}$ is compact.

Appendix: Mathematical notation

In this appendix we gather together various points of mathematical notation used throughout this book. For brevity we make no attempt to give supporting examples here, since they can be found in the main text. Additional material can be found in several books such as Griffiths and Hilton (1982). We have to assume that the reader is willing to accept the convention of terse mathematical layout of definitions and deductions. For brevity we often use the neologism 'iff' for 'if and only if' and denote it symbolically by '$\Leftrightarrow$'. The single-headed arrow '$\Rightarrow$' is occasionally used to mean 'implies' or 'if ... then'.

Sets

The members $x, y, \ldots,$ of a set X are variously referred to as elements or points, and $x \in X$ means that x is a member of X. A convenient mode of defining a set is to use curly brackets in the form.

$$X = \{x \mid x \text{ has property } P\}$$

and the vertical stroke is read 'such that'. If Y is another set, then $Y \subseteq X$ (or $X \supseteq Y$) means that Y is a subset of X; and $X - Y$ denotes the set of those members of X which are not in Y; using curly brackets we have:

$$X - Y = \{x \mid x \in X \text{ and } x \notin Y\}.$$

If X consists of just one element x, then x is called a *singleton*, written $\{x\}$. A set with no elements is called **empty**; for example $\{x \mid x \neq x\}$. Thus if $X \subseteq Y$ then $X - Y$ is empty. Thus $X - Y$ is empty if $X \subseteq Y$. Two sets X, Z are called **equal** if, and only if, both $Z \subseteq X$ and $X \subseteq Z$. (See Griffiths and Hilton (1982), Chapter 1.)

Standard sets and their symbols are as follows:

$$\mathbb{N} \subseteq \mathbb{Z} \subseteq \mathbb{R} \subseteq \mathbb{R}^2$$

where, respectively,

$$\mathbb{N} = \{1, 2, 3, \ldots\} \text{ (set of natural numbers)}$$

$\mathbb{Z} = \{0, \pm 1, \pm 2, \pm 3, \ldots\}$ (set of all integers)

$\mathbb{R}$ = set of all *real* numbers

$\mathbb{R}^2$ = plane of coordinate geometry, whose points are pairs (x, y).

Here x and y lie in $\mathbb{R}$, and $\mathbb{R}$ may be thought of as being represented by the x-axis; hence the above inclusion $\mathbb{R} \subseteq \mathbb{R}^2$. Some standard subsets of $\mathbb{R}$ are:

$$\mathbb{R}^+ = \{x \in \mathbb{R} \mid x \geq 0\}, \qquad \mathbb{R}_> = \{x \in \mathbb{R} \mid x > 0\},$$

and the closed intervals $[a, b] = \{x \in \mathbb{R} \mid a \leq x \leq b)$. Sometimes we may wish to exclude one or other end point, and write, e.g., $[a, b) = [a, b] - [b]$, with similar meanings for $(a, b]$ and (a, b).

In higher dimensions, $\mathbb{R}^n$ denotes the set of all n-tuples $(x_1, x_2, \ldots, x_n)$ where each x_i lies in $\mathbb{R}$. When $n = 3$, $\mathbb{R}^3 = \{(x, y, z) \mid x, y, z \in \mathbb{R}\}$, the usual model of three-dimensional space.

When $\mathbb{R}^2$ is regarded as the Argand diagram of complex numbers, we signal this by using $\mathbb{C}$ instead. Points of $\mathbb{R}^2$, $\mathbb{R}^3$ are often denoted by bold type, as $\mathbf{r} = (x, y)$, and so on, to distinguish them from *scalars* (i.e., elements of $\mathbb{R}$).

For use with a matrix M, we regard $\mathbf{r}$ as a **column** vector, in order to be able to write the matrix product $M \cdot \mathbf{r}$, which is again a column vector. We use the transposition symbol T to write $(x \quad y)^{\mathrm{T}}$ for $\mathbf{r}$ regarded as a column vector. Which version is meant should be obvious from the context, and the row vector is preferred for typographical convenience.

Mappings; functions

A general function is denoted $f: X \to Y$, where X and Y are sets called, respectively, the **domain** and **co-domain**, and f is a rule that tells us how to assign to each $x \in X$ a unique $y \in Y$; we denote y by $f(x)$ and call it the **value** of f at x. The word 'mapping' is often used for 'function', especially if the dimension of the co-domain exceeds 1. We often write $f: X \to Y, x \to f(x)$ when f is being defined by some formula; and we also use the older convention of saying '$f(x, y)$ is a function' when great precision is not needed to spell out the domain of f; here it would be understood to be a subset of $\mathbb{R}^2$.

Some standard functions are as follows. They are defined in detail elsewhere in the book.

abs: $\mathbb{R} \to \mathbb{R}^+$, $\quad \mathrm{abs}(x) = |x|$;

mod: $\mathbb{C} \to \mathbb{R}^+$, $\quad \mathrm{mod}(z) = |z|$;

Pol: $\mathbb{R}^2 - \{0\} \to [0, 2\pi)$, $\quad \mathrm{Pol}(P)$ = the polar angle of $P \in \mathbb{R}^2 - \{0\}$;

$\cos^{-1}$: $[-1, 1] \to [0, \pi]$, $\quad$ (note the domain and co-domain!);

$\sqrt{\ }$: $\mathbb{R}^+ \to \mathbb{R}^+$, $\quad \sqrt{x}$ = positive square root of x;

$\sin^{-1}$: $[-1, 1] \to [-\pi/2, \pi/2]$;

$\tan^{-1}$: $(-\pi/2, \pi/2) \to \mathbb{R}$;

$$\text{ln}: \quad \mathbb{R}_> \to \mathbb{R}, \quad \text{(natural logarithm)};$$

$$\text{exp}: \quad \mathbb{R} \to \mathbb{R}_>, \quad \exp(x) = \mathrm{e}^x \quad \text{(exponential function)}.$$

A particular mapping is the 'identity mapping' on a set X. This is denoted by $\mathrm{id}_X: X \to X$, and is defined by the rule that for each $x \in X$, $\mathrm{id}_X(x) = x$. (In coordinate geometry, this is the function traditionally and confusingly known as '$y = x$'.)

Given functions $f: X \to Y$ and $g: Y \to Z$ we often write $X \xrightarrow{f} Y \xrightarrow{g} Z$; and we can form a new function

$$g \circ f: X \mapsto Z, \qquad x \to g(f(x)),$$

called the **composition** of g with f. In particular

$$g \circ \mathrm{id}_Y = g, \qquad \mathrm{id}_Y \circ f = f.$$

For example, $\ln \circ \exp = \mathrm{id}$. Composition obeys the associative law, in that if also we have $h: Z \to W$, then $h \circ (g \circ f) = (h \circ g) \circ f: X \to W$; because, for every $x \in \mathrm{X}$,

$$(h \circ (g \circ f))(x) = h(g(f(x))) = ((h \circ g) \circ f)(x).$$

If f maps every pair of points to *distinct* points of Y, then f is said to be an **injection**; equivalently we must have: if $x_1 \neq x_2$ in X, then $f(x_1) \neq f(x_2)$ in Y. (Thus, since $|-1| = 1 = |1|$, then abs is not an injection; hence neither is mod.)

On the other hand, if $D \subseteq X$, then the subset $f(D) \subseteq Y$, called the **image** of D by f, is defined by

$$f(D) = (f(a) \mid a \in D\}.$$

If the image of X is the whole of Y then f is said to be **onto**, or a **surjection**.

It is often necessary to show that a function $f: A \to B$ under consideration is *both* an injection and a surjection (then f is called a **bijection**). One way of doing this is to find a function $g: B \to A$ with the property that for all $a \in A$, $b \in B$

$$g(f(a)) = a, \qquad f(g(b)) = b, \qquad (\text{i.e., } g \circ f = \mathrm{id}_A, f \circ g = \mathrm{id}_B).$$

These equations determine g uniquely (given f) so g is denoted by $f^{-1}: B \to A$ and called the **inverse** of f. If $h: B \to C$ is a bijection, so is $h \circ f: A \to C$ and

$$(h \circ f)^{-1} = f^{-1} \circ h^{-1}: C \to R.$$

Applications and examples will be found in the main text.

In algebraic expressions, multiplication is usually denoted by a dot, as with $x \cdot y$; the dot is often omitted, as with matrices: $A\mathbf{z}$, but $A \cdot t\mathbf{z}$ for clarity, rather than $A(t\mathbf{z})$ (which looks like a function). Thus the dot is often used as a form of punctuation, and similarly with brackets, to avoid ambiguity. The special arrow $\mapsto$ is used when indicating a correspondence between *variables*, as in $x \mapsto \mathrm{e}^x$ for the function exp: $\mathbb{R} \to \mathbb{R}_{>0}$.

References

Allgood, K.T. and Yorke, J.A. (1989). Fractal basin boundaries and chaotic attractors. In Devaney, R.L. and Keene, L. (editors) (1989). *Chaos and Fractals: the mathematics behind the computer graphics.* Proc. Symposia in Applied Mathematics, pp. 41–55.

Arnold, V.I. (1992). *Ordinary differential equations.* Springer.

Arrowsmith, D.K. and Place, C.M. (1982). *Ordinary differential equations.* Chapman and Hall.

Artmann, B. (1988). *The concept of number.* Ellis Horwood.

Barnsley, M.F. and Demko, S. (1985). Iterated function systems and the global construction of fractals. *Proc. Roy. Soc. Lond.* A **399**, 243–75.

Baylis, J. (1981). *What is mathematical analysis?* Macmillan.

Bedford, T. and Swift, J. (editors) (1988). *New directions in dynamical systems.* London Mathematical Society Lecture Note Series, 127, Cambridge University Press.

Blows, T.R. and Lloyd, N.G. (1988). Some cubic systems with several limit cycles. *Nonlinearity*, **1**, 000–000.

Blyth, T.S. and Robertson, E.F. (1986). *Linear algebra* (vol 4 of *Essential algebra*). Chapman and Hall.

Boyer, C.B. and Mersbach, U.C. (1989). *Introduction to the history of mathematics.* Wiley.

Broecker, T. and Lander, L. (1978). *Differentiable germs and catastrophes.* London Mathematical Society Lecture Notes, 17, Cambridge University Press.

Brown, A. (1981). Equations for periodic solutions of a logistic differential equation. *J. Austral. Math. Soc.* (Series B), **23**, 78–94.

Bruce, J.W. and Giblin, P.J. (1981). What is an envelope?, *Math. Gazette*, **65**, 186–92.

Bruce, J.W., Giblin, P.J. and Rippon, P.J. (1990). *Microcomputers and mathematics.* Cambridge University Press.

Bryant, V. (1990). *Yet another introduction to analysis.* Cambridge University Press.

Chillingworth, D.J.R. and Furness, P. (1974). Reversals of the earth's magnetic field. In *Dynamical systems – Warwick.* (ed. A. Manning) Springer Lecture Notes in Mathematics 468, pp. 91–8.

Collett, P. and Eckmann, J-P. (1980). *Iterated maps on the interval as dynamical systems.* Birkhäuser.

Cook, M. (1991). Offbeat oscillators. *The Micro-user*, **9**, No. 4.

Cooke, K.L. and Renfrew, C. (editors) (1979). *Transformations: Mathematical approaches to culture change*. Academic Press.

Coppel, W.A. (1966). A survey of quadratic systems. *J. Diff. Eqns*, **2**, 293–304.

Courant, R. (1952). *Differential calculus*, Blackie.

Courant, R. and Robbins, H. (1948). *What is mathematics?* Oxford University Press.

Crilly, T. (1990). Black holes and the gamma function. *Math. Gazette*, **74**, 291–2.

Davis, H. (1962). *Introduction to non-linear and integral equations*, Dover.

Devaney, R.L. and Keene, L. (editors) (1989). *Chaos and fractals: the mathematics behind the computer graphics*. Proc. Symposia in Applied Mathematics. Vol 39, American Math. Soc.

Feynman, R. (1988). *What do you care what other people think?* Unwin.

Garcia-Cobian, R. (1978). Some stochastic models for prediction and control in catastrophe theory. Ph.D. thesis, University of Southampton.

Giblin, P.J. (1982). Parallels. *Math. Gazette*, **66**, 28–30.

Gibson, G.A. (1956). *An elementary treatise on the calculus*. Macmillan.

Gleick, J. (1988). *Chaos: making a new science*. Viking Penguin.

Griffiths, H.B. (1981*a*). *Surfaces*. Cambridge University Press.

Griffiths, H.B. (1981*b*). Cayley's version of the resultant of two polynomials. *Amer. Math. Monthly*, **88**, 328–38.

Griffiths, H.B. (1984). The implicit function theorem – technique versus understanding. *Int. J. Math. Educ. Sci. Technol.*, 541–51.

Griffiths, H.B. (1985). Bifurcations of some plane dynamical systems: and mappings between some elementary catastrophes. In *Differential topology – geometry and related fields, and their applications to the physical sciences and engineering*, Band 76, (ed. G.M. Rassias), Teubner.

Griffiths, H.B. and Hilton, P.J. (1982). *Classical mathematics*. Springer.

Griffiths, H.B. and Howson, A.G. (1974). *Mathematics: society and curricula*. Cambridge University Press.

Griffiths, H.B. and Rand, D. (1977). *An essay on elites and mathematical modelling*. SSRC Report.

Guckenheimer, J. and Holmes, P.J. (1983). *Non-linear oscillations, dynamical systems and bifurcations of vector fields*. Springer.

Hall, H.S. and Knight, S.R. (1942). *Higher algebra*. Macmillan.

Hao Bai-Lin, (1984). *Chaos*. World Scientific.

Hartman, P. (1964). *Ordinary differential equations*. Wiley.

Hénon, M. (1976). A two-dimensional mapping with a strange attractor. *Comm. Math. Phys.*, **50**, 69–77.

Hirsch, M.W. and Smale, S. (1974). *Differential equations, dynamical systems, and linear algebra*. Academic Press.

Holmes, P.J. and Rand, D.A. (1978). The bifurcations of Duffing's equation: an application and catastrophe theory. *J. Sound. Vib.*, **44** (2), 237–253.

Hutchinson, J. (1981). Fractals and self-similarity. *Indiana Univ. J. Math.*, **30**, 713–47.

Kemeny, J.G. and Kurtz, T.E. (1990). *TrueBASIC reference manual*. TrueBASIC Inc.

Klein, F. (1908). *Elementary mathematics from an advanced standpoint: arithmetic, algebra, analysis*. Dover (1939).

Kreyszig, E. (1962). *Advanced engineering mathematics.* Wiley.

Lamb, H. (1907). *Infinitesimal calculus.* Cambridge University Press.

Lanchester, F.W. (1916). Mathematics in Warfare. In Newman (1956), 2138–57.

Li, T.-Y. and Yorke, J.A. (1975). Period three implies chaos. *American Math. Mthly*, **82**, 985–92.

Lotka, A.J. (1956). *Elements of mathematical biology.* Dover.

Malthus, T.J. (1798). An essay concerning population. (ed. A. Flew). Penguin (1982).

Mandelbrot, B. (1982). *The fractal geometry of nature.* Freeman.

Marsden, J.E. and McCracken, M. (1976). *The Hopf bifurcation and its applications.* Applied Math. Series, 19, Springer.

May, R.M. (1976). Simple models with very complicated dynamics. *Nature*, **261**, 459–467.

Maynard Smith, J. (1968). *Mathematical ideas in biology.* Cambridge University Press.

Milne, W.P. (1930). *Higher Algebra.* Arnold.

Mirsky, L. (1955). *Introduction to linear algebra.* Oxford University Press.

Munkres, J. (1975). *Topology.* Prentice-Hall.

Murphy, R.V. (1979). Maximum range problems in a resisting medium. *Math. Gazette*, **63**, 10–16.

Murray, J.D. (1978). Biological and chemical oscillatory phenomena and their mathematical models. *Bull. Inst. Math. Applics.*, **14**, 162–169.

Newman, J.R. (1956). *The world of mathematics.* Simon and Schuster.

Newman, M.H.A. (1961). *Topology of plane sets.* Cambridge University Press.

Oldknow, A.J. (1987). *Microcomputers in geometry.* Ellis Horwood.

Peitgen, H.-O. (editor) (1989). *Newton's method in complex dynamical systems.* Kluwer.

Peitgen, H.-O. and Richter, P. (1986). *The beauty of fractals.* Springer.

Piaggio, H.T.H. (1952). *Differential equations and their applications.* Bell.

Poston, T. and Stewart, I. (1978). *Catastrophe theory and its applications.* Pitman.

Renfrew, C. (1978). Trajectory discontinuity and morphogenesis, the implications of catastrophe theory for archaeology. *American Antiquity* (43), 203–44.

Renton, A. (1983). An iterative matrix method. *Math. Gazette*, **67**, 294–5.

Richardson, L.F. (1935). Mathematics of war and foreign politics. In Newman (1956), pp. 1240–53.

Ross, G. (1990). *Chaos and negative resistance.* Undergraduate project, University of Andrews.

Sandefur, J.T. (1990). *Discrete Dynamical Systems.* Oxford University Press.

Scholl, B. and Thieler, P. (1989). Another real world approach to complex dynamics. In: *Applications and modelling in learning and teaching mathematics.* Ed. by W. Blum et al. Ellis Horwood.

Starbuck, W.K. (1973). *Organisational growth and development.* Penguin.

Stirling, D.G.S. (1990). *Mathematical analysis.* Ellis Horwood.

Thom, R. (1975). *Structural stability and morphogenesis.* Benjamin-Addison Wesley.

Thompson, D'Arcy W. (1961). *Growth and form*, (abridged J.T. Bonner). Cambridge University Press.

Volterra, V. (1931). Leçons sur la théorie mathématique de la lutte pour la vie. Paris.

Webb, J.R.L. (1991). *Functions of several real variables.* Ellis Horwood.

Whitley, D. (1982). *The bifurcations and dynamics of certain quadratic maps of the plane.* Ph.D. thesis, University of Southampton.
Whitley, D. (1983). Discrete dynamical systems in dimensions 1 and 2. *Bull. London Math. Soc.*, **15**, 177–217.
Wicks, K.R. (1990*a*). Hyperspaces and self-similarity. Ph.D. thesis, University of Hull.
Wicks, K.R. (1990*b*). Spiral-based self-similar sets. *Math. Research Reports*, University of Hull, **3**, No. 1.
Wicks, K.R. (1991). *Fractals and Hyperspaces.* Springer Lecture Notes in math. 1492. Springer.
Zeeman, E.C. (1977). *Catastrophe Theory: Selected Papers 1972–77.* Addison Wesley.

List of programs

Index